ADVANCED COMPUTING AND ANALYSIS TECHNIQUES IN PHYSICS RESEARCH

VII International Workshop on Advanced Computing & Analysis Techniques in Physics Research

ACAT 2000

October 16 – 20, 2000

Fermi National Accelerator Laboratory

Batavia, Illinois, USA

SPONSORS

 FERMILAB   U.S. Department of Energy   U.S. National Science Foundation

CORPORATE SPONSORS

http://conferences.fnal.gov/acat2000/

AIHENP workshop series

ADVANCED COMPUTING AND ANALYSIS TECHNIQUES IN PHYSICS RESEARCH

VII International Workshop
ACAT 2000

Batavia, Illinois   16–20 October 2000

EDITORS
Pushpalatha C. Bhat
Matthias Kasemann
Fermilab, Batavia, Illinois

Melville, New York, 2001
AIP CONFERENCE PROCEEDINGS ■ VOLUME 583

Editors:

Pushpalatha C. Bhat
Fermi National Accelerator Laboratory
Mail Station 357, D0 Assy Bldg.
P.O. Box 500
Batavia, IL 60510-0500
USA

E-mail: pushpa@fnal.gov

Matthias Kasemann
Fermi National Accelerator Laboratory
Mail Station 370, FCC 1W
P.O. Box 500
Batavia, IL 60510-0500
USA

E-mail: kasemann@fnal.gov

The articles on pp. 22–30, 51–56, 122–124, 125–127, 133–136, 161–163, 205–207, 220–222, 226–228, 229–231, 247–249, 250–252, 255–257, 261–263, 267–269, 280–284, 287–290, 291–293, 294–296, 307–309, 319–322, 323–325, 326–328, 342–344, 345–347, 348–350, 355–356, and 357–359 were authored by U. S. Government employees and are not covered by the below mentioned copyright.

Authorization to photocopy items for internal or personal use, beyond the free copying permitted under the 1978 U.S. Copyright Law (see statement below), is granted by the American Institute of Physics for users registered with the Copyright Clearance Center (CCC) Transactional Reporting Service, provided that the base fee of $18.00 per copy is paid directly to CCC, 222 Rosewood Drive, Danvers, MA 01923. For those organizations that have been granted a photocopy license by CCC, a separate system of payment has been arranged. The fee code for users of the Transactional Reporting Service is: 0-7354-0023-7/01/$18.00.

© 2001 American Institute of Physics

Individual readers of this volume and nonprofit libraries, acting for them, are permitted to make fair use of the material in it, such as copying an article for use in teaching or research. Permission is granted to quote from this volume in scientific work with the customary acknowledgment of the source. To reprint a figure, table, or other excerpt requires the consent of one of the original authors and notification to AIP. Republication or systematic or multiple reproduction of any material in this volume is permitted only under license from AIP. Address inquiries to Office of Rights and Permissions, Suite 1NO1, 2 Huntington Quadrangle, Melville, NY 11747-4502; phone: 516-576-2268; fax: 516-576-2450; e-mail: rights@aip.org.

L.C. Catalog Card No. 2001093032
ISBN 0-7354-0023-7
ISSN 0094-243X

Printed in the United States of America

CONTENTS

Preface xi
Opening Remarks xv
   M. Witherell
Committees xvii

PLENARY SESSIONS

Information Technology Research: Transforming Our Future 7
   R. Bajcsy
Speaking C++ As a Native 11
   B. Stroustrup
Advanced Analysis Methods in High Energy Physics 22
   P. C. Bhat
Statistical Techniques in High Energy Physics 31
   L. Lyons
The H1 Neural Network Trigger Project 36
   C. Kiesling, B. Denby, J. Fent, W. Fröchtenicht, P. Garda, B. Granado, G. Grindhammer,
   W. Haberer, L. Janauschek, T. Kobler, B. Koblitz, G. Nellen, J.-C. Prévotet, S. Schmidt,
   E. Tzamariudaki, and S. Udluft
Feynman-Diagram Evaluation in the Electroweak Theory with Computer Algebra 45
   G. Weiglein
Grid Computing 51
   I. Foster
Large-Scale Molecular Dynamics Simulations of Materials on Parallel Computers 57
   A. Nakano, T. J. Campbell, R. K. Kalia, S. Kodiyalam, S. Ogata, F. Shimojo,
   P. Vashishta, and P. Walsh

PARALLEL SESSIONS

ARTIFICIAL INTELLIGENCE

Session I Online Applications of Neural Networks

Momentum Reconstruction and Triggering Suggested for the ATLAS Detector 67
   G. Dror and E. Etzion
Neural Network Real-Time Event Selection for the DIRAC Experiment 70
   P. Kokkas, M. Steinacher, L. Tauscher, and S. Vlachos
Intelligent Preprocessing for Neural Networks in the H1 Experiment 73
   J.-C. Prévotet, B. Denby, P. Garda, B. Granado, W. Fröchtenicht, G. Grindhammer,
   L. Janauschek, C. Kiesling, T. Kobler, B. Koblitz, S. Schmidt, B. Tzamariudaki,
   and S. Udluft
An Electronic System for Simulation of Neural Networks with a Microsecond
Real-Time Constraint 76
   A. Chorti, B. Granado, B. Denby, and P. Garda

Session II Applications in Data Analysis I

Selection of W-Pair-Production in DELPHI with Feed-Forward Neural Networks 80
   K.-H. Becks, P. Buschmann, J. Drees, U. Müller, and H. Wahlen
Use of Neural Networks in a Search for Single Top Quark Production at DØ 83
   L. Dudko (for the DØ Collaboration)
A Hybrid Training Method for Neural Energy Estimation in Calorimetry 86
   P. V. M. da Silva, J. M. Seixas, and J. Seixas

Principal Component Analysis for Neural Electron/Jet Discrimination in Highly Segmented Calorimeters 89
    M. R. Vassali and J. M. Seixas

Session III Applications in Data Analysis II

Particle Identification at D∅ Using Neural Networks 92
    D. Chakraborty (for the D∅ Collaboration)
Vertex Reconstructing Neural Network at the ZEUS Central Tracking Detector 95
    G. Dror and E. Etzion
Neural Networks for Higgs Physics 98
    S. Tentindo-Repond, P. C. Bhat, and H. B. Prosper
Top Quark Mass Measurements Using Neural Networks 101
    S. B. Beri, P. C. Bhat, R. Kaur, and H. B. Prosper

Session IV Theoretical Aspects and Other Topics

Clopper-Pearson Bounds from HEP Data Cuts 104
    B. A. Berg
Experimenting with Rule Induction Algorithms in HEP Data Analysis ... 107
    N. Stepanov
Looking for Instanton-Induced Processes at HERA Using a Multivariate Technique Based on Range Searching 110
    T. Carli and B. Koblitz
A Self-Organizing Neural Network for Job Scheduling in Distributed Systems ... 113
    H. B. Newman and I. C. Legrand

INNOVATIVE SOFTWARE ALGORITHMS & TOOLS

Session I Online Monitoring and Controls

Tailorable Software Architectures in the Accelerator Control System Environment ... 119
    I. Mejuev, A. Kumagai, and E. Kadokura
Online Modeling of the Fermilab Accelerators 122
    E. S. McCrory, L. Michelotti, and J.-F. Ostiguy
A Beamline Matching Application Based on Open Source Software 125
    J.-F. Ostiguy

Session II Physics Analysis and Reconstruction Algorithms

Sleuth: A Quasi-Model-Independent Search Strategy for New High p_T Physics ... 128
    B. Knuteson
Fast Tracking in Hadron Collider Experiments 130
    N. Konstantinidis and H. Drevermann
More Performance Results and Implementation of an Object-Oriented Track Reconstruction Model in Different OO Frameworks 133
    I. Gaines and S. Qian

Session III Pattern Recognition Techniques

Simultaneous Tracking and Vertexing with Elastic Templates 137
    A. Haas
Vertex Finding Prior to Tracking in Magnetic Field 140
    Y. A. Yatsunenko

Singular Value Decomposition to Simplify Features Recognition Analysis in Very Large Collection of Images 143
    F. Guillon, D. J. C. Murray, and P. DesAutels
The Hermite Polynomials in Problem of the Searching for Position of the Global Maximum 146
    Y. A. Yatsunenko

Session IV Common Libraries

Event Bookkeeping for CLEO-3 149
    J. J. Thaler
KID-KLOE Integrated Dataflow 152
    I. Sfiligoi (for the KLOE Collaboration)
A Component-Based Approach to Scientific Workflow Management 155
    N. Baker, P. Brooks, Z. Kovacs, J-M. LeGoff, and R. McClatchey

Session V Grid and Distributed Computing Techniques

Protocols and Services for Distributed Data-Intensive Science 161
    W. Allcock, I. Foster, S. Tuecke, A. Chervenak, and C. Kesselman
Simulating Distributed Systems 164
    H. B. Newman and I. C. Legrand

SYMBOLIC PROBLEM SOLVING

Session I Feynman Diagram Algorithms and Tools

HELAC-PHEGAS: Automatic Computation of Helicity Amplitudes and Cross Sections 169
    A. Kanaki and C. G. Papadopoulos
O'Mega: An Optimizing Matrix Element Generator 173
    T. Ohl
A Feynman Graph Selection Tool in GRACE System 176
    F. Yuasa, T. Kaneko, and T. Ishikawa

Session II Symbolic Manipulation via Function Objects

A Library of Function Classes 179
    J. Boudreau, M. Fischler, and P. Maksimović
functionalObjects.h: Using Symbolic Syntax in C++ Programs 182
    R. Nolty
Large-Scale Symbolic Programming with GiNaC 185
    A. Frink, C. Bauer, and R. Kreckel

Session III Symbolic Techniques for Feynman Diagrams

Optimization of Symbolic Evaluation of Helicity Amplitudes 190
    P. S. Cherzor, V. A. Ilyin, and A. E. Pukhov
A Feynman Diagram Analyser DIANA—Graphic Facilities 193
    J. Fleischer and M. Tentyukov
The Tensor Reduction and Master Integrals of the Two-Loop Massless Crossed Box 196
    C. Oleari

Session IV Multi-loop Calculations and Results

Gauge Invariant Classes of Feynman Diagrams and Applications for Calculations 199
    E. E. Boos
A Parallel Version of FORM 3 202
    D. Fliegner, A. Rétey, and J. A. M. Vermaseren
Two-Loop Calculations in the MSSM with FeynArts 205
    S. Heinemeyer

VERY LARGE SCALE COMPUTING

Session I Online Monitoring and Controls

CompHEP-PYTHIA Interface: Integrated Package for the Collision Events Generation Based on Exact Matrix Elements 211
    A. S. Belyaev, E. E. Boos, A. N. Vologdin, M. N. Dubinin, V. A. Ilyin,
    A. P. Kryukov, A. E. Pukhov, A. N. Skachkova, V. I. Savrin, A. V. Sherstnev,
    and S. A. Shichanin
Integration of GRACE and PYTHIA 214
    K. Sato, S. Tsuno, J. Fujimoto, T. Ishikawa, Y. Kurihara, and S. Odaka
Algorithm for Computing Excited States in Quantum Theory 217
    X. Q. Luo, H. Jirari, H. Kröger, and K. Moriarty
Simple Scaling for Faster Tracking Simulation in Accelerator Multiparticle Dynamics ... 220
    J. A. MacLachlan
Adaptive Mesh Simulations of Astrophysical Detonations Using the ASCI Flash Code ... 223
    B. Fryxell, A. C. Calder, L. J. Dursi, D. Q. Lamb, P. MacNeice, K. Olson, P. Ricker,
    R. Rosner, F. X. Timmes, J. W. Truran, H. M. Tufo, and M. Zingale

Session II Analysis Farms and DAQ Systems

Matrix Distributed Processing and FermiQCD 226
    M. Di Pierro
A Terabyte Analysis Machine for SDSS Data 229
    J. Annis, G. Garzoglio, K. Ruthsmandorfer, and C. Stoughton
A Large Linux-PC Farm for Online Event Reconstruction and Data Management at HERA-B 232
    A. Gellrich
Client and Event Driven Data Hub System at CDF 235
    B. Kilminster, K. McFarland, T. Vaiciulis, H. Matsunaga, and M. Shimojima
The COMPASS Computing Farm Project 238
    M. Lamanna

Session III Grid Architectures

A Data Grid Prototype for Distributed Data Production in CMS 241
    M. Hafeez, A. Samar, and H. Stockinger
Object Level Physics Data Replication in the Grid 244
    K. Holtman
SAM for D0—A Fully Distributed Data Access System 247
    I. Terekhov, V. White, L. Lueking, L. Carpenter, H. Schellman, J. Trumbo,
    S. Veseli, and M. Vranicar
The Control Actions Framework 250
    P. Calafiura

POSTER SESSION

EMS: A Framework for Data Acquisition and Analysis 255
    J. M. Nogiec, J. Sim, K. Trombly-Freytag, and D. Walbridge

Effects of Limited Resources in 3D Real-Time Simulation of an Extended ECHO Complex Adaptive System Model 258
    D. M. Dominiak, F. Rinaldo, and M. W. Evans

High Performance Visual Display for HENP Detectors 261
    M. McGuigan, G. Smith, J. Spiletic, V. Fine, and P. Nevski

ROOT OO Model to Render Multi-level 3-D Geometrical Objects via an OpenGL 264
    R. Brun, V. Fine, and F. Rademakers

A C++ Particle Data Table Interface 267
    L. A. Garren

Parallel Computing on a PC Cluster 270
    X. Q. Luo, E. B. Gregory, J. C. Yang, Y. L. Wang, D. Chang, and Y. Lin

Some Advance Methods of Statistical Analysis for the Muon g-2 Experiment at BNL 273
    S. I. Redin (for the g-2 Collaboration)

Analytical Calculation of Heavy Baryon Correlators in NLO of Perturbative QCD 277
    S. Groote, J. G. Körner, and A. A. Pivovarov

Summaries of Recent Computer-Assisted Feynman Diagram Calculations 280
    M. Fischler

WORKING GROUPS AND PANEL DISCUSSIONS

Use of C++ in Scientific Computing

Experiences Reviewing Scientific C++ Code 287
    M. Paterno

Reflections on a Decade of Object-Oriented Programming in Accelerator Physics 291
    L. Michelotti

C++ in Scientific Application: A Case Study 294
    W. E. Brown

Advanced Analysis Environments

The Development of the ROOT Data Analysis System 297
    R. Brun

Analysis Environment Challenges 301
    L. A. Tuura

Large-Scale Simulations

CDF Monte Carlo'2000 307
    P. Murat (for the CDF Collaboration)

Computational Challenges for Large-Scale Astrophysics Calculations 310
    B. Fryxell

DZero Monte Carlo-Simulation 313
    G. E. Graham (for the DZero Collaboration)

Large-Scale Simulations of Clusters of Galaxies 316
    P. M. Ricker, A. C. Calder, L. J. Dursi, B. Fryxell, D. Q. Lamb, P. MacNeice,
    K. Olson, R. Rosner, F. X. Timmes, J. W. Truran, H. M. Tufo, and M. Zingale

CMS Monte Carlo Status, Performance and Future Plans 319
    H. Wenzel (for the CMS Collaboration)

Large-Scale Cluster Surveys and Distributed Computing 323
    J. Annis

Simulation Packages in Accelerator Studies for the ν Factory 326
  V. D. Elvira
ATLAS Simulation: Status, Performance, and Future Plans 329
  F. C. Luehring (for the ATLAS Collaboration)

RAPPORTEUR TALKS

Artificial Intelligence 335
  H. B. Prosper
Status of Neural Network Hardware in High Energy Physics 338
  B. Denby
Innovative Software Algorithms and Tools Parallel Sessions Summary 342
  I. Gaines
Panel Discussion: C++ in Scientific Computing 345
  W. E. Brown, J. Kowalkowski, and M. Fischler
Advanced Analysis Environments—Summary 348
  S. Panacek
Summary: Panel Discussion on Large-Scale Astrophysical Calculations 351
  R. Rosner
Summary of HEP and Accelerator Physics Part of Panel Discussion on Large-Scale Simulations 355
  R. Raja
Worldwide Computing Working Group Summary 357
  I. Gaines
Perspectives on the Workshop Series 360
  M. Werlen and D. Perret-Gallix

Technology Show 367
Participating Companies 369
More of ACAT 2000 in Pictures 375
Workshop Program 389
Workshop Participants 399
ACAT Participants Photograph 403
Author Index 405

Preface

At the turn of the last century, the discovery of the electron, radioactivity, quantum theory and the theory of relativity opened up a new realm of physics that led to an explosive growth in Science and Technology in the twentieth century. Now at the dawn of this century and a new millennium *new physics* might be revealed again! Over the next decade or two, an impressive array of scientific instruments at the Tevatron, RHIC (Relativistic Heavy Ion Collider) and LHC (Large Hadron collider), LIGO (Laser Interferometer Gravitational Observatory) and SDSS (Sloan Digital Sky Survey), to name a few, will usher in the most comprehensive program of study of the fundamental forces of nature and the structure of the universe. Major discoveries are anticipated. But, it is our conviction that the pace of discoveries will be severely impeded unless a concerted effort is made to deploy and employ advanced computing techniques to handle, process and analyze the unprecedented amounts of data. It is against this backdrop that ACAT 2000, the VII International Workshop on "Advanced Computing and Analysis Techniques in Physics Research" was held at Fermilab during October 16-20, 2000.

The first workshop in the series was held in Lyon, France, in 1990 under the name "New Computing Techniques in Physics Research" and was organized by Denis Perret-Gallix (LAPP, Annecy). Following this, the workshop was held in Europe at approximately 18-month intervals. The ACAT 2000 workshop was the first to be held in the US with the updated name and with expanded scope. With the new name and acronym our intention was to emphasize the importance of the advanced analysis techniques in meeting the scientific challenges of the coming decade. The workshop also obtained a new logo which was inspired in part by the T-shirt Richard Feynman helped design for the Thinking Machines Inc. (Feynman worked on the Connection Machine at the Thinking Machines Inc. during the summer of 1983; Physics Today, February 1989).

The workshop followed four main tracks: Artificial Intelligence (neural networks and other adaptive multivariate methods); Innovative Software Algorithms and Tools; Symbolic Problem Solving and Very Large Scale Computing. The workshop covered applications in high energy physics, astrophysics, accelerator physics and nuclear physics. About 200 physicists and computer scientists from all over the world came together to present their work and to discuss new ideas and initiatives. Besides the plenary, parallel and poster sessions, the workshop included working group and panel discussion sessions focused on particular topics – uses of C++, Large Scale Simulations, Advanced Analysis Environments and Worldwide Computing. The working group and panel sessions allowed informal presentations and vigorous and stimulating discussions.

The recent revolution in computer hardware and software is not only transforming our everyday lives but also the way we do science. The key-note address by Dr. Ruzena Baczsy of the U.S. National Science Foundation focused on the transformation that Information Technology is bringing to our society and our lives. Dr. Michael Witherell, Director of Fermilab, in his opening remarks, noted that there is now growing recognition among physicists that we rely heavily on developments in advanced computing technology and that innovative scientists often recognize the need for a revolutionary development before the wider world understands what it is good for. A special Fermilab Colloquium by Dr. Stephen Wolfram gave glimpses of his work on "A New kind of Science," dealing with cellular automata and the evolution of complex systems. Perhaps, Wolfram's ideas could be part of a revolution in the making.

Intelligence on-line and off-line

High energy physics experimenters hope to discover the Higgs boson, Supersymmetry and/or other signals of new physics beyond the Standard Model. Finding signals of new physics may be the veritable case of "finding needles in a hay-stack". Fully multivariate analyses and advanced statistical techniques are crucial to achieve the physics goals. The new generation of experiments will present daunting challenge in data handling at all stages. So, intelligent handling and analysis of data is required both in real-time applications and in off-line data analysis. Artificial intelligence sessions surveyed the progress in the development and applications of adaptive multivariate methods such as neural networks. Prof. John Moody gave a feature talk entitled "Knowledge Discovery through Machine Learning" where he discussed some powerful "learning" algorithms that have been used to discover solutions to difficult problems in a wide range of fields.

Innovation is the key!

The usual topic of *Software Engineering* was renamed *Innovative Software Algorithms and Tools* to focus attention on the need to innovate and to stay on the cutting-edge. A variety of algorithms and tools for data management, processing and analysis were discussed. The world-renowned computer scientist Dr. Bjarne Stroustup, the inventor of C++, gave a feature talk "Speaking C++ as a native" and also served as a panelist in discussions centred around the "Use of C++ in Scientific Computing". His talk explained, by way of several simple but striking examples, how C++ can be used in a much more expressive manner than one commonly finds. Stroustrup, echoing the comments of Mike Witherell, noted that the world is slow to catch on to new ideas. He also emphasized the need for physicists to be involved in the C++ Standards Committees if they wish to influence the further development of that language.

Computations start with symbolism

Symbolic Algebra, which was re-christened *Symbolic Problem Solving* at this workshop, has, over the years, proved to be an effective approach to complex theoretical calculations. The *Schoonship* computer program, developed by one of the early pioneers, Martinus Veltman, helped him and Gerard 't Hooft to develop and verify calculational methods of non-Abelian gauge theory of electroweak interaction. The *Schoonship* program, using symbols, performed algebraic simplifications of the complicated quantum field theoretical expressions. (Veltman and 't Hooft won the 1999 Nobel Prize in physics *for elucidating the quantum structure of the electroweak interaction.*) The importance and efficacy of such symbolic, computer algebra programs to perform symbolic manipulations and precise calculations of physical quantities is now well recognized. There are now many high energy specific as well as commercial, general purpose software packages (such as Maple and Mathematica) available for symbolic calculations. The ACAT 2000 workshop included feature talks by Gaston Gonnet on "Computer Algebra" and Stephen Wolfram on "Mathematica" among others.

Grid for the global village!

The experimental investigations we are embarking on involve thousands of scientists from across the world. Mining the scientific treasures from these experiments, over national and international distances, will present new problems in data access, processing and distribution, and remote collaboration on a scale never encountered in the history of science. "Grid computing" is emerging as the infrastructure that will connect multiple regional and national computational centers, providing data access and creating a universal source of pervasive and dependable computing power. Grid computing was therefore the focus for a whole day at the workshop. Various champions of the grid projects "GriPhyN", "Particle Physics Data Grid (PPDG)" and "European DataGrid", such as Ian Foster (ANL), Paul Avery (Florida), Harvey Newman (Caltech), Miron Livny (Wisconsin), Luciano Barone (INFN), Fabrizio Galgliardi (CERN) and other pioneers of grid and world-wide computing contributed. These were augmented by other plenary, parallel and panel talks on very large scale computing and simulations. Robert Ryne spoke on accelerator physics, Alex Szalay on astrophysics, Paul Mackenzie on lattice QCD calculations and Aiichiro Nakano on molecular dynamics simulations.

Technology Show

A Technology Show was co-sponsored by SGI, Cisco Systems and Fermilab. The show featured the Reality Center for collaborative visualization, IP streaming video, IP telephony, wireless LAN by SGI and Cisco, and hardware and application software exhibits from Wolfram Research, Platform Computing, Objectivity, Kuck & Associates Inc. and Waterloo Maple.

Acknowledgments

We would like to thank Fermilab Director Michael Witherell, our sponsors, the U.S. Department of Energy and the National Science Foundation. We thank Cisco Systems and Silicon Graphics for co-sponsoring the workshop and the Technology show. We also thank the American Physical Society and the European Physical Society for endorsing the workshop. We would like to thank the local and international organizing committees and in particular Denis

Perret-Gallix, Bruce Denby, Mark Fischler and Irwin Gaines for extraordinary help. Our special thanks are due to the Fermilab visual media services headed by Fred Ulrich for help with workshop poster and other graphics, photographs and video streaming. Last but not least, our secretarial staff did a fantastic job in the administrative organization of the workshop; we appreciate their efforts and thank Emily Pahlavan for co-ordinating the efforts and Mari Herrera for helping with the proceedings.

As a final note, we would like to inform the readers that the video-streamed plenary talks are available at the ACAT web site: http://conferences.fnal.gov/acat2000/.

Fermilab   Pushpalatha Bhat
April 9, 2001   Matthias Kasemann

Pushpalatha Bhat   Matthias Kasemann

Opening Remarks

Mike Witherell, Director, Fermilab

"We have wonderful opportunities awaiting particle physics over the next decade. Two technologies are widely recognized as having driven our field from the beginning - accelerators and particle detectors. But there is also growing recognition that we rely on developments in advanced computing technologies. Innovative scientists often recognize the need for a revolutionary development before the wider world understands what it is good for. Once something becomes available, of course, lots of people know what to do with it, as we have learned over the last decade with the World Wide Web. There is a mutual benefit in collaboration between forefront physics research and computing technology. We rely, all over our laboratory (and our community), on continued innovations in the areas being discussed here at this conference."

CHAIRPERSONS OF THE WORKSHOP

Pushpalatha Bhat (Fermilab)
Matthias Kasemann (Fermilab)

ACAT COMMITTEES

International Advisory Committee

Halina Abramowicz	(Tel Aviv Univ.)	Christian Kiesling	(MPI, Munich)
Karl-Heinz Becks	(Wuppertal Univ.)	Paul Kunz	(SLAC, Stanford)
Chris Berger	(RWTH-Aachen Univ.)	Leif Lonnblad	(Lund Univ.)
Pushpalatha Bhat	(Fermilab, Batavia)	Victor Matveev	(INR, Moscow)
Rene Brun	(CERN, Geneva)	Denis Perret-Gallix	(IAC Chair, LAPP, Annecy-le-Vieux)
Bruce Denby	(Versailles Univ.)	Peter Overmann	(Wolfram Research, Inc.)
Jochem Fleischer	(Bielefeld Univ.)	Carsten Peterson	(Lund Univ.)
Raoul Gatto	(Geneva Univ.)	David Quarrie	(LBL, Berkeley)
Gaston Gonnet	(ETHZ, Zurich)	Ettore Remiddi	(Bologna Univ.)
Viacheslav Ilyin	(INP/MSU, Moscow)	Jose Seixas	(UFRJ, Rio de Janeiro)
Fred James	(CERN, Geneva)	Yoshimisu Shimizu	(KEK, Tsukuba)
Robert Jones	(CERN, Geneva)	Dmitri Shirkov	(JINR, Dubna)
Jacques Jousset	(LPC Clermont-Ferrand)	Alexandre Smirnitsky	(ITEP, Moscow)
Toshiaki Kaneko	(Meiji Gakuin Univ. Yokohama)	Ludwig Tauscher	(Basel Univ.)
Andrei Kataev	(INR, Moscow)	Jos Vermaseren	(NIKHEF, Amsterdam)
Setsuya Kawabata	(KEK, Tsukuba)	Monique Werlen	(LAPTH, Annecy-le-Vieux)

Local Organizing Committee

Pushpalatha Bhat	(Fermilab)	Joe Lykken	(Fermilab)
Mark Fischler	(Fermilab)	Ruth Pordes	(Fermilab)
Irwin Gaines	(Fermilab)	Harrison Prosper	(FSU)
Al Goshaw	(Fermilab & Duke U.)	Betsy Schermerhorn	(Fermilab)
Stephen Holmes	(Fermilab)	Marjorie Shapiro	(LBL)
Matthias Kasemann	(Fermilab)	Vicky White	(Fermilab)
Rocky Kolb	(Fermilab & U. Chicago)	John Womersley	(Fermilab)
Lee Lueking	(Fermilab)		

Secretarial Staff

Emily Pahlavan	Susan Deibele	Michelle Lopez	Cynthia Sazama
Mari Herrera	Judy Hentges	Sarah McCook	
Preseta Baldwin	Jo Ann Larson	Patti Poole	

Technical Support

Dane Skow	Tim Doody	Steve Fry	Al Johnson
Dick Adamo	Ed Hagler	David Tang	Jenny Mullins
Gerry Bellendir	Jeff Kallenback	Fred Ulrich	Karen Seifrid
John Bellendir	Jack MacNerland	Diana Canzone	Cindy Arnold
Phillipe Canal	John Urish	Reider Hahn	Ray Fonseca
Keith Coiley	Phil DeMar	Jim Shultz	Larry Thomas

PLENARY SESSIONS

Ruzena Bajcsy
(National Science Foundation)

John Moody
(Oregon Graduate Institute)

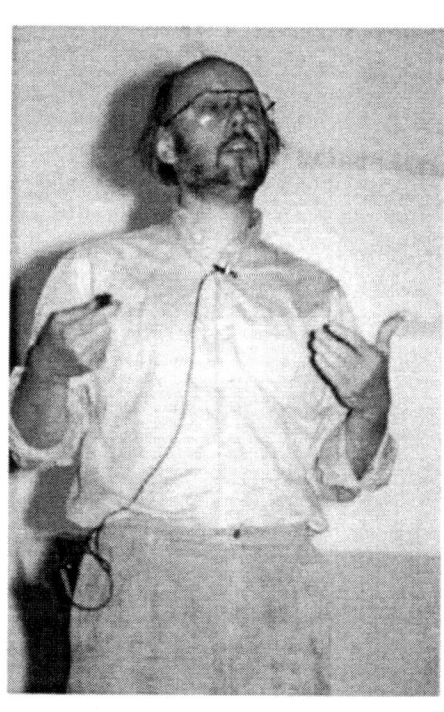
Bjarne Stroustrup
(AT&T Labs-Research)

Paul Mackenzie
(Fermilab)

Bob Aiken
(Cisco Systems)

Rex Tanakit
(Silicon Graphics Inc.)

Yael Maguire
(Massachusetts Institute of Technology)

Pushpa Bhat
(Fermilab)

Louis Lyons
(Oxford University)

C. Kiesling
(Max Planck Institute)

S. Wolfram
(Wolfram Research)

Ian Foster
(Argonne National Laboratory)

Aiichiro Nakano
(Louisiana State Univeristy)

Information Technology Research: Transforming Our Future

Dr. Ruzena Bajcsy

Assistant Director
Computer and Information Science and Engineering Directorate
National Science Foundation

Abstract. The Information Age is transforming our economy and our lives. In its pathbreaking 1999 report to President Clinton, the Presidential Information Technology Advisory Committee (PITAC) outlined the ten crucial ways that new technologies are transforming society in the U.S. It is clear that the Federal government will need to provide the critical R&D investments that will help retain and bolster the U.S. technological lead in the 21st century. These investments will also support efforts to make new technologies and their benefits available to all U.S. citizens.

INTRODUCTION

The Technology Revolution. The Information Age. The New Economy. We are living in exciting times today, and no one is quite sure how to describe them. What is certain however is that rapid advances in computing, information, and computation technologies, particularly over the last two decades, are transforming our world rapidly in ways that are often difficult to comprehend. Like the discovery of fire, the invention of movable type, and the harnessing of electricity, these transformations will profoundly influence how we live our lives in the future—but their effects can be maddeningly unpredictable.

In the U.S., many of these transforming technologies, including the Internet itself, grew out of basic research supported by U.S. Government agencies such as the National Science Foundation (NSF) and the Defense Advanced Research Projects Agency (DARPA). But in recent years, Federal investments in crucial high performance computing and communications research has not kept up with rapid developments in these technologies. To address this problem, the Clinton Administration established, in February 1997, a bipartisan Presidential Information Technology Advisory Committee (PITAC), comprised of technology leaders from business and academia, to advise the U.S. Government how to bolster its sagging investments in critical Information Technology Research and Development (IT R&D).

THE PITAC REPORT

In its pathbreaking 1999 report to the President, *Information Technology Research: Investing in Our Future* [1], the PITAC outlined ten important ways in which technology is transforming modern society. It also issued a series of forward-looking recommendations designed to support the research community and ensure that the benefits of technological education and innovation would be made available to all citizens. In addition, it underlined the continued importance of Government investments in the research and development of leading-edge technologies that might take decades to achieve fruition—precisely the kind of long-term R&D that increasing numbers of private sector businesses, driven by the incessant demands of the competitive global economy, no longer find it economically feasible to support.

The Ten Transformations

In the opinion of the PITAC, the ten National Challenge-level transformations that must be addressed in the 21st century are:

Transforming The Way We Communicate

By now it is abundantly clear that the Internet is the heart of the national and international communications revolution. But the current Internet is already bursting at its technological seams. We must learn to expand, to scale the Internet to anticipate future growth and an ever-increasing hunger for bandwidth. In addition, communications must be simplified and networks must be made more robust. From top scientists to novices who've just purchased their first PC, the Internet experience is far more complicated, far less simple than operating a television, and these reliability and usability problems must be addressed.

Global networking also poses interesting challenges. How do we respect national boundaries and sovereignty? How do we best take advantage of the Internet for sharing information globally? These are questions for which there is no simple answer and no simple solution. But to discover these answers and solutions, we must invest in basic research immediately and continuously.

Transforming the Way We Deal with Information

While the Internet itself is transforming society, the effect of the information it carries may ultimately prove to be even more profound. But the vast treasure trove of information that is now seemingly at our fingertips is often difficult to access and search. How can we improve our interface with this information? Are CRTs and keyboards adequate to the task, or do we need to change the way we think about accessing information? Future information access may be derived from today's experimental multi-modal human computer interaction technologies. In addition to the traditional keyboard, access to information by means of speech, touch, gesture recognition and synthesis will eventually provide equal access to novices and experts, regardless of their physical condition or global location.

Information overload is also a serious problem. Sometimes the problem for a researcher is not discovering where the information is located, but sorting through the huge amount of information that is uncovered. Today's information systems require dramatic improvements in data access methods. Current research issues include network reliability and bandwidth, scalable software support, database structure and retrieval algorithms, robust and secure access, and the maintenance of quality audio and video under heavy network loads. Additional issues, not necessarily involving high end science, but important nonetheless, include copyrights, intellectual property, and business issues.

Transforming The Way We Learn

Education is not keeping pace with today's rapid advances in technology. We must dramatically improve the information infrastructure to transform the way in which the average student or citizen learns. To accomplish this, we must first improve software technologies to enable fast, easy development of new educational materials for the 21st century, and support simplified modification and maintenance of these materials by educators. We must also determine rationally what educational needs can be fulfilled by computing and communications technologies, and what needs can be fulfilled by traditional methods. The ultimate goal is to help all citizens learn and use new technologies effectively in personal and professional lives.

Transforming The Practice Of Health Care

Federal agencies are already working on the problems of a national infrastructure for electronic medical records and health system intranets for data sharing. Chief among the problems to be solved are the issues of privacy and security, and the management of knowledge repositories. Telemedicine is another promising area of research, but we need to solve complex technical problems in order to make telemedicine work for patients *and* physicians. New research is also needed in remote visualization and robotics that will enable expert techniques and skills to be carried simply and reliably to remote or underprivileged areas of the country. But for telemedicine and remote surgery to become a reality, we will need to increase the reliability and stability of today's capricious communications networks. High-reliability, low-latency communications are needed to support sophisticated applications such as telepresence surgery.

Transforming The Nature Of Commerce

The Internet has already changed how business is transacted. It is now clear that leading-edge technologies can get companies closer to customers and reduce paperwork, purchasing costs, and delivery time. But once again, privacy and security are critical topics. International trade relations will become more important in this area.

Transforming The Nature Of Work

It has been estimated that as many as 15 million U.S. workers will become telecommuters over the next decade. This will enhance productivity, and provide organizational flexibility and environmental benefits. To continue transforming the nature of work, high speed networking capacity will be needed, collaborative software technologies will be critical, and privacy and security of communications will once again be crucial. But this transformation involves human factors problems as well. The social and economic implications of workforce and workplace transformation must be studied, and the worker skill base needs to be significantly and continuously bolstered. Lifelong learning will become paramount so that no worker will fear that technology will make him or her obsolete.

Transforming How We Design And Build Things

New technologies can and are increasing productivity, reducing cost of goods sold, improving product quality, longevity, and reliability. Networked computers can allow simultaneous modification of standard products to meet specific customer needs. Leading edge technologies are also helping companies maintain maximum flexibility and reduce design cycle time. But new high-end technologies are needed for concept design, simulation, analysis, and data mining. There is also a need for improvements in planning and scheduling, purchasing, investment, and cost analysis.

Transforming How We Conduct Research

Research problems are becoming more complex and interdisciplinary in nature. Increasingly, high speed computers and networks are enabling discoveries across a broad spectrum of disciplines, and, as a result, innovative methods for collaboration around the globe are being created and promoted. Key technologies in this area include high-end computing for modeling complex physical phenomena; advances in collaborative environments; visualization of complex datasets; innovative data-mining techniques; and management of very large datasets and databases.

Transforming Our Understanding Of The Environment

We must support a substantial increase in research in climate modeling to improve the accuracy of local and regional forecasting; disaster management; and support for national and international energy and environmental policies. Progress in this area depends on improved computational methods requiring order of magnitude increases in computing capability to deal with immense problems of time and space. In addition, researchers will need to design improved numerical methods and algorithms and sophisticated new tools for data storage, management, analysis and visualization, as well as improving software development and testing techniques, and advanced networks for distributed computing.

Transforming Government

We must insure that Federal, state, and local government institutions become more efficient and responsive to all citizens through improved information technologies. Challenges in this area include improvements in systems and data access methods; high performance data storage; and tools to locate and present information. Once again, to accomplish this, robust, reliable, and secure networks are critical. Improvements must be available to *all* citizens, and barriers to access must be surmounted.

PRIORITIES FOR FEDERALLY SUPPORTED RESEARCH:

Priorities for Federally-supported research to address these transformations comprise software, including a close examination of software R&D methods; fundamental research in component technologies; R&D in information capture, management, analysis, and availability; and human and computer interfaces and interactions. To accomplish these aims, the Government will need to make software research a substantive component of every major IT research initiative

In the area of high-end computing, the Government will support R&D in innovative computing technologies and architectures and software for improving high-end performance. Federally supported researchers will also aim to achieve sustained petaops/petaflops performance on real applications via software and hardware. The Government will support this continued science and engineering research via continued acquisitions of high-end systems.

The Government will also focus on crucial networking issues with an emphasis on Scalable Information Infrastructure. Researchers will attempt to more fully understand the behavior of the global-scale network, including collecting and analyzing performance data and modeling and simulating network behavior. Other topics will include the physics of the network, with an emphasis on optical and wireless technologies and bandwidth; R&D on scaling the Internet and creating middleware to enable large-scale systems; and work on large-scale applications and the scalable network services they require. Select Government agencies will also create testbeds to support these efforts.

The Federal government will also address the socioeconomic impact of the Information Age on average citizens who may, at times, feel as if new technologies are passing them by and making their skills obsolete in the world of tomorrow. Government/university/industry partnerships will strive to increase information technology literacy, access, and research capabilities; expand participation of women and minorities in IT careers; and remove barriers to high-bandwidth connectivity for all citizens. Federal technology investments will also aim to accelerate and expand IT education efforts from K-adult, including the vital topic of lifelong learning, thus strengthening the use of IT in education itself.

CONCLUSION

Industry cannot, by its nature, fund long-term research due to shorter-range goals and pressures of running a profitable business in a highly competitive world economy. But it is abundantly clear that if the private sector trend away from long-term, fundamental scientific R&D continues, the flow of innovative ideas that formed the underpinning for the economic miracle of the 1980s and 1990s will soon cease—or at the very least, will slow to a trickle. Under such circumstances, it will be difficult for the U.S. to maintain its technological leadership.

Fortunately, the Federal government is moving to increase funding in the crucial science areas that form the underpinnings of these ten dramatic transformations cited by the PITAC. In their view, the Government's role, ultimately, is to retain and expand U.S. leadership in long-term fundamental IT R&D. The societal gains from this kind of investment are enormous. If we move ahead on the PITAC's vision of the future, our international competitiveness will remain secure; educational levels will be profoundly and dramatically improved for all citizens; and societal problems and dislocations can be addressed, allowing the U.S. to make dramatic advances in the new century—and allow those advances to be shared by all.

REFERENCES

1. The Presidential Information Technology Advisory Committee, *Information Technology Research: Investing in Our Future,* Washington, DC, 1999.

Speaking C++ As A Native[†]

Bjarne Stroustrup

AT&T Labs – Research

Abstract. C++ supports several styles ("multiple paradigms") of programming. This allows great flexibility, notational convenience, maintainability, and close-to-optimal performance. Programmers who don't know the basic native C++ styles and techniques "speak" C++ with a thick accent, limiting themselves to relatively restrictive pidgin dialects. Here, I present language features such as classes, class hierarchies,[*] abstract classes, and templates, together with the fundamental programming styles they support. In particular, I show how to provide generic algorithms, function objects, access objects, and delayed evaluation as needed to build and use flexible and efficient libraries. The aim is to give an idea of what's possible to provide, and some understanding of the fundamental techniques of modern C++ libraries.

INTRODUCTION

What can I tell you about C++ in physics computation? Clearly I can't tell you about physics; I'm not a physicist. Clearly I can't tell you about math, because I haven't practiced it since I got my degree. And clearly I can't tell you about all of the wonderful C++ programs and libraries that you have, because you know them and I don't.

Instead, I'm going to show a lot of little snippets of code and explain why I think they are interesting and useful. Before I do that, I'm going to talk a little about standard C++, assuming that you are not all up on the latest state of the C++ world.

WHAT IS C++?

C++ is a general purpose programming language. It can do more things than any language I've heard deemed "general purpose." It has a bias toward systems programming: you can hack device drivers with C++; it has even been used to program diesel engine fuel injectors. Perhaps most significantly, C++ is a multiparadigm programming language because it's meant to support a variety of ways of expressing yourself.

C++ supports:

- *C-style programming* — C++ is a better C, maintaining C's flexibility and run-time efficiency while improving type checking;

- *data abstraction* — the ability to create types that suit your needs;

- *object-oriented programming* — the idea of programming with class hierarchies and runtime polymorphism; and

- *generic programming* — programming using type parameterization on both data types and algorithms.

Why does C++ support these diverse approaches? Because the most effective styles of programming involve a variety of techniques that people often classify as belonging to different paradigms.

We've had an ISO standard (2) for about 2 years now. While a set of minor clarifications is supposed to be voted in next week,[1] we've been working hard on stability (rather than on changes or new extensions).

This has led to a lot of implementations, and they are converging to the standard. Thus, our ability to write code which is portable across operating systems and machine architectures is improving. Some of these implementations are even free; this is important to grad students and others who like to try new things. C++ works on almost all platforms, including all of the major ones.

[†] This paper is an abridged and edited transcript, prepared by Mark Fischler, Walter Brown, and Bjarne Stroustrup, based on the video recording of the plenary talk.

[*] For lack of space, the discussion of class hierarchy design and the use of abstract classes is missing from this transcript.

[1] The vote has since been held as scheduled: the resolutions approving a Technical Corrigendum passed unanimously. Currently in the hands of the Project Editor, the resulting ISO document will be formally issued shortly (we hope) after he is finished with it.

As a result, there are lots of foundation libraries, lots of scientific libraries, and lots of support for applications of various sorts and for lots of environments. However, here I'm going to show small elegant examples — the building blocks for the programming styles — because you can find just about anything else in a lot of other places.

USING A PROGRAMMING LANGUAGE

Ideals:

- *directness* — represent concepts directly in a program; and

- *independence* — represent independent concepts independently in a program.

If you have some ideas, you want to write them down so that your thoughts are reflected directly in the code. What you want to say, you want to say clearly. If your thoughts are muddled, you are going to get a lousy program. That's a different issue, and there the program language designer can help only marginally: by supporting clear thinking better than woolly, muddled thinking. That's very hard to do without becoming paternalistic and restrictive. C++ invariably errs to the side of allowing you to say more rather than on the side of allowing you to say just what I might consider good.

It should also be possible to represent independent ideas independently. The alternative is a big glob of code that does "everything" for you, but you can't figure out which part is connected to what. To avoid such messes, you try to keep separate concerns and ideas separate, so that if one thing needs to be changed, you can do so without changing lots of apparently unrelated things.

The *class* is the main construct in C++. It is used to express concepts. The *class* plays a lot of roles because there are lots of different kinds of concepts. We can have, for example, value types; function types; constraint declarations; resource handles; node types; interfaces; and many more.

A VALUE TYPE CLASS

One of the simplest examples I've come up with to illustrate some of these ideas is a simple value type, *Range*. A *Range* object holds a value guaranteed to be within specified bounds:

```
void f(Range& r, int n)
{
  try {
    Range v1(0,3,10);
    Range v2(7,9,100);
    v1 = 7;   // ok: 7 is in [0,10)
    int i = v2;  // extract value
    r = 7;    // may throw exception
    v2 = n;   // may throw exception
    v2 = 3;   // will throw exception:
              // 3 is not in [7,100)
  }
  catch(Range_error) {
    cerr << "Range error in f();"
  }
}
```

A value within the bounds is fine. However, you will not succeed if you attempt to enter a value that is not within bounds into a *Range*. In that case, you get an exception. That is, *Range* throws an exception. You can catch such exceptions, as shown above, and possibly recover from the error. That way, you can pass around values guaranteed to be within the specified bounds.

A *Range* is a very simple concept and I can express it quite directly in C++:

```
class Range { // simple value type
  int  value, low, high;

  // invariant: low <= value < high
  void check(int v)
  { if (v<low || high<=v)
      throw Range_error();
  }
public:
  Range(int lw, int v, int hi)
    : low(lw), value(v), high(hi)
    { check(v); }

  Range(const Range& a)
  { low=a.low;
    value=a.value;
    high=a.high;
  }

  Range& operator=(const Range& a)
  { check(a.value); value=a.value; }

  Range& operator=(int a)
  { check(a); value=a; }

  // extract value:
  operator int() const
  { return value; }
};
```

This example embodies the very simplest idea of an object whose value is constrained. Notice the notion of the invariant established in a constructor to make sure that every *Range* object is valid and maintained by every member function. The function *check( )* says that, if the given value is not in bounds, we throw an exception. Each function that sets a value that could be out of bounds *check( )*s it first.

For example, the constructor uses *check( )* when we make an object from a triple (*low, value, high*): If the initial value wasn't in bounds, the object will never be created — the constructor will fail. That's fine because then you don't have an invalid object to get yourself in trouble with later. Assignments, too, *check( )* when needed.

The representation of a *Range* (i.e., the integers *value, low, high*) is *private* and only accessible by the functions declared in class *Range*. Note how construction (initialization) and assignment is specified by the programmer.

GENERALIZING A VALUE TYPE

I said that C++ is there to express ideas directly, but I didn't do quite what I said. While I'd said I was representing *thing*s in a range — and if the *thing* was not in the range you threw an exception — the code used *int*s, not "*thing*s."

The code can actually work for any type of *thing*, provided you can check the invariant that some *thing* is higher than *low* and less than *high*. And so I can generalize by saying that a *Range* is a range over values of type *T*, where *T* is anything that you can meaningfully check a range of. So I rewrote *Range*, not in terms of integers, but in terms of the arbitrary type *T*. This illustrates the C++ *template* concept:[2]

```
// simple value type
template<class T> class Range {
  T   value, low, high;
  // invariant: low <= value < high

  void check(const T& v)
  {   if (v<low || high<=v)
        throw Range_error(); }
public:
  Range(const T& lw,
        const T& v, const T& hi)
    : low(lw), value(v), high(hi)
    {   check(v); }

  Range(const Range& a)
  {   low=a.low;
      value=a.value;
      high=a.high;
  }

  Range& operator=(const Range& a)
  {   check(a.value); value=a.value; }

  Range& operator=(const T& a)
  {   check(a); value=a; }

  // extract value:
  operator T() const { return value; }
};
```

Now we can say that we want a range of integers, or a range of doubles, a range of characters, or even a range of *string*s:

```
Range<int> ri(10, 10, 1000);
Range<double> rd(0, 3.14, 1000);
Range<char> rc('a', 'a', 'z');
Range<string> rs("Algorithm",
    "Function", "Zero");
```

The *string* is the standard library string type. Of course you can compare *string*s: *string* comparison gives lexicographical ordering. It works. For example, here "Function" is between "Algorithm" and "Zero." So this generalizes nicely.

CONSTRAINTS

If you look back at the previous example, I still did not do exactly what I said. I'd said I was going to check ranges for any type *T for which it was meaningful to do comparisons*. But I didn't write that; I defined *Range* for an arbitrary type *T*. The construct *template<class T>* is the good old mathematical "for all *T*." How can we impose the constraint that our objects should have a linear ordering?

You can rely on the compiler to check. Code like this will not compile if you feed it a type for which < or <= doesn't work. However, the error message can be very verbose and cryptic, so let's try to express this constraint directly.

I want to ensure, for the class *Range<T>*, that *T* is comparable and that *T* is assignable. How — in Standard C++ — can I express that? Here is one way:

```
template<class T> struct Comparable {

  // the constraints check:
  static void constraints(T a, T b)
```

---

[2] Templates are often considered new because I didn't invent them until 1988, but the world can be slow to catch on to new ideas.

```
    {  a<b; a<=b; }

  // trigger the constraints check:
  Comparable()
    { void (*p)(T,T) = constraints; }
};

Template<class T> struct Assignable {
  // ...
};

template<class T> class Range
  : private Comparable<T>,
    private Assignable<T> {
  // ...
};

Range<int> r1(1,5,10);   // ok
Range< complex<double> > r2(1,5,10);
  // constraint error: no < or <=
```

I define a little template class *Comparable*, which will compile if *T* fulfills the criteria I defined, namely that if you have two *T*'s, you should be able to compare them using < and <=. The *constraints()* function just checks that constraint. It doesn't do any real work; it just expresses the constraint by exercising the aspect of the type that I am interested in. The constructor makes sure *constraints()* is exercised: it can make a *Comparable<T>* if and only if *constraints()* can be compiled for the type *T*. As you will see, the compiler never actually generates any code for this. I write *Assignable* in the same way.

Notice there are no macros and no magic here. Furthermore, it is pretty minimal:

- I have a single line that names the property I want to check;

- I have a single line that expresses that check; and

- I have a single line that expresses when it's checked.

This is not particularly new; I wrote about constraints in *The Design and Evolution of C++* (6) in 1994, but this is the first time I have been able to express general constraints in small and simple code snippets like this.

Now, when I take a range of *things*, the compiler checks whether *things* are comparable. Compilers can compile this so that it's all a compile-time effort with no run-time effect. So when I talk about representing concepts as classes, it doesn't mean that I have to create objects in the machine representing the concepts and invoke operations on them to get work done. It simply means that I can express my concept and have it work.

Of course *int*s are comparable, so *Range<int>* compiles. Next, we try for a range of double-precision complex numbers, but you can't make a *Range< complex<double> >* because we check *Comparable< complex<double> >*: we try to do a <, but that doesn't work — operator < is not defined for complex numbers — so we get a compile-time error.

A compiler will check anything you do to a template parameter class *T* even if you don't write specific constraints checks. However, if you've ever tried, you know that the error messages leave a lot to be desired. The main point of the *constraints()* technique is to make the constraints explicit. Doing that yields good, specific error messages and, importantly, it allows us to express a very general notion of constraint.

Anything you can say in the language you can check in a constraint. In particular, it is easy to express constraints involving more than one type. For example, if you have a template with three type arguments, *T1*, *T2*, and *T3*, you can say that the result of multiplying a *T1* by a *T2* should be assignable to a *T3* by simply saying *t3=t1*t2* for suitably declared *t1*, *t2*, *t3*. Because you express this in the language itself, rather than in some language designed to express constraints, this technique is actually more powerful than what is found in non-research languages and in most research languages. And still, expressing a constraint is four simple lines of code; no magic is required.

MANAGING RESOURCES

One thing that comes up again and again in my world is that there are a lot of resources to take care of. Memory is the one resource people always talk about, but more critical are things like file handles, thread handles, and sockets. It doesn't actually matter if you can clean up all of your memory if you have left thread handles hanging around, because they own memory you can't clean up.

In general, it's very, very difficult to deal with resources. In practice, however, most resources live in a scope and this is the simple and common case that I'm going to show a solution for.

The general structure of the solution is to acquire a resource by initializing some object that holds it. So we are introducing a class — a resource handle class — that represents the notion of the resource. The class controls access to the resource. Of course classes are good at that kind of control. You can access the resource only by using functions that the handle provides for that — representations are private. Creation is controlled by constructors, copying can be controlled, and final cleanup is provided by destructors. A destructor is a function that is guaranteed to be invoked upon exit from the scope of a variable, and it just releases the resource at that point.

Actually, this technique is the key to exception safety, but I don't have time here to go into that in detail. If you are interested, read Appendix E of *The C++ Programming Language, Special Edition* (5).[3]

To illustrate, I sketch a piece of code I've seen many times in many versions in C and C++ programs: you grab something, you use it, and then you release the resource:

```
void my_fct(const char* p)
{
  FILE* f = fopen(p,"r"); // acquire
  // use f
  fclose(f);              // release
}
```

This is fine as long as you actually get to releasing the resource. If a C program does a *longjump* here you're in trouble; if a C++ code *throws* an exception here you're in trouble. In short, this is very simple but very unsatisfactory code.

The naive fix that everybody uses when they first start playing with exceptions is to wrap a *try* block around the resource's initialization and use:

```
void my_fct2(const char* p)
{
  FILE* f = 0;
  try {
    f = fopen(p,"r");
    // use f
  }
  catch () {  // handle exception
    // ...
  }
  if (f) fclose(f);
}
```

This is fine, but I find that it is only fine if you apply the technique consistently and correctly. Here's what we just did:

- We found a problem in the code, a problem caused by people failing to think things through and take care of all the error conditions.
- Then we solved it by doubling the size of the code and complicating the control structure.

The chance of forgetting something and not getting things right is at least linear with the size of the code. So if I recommended this, I would be recommending a way of

---

[3] If your book doesn't have an appendix E, just go to my home pages (9) and download a version, or augment Bjarne's retirement fund by buying a new copy :-).

dealing with careless errors that doubled the probablity of careless errors. This particular problem actually held back exceptions in C++ for at least a year.

Remember: if I have a concept, I'm suppose to represent it directly by a class. So I create a little handle class to represent my notion of an open file:

```
// in some support library:
class File_handle {
  FILE* p;
public:
  File_handle(const char* pp,
              const char* r)
    :p(fopen(pp,r))
    { if (p==0) throw Bad_file(); }

  File_handle(const string s,
              const char* r)
    :p(fopen(s.cstr(),r))
    { if (p==0) throw Bad_file(); }

  ~File_handle()  // destructor
  { if (p) fclose(p); }

  // access functions:
  // ...
};

void my_fct3(string s)
{
    File_handle f(s,"r");
    // use f
}
```

The constructor creates the handle and opens the file. If *open()* succeeds, all is fine. If *open()* fails, we don't create the handle and exit *my_fct3()* throwing an exception. The destructor releases the resource — here, it closes the file — if you managed to acquire it. The handle class is the kind of stuff you stick in a support library. However, if you have a resource that nobody else has, you have to write a resource manager yourself. That will be maybe ten lines of code that you write once and then use wherever you acquire one of those resources. You typically acquire a kind of resource in many places in your code, but you need to write only one class to handle that safely. On the other hand, if you use the *try*-block approach, you have to get the error handling code right in every case.

The way you now write your code is to create a handle for the file named *s* with read access. Then you use the file through that handle. You don't have to explicitly close the file because that's taken care of by the handle's destructor. So I've simplified the code while making it exception-safe. The chances of making mistakes are now much more limited.

CONTROL ABSTRACTION

There is a related problem that was open, I estimate, for about 20 years. It's trying to deal with the fact that a lot of the code we write is of the form "Do some prefix code, then do the real thing, then do some suffix code." Common examples of that include lock/unlock, transaction-start/transaction-commit, debug-trace-on/off, and "acquire the resource, do the operation, release the resource." If you've written a large program, you'll have code of this general style somewhere.

As shown below, I want to take some arbitrary class X — say one you are going to write tomorrow so I couldn't possibly know what it is — and wrap it so that prefix code and suffix code is implicitly done. When I write *x->count()*, it should translate into *prefix()*, *count()*, *suffix()*, and similarly for other member functions of your class. Further, I shouldn't have to know what *prefix()* and *suffix()* are when I write the wrapper class:

```
void f(X& x)
{
  Wrap<X> xx(x,prefix,suffix);
  int n = xx->count();
    // prefix(); n=x.count(); suffix();
  xx->g(99);
    // prefix(); x.g(99); suffix();
}
```

This is close to what people mean when they say "control abstractions." It does something to the control flow in your program, in a guaranteed and declarative manner.

The constraints on a solution to this problem are that there should be optimal performance: It should be possible to inline *prefix()* and *suffix()*. This is very important if these are, say, assembly code that does lock/unlock. It has to work for pre-existing X's. Oh, and by the way, I can write it in 16 lines of standard C++:[4]

```
template<class T, class Suf>
class Wrap_proxy {
  T* p;
  Suf suffix;
public:
  Wrap_proxy(T* pp, Suf s)
    :p(pp), suffix(s) {}

  ~Wrap_proxy() { suffix(); }

  T* operator->() { return p; }
};
```

---

[4] The example looks longer because some of the lines are artificially wrapped to fit the two-column format.

```
template<class T, class Pre, class Suf>
class Wrap {
  T* p;
  Pre prefix;
  Suf suffix;
public:
  Wrap(T& x, Pre pref, Suf s)
    :p(&x), prefix(pref), suffix(s) {}

  Wrap_proxy<T,Suf> operator->()
  { prefix();
    return Wrap_proxy<T,Suf>(p,suffix);
  }
};
```

This wraps an object by storing away the prefix and the suffix and a pointer to the object. Whenever you call the resulting wrapper object using operator arrow, the prefix code is invoked, then a proxy is created and the proxy's *operator->()* is called. When you are finished with the proxy, it is of course destroyed, calling the suffix code. This works. I present it here to show that

- you can do some control abstraction in C++, and
- the range of notions you can represent as a class is much wider than most people are willing to believe.

I describe the wrapper further in a paper (8).

GENERIC PROGRAMMING

Now I'm going to explain some of the basics of generic programming as it is represented in the C++ standard library. The first idea is that you can make yourself a lot of useful containers, such as *vector<T>*, *list<T>*, and *map<K,V>*. Since the standard library has all these and more, and since these containers' quality is really quite good, you can just use them without having to write them yourself.

Further, there are some very common things — you can find them in Knuth or Sedgwick — that we frequently do to all kinds of containers:

- find an element in a container,
- sort a container,
- perform an operation on each element of a container,
- remove elements that meet a given criteria from a container,
- copy a container.

You don't want to hand-code these algorithms each time you use them. That would just be a waste, a nuisance, and a well-known source of bugs. People don't write their own sorts anymore, except for the few people who actually have a chance of getting them right. Similarly with the other basic algorithms, so they're provided in the standard library.

And we don't want to repeat the code for each algorithm for each container. That would be a nuisance and a maintenance hazard.

The standard library is organized as a framework of containers and algorithms. This organization is the work of Alex Stepanov. The problem was "how do you provide an algorithm over a set of containers that includes the Standard ones as well as those you have defined yourself?" The key idea is to say that any container can be seen as a sequence of elements. A sequence has a beginning and an end, and if you have access to an element you can get to the next element. That's all there is to it.

```
begin                     end
  |                        |
  v                        v
elem -> elem -> ... -> 0
```

Something that refers to an element of a sequence is called an *iterator*. The obvious C-style notation is that ++ for an iterator means "refer to the next element" and * means "get the value of the element referred to." This sequence notion is very general: it covers vectors, it covers lists, it covers trees. While the implementation of these notions may be different, the semantics of getting to the next element and getting the value of the next element are independent of what kind of data structure you are talking about as long as you can view the elements as part of a sequence.

Here is some pseudo-code expressing what we want to do: copy a sequence from begining to its end onto output, find the value in the sequence, and count the number of occurrences of the value in the sequence; we want to make it into real code:

```
// copy sequence to output:
copy(begin,end,output)

// find value in sequence:
find(begin,end,value)

// count occurrences of value
// in sequence:
count(begin,end,value)
```

One of the ideals of programming is the idea of direct representation of ideas in code. Given this pseudocode, what is the smallest step we could do to turn it into real code? Well, that would be doing nothing. We can't quite do that in C++: we have to put a semicolon after each expression to make it a statement.

We are also getting close to the other major ideal here: to represent independent concepts independently in the code. Notice what we have kept independent here.

- *Container type*: When we look at elements, we don't have to know which kind of container we are looking at. There's the notion that a container should have a beginning and an end, and — given its notion of begin and end — we can start at the beginning and examine each element until we reach the end. That's all that's required of a container.

- *Element type*: Element types are independent of the container types. A type is not required to be part of some class hierarchy to be used for elements in a container. The container notion does not intrude on the notion of an element.

- *Algorithm*: We separate algorithms from the containers. An algorithm need not be a member of a (container) class.

- *Comparison criteria*: When we do anything interesting with algorithms, we have comparison criteria, polices, and such. Each can be independently specified.

We can vary these four things (containers, elements, algorithms, and policies) independently. This is what allows the standard library to be five or six thousand lines of code, yet to do more than many libraries 20 times its size.

Let me show you some code for a simple linear search to find, in the sequence from *first* to *last*, the value *val*:

```
template<class In, class T>
// find val in sequence [first,last):
In find(In first, In last, T val)
{
   while (first!=last && *first!=val)
      // while we haven't reached the end
      // and haven't found what we seek
      ++first;  // carry on
   return first;
}
```

This is real code: The standard library looks like this. We've parameterized *find()* so that we don't need to know which kind of iterator is used to represent the sequence. The type of element is another parameter.

So, we go through a loop until we have reached the end or found what we are looking for. As long as we haven't reached the end and as long as we haven't found the value we are looking for, we make the iterator refer to the next element and try again.

You may or may not like C or C++ syntax, but this is colloquial. If you want to deal with this class of languages, you'd better get used to it. Familiarity is often confused with what is natural. I don't think I'm doing that: this notation is not natural, but it's familiar to a lot of people. People can come to love it; I'm not sure they should, but they do.

What we can do now is to take a vector of integers and, say, apply *find()* to it for some value *x*, Did we hit the end? If so, *x* wasn't there; otherwise, we found *x*:

```
void f(vector<int>& v, int x)
{
  vector<int>::iterator p
    = find(v.begin(), v.end(), x);
  if (p != v.end())
    { /* we found x */ }
  // ...
}
```

Since this is a vector, the iterator is almost certainly implemented as an ordinary pointer. So ++*first* simply makes *first* point to the next element in the vector. It's a standard machine instruction that adds a constant to a pointer. That's simple and efficient. Looking for the value **first* means dereference a pointer. If you measure this code, you will find that it's optimal; you cannot write better-performing code in C.

Now, let's try with a list of strings. I try to find the string *s* in it, using *find()*:

```
void f(list<string>& lst, string s)
{
  list<string>::iterator q
    = find(lst.begin(), lst.end(), s);
  if (q != lst.end())
    { /* we found s */ }
  // ...
}
```

An iterator for a list is unlikely to be implemented as a pointer to an element. It's going to be a pointer to some kind of link node. When we do a comparison here, it compares two link nodes. That's fine; that's still a simple and efficient pointer comparison. When I dereference — when I want to get a value — I grab into that node to extract its value field. When I increment the list iterator to get to the next element, I indirect through a "next field" to the link node for the next element. Again, you can see that this is exactly the code you would have hand written in any language you care to use: C, assembler, C++, whatever.

When we get out of the loop, we've found *s* or we've reached the end. The *find()* algorithm is really basic. However, there's lots and lots of code like that in the C++ standard library. It's deceptively simple, but it is fast and it is general.

Looking for a specific value is a special case of looking for something that meets some criteria. In my work, I more often look for something that fulfills a predicate *P*. That is, I'm not looking for a specific value such as 7, I'm looking for a value less than a threshold, or higher than a threshold, or something like that. So, I want to specify a predicate, something that express my criteria:

```
template<class In, class Pred>
In find_if(In b, In e, Pred p)
{
  while(b!=e && !p(*b))
      // while we haven't reached the end
      // and haven't found what we seek
    ++b; // carry on
  return b;
}
```

We just replace the earlier "not equal to" by "not meeting my criteria," and all of the code works again. Of course the *find()* function is just a simplification of *find_if()* where *P* is "equals." Here, I look in a vector of strings *v* for a string *"foo"*, using a predicate *less than "foo"*:

```
void f(vector<string>& v)
{
  vector<string>::iterator p
    = find_if(v.begin(), v.end(),
              Less_than<string>("foo"));
  if (p != v.end())
    { /* found: *p < "foo" */ }
  // ...
}
```

We go through the vector, from the beginning to the end, looking for something that's *less than "foo"*. If we didn't reach the end, we found something that meets that criteria and p now points to an element that did. This generates what I would consider the obvious code.

We can use *find_if()* for a list of records, where we want to check that the name field in the record is equal to that of a record that I'm interested in. For a lot of data processing that is exactly what you need: you check a notion of equality which is not the equality of the value of the record, it's the equality of some field of the record. Here, some notion of name-equality is used:.

```
void f(list<record>& lst,
       const Record& my_rec)
{
```

```
    list<Record>::iterator q
      = find_if(lst.begin(), lst.end(),
              Name_eq(my_rec));
    if (q != lst.end()) {
      // found: *q has same key as my_rec
    }
    // ...
}
```

FUNCTION OBJECTS

I have illustrated a general form of flexibility. *Name_eq* is the archetype of a predicate: it holds a value that you compare against. That value is stored when you construct the *Name_eq* object, and *operator()* — the application operator — simply does the comparison:

```
class Name_eq {
  const string s;
public:
  Name_eq(const Record& r)
    : s(r.name) {}

  static bool
  operator() const (const Record& r)
  { return r.n == s; }
};
```

We use *Name_eq* like this:

```
void f(list<record>& lst,
       const Record& my_rec)
{
  // ...
  find_if(lst.begin(), lst.end(),
          Name_eq(my_rec));
  // ...
}
```

For each element in *lst*, the predicate objects created by *Name_eq(my_rec)* is invoked. That function object compares the name field of the current element with the copy of name field of *my_rec* that was stored away by *Name_eq(my_rec)*.

Here is an archetypal function object. Such an object has a state that is established when you construct that object and that is used (in the application operator, *operator()*) as you go along:

```
template<class S> class F {
  S s;   // state
public:
  F(const S& ss) : s(ss)
  { /* establish initial state */ }

  void operator() (const S& ss)
  { /* do something with ss to s */ }

  // reveal state:
  operator S() { return s; }
};
```

This is very general. It is more general than a function because a function cannot be initialized to work against a contained state: A function object, in contrast, can carry state and you can extract the state from it. By parameterizing algorithms with such function objects you can express arbitrary predicates and policies.

It's quite common to pass a function object along, updating its state by an operation on each element of a container. The simplest example of that is to take a sum: you initialize a sum object to zero in its constructor — this would be the state, here a numeric value. As you go along, you add elements of the container to that value. When you are finished, you extract the resulting value sum from the sum object.

Interestingly, function objects also run faster than equivalent functions because little function objects inline better than functions. The reason is that when you pass a function you are passing a pointer to function and optimizers are not very good at dealing with pointers. On the other hand, if you pass a function object, you're passing an object rather than a pointer; when you do the operation on the object, you have the object, you have the function, and inlining is easy.

This is the reason that the generic general *sort()* in C++ often runs several times faster than *qsort()* in C. I have measured it from 2 to 7 times faster on things like floating-point numbers and simple strings. It's not really magic, these generic programming techniques just fit better with compiler technology than do C-style parameterization with pointers to functions.

DELAYED EVALUATION

You can use function objects directly. However, sometimes you'd like to use "the natural notation." That often means using operators like +, -, and *. In particular, we often want to express vector and matrix manipulation using conventional notation, e.g. *v=m*v2+v*. In addition, we want to evaluate such experessions without using temporaries, and without having expensive function calls compromise your run-time performance. The point of the following example is partly to avoid temporary values and partly to show you how to get the "natural" notation without overhead.

```
Matrix m;
```

```
Vector v, v2, v3;
// ...
v = m * v2 + v3;
```

The basic implementation idea is to generate a single function, *mul_add_and_assign(v,m,v2,v3)*, that knows that it's supposed to multiply, add, and assign. If this is on some form of CRAY you can write very beautiful code vectorizing such compound operations, but given only *v=m*v2+v3*, compilers are generally not smart enough to vectorize without help from the programmer. To help, we write something like this:

```
struct MV { // object representing
            // the need to multiply
  Matrix* m;
  Vector* v;
  MV(Matrix& mm, Vector& vv)
    : m(&mm), v(&vv) {}
};

MV operator * (const Matrix& m,
               const Vector& v)
{ return MV(m,v); }

MVV operator + (const MV& mv,
                const Vector& v)
{ return MVV(mv.m,mv.v,v); }

v = m*v2+v3;
  // v = MVV(MV(m,v2),v3);
  // mul_add_and_assign(m,v2,v3,v);
```

We make a little function object *MV* that simply keeps track that it has seen an *m* and a *v*. This represents the notion that *m* wants to be multiplied by *v*. We have operator *, given a matrix and a vector, make one of these *MV* objects. *MV(m,v)* expresses the notion that *m* would like to be multiplied by *v*. We do a similar thing for operator +: if you get an *MV* and a vector, it creates an *MVV* object that holds the matrix and the two vectors. So when we execute *m*v2+v3*, we just construct little objects until we have *MVV(MV(m,v2),v3)* — unravelling the expression collecting information — and in the end we use the collected information to generate *mul_add_and_assign(v,m,v2,v3)*.

The above example collects references to matrices and vectors. I chose matrices and vectors for this example because I know that lots of people must use large vectors and matrices. Copying a 10000x10000 matrix is expensive to most people. Another example of the delayed evaluation technique is to collect the value, the format, and an output stream so that when all of these things are together I can output the value with the right format onto the stream. Again, this relies on function objects. The whole thing is done by creating little function objects to hold the information until you got to the final function. Inlining is very important. So is pass-by-value, because these function objects get passed along, then the optimizer gets them, and they just disappear.

Function objects tend to be templates. An example here would be a matrix of *double*s stored densely: *Matrix<double,Dense>*. Most current C++ vector and matrix libraries work with these techniques, so they generate fast code. This is why many of you have seen graphs comparing the performance of Fortran and C++, with C++ winning. If you haven't seen such graphs, look for links on my C++ page. That was "known" to be impossible, but it's always nice to disprove a myth. These libraries' vector and matrix classes have little "policy objects" associated with them so a matrix is not just a matrix of *double*s, it's also something that controls, for further optimization, the way elements are stored and accessed. These "policy objects" controls need only be seen by expert users who care.

CLASS HIERARCHIES

I've spoken about C++ at length, giving a variety of examples, yet I haven't shown a single class hierarchy. According to some people's definition of OO, this means I haven't yet talked about Object-Oriented Programming. I should do so, because OOP is important and because some of the most interesting and important uses of C++ are in application domains that use class hierarchies effectively. However, object-oriented programming is the use of C++ that people know best — at least they think they know best. So here I have emphasized the other programming styles.

I think one of the keys to modern C++ is lots of little objects, as opposed to huge hierarchies. One of the reasons that hierarchies get large and massive is that you throw too much into them. Little objects representing policies, values, constraints, etc., are very useful and can provide generality, flexibility, and efficiency. In particular, "little objects" can be used to design leaner hierarchies by not relying exclusively on facilities represented within a hierarchy.

I do not want to be misunderstood: class hierarchies and their use in object-oriented programming are important. Lack of space, unfortunately, keeps me from describing this last piece of my talk in this transcript (but it is in the video of my talk that you can view from the ACAT2000 conference website).

SUMMARY

Try to think of C++ as a new language. A lot of you have used it for a long time; you will know that there are techniques that didn't work a few years ago when you last tried. A lot of these now work.

This is a good time to be adventurous because the standard is out, the compilers are starting to support the standard, not just in language features but also in terms of efficiency. On the other hand, of course, be careful! Not every technique works for every project and for every group of people. But this is a good time to start to see what concepts you can express more directly and more efficiently than before.

For those of you who are beginning with C++, please remember that C++ is not just C with a few useless and inefficient bits added; you can write cleaner, shorter, and faster-running code in C++ than in C if you know how. An example is sorting: the general, generic, and typesafe *sort()* in the C++ standard library is not just easier to use than the C-style *qsort()*, it is often several times faster. And C++ is not just class hierarchies, there is a lot more to it. A lot of modern C++ techniques are focussed on templates, containers, and function objects.

Prefer the C++ standard library style to traditional C style; it is simply easier to express ideas using *vector*, *list*, and *string*, rather than with arrays, pointers, and casts. If you are not careful, you can get overhead in both cases. You have to understand things to write good code; you can't just blindly plow along in either the C++ style or the C style and expect to produce efficient and maintainable code.

FOR MORE INFORMATION

There's a lot of reference material available. You can look at my "Third Edition" book (5) — the "Special Edition" is the hardcover version — or you can get the Standard itself (2) via the web. My *Special Edition* was updated last year from my *Third Edition*: I added another 100 pages, corrected many errors, and clarified numerous issues. I'm now confident enough to offer $16 for every new bug reported to me. I haven't yet been ruined.

The Design and Evolution of C++ (6) is for people who are interested in why things are the way they are. Answers to many "why?" questions about the design of C++ can be found there. It is the closest thing we have for a rationale for the design of C++.

I'm the editor of Addison Wesley's *C++ In Depth* series. I'll mention two of the books here. One is Herb Sutter's book on exceptions (10), which gives a lot of exercises and discussions, going into greater detail about exception handling techniques than I do in my *Special Edition*. Another is Andy Koenig and Barbara Moo's book called *Accelerated C++* (3) which basically is a tutorial on modern C++; it is probably the first such introduction. It introduces templates four chapters before it introduces pointers. This gives you an idea of how much the world has changed.

There are some papers on the web. In (7), I do a microanalysis of some very simple C and C++ examples used in education. The results were good enough — from a C++ perspective — to cause a firestorm of letters to the editor when it was published last year. I consider it non-controversial. The code is available on my web site (9) so you can run it yourself.

There are many useful links on my C++ page (www.research.att.com/ bs/C++.html). In particular, the ACCU site (1) has many useful book reports. These reviews are done by professionals and are reasonably unbiased — as opposed to most reviews that you find on the web. Many of my favorite links can be found on my home pages: FAQ's, the standard itself, compilers, garbage collectors, papers, book chapters, etc.

REFERENCES

1. Association of C and C++ Users. www.accu.org
2. International Standard Organization, *The C++ Programming Language*, 1998.
3. Koenig, Andrew and Barbara Moo, *Accelerated C++*, Addison Wesley Longman, 2000.
4. Stroustrup, Bjarne, *Learning Standard C++ as a New Language*, C/C++ Users Journal. pp 43-54. May 1999.
5. Stroustrup, Bjarne, *The C++ Programming Language, Special Edition*, Addison Wesley Longman, 2000.
6. Stroustrup, Bjarne, *The Design and Evolution of C++*, Addison Wesley Longman, 1994.
7. Stroustrup, Bjarne, *Why C++ isn't Just an Object-oriented Programming Language*, Addendum to OOPSLA'95 Proceedings. OOPS Messenger. October 1995.
8. Stroustrup, Bjarne, *Wrapping C++ Member Function Calls*, The C++ Report. June 2000, Vol 12/No 6.
9. Stroustrup, Bjarne, www.research.att.com/~bs
10. Sutter, Herb. *Exceptional C++*, Addison Wesley Longman, 2000.

Advanced Analysis Methods in High Energy Physics

Pushpalatha C. Bhat

Fermi National Accelerator Laboratory, P.O.Box 500, Batavia, IL 60510

Abstract. During the coming decade, high energy physics experiments at the Fermilab Tevatron and around the globe will use very sophisticated equipment to record unprecedented amounts of data in the hope of making major discoveries that may unravel some of Nature's deepest mysteries. The discovery of the Higgs boson and signals of new physics may be around the corner. The use of advanced analysis techniques will be crucial in achieving these goals. I will discuss some of the novel methods of analysis that could prove to be particularly valuable for finding evidence of any new physics, for improving precision measurements and for exploring parameter spaces of theoretical models.

"A reasonable man adapts himself to the world.
An unreasonable man persists to adapt the world to himself.
So, all progress depends on the unreasonable one."
-Bernard Shaw.

INTRODUCTION

The CDF and DØ experiments are preparing for a new and possibly a decade-long run at the upgraded Fermilab Tevatron. A new generation of accelerators and detectors are on the horizon. In the coming decade, we hope to discover the Higgs boson and find evidence of physics beyond the Standard Model (SM) such as Supersymmetry or Technicolor, or something completely unexpected. In order to achieve the goals of the high energy physics (HEP) community, I believe it is crucial that advanced and optimal data analysis methods be used both on-line and off-line [1,2].

In our quest to understand the universe, we continually experiment, analyze observations, interpret results and update our knowledge. In high energy physics, there was a time when we exposed nuclear emulsion targets to particle beams or took bubble chamber photographs of interactions and "recorded" data from scans off-line. In the not-so-distant past, we could afford the luxury of writing data to storage media based on simple interaction criteria and organize, reduce and analyze data completely off-line. But, as our knowledge-base increased, and as we began to address more complex problems, looking for extremely rare processes at higher beam energies and higher luminosities, it became necessary to sift through large amounts of data on-line before selected data were written out. Each new generation of experiments is more demanding than the previous in terms of data handling; the rates of interactions and the number of detector channels to be read-out often grow by orders of magnitude. Finding the signals of new physics become a veritable case of "finding needles in a hay-stack". So new paradigms and new technologies need to be identified, developed and adopted.

INTELLIGENT DETECTORS

Today, data analysis in HEP experiments starts when a high energy event occurs. The electronic data from the detectors need to be transformed into useful "physics" information in real-time. One can envision that the calorimeter, for instance, can have "intelligence" close to its electronic read-out so that the clustering and energy measurements are readily available. Such information from different sub-detectors can be used to extract event features, such as number of tracks, high transverse momentum (p_T) objects and object identities. The features are then used to make a decision about whether the event should be recorded. So we need to build intelligent detectors and smart triggers! Feature extraction, classification or particle identification can be accomplished using algorithms implemented either in specialized hardware (neural network chips, for example) or in conventional hardware such as Field Programmable Gate Arrays (FPGAs) or Digital Signal Processors (DSPs). The H1 experiment at HERA, for example, has used neural network hardware in its Level-2 trigger. This has been operated successfully since 1996 and has been crucial for the rich physics results from H1. The project has been discussed in detail in these proceedings by Chris Kiesling [2]. Innovative data management on-line, and the use of smart algorithms encoded in trigger hardware would be beneficial in meeting the demands of data handling and analysis on-line. Use of expert and fuzzy logic systems in controls and monitoring of detector electronics is an area that has not received much attention and needs to be explored.

OPTIMAL ANALYSIS METHODS

My golden rule for an optimal analysis is this:

*"Keep it simple.
As simple as possible.
Not any simpler."* - *Einstein*

Most data analysis tasks such as charged particle tracking, particle identification and signal/background discrimination, fitting, parameter estimation, functional approximation (deriving various correction and rate functions) and data exploration, normally involve several measured quantities or 'feature variables". To obtain the best possible results it is necessary to make maximal use of information in the data and hence employ optimal multivariate methods of analysis [1,3].

The power of multivariate methods in discrimination tasks can be illustrated by the following simple example. In Fig.1, I have shown distributions of two observables x1 and x2 arising from two bi-variate Gaussians. One sees considerable overlap of the two classes in the one-dimensional projections (Fig. 1(a,b). But if one examines the data in 2-dimensions, one sees that the two classes of events are separable (Fig.1(c)). A Fisher linear discriminant, an appropriate linear combination of x1 and x2 plotted in Fig. 1(d), can provide a clear separation of the two classes.

In real life examples, decision boundaries between classes are more complicated and require the use of more sophisticated, more flexible, non-linear methods

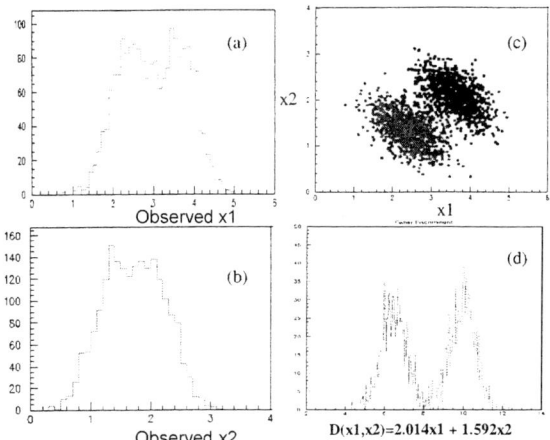

FIGURE 1. (a,b) Distributions of two hypothetical observables x1 and x2 arising from a mixture of two classes of events. (c) Original 2D distribution for the two classes of events and (d) Fisher linear discriminant that provides a mapping to 1-dimension in a way that cleanly separates the two classes.

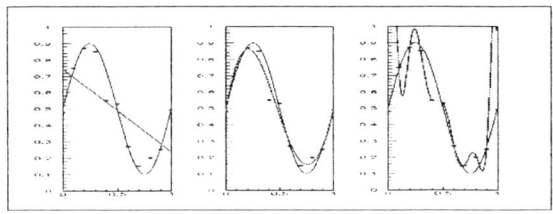

FIGURE 2. Results of fitting a data set (shown by points) with a 1st, 3rd, and 10th order polynomial (plots from left to right). The generator function $f(x) = 0.5 + 0.4\sin(2\pi x)$ is superposed in each case.

to calculate them. But, I want to emphasize that, as is true of all methods, it is important to make an appropriate choice of model complexity. A highly flexible model with lots of parameters will over-fit the data. This is illustrated by an example of polynomial fitting shown in Fig.2. The smooth curve is the parent function $f(x) = 0.5 + 0.4\sin(2\pi x)$. The data points are generated by adding random noise. Having either too few or too many parameters to fit the data yields a model that provides a poor representation of the underlying parent function.

Since event classification (or discrimination) is one of the most common tasks we deal with in high energy physics, I will concentrate on that topic in the rest of this section.

Optimal discrimination minimizes the probability of mis-classification. The traditional procedure of choosing and applying cuts on one event variable at a time is rarely optimal in that sense. However, given a set of event variables (denoted by a vector **x**), if correlations exist between them, optimal separation can *always* be achieved if one treats the variables in a fully multivariate manner. The optimal way to partition a multidimensional space populated by two classes of events 's' and 'b', for example, is to apply a cut on the ratio of the probabilities,

$$r(\mathbf{x}) = \frac{p(s|\mathbf{x})}{p(b|\mathbf{x})} = \frac{p(\mathbf{x}|s)p(s)}{p(\mathbf{x}|b)p(b)}, \quad (1)$$

where $p(\mathbf{x}|s)$ and $p(\mathbf{x}|b)$ are the class conditional probabilities, that is, probability density functions for signal and background, respectively; $p(s)$ and $p(b)$ are the prior probabilities.

The posterior probability for the desired class 's' then becomes,

$$p(s|\mathbf{x}) = \frac{r}{1+r} = \frac{p(\mathbf{x}|s)p(s)}{p(\mathbf{x}|s)p(s) + p(\mathbf{x}|b)p(b)}. \quad (2)$$

The discriminant 'r' is called the Bayes discriminant. The problem of discrimination, then, mathematically reduces to that of calculating the Bayes discriminant $r(\mathbf{x})$ or the class conditional probabilities. I should note here that algorithms such as neural networks, interestingly, can directly yield the posterior probability $p(s|\mathbf{x})$.

In general, when many classes C_k ($k=1,\ldots,N$) are present, the Bayes posterior probability is written as,

$$p(C_k|\mathbf{x}) = \frac{p(\mathbf{x}|C_k)p(C_k)}{\sum p(\mathbf{x}|C_k)p(C_k)}. \quad (3)$$

The Bayes rule for classification is to assign the object to the class with highest posterior probability.

PROBABILITY DENSITY ESTIMATION

I will briefly describe a few popular multivariate methods, most of which are probability density estimators.

Histogramming

The problem of probability density estimation in principle can be solved quite simply! One would merely histogram the multivariate data **x** in **M** bins in each of the d feature variables. The fraction of events (points) that fall within each bin yields a direct estimate of the density at the value of the feature vector **x**, say at the center of the bin. The bin width (and therefore the number of bins M) has to be chosen such that the structure in the density is not washed out (due to too few bins) and the density estimation is not too spiky (due to too many bins). The serious disadvantage of the histogramming method is that the total number of bins required grows like $\mathbf{M}^d$ (referred to as Bellman's curse of dimensionality). We would require a huge number of data points or else most of the bins would be empty leading to an estimated density of zero for those bins. The other issue is that the variables are generally correlated, and tend to be restricted to a sub-space of lower dimensionality, referred to as *intrinsic dimensionality*. Clearly, this method is inadequate for high dimensional data. There are better and more efficient methods for density estimation.

Kernel-based Methods

These methods sample neighborhoods of data points to provide probability densities. Let us take the simple example of a hypercube of side h as the kernel function in a d-dimensional space. The method consists of placing such a hypercube at each data point $\mathbf{x_n}$, counting the number of data points that fall within the hypercube and dividing that by the volume of the hypercube and the total number of data points, i.e.,

$$\tilde{p}(\mathbf{x}) = \frac{1}{N}\sum_{n=1}^{N}\frac{1}{h^d}H\left(\frac{\mathbf{x}-\mathbf{x}_n}{h}\right), \quad (4)$$

where N is the total number of data points, and

$H(u)=1$ if **x** is in the hypercube, 0 otherwise.

The method is akin to histogramming, but with overlapping bins (hypercubes) this time placed around each data point. Smoother and more robust density estimates can be obtained by using smooth functional forms for the kernel function. A common choice is a multivariate Gaussian,

$$H(u) = \frac{1}{(2\pi)^{d/2}} \exp\left(-\frac{\|\mathbf{x} - \mathbf{x}^n\|^2}{2h^2}\right), \quad (5)$$

where the width of the Gaussian h acts as a smoothing parameter to be chosen appropriately for the problem.

If the kernel functions satisfy,

$$H(u) \geq 0; \int H(u)du = 1 \quad (6)$$

then, the estimator satisfies

$\tilde{p}(\mathbf{x}) \geq 0$ and $\int \tilde{p}(\mathbf{x})d\mathbf{x} = 1$, as required.

The PDE method[4], used at D∅ in the measurement of the top quark mass using dilepton events [5], is an example of such a kernel-based method.

K-Nearest Neighbor Method

In the kernel-based approach, the parameter h is a constant and consequently the density estimation can be over-smoothed in some regions and spiky in some others. This problem is addressed in the K-nearest-neighbor approach. In this case, we place a kernel, say a hypersphere, at each data point $\mathbf{x}$ and instead of fixing its volume V and counting the number of data points that fall within it, we vary the volume (i.e., the radius of the hypersphere) until a fixed number of data points are within the volume. Then, the density is calculated as,

$$\tilde{p}(\mathbf{x}) = \frac{K}{NV}. \quad (7)$$

A classification criterion can be directly obtained in the K-nearest-neighbor approach as follows: if there are N_k points belonging to class C_k and N points in total, so that $\sum_k N_k = N$, then the class conditional probabilities can be written as

$$p(\mathbf{x}|C_k) = \frac{K_k}{N_k V}, \quad (8)$$

where K_K is the number of points in volume V for class C_K.

The prior probability,

$$p(C_k) = \frac{N_K}{N}. \quad (9)$$

The Bayes posterior probability is;

$$p(C_k|\mathbf{x}) = \frac{p(\mathbf{x}|C_k)P(C_k)}{p(\mathbf{x})} = \frac{K_k}{K}. \quad (10)$$

This yields the following algorithm: a new feature vector $\mathbf{x}$ should be assigned to the class C_k that has the most representatives in the volume of the hypersphere.

The contribution from Carli and Koblitz [6] at this workshop is an example of this method.

Adaptive Mixtures

The method of adaptive mixtures (AM; also called the mixture model) is a variant of the Kernel-based approach where the density estimate is obtained by a linear combination of an adjustable number of basis functions or component densities $p(\mathbf{x}|j)$,

$$\tilde{p}(\mathbf{x}) = \sum_{j=1}^{M} p(\mathbf{x}|j)p(j), \quad (11)$$

where M is typically far less than the number of points N, and the coefficients $p(j)$ are the mixing parameters. The most common functional form assumed for the component densities is a multivariate Gaussian,

$$p(\mathbf{x}|j) = \frac{1}{(2\pi)^{d/2}|\Sigma_j|^{1/2}} \times \exp\left[-\tfrac{1}{2}(\mathbf{x}-\boldsymbol{\mu}_j)^T \Sigma_j^{-1}(\mathbf{x}-\boldsymbol{\mu}_j)\right] \quad (12)$$

where μ_j is the mean and Σ_j is the covariance matrix. The adaptive mixtures algorithm would incorporate rules for adding or deleting components and for adjusting μ_j and Σ_j.

The mixture models or the method of mixtures have been used quite extensively in the statistical community. These traditional applications assume that the data came from a mixture of a given number of components, where as in AM this assumption is not made.

Neural Networks

Even though the concepts of neural networks were inspired from biology, the algorithms have deep statistical underpinnings. Neural network algorithms have emerged as powerful and flexible methods for a variety of multivariate data analysis applications. Feed-forward neural networks, also known as multilayered perceptrons, are the most popular and widely used. The output of a feed-forward neural network trained by minimizing, for example, mean square error function, directly approximates the Bayesian posterior probability $p(s|\mathbf{x})$ (Eq. 2) [7] without the need to estimate the class-conditional probabilities separately. A schematic of a feed-forward neural network (NN) is shown in Fig. 3. Such networks provide a general framework for estimating non-linear functional mappings between a set of input variables $\mathbf{x} (\equiv (x_1, x_2, x_3, ... x_k))$ and an output variable $O(\mathbf{x})$ (or a set of output variables) without requiring a prior mathematical description of how the output formally depends on the inputs. The mapping involves transforming the input variables with an arbitrary number of adaptive non-linear functions. The output in the simple example shown in Fig. 3 can be written as,

$$O(\mathbf{x}) = g\left(\sum_i w_j h_j + \theta_i\right) \equiv p(s|\mathbf{x}), \quad (13)$$

with $\quad h_j = g\left(\sum w_{jk} + \theta_j\right),$

and where g is a non-linear "activation" function normally taken as a logistic sigmoid

$$g(a) = \frac{1}{1+e^{-a}}. \quad (14)$$

These "hidden" transformation functions g, or more precisely the weights w_j and w_{jk} and the thresholds (not shown in the figure) θ_i and θ_j adapt themselves to the data as part of the "training" process of the neural network. The number of such parameters need to grow only as the complexity of the problem grows. The parameters are determined by minimizing an error function, usually the mean square error between the actual output O^p and the desired (target) output t^p,

$$E = \frac{1}{2N_p}\sum_{p=1}^{N}\left(O^p - t^p\right)^2, \quad (15)$$

with respect to the parameters. Here, p denotes a feature vector or pattern. The stochastic optimization algorithms used in learning enable the model to be improved a little bit for each data point in the training sample. The Bayes discriminant in terms of the NN output is

$$r(\mathbf{x}) = \frac{O(\mathbf{x})}{1-O(\mathbf{x})}. \quad (16)$$

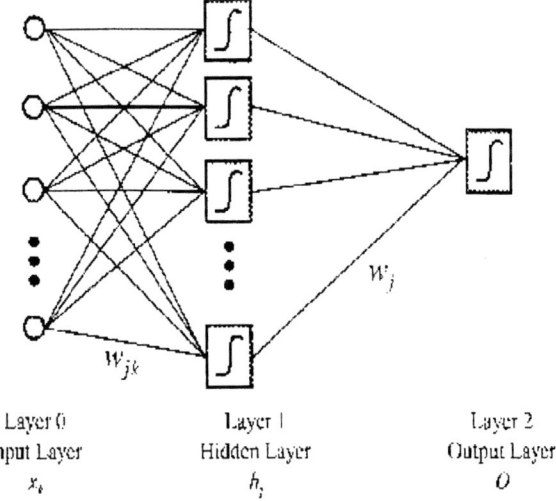

FIGURE 3. A schematic representation of a three layer feed-forward neural network.

Neural networks, apart from being universal approximators (i.e., they approximate probability densities or posterior probabilities to arbitrary accuracy), provide a very practical tool because of the relatively small computational times required in their training (in a majority of applications in HEP). The fast convergence as well as the robustness in supervised learning of multilayer perceptrons are due to efficient and powerful algorithms developed in recent years.

Good generalization, that is good predictions for new inputs, is controlled by model complexity as we discussed in the example of polynomial curve fitting in the previous section. The traditional approaches used to control model complexity are *structure stabilization* (optimizing the size of the network) and *regularization*. In the former one starts with large networks and *prunes* connections or starts with small networks and adds units/neurons as necessary. In regularization, one penalizes complexity by adding a penalty term to the error function.

There are many new and sophisticated approaches to achieve good generalization. It is important to note here that the generalization error (g.e.) of an NN can be decomposed into the sum of the bias-squared (b^2) plus the variance (σ^2), i.e., the generalization error,

$$g.e. \equiv \sqrt{b^2 + \sigma^2}.$$

The goal is to minimize the g.e., that is, finding the best compromise between bias and variance. Ensembles of networks, such as committees or stacks, can be used to control bias and variance [8].

Bayesian learning of network parameters can in principle handle networks of arbitrarily high complexity without over-fitting. Bayesian networks also provide a rigorous way to assign errors to network predictions [8].

Aside from the MLP, there are other neural network types which are potentially useful in some applications in HEP. One of them is the self-organizing map (SOM). This is an unsupervised technique and appears to be an excellent tool for model-independent data exploration. It maps input space on to a low-dimensional (usually 2-D) regular grid that can be used to visualize and explore properties of the data. Given models for background processes, one could use it in a manner similar to the program "Sleuth" developed at DØ to search for new physics [9].

More detailed accounts and discussions of neural networks and other methods can be found in ref. [8]

PHYSICS ANALYSIS EXAMPLES

I describe here a couple of example physics analyses to illustrate the power of multivariate methods.

Top Quark Mass Measurement

The top quark mass measurement was one of the most important results from Run I of the Tevatron collider experiments. Since the DØ experiment did not have a silicon vertex detector (SVX) and used only soft muon tagging for b-jet identification, the b-tagging efficiency was only 20% in the lepton + ≥ 4-jets channel ($t\bar{t} \rightarrow Wb W\bar{b} \rightarrow l\nu b q\bar{q}\bar{b}$ process) compared to approximately 53% at CDF which had the ability to tag b-jets with its SVX.

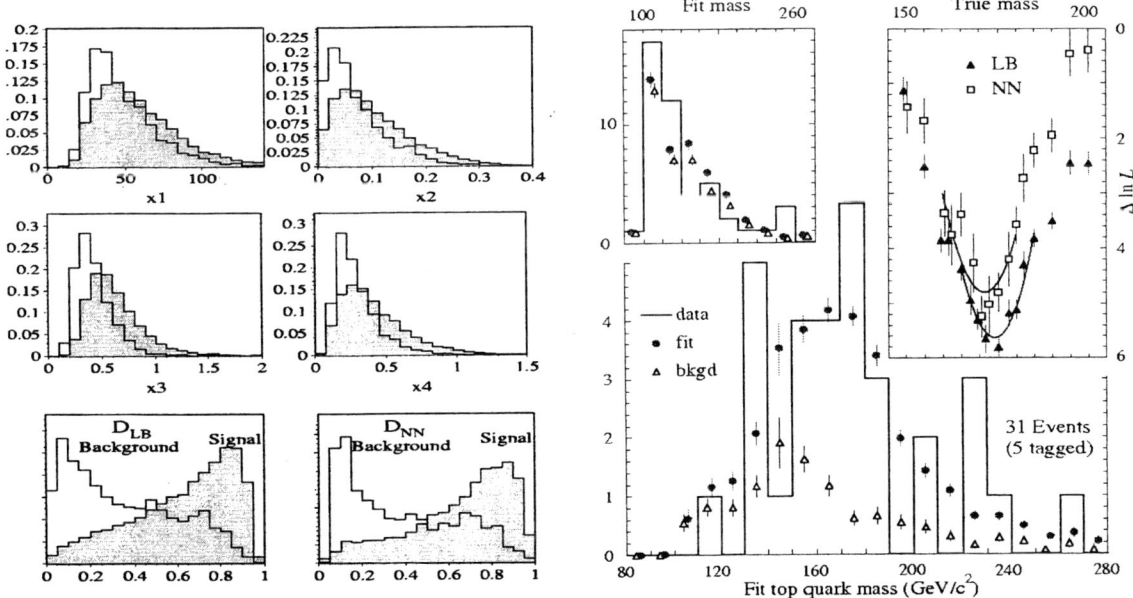

FIGURE 4. Distributions of discriminant variables x_1, x_2, x_3, x_4 (see [10] for definitions) and the final multivariate discriminant D for signal (filled histograms) and background. All histograms are normalized to unity. (Right) The fitted mass distribution for events in the signal-rich sample. Left inset: The same for events in background–rich sample. Right inset: The relative log likelihood functions for the two methods. For details of the analysis, see ref. [11].

Nonetheless, DØ was able to measure the top quark mass with a precision approaching that of CDF, by using multivariate techniques for separating signal and background while minimizing the correlation of the selection with the top quark mass.

Two multivariate methods, (1) a modified log-likelihood technique (LB method) and (2) a feed forward neural network (NN method), were used to compute a signal probability $D \equiv p(top \mid \mathbf{x})$ for each event, given data $\mathbf{x}$. A likelihood fit, based on a Bayesian method [4], of the data to discrete sets of signal and background models in the $[p(top \mid \mathbf{x}), m_{fit}]$ plane was used to extract the top quark mass. (m_{fit} is the fitted mass for each event from a kinematic fit to the $t\bar{t}$ hypothesis.) The distributions of variables and the results of the fits are shown in Fig. 4. Combining the results of the fits from the two methods, DØ measures $m_t = 173.3 \pm 5.6$ (stat) ± 5.5 (syst) GeV/c² [11].

Discovering the Higgs Boson

In the SM framework, a global fit to the electroweak precision data, including the directly measured top quark and W boson masses, yields a Higgs boson mass of $M_H = 107^{+67}_{-45}$ GeV/c² and a 95% C.L. upper limit of 225 GeV/c² [12]. In broad classes of supersymmetric (SUSY) theories, the mass m_h of the lightest CP-even neutral Higgs boson h is constrained to be less than 150 GeV/c² [13]. In the minimal supersymmetric SM (MSSM), $M_h < 130$ GeV/c² and there are tantalizing hints of a 115 GeV/c² Higgs boson from the recently completed LEP experiments [14]. These intriguing indications of a low-mass Higgs boson motivated studies of strategies that maximize the potential for its discovery at the upgraded Tevatron [15]. Our study of the Higgs discovery potential focused on a standard model Higgs boson in the mass range 90 GeV/c² < M_H <130 GeV/c² that would be produced via the processes,

$$p\bar{p} \to WH \to l\nu b\bar{b}, \quad p\bar{p} \to ZH \to llb\bar{b}, \nu\bar{\nu}b\bar{b}.$$

The dominant backgrounds in these channels come from $Wb\bar{b}, WZ, t\bar{t}$ and single top processes. We have shown that a neural network analysis could yield a 5σ discovery for $100 \leq M_H \leq 130$ GeV/c² with only half the integrated luminosity needed for a conventional analysis. Fig. 5 shows the neural network distributions for signal Monte Carlo events with M_H =110 GeV/c² compared with the specified backgrounds, for a set of seven input variables. (For details, see ref. 15). A plot of the required integrated luminosity for a 5σ observation is also shown in Fig. 5. For a 110 GeV/c² Higgs boson, if 10% systematic uncertainties are assumed, CDF and DØ would require about 13 fb⁻¹ for independent 5σ discovery. Our study shows that with 20 fb⁻¹, a 3-5σ observation of a neutral Higgs boson is possible at the Tevatron for masses with $M_H \leq 130$ GeV/c².

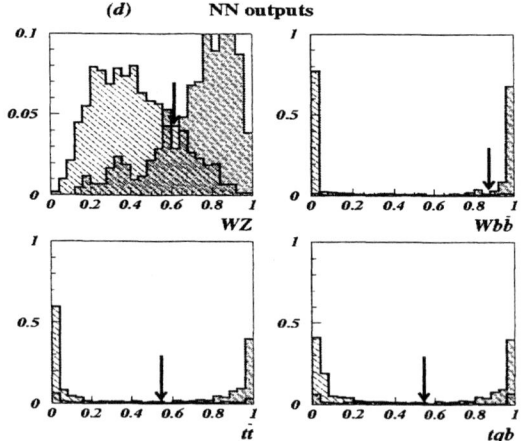
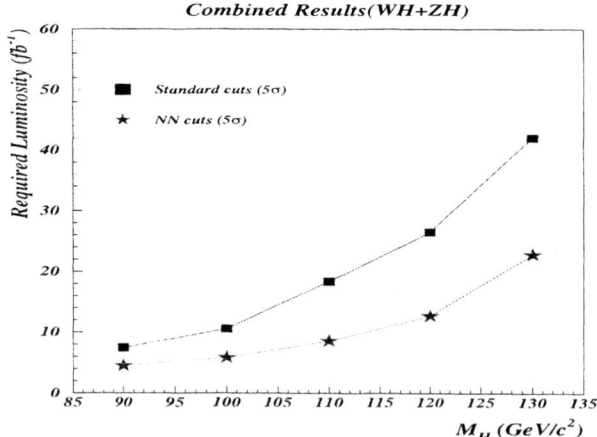

FIGURE 5. (Left) Neural network distributions for WH signal (M_H =110 GeV/c²; heavily shaded histograms) compared with backgrounds $Wb\bar{b}, WZ, t\bar{t}$ and single top. (Right) Comparison of the required integrated luminosities for a 5σ observation in the CDF and DØ experiments for NN and conventional cuts (WH and ZH channels combined).

EXPLORING MODELS

Physicists are becoming increasingly convinced of the value of Bayesian reasoning as a powerful way of extracting information from data and of updating knowledge upon arrival of new data. The Bayesian approach provides a well-founded mathematical procedure to compute the conditional probability of a model (or a hypothesis) and therefore to do straight-forward and meaningful model comparisons. It also allows treatment of all uncertainties in a consistent manner. We have applied these ideas in two analyses (1) fitting binned data to one or more multi-source models [16] which was eventually used in the top quark mass measurement at DØ and (2) the extraction of the solar neutrino survival probability [17] as a function of neutrino energy, using data and solar neutrino model predictions. These practical applications illustrate the usefulness of Bayesian methods in data analysis.

The Bayesian approach provides a systematic way of extracting probabilistic information for each parameter of a model, say for example, a particular SUSY model, via marginalization over the remaining parameters. This probabilistic approach to model exploration could prove to be extremely fruitful. We are studying this approach in the search for supersymmetric Higgs boson predicted by the SO(10) model [18].

CONCLUSIONS

The discovery of the Higgs boson and signals of new physics beyond the SM may be just around the corner in Run-II at the Fermilab Tevatron. Somewhat later, the experiments at the Large Hadron Collider at CERN will enable us to probe physics at the TeV scale. We are entering an exciting era with lots of optimism and hope. The physics pursuits are extremely challenging, even daunting!

Use of optimal analysis methods will have to become routine in order to achieve the high energy physics goals for the coming decade. These methods, particularly neural network techniques have already made an impact on discoveries and precision measurements and I believe that they will be the methods of choice for future analyses.

ACKNOWLEDGMENTS

My research is supported in part by the U.S. Department of Energy under contract number DE-AC02-76CH03000.

REFERENCES

1. For some examples not covered in the paper, see the talk video at http://conferences.fnal.gov/acat2000/;

   P.C. Bhat, "New Directions in Data Analysis" to be published in *Int. J. Mod. Phys.* (2001), *Proceedings of the Division of Particles and Fields of the American Physical Society, August 2000, Columbus, Ohio*.

2. C. Kiesling, these proceedings. For talk video see http://conferences.fnal.gov/acat2000/.

3. P. C. Bhat (DØ collaboration), *Proceedings of the 10th Topical Workshop on Proton-Antiproton Collider Physics*, Batavia, Illinois (AIP, Woodbury, NY 1995), p. 308; P. C. Bhat (DØ collaboration), *Proceedings of the 8th Meeting of the the Division of Particles and Fields of the American Physical Society, Albuquerque, New Mexico* (World Scientific, NJ 1994), p. 705; P. C. Bhat et al., *Proceedings of the DPF summer study in High Energy physics, Snowmass, Colorado* (1990), p. 169.

4. L. Holmström, R. Sain and H. E. Miettinen, Comput. Phys. Commun. **88**, 195 (1995).

5. DØ Collaboration, *Phys. Rev. Lett.* **80**, 2063 (1998).

6. T. Carli and B. Koblitz, these proceedings.

7. D. W. Ruck et al., *IEEE Trans. Neural Netw.* **1**, 296 (1990); E. A. Wan, *ibid*, **1**, 303 (1990); E. K. Blum and L. K. Li, *Neural Networks* **4**, 511 (1991).

8. C.M. Bishop, *Neural Networks for Pattern Recognition* (Clarendon, Oxford, 1998); R. Beale and T. Jackson. *Neural Computing: An Introduction* (Adam Hilger, New York, 1991), P. C. Bhat and H. B. Prosper, *Multivariate Analysis in High Energy Physics* (in preparation; to be published by World Scientific).

9. B.Knuteson, these proceedings, and references therein

10. The four variables were: $x_1 \equiv E_T$, the missing transverse energy in the event, $x_2 \equiv A$, the event aplanarity, $x_3 \equiv (H_T - E_T^{jet1})/H_z$ where $H_T = \sum E_T$ of all selected jets, and $H_z = \sum |E_z|$ of all objects in the event, E_z being the longitudinal

momentum. $x_4 \equiv \Delta R_{jj}^{min} \cdot E_T^{min} / \left(E_T^l + E_T \right)$ where ΔR_{jj}^{min} is the minimum ΔR (the distance between jets in $\eta - \varphi$ space) of the six pairs of four jets and E_T^{min} is the smaller jet E_T from the minimum ΔR pair.

11. P.C. Bhat, H.B. Prosper and S.S. Snyder, *Int. J. Mod. Phys.* **13** 5113 (1998); DØ Collaboration, *Phys. Rev. D.* **58**, 052001 (1998); CDF Collaboration, *hep-ex/0006028*.

12. J. Erler and P. Langacker, *Proceedings of the 5th International WEIN Symposium: A Conference on Physics beyond the Standard Model, Santa Fe New Mexico,* (1998); e-print hep-ph/9809352.

13. See, for example, S. Martin, in *Perspectives on Supersymmetry*, edited by G. L. Kane (World Scientific, Singapore, 1998); e-print hep-ph/9709356 v3, 1999..

14. P. Igo-Kemenes, Talk at CERN LEPC meeting, 3, Nov 2000; see also http://lephiggs.web.cern.ch/LEPHIGGS/talks/pik_lepc_nov3_2000.ps

15. P.C. Bhat, R. Gilmartin and H.B. Prosper, *Phys. Rev. D* **62** 074022 (2000); Run II Higgs Working Group Report, *hep-ph 0010338*.

16. P.C. Bhat, H.B. Prosper and S.S. Snyder, *Phys. Lett.* **B407**, 73 (1997).

17. C.M. Bhat *et. al., Phys. Rev. Lett.* **81** 5056 (1998).

18. H. Baer, et al., hep-ph/0005027.

Statistical Techniques in High Energy Physics

Louis Lyons

Nuclear and Astrophysics Laboratory, Keble Rd, Oxford, UK

Abstract. Rather than attempting to cover a wide range of statistical problems, I shall concentrate on four topics: 1) The argument between Bayesians and Frequentists, 2) A paradox in comparing data with two hypotheses; 3) The CLs method used in the search for the Higgs at CERN, 4) The MLBZ method of using data to estimate some of the systematic effects in measuring the W boson's mass.

BAYES VERSUS FREQUENTISM

The Confidence Limits Workshops[1,2] earlier this year brought into sharp focus the differences between the Bayes and Frequentist approaches.

Bayesians start from Bayes' Theorem

$$P(A \text{ and } B) = P(A|B) * P(B) = P(B|A) * P(A) \quad (1)$$

where $P(A|B)$ means the probability of A happening, given that B has happened. This is completely uncontroversial when used in situations where A and B describe certain events (in the statistical sense) e.g. for a high energy $\overline{pp}$ interaction, A and B are the production of a W boson and a top quark respectively. However Bayesians use eqn (1) for A = hypothesis (or the value of a parameter) and B = data whence (1) can be rewritten

$$P(parameter|data) \; \alpha \; P(data|parameter) * P(parameter) \quad (2)$$

i.e. (posterior prob) α (likelihood fn) * (prior). Thus after performing an experiment, the posterior knowledge about a parameter is obtained by combining one's prior knowledge with the likelihood function, as deduced from the experimental data.

To frequentists, this is anathema because they would object to making a probability statement about the value of a physical parameter. For them, the value of α_s either is between 0.115 and 0.120, or it is not (even if we do not know which), and it does not make any sense to ascribe a probability to this.

Furthermore, to deduce the Bayesians' posterior probability requires a functional form for the prior; there is some arbitrariness in how this should be chosen. It is tempting to try to assign an "uninformative prior", e.g. one which is flat in the parameter, p. However, flat in p is not the same as flat in p^2 or $\sqrt{p}$ or $\log(p)$, and it is not usually clear which is best. A frequentist would want the result of an experiment to be independent of such arbitrary choices.

Bayesians, on the other hand, would interpret $P(parameter|data)$ or $P(parameter)$ not so much as a probability in the classical sense, but more as a degree of belief, and then quote a "credible interval" for the parameter, p. Concerning priors, they would either justify the use of subjective priors as expressing genuine differences of knowledge of different experimentalists; or they would attempt to find priors with some theoretical justification. They would further argue that the Bayesian approach most closely resembles the way scientists make decisions; in deciding what research to pursue, personal judgments play an important role. Another effect of the prior is that unphysical values of a parameter are excluded; this is not necessarily so in frequentist approaches.

Frequentists construct confidence intervals without invoking $P(parameter|data)$ or $P(parameter)$, and hence do not require a prior. Their method is simply to use $P(data|parameter)$ to construct a probability interval for the data (i.e. the result of the experiment) for each value of p. Thus the shaded band of Fig. 1 shows the likely result of the experiment for each p. Then for the given result of a particular experiment, the confidence belt for p is given by where a vertical line at the experimental result cuts the shaded band (see Fig. 1).

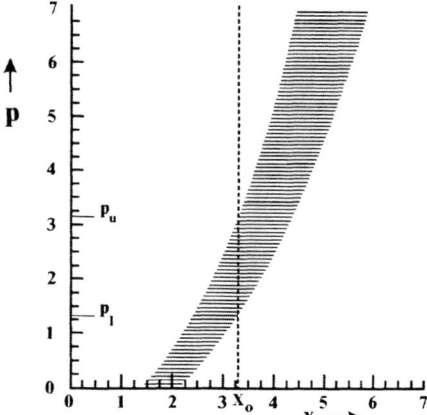

FIGURE 1. Confidence belt. p is a theoretical parameter and x is an observation. For example, p could be the temperature at the centre of the Sun, and x the production rate for events in a solar neutrino detector. For a given p, a band in x is calculated such that the probability of observing x in this range is 90%. As p takes on all possible values, the shaded region is obtained; this gives the likely values of x for any p. For a given experiment observing x_0, the range p_l to p_u contains those values of p for which x_0 was a likely observation.

The frequentist thus produces a statement such as

$$p_l < p < p_u \text{ at 90\% confidence.} \quad (3)$$

This is based just on the data, without any preconceived ideas about the relative probabilities of different values of p. In eqn (3), the true value of p is unknown but regarded as a constant, while the known p_l and p_u are regarded as random variables; eqn (3) is thus a statement about the probability of the random p_l and p_u containing the unknown p. In a similar statement by Bayesians, p is regarded as a random variable with a probability distribution, and p_l and p_u are constants.

For accurate measurements where the errors are approximately Gaussian (e.g. M_Z, the mass of the Z as determined at LEP, which is 91188±2 MeV), the two approaches give the same results [3]. This is the case where, for the Bayesian, "the data overshadows the prior". The functional form of the prior is then unimportant, because any reasonable prior will be virtually constant over the small mass range of interest. In contrast, in situations where we are dealing with limits, the relevant range extends down to zero, and so here the form of the prior can be very important. This is why the Frequentist-Bayes argument is more relevant in limit situations than for accurate measurements.

Thus Narsky [4] showed how variable upper limits could be, depending on the exact way they are determined. He considered the example of the observation of a given number of events, assumed to be Poisson distributed, when the expected background was b. For six different methods applied to n=3 and b=1.0, upper limits ranged from 0.3 to 3.3. When quoting upper limits, it is thus crucial to explain clearly how they were deduced.

One of the problems apparent from the Workshops was the confusion between P(hypothesis|data) and P(data|hypothesis). This should be clarified by the example of considering a certain unseen person, who is hypothesised to be either male or female. The data is whether or not they are pregnant. For random human beings,

$$P(\text{pregnant}|\text{female}) \sim 3\%$$

whereas P(female|pregnant) is considerably larger.

A specific example of a limit calculation discussed at the Workshops is given in 'CERN Higgs Search' below.

CHOOSING BETWEEN HYPOTHESES

There is a well-known paradox connected with parameter estimation and hypothesis testing [5].

Imagine that a χ^2 method is being used on a histogram of 100 bins, to estimate the value of a single parameter p. Assume that p_0, the best value of the parameter gives a minimum χ^2 of 90. Then the error σ on p is given in terms of $p_1 = p_0 + \sigma$ the value of p for which χ^2 increases to 91.

Now consider p_2, another value of p, for which $\chi^2 = 115$. Is this value satisfactory?

The probability of $\chi^2 = 115$ for 99 degrees of freedom is not unreasonable. i.e. we would not want to exclude p_2. However, given that σ is such that $\chi^2(p_0+\sigma) = 91$, in the usual parabolic approximation, a χ^2 of 115 corresponds to a 5σ effect i.e p_2 is "completely" excluded. So which are we supposed to believe?

The answer is that, although in general a value of 115 for χ^2_{min}, the minimum value of a χ^2 variable, based on 99 degrees of freedom is satisfactory, in the case where χ^2_{min} is 90, a value of $\chi^2(p)$ of 115 is not.

It is the **difference** in χ^2 of the two hypotheses ($p=p_0$ or $p=p_2$) which is relevant in discriminating between them.

This approach is thus important for reducing "errors of the second kind" i.e. accepting a hypothesis when it is in fact wrong. It can be used for discriminating between whether atmospheric neutrino data is more consistent with ν_μ oscillating to ν_τ or to $\nu_{sterile}$; or which set of parton distributions is consistent with jet production in high energy $p\aleph$ collisions. A more elaborate example is provided below.

CERN HIGGS SEARCH

The search for the Higgs at the LEP Collider at CERN is based on the CL_s method [6]. Basically the problem is to try to use the data to distinguish between two hypotheses: A) Background processes produced according to the Standard Model (S.M.) without the Higgs (or with a Higgs that is too heavy to be accessible); or B) S.M. plus Higgs of a certain mass.

For each Higgs mass, the aim of the procedure is to make one of the following choices: i) The data are inconsistent with B at some level (e.g. 5%), and hence a Higgs of that mass is excluded. ii) The data are inconsistent with A at some level (e.g. equivalent to a one-sided 5σ effect) and are more consistent with B. iii) The data cannot either exclude or confirm a Higgs of that mass.

From the data, a test statistic X is constructed. This is in fact the likelihood ratio for the two hypotheses. It involves not only the number of events, which of course is expected to be larger for B than for A, but other kinematic variables e.g. mass of Higgs candidates, whether there are b-hadrons in the possible Higgs decay, etc.

For a given mass Higgs, Monte Carlo simulation is used to predict the expected distributions of X for hypotheses A and B. (Because the production cross-section of a S.M. Higgs of a given mass is well-defined, these predictions involve no free parameters.) Then X from the data is compared with each distribution to decide whether it is possible to favour one hypothesis over the other. For distributions that are well-separated, this is relatively easy. For more realistic situations where there is some overlap (see Fig. 2(a)), a numerical procedure is required.

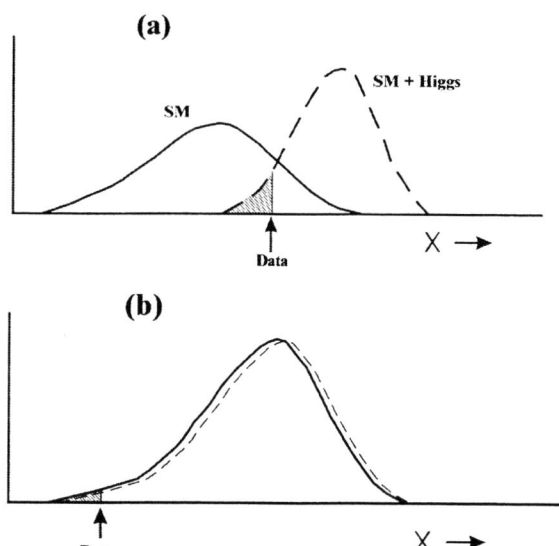

FIGURE 2(a). Predictions from simulation for a particular data statistic X, for Standard Model background only (solid curve) and SM plus Higgs of a particular mass (dashed curve). The shaded area gives CL_{bs}, the fractional area of the dashed curve that is more background-like than the data. CL_b is the corresponding fractional area on the SM curve to the left of the data. Finally CL_s is defined as CL_{bs}/CL_b.
FIGURE 2(b). For a very heavy Higgs that is barely produced, the solid and dashed curves become almost indistinguishable. If the data fluctuates downwards, CL_{bs} is small, but CL_s is still close to unity. The advantage of CL_s is that it prevents the exclusion of such a Higgs.

In order to see whether exclusion of Higgs is possible for the given value of X for the data, CL_{bs} is defined as the fractional area of the simulated "SM + Higgs" distribution to the left of X (i.e. more background-like). The usual frequentist approach is to exclude this hypothesis if CL_{bs} is less than some preset level (say 5%). However the CERN Higgs group want to avoid the situation in which, for a heavy Higgs which it is barely possible to produce (see Fig. 2(b)), a downward data fluctuation could result in CL_{bs} being small enough to exclude the Higgs, even though the experiment has no sensitivity to it; this would happen at the 5% probability level. So a more conservative approach is adopted by defining

$$CL_s = CL_{bs}/CL_b$$

where CL_b is the fractional area of the "background only" distribution to the left of the data X. Thus CL_s is the ratio of two confidence levels, rather than itself being a confidence level.

The data is then said to exclude the Higgs of that mass if CL_s is below a certain cut e.g. 5%. Since CL_s is forced to be larger than CL_{bs}, the cut on CL_s is conservative in its coverage i.e. the Higgs hypothesis will be excluded when the Higgs is really there in not more than 5% of experiments. Then in the situation of Fig. 2(b), CL_s is close to unity, and the CL_s method will never be able to exclude a Higgs to which an experiment has no sensitivity.

Although a relatively modest level (95%) is chosen for exclusion, in order to be sure of a discovery claim, a higher degree of confidence is required. The criterion is that CL_b is required to be very close to unity: $1- CL_b$ is below 5.6×10^{-7}. Since the probability of this happening for a random fluctuation of the background is so small, it is not deemed necessary to adopt the analogy with the exclusion procedure, and to divide by $1-CL_{bs}$.

Fig. 3 shows CL_s and CL_b for recent data [7]. These exclude Higgs masses up to 113.6 GeV, and there is a signal-like effect in CL_b around 115 GeV, but not at the required discovery level.

MLBZ

Systematic errors are much more problematic than statistical ones. Here we discuss a novel approach to trying to estimate realistically some of the systematic errors in the determination of M_W, the mass of the W boson, produced in the reaction

$$e^+e^- \rightarrow W^+W^- \rightarrow 4 \text{ jets} \qquad (4)$$

at the CERN LEP Collider at centre of mass energies around 200 GeV.

Some of the potential sources of error are associated with the estimated jet energies and directions. These include the detector resolution, and the way the fragmentation of quarks and their subsequent hadronisation are described. The traditional method is to use simulation techniques to estimate these, but as ever the question is how realistic these simulations are. The Mixed Lorentz Boosted Z^0s (MLBZ) approach [8] instead uses data to check these effects.

It relies on the fact that the four jets in reaction (4) are rather similar to those of the two jets in

$$e^+e^- \rightarrow Z^0 \rightarrow 2 \text{ jets} \qquad (5)$$

at centre of mass energies around $M_Z = 91$ GeV. Thus if we take two Z^0 2-jet events, and boost them in opposite directions with a Lorentz boost β corresponding to reaction (4) at a given LEP energy,

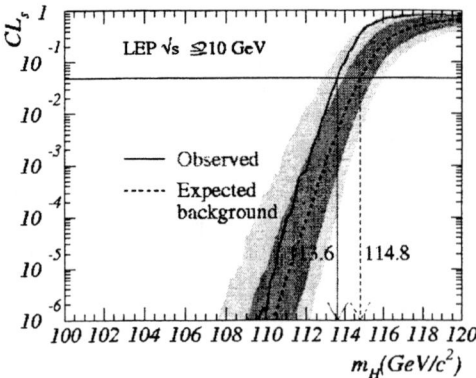

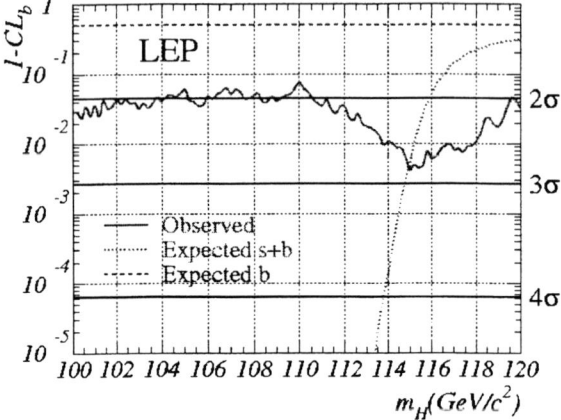

FIGURE 3. Search for the Higgs at LEP. CL_s as a function of M_H is used to see whether it is possible to exclude the Higgs at that mass. The combined data from the four LEP experiments has CL_s (solid curve) below 5% for masses up to 113.6 GeV, and so masses below that are excluded. The median expected value (dashed curve) passes through 5% at 114.8 GeV. In the lower diagram, $1-CL_b$ is used to see whether there is evidence of the Higgs; the data is shown as the solid curve. The expectation for background only is the horizontal dashed line at 0.5, while that for 'background + Higgs' is the dotted curve. The data shows an effect at slightly below 3σ significance for M_H around 115 GeV. The total luminosity is such that a 5sigma signal is not expected for M_H above 113 GeV.

the event configuration will resemble that for reaction (4), albeit at a slightly higher energy. (Alternatively, the Z^0 jets can be scaled down in energy, so as to correspond to M_W). The same analysis procedure as used to extract M_W from the real 4-jet events is then used on MLBZ events, constructed either from real data or from simulated events of reaction (5). By comparing the mass shifts observed between real and simulated MLBZ events, the reliability of the simulation to predict detector, fragmentation and hadronisation effects can be checked.

Sophistications include reweighting MLBZ events to allow for the different angular distributions of the jets in reactions (4) and (5); to incorporate the finite width of the bosons in (4), compared with all Z^0 in (5) being produced at the same centre of mass energy; to correct for the fact that Z^0 decays include b jets, while W decays do not; etc.

Of course the MLBZ technique does not include all systematics (e.g. initial state radiation, colour reconnection, etc.) but for those that are, it provides a more direct and powerful way of estimating them. Thus the DELPHI systematic errors calculated by MLBZ have now been explored to a level of precision which is more than a factor of 4 better than was previously possible using simulation.

The method can be extended to deal with aspects of W physics other than just M_W (e.g. triple gauge couplings); and with other reactions e.g.

$$e^+e^- \to W^+W^- \to 2 \text{ jets} + \text{lepton} + \text{neutrino}$$
$$e^+e^- \to Z^0 Z^0$$
$$e^+e^- \to Z^0 + \text{Higgs}$$

CONCLUSION

There continue to be interesting statistical analyses to perform in High Energy Physics. If you are aware of any challenging problems (and especially if you also know how to solve them!), please let me know.

REFERENCES

1. Workshop on Confidence Limits, CERN, 27-28 January 2000, CERN Yellow Report 2000-005.

2. Workshop on Confidence Limits, Fermilab, 27-28 March 2000, http://conferences.fnal.gov/cl2k/.

3. R. Cousins, Am. J. Phys. **63**, 398 (1995).

4. I. Narsky, Nuclear Instruments and Methods **A450**, 444 (2000).

5. L.Lyons, 'Selecting between two hypotheses', Oxford preprint OUNP-99-12.

6. A.L. Reed, ref.1) page 81; W. Murray, ref.2).

7. P. Igo-Kemenes, Talk at CERN LEPC meeting, 3rd Nov 2000; see also http://lephiggs.web.cern.ch/LEPHIGGS/talks/pik_lepc_nov3_2000.ps.

8. N. Kjaer and M. Mulders, 'Mixed Lorentz Boosted Z^0s', DELPHI 2000-51 CONF 366

The H1 Neural Network Trigger Project

C. Kiesling[1],*, B. Denby[2], J. Fent[1], W. Fröchtenicht[1], P. Garda[2], B. Granado[2],
G. Grindhammer[1], W. Haberer[1], L. Janauschek[1], T. Kobler[1], B. Koblitz[1], G. Nellen[1],
J.-C. Prevotet[2], S. Schmidt[1], E. Tzamariudaki[1], S. Udluft[1]

[1] *Max-Planck Institut für Physik, München, Germany*
[2] *Laboratoire des Instruments et Systemes, Universite Pierre et Marie Curie, Paris VI, France*

Abstract. We present a short overview of neuromorphic hardware and some of the physics projects making use of such devices. As a concrete example we describe an innovative project within the H1-Experiment at the electron-proton collider HERA, instrumenting hardwired neural networks as pattern recognition machines to discriminate between wanted physics and uninteresting background at the trigger level. The decision time of the system is less than 20 microseconds, typical for a modern second level trigger. The neural trigger has been successfully running for the past four years and has turned out new physics results from H1 unobtainable so far with other triggering schemes. We describe the concepts and the technical realization of the neural network trigger system, present the most important physics results, and motivate an upgrade of the system for the future high luminosity running at HERA. The upgrade concentrates on "intelligent preprocessing" of the neural inputs which help to strongly improve the networks' discrimination power.

INTRODUCTION

It was here at Fermilab, in the early nineties, where the first applications of hardwired neural networks in high energy physics were initiated. The goal of these projects was to construct fast, intelligent triggers for HEP experiments. At that time, only a few neuromorphic chips were available, mostly in the form of analog devices. The early Fermilab applications all used such analog chips. Since then a number of digital instances have been developed, opening up the field for concrete neural trigger applications, which can be bit-precisely simulated offline for controlled determination and monitoring of trigger efficiencies. In these applications it has become clear that the digital approach is essential for further progress of the neural technology in high energy physics.

While many successful neural applications have emerged during the past years in the field of offline analysis (see, e.g., reports at this conference such as (1, 2)), the trigger area is still highly sensitive and new methods such as neural networks, although understood theoretically in great detail, should be prepared for quite some skepticism. Nevertheless, a number of successful projects have emerged and are turning out now new physics results, hard to obtain with the more "classical" methods of triggering. The main subject of this article is to show, giving the concrete example of the H1 neural network trigger, that the neural method in the trigger area is now firmly established, provides new physics information, and certainly has a splendid future in areas where fast and efficient pattern recognition algorithms are mandatory.

NEUROMORPHIC HARDWARE AND EARLY HEP APPLICATIONS

Neural networks have traditionally been studied in a simulation approach on standard serial computers, where the inherent parallelism of the neural method is not exploited. These more academic applications are usually not time-critical and can afford an execution speed at the "millisecond" scale (we speak here about the computations of a multi-layer feed-forward net in its "recall" step. The training of networks, which is done beforehand in the standard backpropagation supervised mode, needs many order of magnitude more execution time, depending on the amount of training data available and on the complexity of the network architecture). In the early 90's first neuromorphic hardware (massively parallel processor architectures) became available from industry so that high speed ("microsecond scale") realtime applications, such as trigger processors, could be seriously envisaged. These applications are today the most demanding concerning speed of execution. It was recognized quite early (see, e.g., (9)) that pattern recognition tasks, such as event

---

* Author's e-mail: cmk@mppmu.mpg.de

Table 1. Major developments in neuromorphic hardware. The performance of these multiprocessor chips is indicated by the execution time of a specific multi-layer perceptron. The corresponding network topology is indicated by $i \times h \times o$, where i is the number of inputs, h the number of hidden nodes, and o the number of output nodes. For the digital devices typical execution times are about 5 to 10 μs. The number of bits available for the input data and the weights is indicated. For the accumulation of the neural activity the precision is usually augmented by at least 8 bits.

Analog Devices					
Chip name	manufacturer	year	# of proc.	net topology	exec time
ETANN (3)	Intel Corporation (USA)	1991	64	$64 \times 64 \times 4$	8 μs
NeuroClassifier (4)	Univ. Twente (NL)	1994	6	$70 \times 6 \times 1$	50 ns
Digital Devices					
Chip name	manufacturer	year	# of proc.	clock	precision
CNAPS (5)	Adaptive Solutions (USA)	1993	64	20 MHz	8/16
MA16 (6)	Siemens (D)	1994	16	50 MHz	16/16
Totem (7)	Uni. of Pisa (I)	1994	32	30 MHz	16/8
SAND1 (8)	KfK Karlsruhe (D)	1995	4	50 MHz	16/16

selection at the trigger level, could develop into very fruitful applications of neural networks.

Table 1 gives an overview of the major developments in the field of dedicated neural processor systems. The first large-scale industrial production came from Intel Corporation (3), the famous *Electrically Trainable Artificial Neural Network* ("ETANN") chip. This chip was of the analog variety and was able to carry out in parallel 64 multiply/add operations in less than 5 μs. The weights were stored on chip in an analog way with a precision of roughly 4 bits. The ETANN was used first in a test experiment (10) at Fermilab, where a rather involved neural approach was realized: From the space coordinates in a three-layer driftchamber, originating from muons leaving a beam dump, two parameters in 16 bit precision were requested from a $12 \times 64 \times 64$ network, namely the origin and the direction of the muon. Since the driftchamber can only provide coordinates in one plane, the two parameters translate into intercept and slope of the particle trajectory projected onto the plane transversely to the incident beam. The drift time was converted to a voltage as analog input to the ETANN. Each output node of the network was trained offline to represent a bit of each of the two 32-bit quantities. Compared to standard track reconstruction the network behaved extremely well (10), reaching the offline results within a factor of 2 in precision. The main strength of this test experiment, however, was to show that a particle trajectory could be determined with good accuracy at the few microsecond scale, so that serious trigger applications could be imagined.

Another approach, also tried out at Fermilab, was a more classical pattern recognition task, namely shower shape evaluation, discriminating electrons from hadrons in the electromagnetic calorimeter of the CDF detector (11). Here, the network was of a constructed type, rather than trained, using the 5 by 5 array of calorimeter cells around the shower maximum as inputs. The weights were calculated based on isolation and energy sharing criteria in the array. The intention of the neural approach was to trigger on electrons coming from semileptonic decays of B-mesons.

Although demonstrating impressively the validity of the neural ideas for triggering, the main drawback of these approaches, finally preventing any further trigger development, was the delicate analog nature of the ETANN chip: It had to be trained in a cumbersome way to maintain its properties (the preloaded weights, thresholds) over time and it furthermore was suffering from the temperature instabilities inherent to most analog devices.

With the availability of digital neuromorphic chips (see Table 1) a number of real trigger applications emerged: At CERN the WA92 experiment (12) realized an electron trigger for charm events based on the Siemens MA16 chip, replacing the ETANN in their first prototype. In nuclear physics a neutrino oscillation experiment (13, 14) at the Chooz reactor utilized the CNAPS chip as a fast trigger to discriminate neutrino interactions from background cosmic events: The signature of an inverse β-decay $\bar{\nu}_e p \rightarrow e^+ n$ is a prompt signal from the annihilation photons, followed by a delayed large signal from neutron capture with a typical time difference of about 2 μs. The neural network was trained to reconstruct, from a set of digitized PMT time and energy signals, the position and energy of the event within the fiducial volume of the target. The neural trigger was running at the second level, following a first level trigger based on PMT pulse height discrimination. Using the CNAPS, the task is performed in about 150 μs, allowing for much higher rates at the first level and consequently much lower thresholds than the standard Chooz trigger. Several other applications of the neuromorphic processors presented in Table 1 have been reported (see, e.g., (15) and the proceedings of previous conferences of this series (16)).

One should realize, however, that the number of concrete running applications using neuromorphic hardware are still scarce. This is mainly due to the fact that most of the classical large-scale industrial applications, such as online character recognition (OCR) do not require execution times at the microsecond scale. In these fields modern Pentium-type serial computers are fast enough. So it is not too surprising that most neuromorphic developments by industry have died out these days (ETANN, MA16, CNAPS, just to name the most prominent of them). As time progresses, however, and the problems to tackle will become more demanding, either concerning execution time or complexity of the problem, the parallel neuromorphic way seems the only viable solution: In the field of realtime triggering or pattern recognition, neural networks play, due to their inherent parallelism of computation, a unique role concerning speed, discrimination power, and adaptability to highly complex tasks. In this sense we foresee a renaissance of neuromorphic hardware in the future (see, e.g., the Silicon Brain Project (17)).

THE H1 NEURAL NETWORK TRIGGER

In the following we will describe in more detail a very ambitious neural trigger project, operational in the H1 experiment at the electron-proton collider HERA at DESY. The H1 neural network trigger project presently is the largest such application in high energy physics and is running successfully now since a number of years. For the future high luminosity running at HERA an upgrade of the neural preprocessing is underway, which aims at constructing, in hardware, physically motivated "intelligent" variables.

Detector and Trigger Scheme

A modern, large particle detector system such as H1 at HERA is built with the intention to serve as a general purpose facility and be prepared to be sensitive for the expected physics as well as potentially new phenomena. To this end a large variety of detection principles are employed, covering efficient detection and measurement of hadronic particles, photons and leptons (electrons and muons). For charged particle detection H1 uses an arrangement of drift and proportional chambers surrounding the interaction point (central tracking detector) and the forward (proton direction) region (forward tracker). For particle energy measurement the tracker system is surrounded by a central liquid argon calorimeter and two warm calorimeters to cover the extreme forward ("plug") and backward ("SpaCal") parts of the solid angle. The calorimeter is enclosed by a superconducting solenoid with diameter of about 6 m, followed by an instrumented magnetic flux return yoke for muon measurement. In the forward direction, a warm toroidal magnet, sandwiched between sets of drift chambers, allows additional measurement of high energy muons. For photoproduction reactions and for luminosity measurements several electron and photon taggers are installed upstream of the proton beam. In the downstream part Roman pots and a neutron calorimeter are installed. Further details on the H1 detector are given elsewhere (18).

For triggering the apparatus, H1 has installed a scheme of three levels, two hardware levels and one software level ("level 4"). An intermediate software level ("level 3") is provided, but not used at present. At level 1, each of the detector components (subdetectors) provides a set of triggers to a central trigger box, where they can be subjected to simple coincidence logic. Since the HERA bunch cross frequency is 10.4 MHz (corresponding to a time interval of 96 ns between possible interactions), the level 1 triggers must be able to derive a trigger decision each 96 ns. This is achieved by storing the information for each bunch cross (BC) in a digital pipeline for a maximum of 30 BC's. After about 2.3 μs (24 BC's) a level 1 trigger decision is formed and the information is transferred to the level 2 systems. At this point the primary deadtime starts, no further triggers can be accepted until a fast clear signal from the level 2 trigger system has been issued rejecting the event. When the event is accepted by the level 2, the detector readout is initiated and the full event information is sent to the level 4 processor farm, where a full event reconstruction is performed and the final event decision is taken.

For the level 2 hardware trigger the decision time is limited to 20 μs in order to digest a maximum of 1-2 kHz from level 1 while keeping the deadtime below 2 %. At level 2, the information from all level 1 processors is available, so that "intelligent" use of this information is possible, exploiting the correlations among the various trigger quantities. The output of the level 2 trigger must not exceed 100 Hz which is the maximum rate for the level 4 RISC processor farm. The output rate of level 4 is dumped to permanent tape and is limited to about 10 Hz.

The Need for Complex Triggers

Since its start in 1992 the HERA ep collider has constantly improved its performance. Already during the first years of operation it became clear that many high cross section physics reactions, most importantly low Q^2 processes and photoproduction could not be triggered with their full rate due to the competing background (dominated by beam-gas reactions). More severe level 1 conditions could not help out due to substantial loss in physics efficiency. The "natural" way out was to scale down ("prescale"), sometimes quite heavily, the respective level 1 triggers designed for these physics reactions.

In 1995 H1 has tried to improve on this unsatisfactory situation by commissioning two hardware systems at the second trigger level, one of them a novel digital neural network trigger based on the CNAPS chip. The level 2 triggers have access to the information provided by the various subdetectors at level 1 and can use this information in an "intelligent" way.

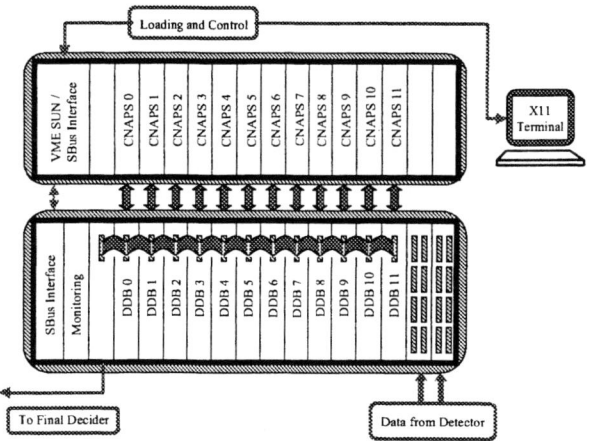

FIGURE 1. *Layout of the H1 Neural Network Trigger system ("L2NN"). Each network processor (CNAPS) is associated with a preprocessing module (DDB) preparing the net inputs individually for its companion CNAPS board. The system is steered by a VME SPARCstation with remote accessibility.*

Architecture of the Neural Network Trigger

The level 2 trigger system has the task to be able to discriminate all possible physics reactions. This request has a very important implication for the neural architecture in the H1 level 2 neural trigger system: Our investigations have shown that *small* nets trained for *specific* physics reactions, working all in parallel, are more efficient compared to a single larger net trained on all possible physics reactions simultaneously. Most importantly, putting these nets to a real trigger application, the degree of modularity is extremely helpful when a new trigger for a new kind of physics reaction is to be implemented: there is no need to retrain the other nets, the new physics net is simply added to the group of the others.

For the network computations (matrix-vector multiplication and accumulation) the CNAPS chip by Adaptive Solutions (5) was used. For the preparation of the input quantities and their interfacing to the CNAPS chip dedicated hardware has been built at the Max-Planck-Institute (preprocessing hardware, see below). Due to the high flexibility of programming the CNAPS chip, arbitrary algorithms can be realized, provided they fit in the latency of level 2 (L2). In the L2 Neural Network Trigger of H1 three different algorithms are used at present:

- Feed Forward Networks: these networks have a fully connected three-layer structure with one input and one hidden layer (with a maximum of 64 nodes each), and one output node. The input layer is fed with the components of a vector $\vec{x}$ spanning the "trigger space", prepared by our preprocessing hardware.

The output y is used as a discriminator to make the trigger decision. Details on the network architectures chosen and on the methods to determine the weights $\vec{w}$ are found elsewhere (19).

- Background encapsulators: To evade possible bias in selecting a specific physics class for training against the background, various approaches such as self-organizing networks or radial basis function networks for encapsulating the background are under study.

- Constructed Nets: For some simple, low-complexity applications a topological correlator, exploiting the fast matrix-vector multiplication hardware of the CNAPS chip, is used (20).

The strategy of using the networks is the following: Each of the networks is trained for a specific physics channel and is coupled to a set of level 1 subtriggers, particularly efficient for that channel. Because the level 1 subtriggers are sufficiently relaxed to be efficient, their rate is usually unacceptably high. The level 2 trigger therefore has the task to reduce the excess background rate in the subtrigger set while keeping the efficiency for the chosen physics channel high. At present, 12 networks are running in parallel, mostly optimized for production of vector mesons, which are difficult to separate from the background at level 1. Typical rate reductions are between a factor of 5 to over a few hundred.

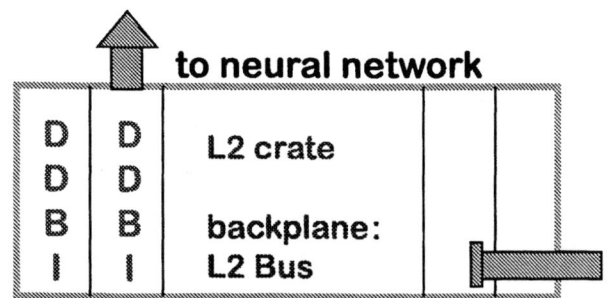

FIGURE 2. *Structure of the DDB-crate which manages the data stream from the level 1 trigger processors to the neural networks: The cable at the right signifies the data from the various subdetectors, which are distributed over a 128bit wide backplane ("L2 bus"). The DDBs pick up the information suited for their respective companion net, prepare the preprocessed net input quantities, and send them to the CNAPS boards.*

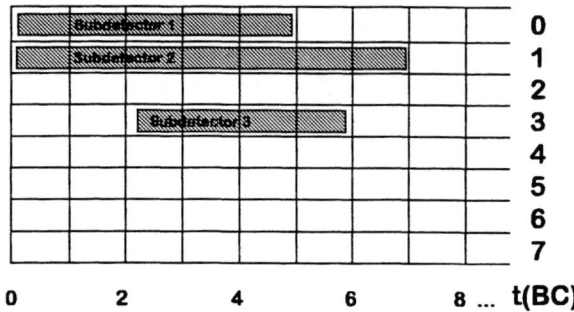

FIGURE 3. *Details of the data flow on the backplane of the DDB-crate: The data from each subdetector are transmitted as a sequence of 16 bit words on one of the 8 subbuses of the L2 bus (gray areas). In the figure three subdetector data streams are shown. The total amount of subdetectors supplying data for L2 is eight at present. The DDBs are programmed to pick up the respective set of items from the subbuses foreseen for preprocessing for the companion neural network.*

The Trigger Hardware

According to the principles described above, the hardware realization for the network trigger is chosen (see Fig. 1): Receiver cards collect the incoming trigger information of the various subdetectors and distribute them via a 128 *bit* wide L2 bus to the preprocessing units, called *Data Distribution Boards* (DDB). Each DDB is able to pick up a freely choosable set of items from the L2 data stream. It performs some basic operations on the items (e.g bit masking, summing) and provides an input vector of maximally 64 8*bit* words for one CNAPS/VME board. Controlling and configuring of the complete system is done by a THEMIS VME SPARCstation, which is located in the crate housing the CNAPS/VME boards.

The CNAPS board

The algorithms calculating the trigger decision are implemented on a VME board housing the CNAPS chip, which is a parallel fixed-point computer in SIMD architecture. The CNAPS-1064 chip (also called *array*) houses 64 processor nodes (PN). Up to eight chips (512 PN's in total) can be combined on one board. A PN is a processor for itself except that it shares the instruction unit and I/O buses with all other PNs. The instruction unit, the sequencer chip CSC-2, is responsible for the command and data flow. The commands are distributed via a 32 *bit* PN command bus. The 8 *bit* wide input and output buses are used for the data transfer to and from the CNAPS arrays. A direct access to these I/O buses is realized with a mezzanine board developed by us. Through this board the input vector is loaded into the CNAPS chip and the trigger result is sent back to the DDB without significant time delay. For synchronization reasons the CNAPS boards are driven with an external clock at 20.8 MHz (2 times the HERA clock frequency of 10.4 MHz).

The main internal parts of the PNs are arithmetic units like adder(32 *bit*) and multiplier(24 *bit*), logic unit, register unit, 4K memory and a buffer unit. All units are connected via two 16 *bit* buses. Calculations are done in fixed-point arithmetic. The sigmoidal transfer function is implemented via a 10 *bit* look-up table. A feed forward step with 64 inputs, 64 hidden nodes and 1 output node can be computed in 8 μs at 20.8 MHz, or in 166 clock cycles. To get the same result with a single conventional CPU one would have to clock it at several GHz.

The Data Distribution Board (DDB)

The Data Distribution Board resides in a special *"L2 VME crate"* equipped with the L2 Bus, an 8 times 16 bit parallel data bus (see figs. 2, 3) running with the HERA clock in an interleaved mode, yielding an effective 20 MHz transfer rate. For each subdetector the level 1 data are sent serially onto one of the eight subbuses of the L2 backplane. For system control purposes, a special monitor board ("spy") with an independent readout of the data transmitted over the L2 bus is residing in the same crate. The data is a heterogeneous mixture from different subdetectors, e.g., calorimetric energy sums, tracker vertex histograms, tracker rays (bits in the $\theta - \phi$ plane), bit-coded muon hit maps, etc.

On the DDB, the L2 data received are passed through a data type selection where they can be transformed (e. g. split into bytes or single bits) using look-up tables (*LUT*).

After bit splitting, several preprocessing algorithms like summing of bits and bytes, bit selections or functions (again realized via LUT's) can be applied. The data may also be sent unchanged to a selection RAM, where the input vector for the neural network computer is stored. Through the use of XILINX 40XX chips the hardware can be flexibly adapted to changes, e.g., for new data formats in the received input. Using selection masks the data are transmitted via a parallel data bus to a mezzanine receiver card directly connected to the local data bus on the CNAPS board.

The complete system with the 2 VME crate is shown in Fig. 4, as it is installed in the H1 experiment (on top the CNAPS crate with the VME control computer, below the DDB crate housing the L2 data bus). Further details on the neural trigger system can be found elsewhere (21).

SOME PHYSICS RESULTS WITH THE NEURAL TRIGGER

With the setup described above various physics channels have been investigated, most importantly elastic and inelastic vector meson photoproduction (ρ, ϕ and J/ψ and). These reactions are difficult to trigger on efficiently since their event signatures are very close to the high rate background (few low momentum tracks in the main detector). Figure 5 shows the trigger rates into 10 different neural networks (red lines), the output rates (blue lines and the rate reduction factor (green lines) for a typical HERA day. As example, for J/ψ elastic photoproduction three different nets are installed (box 2, 4 and 6), depending on the photon-proton center of mass energy with correspondingly quite different detector signatures. Typical rate reductions are between a factor of 5 (network 2) and over 200 (network 6). With these networks a sizeable sample of elastic J/ψ photoproduction events was triggered and led to published cross section measurements shown in Fig. 6. Most remarkable is the fact that the cross section measurements, with the help of the neural trigger technique, could be extended to the highest HERA center of mass energies, where so far no data have existed. An important physics message is extracted from these data: The standard Regge expectation is not in agreement with the data, in contrast to the QCD prediction. The validity of the QCD calculations rely on the fact that the J/ψ meson is heavy enough to justify the perturbative approach.

FIGURE 4. *View of the H1 Neural Network Trigger system (L2NN). Each network processor (CNAPS, upper crate) is connected with a preprocessing module (DDB, lower crate). The cables at the lower right come from the subdetectors, the ones at the lower left carry the decision bits from the neural networks.*

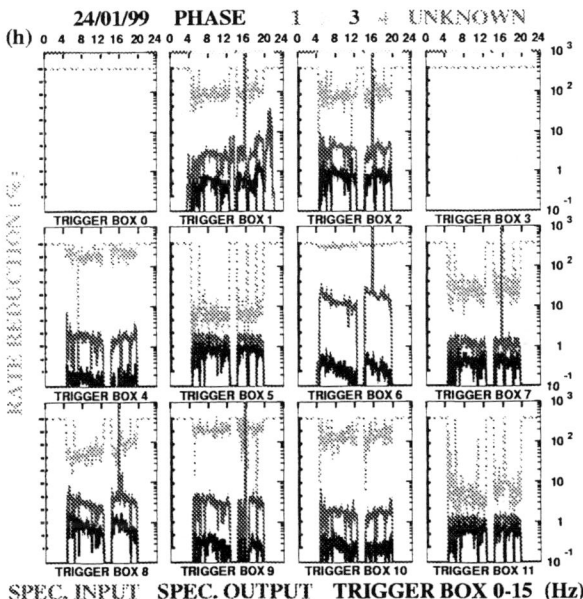

FIGURE 5. *Online rate measurements for the twelve neural trigger boxes operational during the run period of early 1999 (see text).*

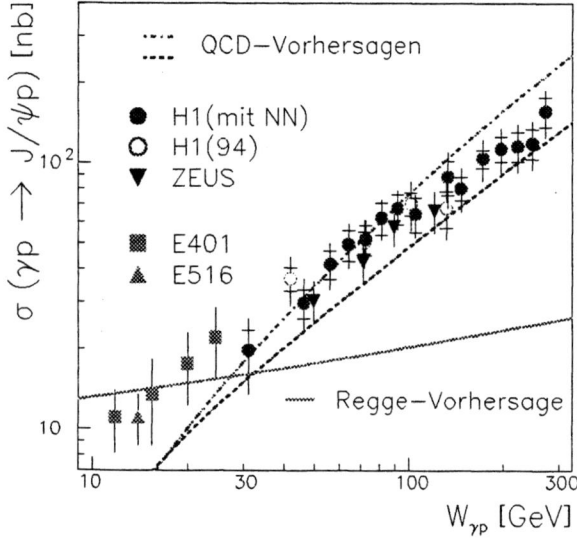

FIGURE 6. Measurements of the total cross section for exclusive photoproduction of J/ψ mesons as function of the photon-proton center of mass energy W. The H1 points are obtained with triggers from three different neural networks, depending on the first level triggers and the observed final state topology. The measurements are compared to those from ZEUS and to some fixed-target experiments at lower energy from Fermilab.

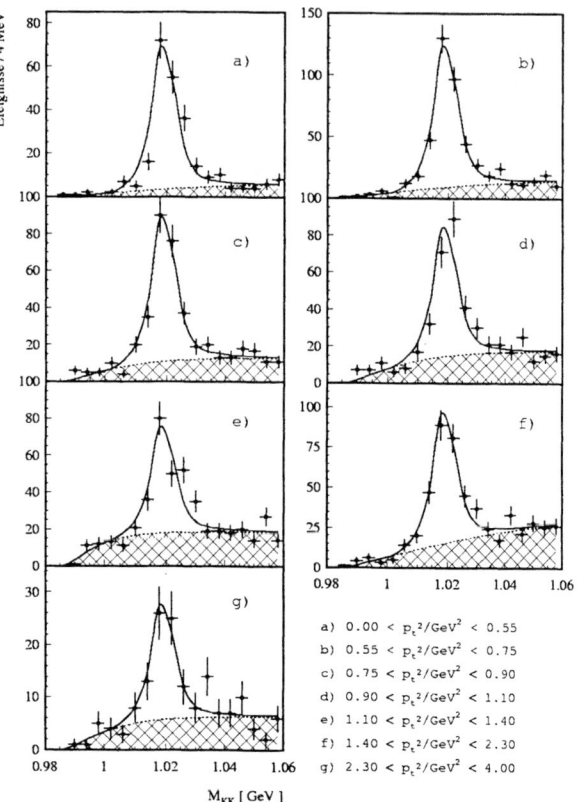

FIGURE 7. Invariant mass distributions from the two final state particles in events from elastic photoproduction of ϕ-mesons, triggered by a neural network trigger. The various distributions correspond to different bins in transverse momentum of the kaon pair. Clear peaks at the ϕ mass are observed.

Another example, close to the final analysis, is the photoproduction of ϕ mesons (see the spectra from the K^+K^- invariant mass spectrum in Fig. 7). This channel could only be triggered with the help of the neural technique.

NEW PREPROCESSING ALGORITHMS: THE DDB2

HERA will move on to improve its luminosity by a factor of 5 in the year 2001. This increase will be achieved mainly through focusing magnets inserted close to the interaction regions of the collider experiments. Since the output rate to tape is to be limited to less than 10 Hz, the H1 trigger system will face the challenge of increased rejection power compared to the present situation. As a consequence of this a new preprocessing hardware is being built for the neural network trigger with the aim to provide the network with more physics-oriented quantities.

The idea behind the improved preprocessing is that the "trivial" part of the correlations in the trigger data are determined, namely the association of information from the various subdetectors linked by the passage of particles or jets, as sketched in Fig. 8. In this way physical objects are formed, defined by their topological vicinity in the various parts of the detector. With the new preprocessing the full granularity of the level 1 trigger information is exploited, but the input data volume to the networks is limited to the physically relevant information.

The central algorithm of the preprocessing is clustering in the various subdetectors, with subsequent matching in the angular coordinates θ and ϕ. To be specific, the following steps will be performed in dedicated hardware, making extensive use of fast modern FPGA technology (for a more detailed discussion see (22)):

Cluster Algorithms: The energy depositions in both the electromagnetic and the hadronic parts of the LAr calorimeter will be clustered, using the highest available granularity (656 "trigger towers" or TTt's), summing nearest neighbors around a local maximum. Before the clustering, look-up tables will perform arbitrary transformations on the TT energies (e.g., transverse energy, taking into account the polar angle θ). Double counting of touching towers is explicitly avoided by the algorithm.

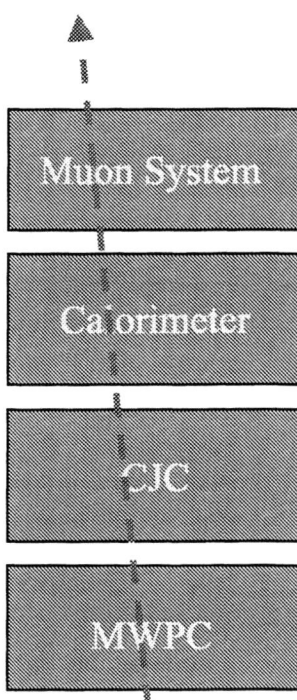

FIGURE 8. *Principle of the preprocessing algorithm for the DDB 2: Particles or jets traverse the various subdetectors of the H1 experiment. The clusters found in each subdetector are correlated in the angular variables θ and ϕ.*

For each cluster found, the total energy $Etot$, the center $Ecent$ with its $\theta - \phi$ values, the ring $Ering$ and the number $nhit$ of towers containing energy above a suitable threshold are stored.

Bit fields (such as hit maps from the MWPCs, the drift chambers or the trigger cells of the SpaCal calorimeter) will also be subjected to a cluster algorithm: In this case a preclustering will be performed, summing all immediate neighbors to a given "seed" bit. After this preclustering, which results in a "hilly" $\theta - \phi$ plane, the same algorithm as for the calorimeter is executed.

Due to their coarse granularity, the trigger information from the muon chambers will not be clustered, but the $\theta - \phi$ information of the muon hits is stored.

Matching: Based on common angular coordinates (θ and ϕ) the clusters from the various layers of the detector will be gathered in physical objects, forming a vector the components of which represent the list of cluster quantities determined in the previous clustering step. In addition, the objects will be ordered in magnitude according to chosen components. Three parallel sorting machines are foreseen, delivering arrays of vectors sorted according to three pre-determined vector components. Ordered input was found very helpful in speeding up the learning and in improving the discrimination power.

Net Inputs: The input to the networks are the sorted vector components (or a freely choosable subset, depending on physics reaction considered). Due to the serial clocking-in of data into the CNAPS chip, a limit of about 8 to 10 objects (with 8 components each) as net input is imposed, which should be quite sufficient for the physics applications considered at present, such as heavy flavor production, both open and hidden, jets in photoproduction, low Q^2 reactions and other physics channels depositing mostly low energy in the calorimeter.

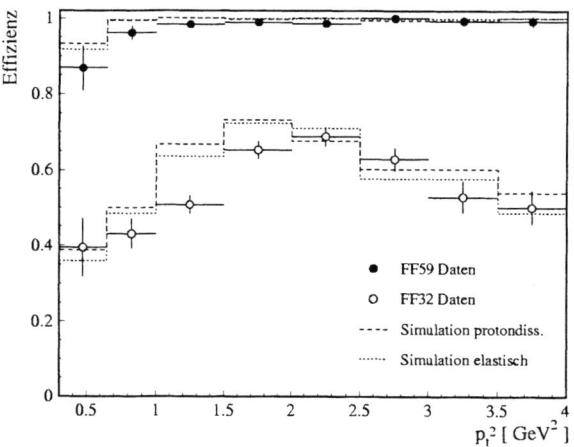

FIGURE 9. *Efficiency for a network trained to select elastic photoproduction of ϕ-mesons, using inputs obtained from the present DDB compared to the efficiency with quantities supplied by the DDB 2 (upgraded preprocessing). The improved efficiency by the DDB 2 is obvious (24).*

Post processing: A final step is considered which determines some physical quantities from the vector components, such as cluster counting, angular differences etc. The exact specifications for the post-processing is still under investigation, studying specific physics reactions.

As an example for the increased selection power of the DDB 2 with respect to the less sophisticated DDB ("DDB 1") presently operating, the elastic photoproduction of ϕ-mesons has been studied using the neural network trigger. This reaction could be observed in H1 for the first time (23). The performance of the network based on DDB 1 quantities is shown in Fig. 9. Using in a simulation the new physics objects (24) provided by the DDB 2, the selection efficiency - at constant background rejection - could be increased by more than 40 percent (see Fig. 9).

CONCLUSIONS

We have presented the principles and the hardware realization for a second level neural network trigger, operational in the H1 experiment at HERA as a global event decision machine since summer of 1996. Based on commercially available, massively parallel digital ULSI neural network chips, a 20 μs decision time is achieved for the network trigger. The network inputs are derived from the trigger information provided by the various level 1 trigger systems and are preprocessed by custom-designed hardware (DDB 1). The system has demonstrated its physics potential by largely increasing the statistics for vector meson production, producing the first physics result from a neural network trigger. The neural trigger also allowed observing elastic ϕ photoproduction for the first time in H1.

For the data taking with the upgraded HERA machine higher selectivity of the neural trigger is needed, and an improved preprocessing unit is under development (DDB 2), performing a quasi physics analysis at the trigger level: Physical objects ("particles" , "jets") are created from the full granularity of the level 1 trigger processors, extracting the physically relevant information while keeping the data volume into the networks manageable. The proposed unit was subjected to test reactions, demonstrating its superior selection power compared to the present DDB 1.

The fact that industry has discontinued the production of neuromorphic hardware is not a sign of lost interest in neural network realtime applications. It is just a reaction to the market which in most of its applications is satisfied with a time response within fractions of seconds, easily obtainable with present-day Pentium type computers. For the visible future realtime applications in physics experiments (high energy physics, astrophysics etc.) trigger processors will have to provide response times in the microsecond range, only obtainable with parallel computing architectures. Developments in this area are underway in the academic world (see, e.g., (25)). We are convinced that these efforts will not only be beneficial for science but also leave their impact for future commercial applications.

REFERENCES

1. P. Bhat, "Advanced Analysis Techniques", Invited Talk, these proceedings.

2. H. Prosper, "Artificial Intelligence - Offline Applications", rapporteur talk, these proceedings.

3. ETANN 80170NX Chip, Specification Booklet, Intel Corporation, 1991.

4. Neuro-Classifier Chip: P. Masa and H. Wallinga, University of Twente, Internet: www.ice.el.utwente.nl/Finished/Neuro/NeuroClassifier.htm

5. CNAPS Release Notes 2.0, Adaptive Solutions, Inc. , Beaverton Or. (1993).

6. U. Ramacher et al., "Design of a First Generation Neurocomputer", in: VLSI Design of Neural Networks, eds. U. Ramacher and U. Rückert, Kluwer Academic Publishers, 1991.

7. G. Anzellotti et al., in "New Computing Techniques in Physics Research", eds. B. Denby & D. Perret-Gallix, World Scientific, Singapore 1995.

8. SAND1 chip, Forschungszentrum Karlsruhe and Institut für Mikroelektronik, Stuttgart. Internet: fuzzy.fzk.de/ fischer/sand/index.html

9. B. Denby, Comp. Phys. Comm. 49 (1988) 429.

10. C. Lindsay et al., Nucl. Instr. Meth. A317 (1992) 346.

11. M. Badgett et al., "A Neural Network Calorimeter Trigger Used in CDF", Proceedings of the 1992 IEEE Nuclear Science Symposium, Orlando, Florida.

12. A. Baldanza et al., in "New Computing Techniques in Physics Research III", eds. K.-H. Becks, D. Perret-Gallix, World Scientific, Singapore, 1994.

13. M. Apollonio et al., Phys. Lett. B420 (1998) 397

14. A. Baldini et al., Nucl. Inst. Meth. A389 (1997) 144

15. B. Denby, Comp. Phys. Comm. 119 (1999) 219.

16. Proceedings of the Workshop on New Computing Techniques in Physics Research IV, eds. B. Denby, D. Perret-Gallix, World Scientific, Singapore, 1995.

17. The Silicon Brain Initiative, Irvine Sensors Inc., Internet: www.irvine-sensors.com

18. I. Abt et al. "The H1 detector at HERA", Nuclear Instruments and Methods in Physics Research A 386 (1997) 310.

19. S. Udluft et al., in "New Computing Techniques in Physics Research VI", eds. G. Athanasiu & D. Perret-Gallix, Parisianou, Athens, 2000, p. 166.

20. J. Möck, "Untersuchung diffraktiver J/Ψ-Ereignisse im H1-Experiment bei HERA und Entwicklung neuronaler Triggeralgorithmen", PhD thesis, TU München (1997).

21. J.H. Köhne et al., Nuclear Instruments and Methods in Physics Research A 389 (1997) 128.

22. J.-C. Prevotet et al., "Intelligent preprocessing for Neural Networks in the H1 experiment", these proceedings.

23. F. Gaede, *Exklusive Produktion von ϕ-Mesonen in ep-Streuung am H1-Experiment bei HERA*, PhD thesis, Christian-Albrechts-Universität zu Kiel (1997)

24. S. Udluft, *Protondissoziative Photoproduktion von ϕ-Mesonen am H1-Experiment bei HERA*, PhD thesis, Ludwig-Maximilians-Universität, München (2000)

25. B. Granado, L. Lacassagne and P. Garda, "Maharadja: A candidate system for RBF networks with the Mahalanobis distance in HEP", in "New Computing Techniques in Physics Research VI", eds. G. Athanasiu & D. Perret-Gallix, Parisianou, Athens, 2000, p. 80.

Feynman-diagram evaluation in the electroweak theory with computer algebra

G. Weiglein

CERN, TH Division, CH-1211 Geneva 23, Switzerland

Abstract. The evaluation of quantum corrections in the theory of the electroweak and strong interactions via higher-order Feynman diagrams requires complicated and laborious calculations, which however can be structured in a strictly algorithmic way. These calculations are ideally suited for the application of computer algebra systems, and computer algebra has proven to be a very valuable tool in this field already over several decades. It is sketched how computer algebra is presently applied in evaluating the predictions of the electroweak theory with high precision, and some recent results obtained in this way are summarized.

INTRODUCTION

The electroweak and strong interactions of elementary particles are very successfully described by quantized gauge field theories. The quantized nature of these theories manifests itself via corrections beyond the lowest order in the perturbative expansion, which is based on Feynman diagrams. The evaluation of higher-order Feynman diagrams (which are called loop diagrams) is a technically very complicated but on the other hand algorithmic procedure. The development of computer-algebra systems was boosted by the demand for this kind of applications in particle physics, and *Schoonschip* (1) was one of the first implementations of a powerful computer-algebra program. Further examples of computer-algebra systems that have their roots in particle physics are *Reduce* (2), *Macsyma* (3), *Mathematica* (4), and *FORM* (5).

Computer-algebra systems allow to perform symbolic manipulations and algebraic calculations without round-off errors. They are equipped with a number of built-in algorithms which provide the basis for the user to implement his/her own algorithms for handling specific problems. Present-day computer-algebra systems furthermore possess capabilities for communicating with external programs, e.g. with routines for numerical evaluation, text processing and graphics, or with other computer-algebra programs.

Examples of computer-algebra systems being widely used at present in high-energy theory are *Mathematica* (4), *Maple* (6) and *FORM* (5). *Mathematica* and *Maple* are general-purpose programs containing a large number of built-in functions (and many additional software packages are available). These programs offer capabilities for both symbolic and numerical computations, support graphical display, and possess user-friendly interactive platforms. The application of these systems to problems in high-energy physics involving expressions with a huge number of terms can be limited by the computing speed or by memory problems. The latter applies in particular to non-local operations, like e.g. factorization, which require to have all terms of an expression available within the physical memory of the computer. *FORM*, on the other hand, is a program that was specifically optimized for handling very large expressions. It is less user-friendly than *Mathematica* and *Maple*, containing much fewer built-in operations and allowing only non-interactive execution. For recent developments concerning the parallelization of *FORM*, see Ref. (7).

PERTURBATIVE EVALUATION OF GAUGE THEORIES WITH COMPUTER-ALGEBRAIC METHODS

The concept of treating interactions as a perturbation to a free field theory and performing an expansion in the coupling constants leads to a description of scattering processes in terms of Feynman diagrams. The lowest-order prediction, corresponding to the classical limit, for a process with a certain number of external particles (e.g. a $2 \to 2$ scattering process) is obtained from the sum of the connected diagrams containing the lowest possible power of the coupling constants (which enter via the interaction vertices). These are in general tree-level diagrams. In higher-order diagrams additional interaction vertices give rise to closed loops of propagators, for which an integration over the internal momenta has to be performed.

The prediction for a scattering process of certain fields, assigned to the external legs, and a specified number of loops can be obtained via an algorithmic procedure. In a first step, all topologically different diagrams (for which in renormalizable theories only 3-point and 4-point interaction vertices are possible) have to be generated. Inserting the fields of the model under consideration into the topologies in all possible ways leads to the Feynman diagrams. The Feynman rules translate these graphical representations into mathematical expressions.

Since the loop integrals in general lead to divergences, the expressions need to be regularized (i.e. made mathematically meaningful). In a renormalizable theory the divergences can be absorbed into a redefinition of the parameters of the theory. The renormalization is furthermore necessary in order to fix the physical meaning of the parameters order by order.

The evaluation of the Feynman amplitudes involves a treatment of the Lorentz structure of the amplitude, calculation of Dirac traces etc. At the one-loop level it is possible to reduce all tensor integrals to a set of standard scalar integrals, which can be expressed in terms of known analytic functions. As a consequence, with the existing techniques a wide class of processes with up to four external legs can be evaluated at the one-loop order in massive gauge theories (for a discussion of the technical problems occurring in one-loop processes with six external legs, see e.g. Ref. (8)).

In contrast to the one-loop case, no general algorithm exists so far for the evaluation of two-loop corrections in the electroweak theory. The main obstacle in two-loop calculations in massive gauge theories is the complicated structure of the two-loop integrals, which makes both the tensor integral reduction and the evaluation of scalar integrals very difficult. In general the occurring integrals are not expressible in terms of polylogarithmic functions. For the evaluation of some types of integrals that do not permit an analytic solution numerical methods and expansions in their kinematical variables have been developed.

Applying the appropriate on-shell conditions to the external legs one obtains the S-matrix element from the sum of all contributing Feynman amplitudes. Squaring it and performing the phase space integrations one finally arrives at predictions for cross sections and life times.

Computer-algebraic methods can facilitate most of the above-mentioned steps. Besides benefits from automation, a computer-algebraic treatment is also useful for verifying the correctness of the different steps of a certain calculation. In particular, results obtained at the algebraic level (before inserting specific numerical values for the parameters) are well suited for highly non-trivial checks, e.g. with respect to their UV- and IR-finiteness, gauge-parameter independence, and the validity of Slavnov–Taylor identities. As an example, in Ref. (9) a Slavnov–Taylor for the two-loop Z-boson self-energy in the electroweak Standard Model (SM) has been verified by showing that the results of about 4000 Feynman diagrams add up to zero algebraically.

As indicated by the above example, powerful computer-algebraic tools are very useful for calculations (in particular of higher-order corrections) in the SM, since the large number of different fields in the SM gives rise to a large number of contributing Feynman diagrams (at the one-loop level typically $O(10^2)$, at the two-loop level $O(10^3)$), and the massiveness of the fields makes the evaluation of the loop diagrams very complicated in general. The technical complications are even higher in extensions of the SM. In the Minimal Supersymmetric Standard Model (MSSM) the duplication of the number of fields compared to the SM leads to a plethora of possible interaction vertices and consequently to a large increase in the number of diagrams contributing at a certain order. In QCD, on the other hand, computer-algebraic tools are particularly valuable for multi-loop applications. In Ref. (10), for instance, the four-loop β function of QCD has been calculated. This required the computation of about 50000 diagrams, showing clearly the need for a high degree of automation. Similarly, for tree-level processes with many particles in the final state thousands of diagrams can contribute and algebraic methods can be useful for obtaining compact and numerically efficient representations, see e.g. Ref. (11).

Examples of computer-algebra based collections of program packages presently used for higher-order calculations in the electroweak thory and QCD are (where the different programs in each collection mostly use common syntax and can be linked together)

(i) *FeynArts* (12), *FeynCalc* (13), *FormCalc* (14), *TwoCalc* (9, 15), *LoopTools* (14), *s2lse* (16),

(ii) *GEFICOM* (17), *QGRAF* (18), *MATAD* (19), *MINCER* (20),

(iii) *DIANA* (21), *QGRAF* (18), *ON-SHELL2* (22),

(iv) *xloops* (23), *GiNaC* (24).

FeynArts, *FeynCalc*, *FormCalc* and *TwoCalc* are written in *Mathematica* (*FormCalc* is partially written in *FORM*). *FeynArts* is a program for generating all Feynman amplitudes contributing to a certain process to a given order in *Mathematica* format and for drawing the corresponding Feynman diagrams. As a feature of particular importance for higher-order calculations in the electroweak theory, *FeynArts* generates not only the unrenormalized diagrams at a given order but also the counterterm contributions at this order and the counterterm diagrams needed for the subloop renormalization. The model files for the electroweak SM (including the Feynman rules for the background-field formulation of the SM (25)) and QCD are predefined in *FeynArts*. In applications to other models, e.g. the two Higgs-doublet model (26), the MSSM (27, 28, 29, 30) or chiral perturbation theory (31), the appropriate model file has to be provided by the user.

FeynCalc and *FormCalc* are programs (using the *FeynArts* syntax) for algebraically evaluating one-loop diagrams in the electroweak theory and QCD with up to four external legs in a highly automatized way. *FormCalc* internally uses an interface to *FORM*, which is used for the memory- and time-intensive parts of the calculation. *FormCalc* can directly be linked to *LoopTools*, which contains routines for the numerical evaluation of scalar one-loop integrals and one-loop tensor coefficients. *LoopTools* is based on the *FF* (32) package and provides a *Fortran* and a *C++* library.

As mentioned above, much less tools are available for two-loop calculations in massive gauge theories compared to the one-loop case. The program *TwoCalc* is based on an algorithm for the tensor reduction of general two-loop 2-point functions, which extends the algorithm for the tensor reduction of one-loop integrals (33). *TwoCalc* can be used for an automatic reduction of Feynman amplitudes for two-loop self-energies with arbitrary masses, external momenta, and gauge parameters to a set of standard scalar integrals. It can directly be linked to the program *s2lse*, which is written in *C++* and performs the evaluation of the scalar two-loop 2-point integrals by means of one-dimensional integral representations in terms of elementary functions, which allow a fast numerical evaluation with high precision.

The program collection *GEFICOM*, *QGRAF*, *MATAD*, and *MINCER* is mainly used for calculations in QCD and for the evaluation of QCD corrections to electroweak observables. *GEFICOM* acts as the master program that calls the other packages. It contains *Mathematica* and *Fortran* routines as well as elements written in the script languages *AWK* and *PERL*. For details of *GEFICOM* and some examples of its applications, see Ref. (17).

The *Fortran* program *QGRAF* is an efficient generator for Feynman diagrams. As output the diagrams are encoded in a symbolic notation. Being optimized for high speed, *QGRAF* is particularly useful for applications involving a very large number (i.e. $O(10^4)$) of diagrams. Within *GEFICOM*, the evaluation of the diagrams proceeds by performing expansions in their kinematical variables. The resulting integrals are then computed with *MINCER* and *MATAD*. The program *MINCER* performs the computation of integrals up to three-loop order where all lines are massless and only one external momentum is non-zero. It makes in particular use of integration-by-parts methods (34). While its original version was written in *Schoonschip*, the present version of *MINCER* is realised in *FORM*. The *FORM* program *MATAD* was designed for the computation of vacuum integrals up to three-loop order which contain only one mass scale (i.e. their propagators are either massless or carry a common mass).

The *C* program *DIANA* is designed as a master program for higher-order calculations, i.e. it calls the necessary subprograms for a specific computation. It reads the output of *QGRAF* and can produce a graphical representation for the diagrams if the relevant topologies are pre-defined by the user. For the calculation of the diagrams *FORM* programs are called, e.g. the package *ONSHELL2* which can be used for the calculation of single-scale two-loop 2-point functions (diagrams with only one non-zero mass in the internal lines and the external momentum on the same mass shell).

xloops is a *Maple* package for calculating certain one-loop and two-loop diagrams in the electroweak SM, which is linked to *C++* routines for numerical integration of loop integrals. The symbolic part of *xloops* is planned to be based in the future on *GiNaC*, which is a specifically designed framework written in *C++*.

EXAMPLES OF HIGHER-ORDER RESULTS IN THE SM AND THE MSSM

In the following some examples are sketched of recent higher-order results obtained with *FeynArts*, *TwoCalc* and *s2lse* in the electroweak SM and the MSSM. Within the SM, higher-order calculations are necessary for the comparison of the theory predictions with the experimental results for electroweak precision observables like M_W, $\sin^2 \theta_{eff}$ etc. which have meanwhile reached an accuracy of better than 1×10^{-3} (35). The precision tests of the SM allow in particular to set constraints on the mass of the Higgs boson, which is the last missing ingredient of the SM and plays a crucial role for a consistent description of massive particles.

In Ref. (37) the currently most accurate prediction for the W-boson mass, M_W, within the SM has been obtained. It contains it particular the complete fermionic contributions at the two-loop level, which are treated exactly, i.e. without an expansion in the top-quark or the Higgs-boson mass. The result for M_W is shown in Fig. 1 as a function of the Higgs-boson mass, M_H. It is compared with the current experimental value for M_W (35). The present 95% C.L. lower bound on M_H from the direct search at LEP of $M_H = 113.5$ GeV (36) is also indicated. The plot shows the well-known preference for a light Higgs boson within the SM. Confronting the theoretical prediction (allowing a variation of the top-quark mass, m_t, which at present dominates the theoretical uncertainty, within 1σ) with the 1σ region of M_W^{exp} and the 95% C.L. lower bound on M_H, one finds that at the present level of accuracies the 1σ regions do no longer overlap.

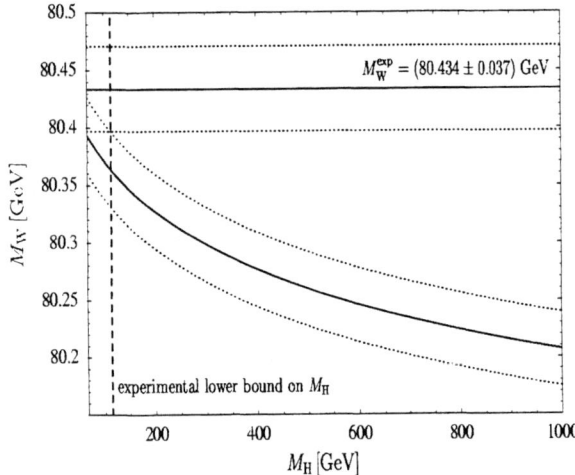

FIGURE 1. The prediction for M_W as a function of M_H for $m_t = 174.3 \pm 5.1$ GeV is compared with the current experimental value, $M_W^{exp} = 80.434 \pm 0.037$ GeV (35), and the experimental 95% C.L. lower bound on the Higgs-boson mass, $M_H = 113.5$ GeV (36).

By comparing the SM predictions for the precision observables with those of extended models, it can be investigated whether the data allow a distinction between different kinds of possible models. In Fig. 2 the predictions for M_W in the SM and the MSSM are shown as a function of m_t. The MSSM prediction contains the dominant SUSY contributions of $O(\alpha \alpha_s)$ to the ρ parameter (27). The allowed region in the SM corresponds to varying M_H in the interval $113 \text{ GeV} \leq M_H \leq 400 \text{ GeV}$, while in the region of the MSSM prediction the SUSY parameters are varied, taking into account the constraints from direct searches for SUSY particles. As indicated in the figure, the predictions in the SM and the MSSM give rise to two bands with only a relatively small overlap region. This region corresponds to the SM with a light Higgs, $M_H \lesssim 130$ GeV, and to the MSSM with heavy superpartners, whose virtual contributions decouple from the electroweak precision observables.

The predictions for M_W in the SM and the MSSM are confronted in Fig. 2 with the current experimental accuracies of M_W and m_t (LEP2/Tevatron, outermost ellipse) and with the prospective accuracies at the LHC and a future Linear Collider (LHC/LC) and at a high-luminosity Linear Collider running in a low-energy mode on the Z-boson resonance and the W-pair threshold (GigaZ). As can be read off from the figure, the data on M_W and m_t presently show a slight preference for the MSSM over the SM, which however statistically is not very significant. The figure shows that the next generation of colliders, in particular a Linear Collider in the GigaZ mode, promises an enormous improvement in the experimental accuracies of M_W and m_t (and furthermore also for $\sin^2\theta_{eff}$) which will allow to test the electroweak theory with unprecedented sensitivity (38).

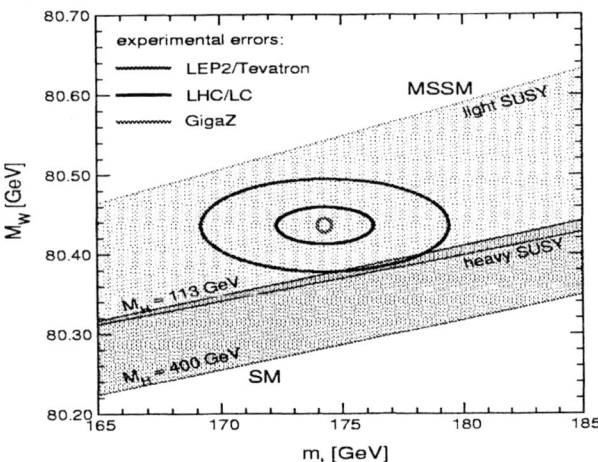

FIGURE 2. The theoretical prediction for M_W within the SM and the MSSM in comparison with the current experimental accuracies (LEP2/Tevatron) and the prospective accuracies at the LHC/LC and at GigaZ.

Besides the indirect constraints from electroweak precision tests, supersymmetric models provide a very stringent direct test since they predict the existence of a relatively light Higgs boson, whose mass can be calculated from the other parameters of the model.

In Ref. (28) a Feynman-diagrammatic result has been obtained for the dominant two-loop contributions to the masses of the neutral $\mathcal{CP}$-even Higgs bosons in the MSSM. The algebraic result obtained with *FeynArts* and *TwoCalc* has been converted into *Fortran* code and has been implemented into the program *FeynHiggs* (39).

While at the tree-level the lightest $\mathcal{CP}$-even Higgs boson in the MSSM is bounded to be lighter than the Z-boson mass, this bound is shifted upwards to $m_h \lesssim 135$ GeV taking into account corrections up to the two-loop order. The highest possible values for m_h are obtained for large values of $\tan\beta$, the ratio of the vacuum expectation values of the two Higgs doublets of the MSSM, large values of the mass of the $\mathcal{CP}$-odd Higgs boson, and a large mixing between the superpartners of the top quark. Comparing the theoretical prediction for the upper bound on m_h as a function of $\tan\beta$ with the experimental exclusion limits obtained at LEP2, it is possible to derive constraints on $\tan\beta$. This is shown in Fig. 3, where the excluded region results from combining the data of the four LEP experiments (40), and the upper (and lower) bound within the MSSM (indicating the boundary to the "theoretically inaccessible" region) has been obtained with *FeynHiggs*.

The upper plot shows the case of the so-called "m_h^{max} benchmark scenario" (41), in which the MSSM parameters (for fixed values of $m_t = 174.3$ GeV and the SUSY scale $M_{SUSY} = 1$ TeV) are chosen such that m_h as a function of $\tan\beta$ takes its maximal values. From the intersection of the experimentally excluded region with the boundary to the theoretically inaccessible region one finds an excluded region of $0.5 < \tan\beta < 2.3$ within this scenario. In the lower plot the MSSM parameters have been chosen according to the "no-mixing benchmark scenario" (41), which differs from the m_h^{max} scenario in that vanishing mixing in the scalar top sector has been assumed. In this case a much wider region of $\tan\beta$ values, up to about $\tan\beta \approx 8$, can be excluded for $m_t = 174.3$ GeV and $M_{SUSY} = 1$ TeV.

CONCLUSIONS AND OUTLOOK

The development of powerful computer-algebra systems was triggered by applications in high-energy physics. Computer-algebra tools have extensively been used in this field already for several decades, and many of todays calculations would not have been feasible without computer algebra. A brief overview of computer-algebraic methods for the perturbative evaluation of gauge theories has been given, and some examples have been discussed of recent higher-order results obtained in the electroweak Standard Model and its Minimal Supersymmetric extension.

The use of modern computer-algebra programs goes beyond their application as tools for certain steps of the calculations. As indicated by the above examples of different collections of programs for precision calculations within the theory of the electroweak and strong interactions, an efficient communication of computer-algebra systems with other program components is of particular importance. These external programs can be packages for numerical evaluations, text processing tools, data bases, expert systems, but also other computer-algebra programs being particularly well suited for certain sub-parts of the problem. In order to facilitate this kind of communication, the need for a certain degree of standardization for the integration of program parts and the data transfer between different systems will become more pronounced in the future.

Accordingly, future improvements of soft- and hardware promise a further extension of the applicability of computer-algebra systems in two different ways. On the one hand they could allow highly sophisticated calcula-

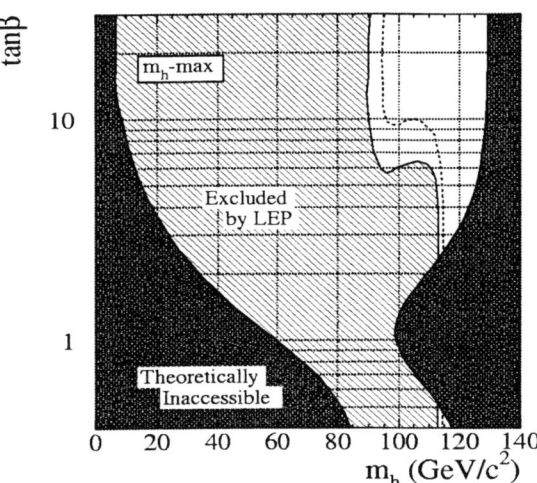

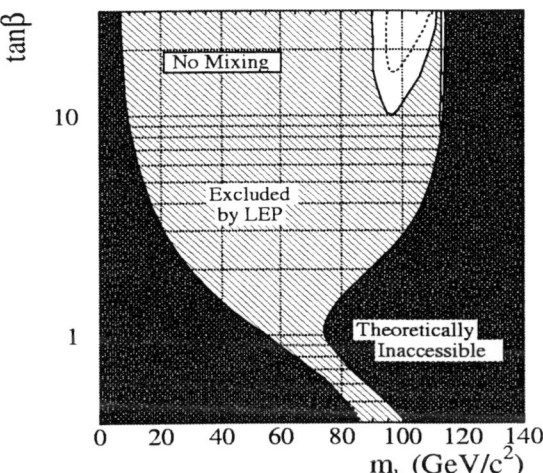

FIGURE 3. The 95% C.L. bounds on m_h in the m_h^{max} and the no-mixing benchmark scenarios obtained from combining the data of the four LEP experiments are compared with the upper bound on m_h within the MSSM (40).

tions which go beyond the scope of present capabilities. On the other hand computer-algebra systems could more and more become parts of general problem-solving environments, where different components are integrated in such a way that the different parts of calculations and the tasks of text processing, graphical representation etc. are handled in the most efficient way.

ACKNOWLEDGMENTS

The author thanks P. Bhat, the other organizers of ACAT 2000, and the Fermilab Theory Group for the invitation and their kind hospitality during his stay at Fermilab.

REFERENCES

1. M. Veltman, *Schoonschip*, CERN preprint, 1967.
2. A.C. Hearn, *Reduce User's Manual*, (The Rand Corp., Santa Monica, CA, 1985).
3. J. Moses, *ACM SIGSAM Bull.*, **8** (1974) 105.
4. S. Wolfram, *Mathematica: A system for Doing Mathematics by Computer*, (Addison-Wesley, Reading, Massachusetts, 1988).
5. J.A.M. Vermaseren, *Symbolic Manipulation with FORM*, (Computer Algebra Netherlands, Amsterdam, 1991); J.A.M. Vermaseren, these proceedings.
6. B.W. Char, K.O. Geddes, M.W. Gentleman and G.H. Gonnet, in *Springer Lecture Notes in Computer Science 162*, (Springer, Berlin, 1983), p. 101.
7. D. Fliegner, A. Rétey and J.A.M. Vermaseren, TTP-99-15, hep-ph/9906426; TTP00-13, hep-ph/0007294.
8. A. Vicini, *Acta Phys. Pol.* **B 29** (1998) 2847.
9. G. Weiglein, R. Scharf and M. Böhm, *Nucl. Phys.* **B 416** (1994) 606.
10. T. van Ritbergen, J.A.M. Vermaseren and S.A. Larin, *Phys. Lett.* **B 400** (1997) 379.
11. T. Ohl, these proceedings, IKDA-2000-30, hep-ph/0011243.
12. J. Küblbeck, M. Böhm and A. Denner, *Comput. Phys. Commun.* **60** (1990) 165; T. Hahn, KA-TP-23-2000, hep-ph/0012260.
13. R. Mertig, M. Böhm and A. Denner, *Comput. Phys. Commun.* **64** (1991) 345.
14. T. Hahn, *Nucl. Phys.* **B (Proc. Suppl.) 89B** (2000) 231.
15. G. Weiglein, R. Mertig, R. Scharf and M. Böhm, in *New Computing Techniques in Physics Research 2*, ed. D. Perret-Gallix (World Scientific, Singapore, 1992), p. 617.
16. S. Bauberger, F.A. Berends, M. Böhm and M. Buza, *Nucl. Phys.* **B 434** (1995) 383; S. Bauberger, F.A. Berends, M. Böhm, M. Buza and G. Weiglein, *Nucl. Phys.* **B (Proc. Suppl.) 37B** (1994) 95; S. Bauberger and M. Böhm, *Nucl. Phys.* **B 445** (1995) 25.
17. R. Harlander and M. Steinhauser, *Prog. Part. Nucl. Phys.* **43** (1999) 167.
18. P. Nogueira, *J. Comput. Phys.* **105** (1993) 279.
19. M. Steinhauser, DESY 00-124, hep-ph/0009029.
20. S.G. Gorishny, S.A. Larin and F.V. Tkachov, INR P-0330 (1984); S.G. Gorishny, S.A. Larin, L.R. Surguladze and F.V. Tkachov, *Comput. Phys. Commun.* **55** (1989) 381; S.A. Larin, F.V. Tkachov and J.A.M. Vermaseren, NIKHEF-H/91-18.
21. M. Tentyukov and J. Fleischer, *Comput. Phys. Commun.* **132** (2000) 124; J. Fleischer and M. Tentyukov, these proceedings, hep-ph/0012189.
22. J. Fleischer and M.Yu. Kalmykov, *Comput. Phys. Commun.* **128** (2000) 531.
23. L. Brücher, J. Franzkowski and D. Kreimer, *Comput. Phys. Commun.* **115** (1998) 140.
24. C. Bauer, A. Frink, R. Kreckel, MZ-TH-00-17, cs/0004015; these proceedings.
25. A. Denner, S. Dittmaier and G. Weiglein, *Nucl. Phys.* **B 440** (1995) 95.
26. W. Beenakker, A. Denner and A. Kraft, *Nucl. Phys.* **B 410** (1993) 219.
27. A. Djouadi, P. Gambino, S. Heinemeyer, W. Hollik, C. Jünger and G. Weiglein, *Phys. Rev. Lett.* **78** (1997) 3626; *Phys. Rev.* **D 57** (1998) 4179.
28. S. Heinemeyer, W. Hollik, and G. Weiglein, *Phys. Rev.* **D 58** (1998) 091701; *Phys. Lett.* **B 440** (1998) 296; *Eur. Phys. Jour.* **C 9** (1999) 343.
29. A. Arhrib and G. Moultaka, *Nucl. Phys.* **B 558** (1999) 3.
30. S. Heinemeyer, these proceedings.
31. U. Bürgi, *Nucl. Phys.* **B 479** (1996) 392.
32. G.J. van Oldenborgh and J.A.M. Vermaseren, *Z. Phys.* **C 46** (1990) 425; G.J. van Oldenborgh, NIKHEF-H/90-15.
33. G. Passarino and M. Veltman, *Nucl. Phys.* **B 160** (1979) 151.
34. F.V. Tkachov, *Phys. Lett.* **B 100** (1981) 65; K.G. Chetyrkin and F.V. Tkachov, *Nucl. Phys.* **B 192** (1981) 159.
35. A. Gurtu, talk given at ICHEP 2000, Osaka, July 2000, to appear in the proceedings.
36. P. Igo-Kemenes, talk given in the LEPC open session (Nov. 3, 2000) on behalf of the LEP Higgs Working Group, lephiggs.web.cern.ch/LEPHIGGS/talks.
37. A. Freitas, W. Hollik, W. Walter and G. Weiglein, *Phys. Lett.* **B 495** (2000) 338; A. Freitas, S. Heinemeyer, W. Hollik, W. Walter and G. Weiglein, *Nucl. Phys.* **B (Proc. Suppl.) 89B** (2000) 82; CERN-TH/2001-018, hep-ph/0101260.
38. J. Erler, S. Heinemeyer, W. Hollik, G. Weiglein and P.M. Zerwas, *Phys. Lett.* **B 486** (2000) 125; S. Heinemeyer and G. Weiglein, hep-ph/0012364.
39. S. Heinemeyer, W. Hollik and G. Weiglein, *Comput. Phys. Commun.* **124** (2000) 76; hep-ph/0002213.
40. The LEP working group for Higgs boson searches, S. Andringa et al., ALEPH 2000-074 CONF 2000-051, DELPHI 2000-148 CONF 447, L3 Note 2600, OPAL Technical Note TN661.
41. M. Carena, S. Heinemeyer, C. Wagner and G. Weiglein, CERN-TH/99-374, hep-ph/9912223.

Grid Computing

Ian Foster

Mathematics and Computer Science Division, Argonne National Laboratory
Department of Computer Science, The University of Chicago

Abstract. The term "Grid Computing" refers to the use, for computational purposes, of emerging distributed Grid infrastructures: that is, network and middleware services designed to provide on-demand and high-performance access to all important computational resources within an organization or community. Grid computing promises to enable both evolutionary and revolutionary changes in the practice of computational science and engineering based on new application modalities such as high-speed distributed analysis of large datasets, collaborative engineering and visualization, desktop access to computation via "science portals," rapid parameter studies and Monte Carlo simulations that use all available resources within an organization, and online analysis of data from scientific instruments. In this article, I examine the status of Grid computing circa 2000, briefly reviewing some relevant history, outlining major current Grid research and development activities, and pointing out likely directions for future work. I also present a number of case studies, selected to illustrate the potential of Grid computing in various areas of science.

INTRODUCTION

The term "computational Grid" or simply "Grid" refers to a new class of infrastructure designed to enable resource sharing among geographically distributed, typically multi-institutional communities [10]. Much as the Internet and Web have reduced barriers to the exchange of information, Grid protocols and services are intended to facilitate remote access to and the coupling of computers, storage systems, display devices, and people, regardless of physical location.

Grid concepts originated in scientific and engineering computing, motivated by the ever-increasing focus on multi-disciplinary, collaborative research and computationally oriented approaches to scientific problems. We are currently at an interesting juncture in the development and application of the technology. Five years of prototyping studies have demonstrated the promise of new Grid-based problem-solving approaches and created an extensive knowledge base concerning tools and techniques. Spurred by the success of these studies, significant Grid deployment efforts have started, in which key infrastructure elements are widely deployed, and major scientific communities have started to retool to embrace the use of Grid technologies. The next several years should see significant success stories as well as refinements in our understanding of the technology.

HISTORICAL ANTECEDENTS

We can trace the historical antecedents of today's Grid computing back to the earliest days of computer networking: after all, the original goal of work on projects such as ARPANET and Multics was to enable remote access to computers. Distributed systems research also has a long and distinguished history of both fundamental contributions and practical application, and contributes much to the body of technology on which Grids are constructed.

Interest in the more ambitious applications considered here really developed with the deployment of high-speed experimental gigabit testbeds in the late 1980s. For example, the CASA testbed linked Caltech, the Jet Propulsion Laboratory, Los Alamos National Laboratory, and the San Diego Supercomputer Center with a dedicated high-speed network and then demonstrated that it was possible, via a combination of specialized algorithms and protocol refinements, to achieve significant speedups for a variety of distributed applications [14, 15]. However, these and other related experiments did not attempt to put in place any protocols or services designed to facilitate sharing: resources were scheduled manually.

The concept of a Grid infrastructure distinct from that of the underlying Internet emerged as a result of the I-WAY project in 1995, in which many of the nation's high-speed networks and supercomputers were interconnected to provide a powerful, although short-lived, application testbed [6]. An important part of this project was the creation of a software infrastructure that provided a uniform authentication, scheduling, and information service [7]. Both the I-WAY and its software infrastructure proved remarkably effective, supporting some 60 application groups in a wide variety of domains.

Inspired in part by the success of the I-WAY effort, a number of new Grid-related initiatives emerged in the late 1990s, including the Globus project [9] and its Globus Ubiquitous Supercomputing Testbed Organization [4], the National Computational Science Alliance's National Technology Grid [17], and most recently the NASA Ames Information Power Grid [11] and DOE ASCI Grid. These and other related efforts are engaged in prototyping and deploying a national-scale Grid infrastructure designed to support next-generation applications.

GRID TECHNOLOGIES

The experience gained during the I-WAY experiment and in subsequent research, development, and deployment activities has resulted in a good understanding of the technologies required to support advanced applications. These technologies build heavily on concepts and standards developed within the Internet community but extend their scope in three respects, to address (1) remote access to and sharing of the end-systems (computers, storage systems, etc.) that Grid applications must deal with, (2) the need for extremely high performance often encountered in Grid applications, and (3) the truly distributed, multi-party—rather than client-server—interactions often encountered in Grid applications. Key ideas underlying ``Grid architecture'' include the following:

- *Grid protocols.* Much as the Internet Protocol defines a ``lingua franca'' that allows disparate devices to exchange information, the Grid architecture defines a set of protocols for resource management, data access, and resource discovery. The Grid Security Infrastructure (GSI) defines SSL-based protocols for authentication, authorization, and delegation in multi-institutional settings. The Grid Resource Allocation and Management (GRAM) protocol builds on GSI to provide for authentication, authorization, job submission, and computation management, hence enabling secure remote job submission. (The Secure Shell protocol represents another access mechanism.) The LDAP-based Grid Resource Information Service (GRIS) protocol provides for both remote enquiry about resource state and registration with index server(s) that provide resource discovery services. The FTP protocol is used for data access, with a combination of standardized but not widely used features (e.g., GSS support for security) and extensions used to address specific concerns of Grid environments.

- *Resource managers.* The definition of this standard protocol suite makes it easy to define what we mean by ``Grid-enabled resources'': any resource that supports the protocols just listed is Grid-accessible. The ``resource managers'' that implement the protocols can vary greatly in sophistication. For example, a simple ``storage resource manager'' (SRM) might be just a GSI-enabled version of a standard FTP server, while a more sophisticated SRM might provide additional space reservation, request queuing, request time estimation, and other functions.

- *Grid services.* The definition of standard resource access protocols makes it possible to define a variety of higher-level services concerned with resource sharing across virtual organizations. For example, Grid Information Index Services (GIIS) leverage GRIS capabilities to provide resource discovery services for a collection of distributed resources, while replication services support the replication of data across multiple storage systems and network caches.

- *Grid tools.* The broad deployment of Grid protocols and services makes it feasible to create Grid tools, i.e., tools designed to ease the development of various classes of application. Examples of such tools include Condor-G [13] and Nimrod-G [1], which support high-throughput computations involving large numbers of independent tasks; tools for data-intensive computing, such as replica management tools for creating and selecting from among data replicas; collaborative environment tools, for managing the sharing of state information by large communities; ``Science Portal'' tools for creating desktop gateways to Grid resources; and distributed computing tools such as MPICH-G for writing applications that exploit Grid resources.

This layered architecture enables individual resources and sites to participate in Grid applications with relatively little effort (resources just need to speak a few simple protocols). At the same time, broad deployment of protocols and services greatly simplifies the task of developing higher-level tools and applications.

PRODUCTION GRIDS

Until recently, a significant obstacle to Grid computing was the fact that the protocols and services were not deployed in any persistent manner and hence could not be taken for granted. Several programs have been launched in the last year aimed at the creation of "production Grids:" that is, Grid infrastructures that are persistent, are supported, and that extend to a significant number of resources found interesting by their target user community. These different efforts each have a different focus, but all build on the core Grid services identified in the preceding section.

The U.S. National Science Foundation (NSF)'s Partnerships for Advanced Computational Infrastructure (PACIs), the NCSA Alliance, headed up by NCSA in Illinois and NPACI, headed up by SDSC in California, both have ambitious Grid deployment activities underway, with the term "National Technology Grid" being used as an umbrella term for the target environment.

NCSA Alliance activities center around its Virtual Machine Room (VMR), Access Grid (AG), and Science Portal projects. The VMR is intended to link the major computational resources across the Alliance into a single integrated computing system, with uniform access mechanisms as well as specialized resource discovery and brokering services. At the time of writing, a first version of the VMR is operational: selected Grid services are in place and various support services (Help Desk, Certificate Authority) have been established. However, much remains to be done before the long-term goal of truly seamless access to the Alliance's resources is achieved.

Work at NPACI emphasizes the deployment of Grid services on NPACI resources and the creation of Science Portals as a means of facilitating access. Their HotPage system is a nice example of a Web-based Portal and in its latest incarnation makes extensive use of Grid services for remote access and monitoring.

NASA's Information Power Grid (IPG) project was started in 1998 with the ambitious goal of integrating Grid computing into the practice of science and engineering within NASA. Based at NASA Ames, the IPG is simultaneously creating a production Grid infrastructure and starting new initiatives relating to tools and applications. Persistent Grid services and associated accounting, support, and Certificate Authority services are in place at four NASA site: NASA Ames, Glenn, Goddard, and Langley.

The U.S. Department of Energy's Office of Defense Program's ASCI program is developing a tri-lab Grid as part of its DISCOM project. DISCOM builds once more on the Globus software used in the other efforts listed here, but to comply with DOE security policies relies on Kerberos for authentication. The switch to Kerberos is straightforward because the Globus Toolkit uses the Generic Security Services (GSS) API for all authentication operation.

Reseachers within the U.S. Department of Energy's Office of Science have proposed the development of a DOE Science Grid that would link major DOE Science laboratories and collaborators, providing both standard Grid services and advanced experimental services such as network quality of service. Some preliminary work has been done in this area, with the creation of the Globus Advance Reservation Network Testbed (GARNET) linking systems at ANL, LBNL, and other institutions.

APPLICATIONS

We review briefly a number of applications chosen to be representative of the wide range of areas in which Grid concepts are being pursued.

Collaboration. CAVERNsoft [12] and Access Grid [5] represent two different technologies concerned with enabling large-scale collaborative work by geographically distributed communities. CAVERNsoft emphasizes support for collaborative manipulation of shared virtual spaces within immersive virtual reality environments, while Access Grid is concerned with communication and information sharing in the context of large-scale (wall-sized) shared display systems.

Tele-instrumentation. Foster et al. [8] describe a system for the collaborative, online analysis of experimental data from the Advanced Photon Source, a high-brilliance X-ray source. This system uses Grid protocols and services to acquire dynamically the supercomputer resources required for online reconstruction of APS data and the advanced visualization systems needed for subsequent incremental dissemination of reconstructed data to remote collaborators. The result is that a batch-mode scientific instrument is transformed into an interactive device.

Portals. HotPage and ECCE' are just two examples of the many "Portals" and "problem solving environments" that have been developed that enable desktop access to Grid resources. In the HotPage portal, the focus is on providing uniform Web access to the diverse supercomputer resources of a national collaboration, the National Partnership for Advanced Computational Infrastructure; in the ECCE' problem solving environment for computational chemistry, the remote supercomputers are essentially invisible, being called upon by ECCE' when a user requests a computationally demanding calculation.

Distributed computing. A Caltech-based group has used Grid services to assemble a total of 13 supercomputers at 11 sites that were together used to perform a record-breaking distributed interactive simulation (DIS) computation [16]. An Argonne-Iowa-Northwestern-Wisconsin group used the Condor-G system to solve a challenging open problem in numerical optimization, accumulating a total of 96,000 node hours during a seven-day period, peaking at over 1000 processors at 7 sites worldwide. In both cases, the Grid increased resources available to researchers by an order of magnitude.

FUTURE DIRECTIONS

We can expect the future to see rapid and sometimes startling progress in Grid technologies, with major advances and changes occurring not only within the scientific and engineering communities that have pioneered many of these ideas, but also within the commercial space.

In science and engineering, one major focus for many researchers over the next several areas will be the development of the technology base required to support large-scale data-intensive computing in distributed environments. One significant driving force for this work will be the large quantities of data to be produced by frontier science experiments (e.g., at the Large Hadron Collider at CERN) as well as major simulation efforts (e.g., in climate). Communities of thousands of scientists, distributed globally and served by networks of varying bandwidths, need to extract small signals from enormous backgrounds via computationally demanding analyses of datasets that will grow from the 100 Terabyte to the 100 Petabyte scale over the next decade. The computing and storage resources required will be distributed, for both technical and strategic reasons, across national centers, regional centers, university computing centers, and individual desktops.

Numerous research groups and projects are pursuing the design, development, and application of so-called "Data Grid" technologies. For example, the SDSC Storage Resource Broker (SRB) [3] provides an integrated framework for metadata management, data access, and analysis; the Globus Data Grid project is developing basic mechanisms for high-speed transport, replica management, and other functions [2]. Projects such as the Earth System Grid, Particle Physics Data Grid, and European Union Data Grid projects are developing and applying key technologies.

The Grid Physics Network (GriPhyN) project (www.griphyn.org) is another large effort focused on realizing the technical concept of "virtual data." GriPhyN uses the term *virtual data grid* as a unifying concept to describe the new technologies required to support such next-generation data-intensive applications. We use this term to capture the following unique characteristics:

- A virtual data grid has *large extent*—national or worldwide—and *scale*, incorporating large numbers of resources on multiple distance scales.

- A virtual data grid is more than a network: it layers sophisticated *new services* on top of local policies, mechanisms, and interfaces, so that geographically remote resources can be used in a coordinated fashion.

- A virtual data grid provides a new degree of *transparency* in how data-handling and processing capabilities are integrated to deliver data products to end-user applications, so that requests for such products are easily mapped into computation and/or data access at multiple locations. (This transparency is needed to enable optimization across diverse, distributed resources, and to keep application development manageable.)

These characteristics combine to enable the definition and delivery of a potentially unlimited virtual space of data products derived from other data. In this virtual space, requests can be satisfied via direct retrieval of materialized products and/or computation, with local and global resource management, policy, and security constraints determining the strategy used. The concept of *virtual data* recognizes that all except irreproducible raw experimental data need 'exist' physically only as the specification for how they may be derived. The grid may instantiate zero, one, or many copies of derivable data depending on probable demand and the relative costs of computation, storage, and transport. In high-energy physics today, over 90% of data access is to derived data. On a much smaller scale, this dynamic processing, construction, and delivery of data is precisely the strategy used to generate much, if not most, of the web content delivered in response to queries today.

In the commercial space, the rise of the Application Service Provider (ASP) and of Internet Computing—that is, the exploitation for computational purposes of the millions of often idle CPUs on the Internet—are two significant trends. ASPs seem likely to have a revolutionary effect on many aspects of the computer industry, transforming today's business relationships so that users interact exclusively with ASPs rather than with the software and hardware vendors they have dealt with in the past. This trend has the potential to reduce significantly current barriers to the use of advanced simulation technologies, if only user interface and support issues can be addressed. Current activities in the science and engineering arena concerned with ``Science Portals'' can be viewed as precursors of this trend.

Finally, Internet Computing has risen to prominence as a result firstly of the several ``volunteer'' efforts that have delivered large number of cycles to various highly parallel problems (e.g., SETI@Home) and more recently as a result of commercial endeavors such as Distributed.Net and Entropia.Com. While it is too early to say what range of applications will prove amenable to highly distributed execution, the possibilities for a transforming impact on science as a result of order-of-magnitude reductions in computing costs are significant.

ACKNOWLEDGMENTS

I gratefully acknowledge discussions with many colleagues on these topics, in particular Carl Kesselman, Steven Tuecke, Bill Johnston, and Rick Stevens. This work was supported in part by the Mathematical, Information, and Computational Sciences Division subprogram of the Office of Advanced Scientific Computing Research, U.S. Department of Energy, under Contract W-31-109-Eng-38; by the Defense Advanced Research Projects Agency under contract N66001-96-C-8523; by the National Science Foundation; and by the NASA Information Power Grid program.

BIBLIOGRAPHY

1. Abramson, D., Sosic, R., Giddy, J. and Hall, B. Nimrod: A Tool for Performing Parameterised Simulations using Distributed Workstations. in *Proc. 4th IEEE Symp. on High Performance Distributed Computing*, 1995.

2. Allcock, B., Bester, J., Chervenak, A.L., Foster, I., Kesselman, C., Nefedova, V., Quesnel, D. and Tuecke, S., Efficient Data Transport and Replica Management for High-Performance Data-Intensive Computing. in *Mass Storage Conference*, (2001).

3. Baru, C., Moore, R., Rajasekar, A. and Wan, M. The SDSC Storage Resource Broker. in *Proceedings of CASCON'98 Conference*, 1998.

4. Brunett, S., Czajkowski, K., Fitzgerald, S., Foster, I., Johnson, A., Kesselman, C., Leigh, J. and Tuecke, S. Application Experiences with the Globus Toolkit. in *Proc. 7th IEEE Symp. on High Performance Distributed Computing*, 1998, 81-89.

5. Childers, L., Disz, T., Olson, R., Papka, M.E., Stevens, R. and Udeshi, T. Access Grid: Immersive Group-to-Group Collaborative Visualization. in *Proceedings of the Fourth International Immersive Projection Technology Workshop 2000*, to appear.

6. DeFanti, T., Foster, I., Papka, M., Stevens, R. and Kuhfuss, T. Overview of the I-WAY: Wide Area Visual Supercomputing. *International Journal of Supercomputer Applications*, 10 (2). 123-130.

7. Foster, I., Geisler, J., Nickless, W., Smith, W. and Tuecke, S. Software Infrastructure for the

I-WAY Metacomputing Experiment. *Concurrency: Practice & Experience, 10* (7). 567--581.

8. Foster, I., Insley, J., Laszewski, G.v., Kesselman, C. and Thiebaux, M. Distance Visualization: Data Exploration on the Grid. *IEEE Computer, 32* (12). 36-43.

9. Foster, I. and Kesselman, C. Globus: A Toolkit-Based Grid Architecture. in *The Grid: Blueprint for a Future Computing Infrastructure*, 1998, 259-278.

10. Foster, I. and Kesselman, C. (eds.). *The Grid: Blueprint for a Future Computing Infrastructure*, 1999.

11. Johnston, W.E., Gannon, D. and Nitzberg, B. Grids as Production Computing Environments: The Engineering Aspects of NASA's Information Power Grid. in *Proc. 8th IEEE Symp. on High Performance Distributed Computing*, 1999.

12. Leigh, J., Johnson, A. and DeFanti, T.A. CAVERN: A Distributed Architecture for Supporting Scalable Persistence and Interoperability in Collaborative Virtual Environments. *Virtual Reality: Research, Development and Applications, 2* (2). 217-237.

13. Litzkow, M. and Livny, M. Experience With The Condor Distributed Batch System. in *IEEE Workshop on Experimental Distributed Systems*, 1990.

14. Lyster, P., Bergman, L., Li, P., Stanfill, D., Crippe, B., Blom, R., Pardo, C. and Okaya, D. CASA Gigabit Supercomputing Network: CALCRUST Three-dimensional Real-Time Multi-Dataset Rendering. in *Proc. Supercomputing '92*, 1992.

15. Messina, P. Distributed Supercomputing Applications. in Kesselman, I.F.a.C. ed. *The Grid: Blueprint for a Future Computing Infrastructure*, 1998, 55-73.

16. Messina, P., Brunett, S., Davis, D., Gottschalk, T., Curkendall, D., Ekroot, L. and Siegel, H. Distributed Interactive Simulation for Synthetic Forces. in *Proceedings of the 11th International Parallel Processing Symposium*, 1997.

17. Stevens, R., Woodward, P., DeFanti, T. and Catlett, C. From the I-WAY to the National Technology Grid. *Communications of the ACM, 40* (11). 50-61.

Large-Scale Molecular Dynamics Simulations of Materials on Parallel Computers

Aiichiro Nakano,[a] Timothy J. Campbell,[a,b] Rajiv K. Kalia,[a] Sanjay Kodiyalam,[a] Shuji Ogata,[a,c] Fuyuki Shimojo,[a,d] Priya Vashishta,[a] and Phillip Walsh[a]

[a]*Concurrent Computing Laboratory for Materials Simulations*
Department of Computer Science, Department of Physics & Astronomy
Louisiana State University, Baton Rouge, LA 70803, USA
[b]*Logicon Inc., Naval Oceanographic Office Major Shared Resource Center, Stennis Space Center, MS 39529, USA*
[c]*Department of Applied Sciences, Yamaguchi University, Ube 755-8611, Japan*
[d]*Faculty of Integrated Arts and Sciences, Hiroshima University, Higashi-Hiroshima 739-8521, Japan*

Abstract. Scalable space-time multiresolution algorithms implemented on massively parallel computers enable large-scale molecular dynamics simulations involving up to a billion atoms, which are applied to the study of nanosystems of great technological importance. These include sintering, structure, and mechanical properties of nanostructured ceramics and nanocomposites, structural transformation in semiconductor nanocrystals, nanoindentation, and oxidation of metallic nanoparticles.

INTRODUCTION

Advanced materials and devices with nanometer grain/feature sizes are being developed to achieve higher strength and toughness in ceramic materials and greater speeds in electronic devices. Below 100 nm, however, continuum description of materials and devices must be supplemented by atomistic descriptions [1]. Current state-of-the-art atomistic simulations involve 1 million to 1 billion atoms. Scalable multiresolution algorithms that enable these large-scale simulations are described. We discuss the application of these algorithms to molecular dynamics simulations of various nanosystems of great technological importance.

SCALABLE ATOMISTIC SIMULATION ALGORITHMS

We have developed a suite of scalable simulation, parallel-computing, and data-management algorithms [2,3]. Billion-atom molecular dynamics (MD) simulations have been demonstrated [4] on massively parallel computers using: i) space-time multiresolution algorithms; ii) wavelet-based adaptive load balancing in curvilinear computational space; and iii) spacefilling-curve-based adaptive data compression for scalable I/O.

We have also developed [4] scalable algorithms for quantum-mechanical (QM) calculations based on the density functional theory (DFT) [5], thereby demonstrating that 20,000-atom DFT-based QM calculations are feasible on massively parallel computers.

Space-time Multiresolution Molecular Dynamics

The MD approach obtains the phase-space trajectories (positions and velocities of all atoms at all time) by numerically integrating coupled ordinary differential equations. The dynamics is encoded in the interatomic potential energy, $E_{MD}(\mathbf{r}^N)$, which is a function of the positions of all N atoms, $\mathbf{r}^N = \{\mathbf{r}_1, \mathbf{r}_2, ..., \mathbf{r}_N\}$, in the system. In our many-body interatomic potential scheme, $E_{MD}(\mathbf{r}^N)$ is an analytic function of relative positions of atomic pairs and triplets [6].

We have developed highly efficient, multiresolution algorithms to carry out large-scale MD simulations on parallel computers. The most compute-intensive problem in an MD simulation is the $O(N^2)$ computation of the electrostatic energy for N charged atoms. We use the Fast Multipole Method (FMM) by

Greengard and Rokhlin to reduce the complexity to $O(N)$ by computing the electrostatic field recursively on an octree [7]. To extend the simulated time scales, we combine the FMM with a multiple time-scale (MTS) method that applies different force-updating schedules for different force components [8].

We have implemented this multiresolution molecular-dynamics (MRMD) algorithm on a number of parallel computers (1,280-processor IBM SP3, 1,088-processor Cray T3E, and 512-processor SGI Origin supercomputers as well as on our 166-node PC and 64-node Digital Alpha clusters) using spatial decomposition. The MRMD algorithm is highly scalable [4]: For a 1.02-billion-atom silica system, one MD step takes only 26.4 seconds on 1,024 Cray T3E processors. For this system, the parallel efficiency is 0.97.

Wavelet-based Adaptive Load Balancing

To overcome the load-imbalance problem associated with parallel computation of irregular atomic distribution, we have developed an adaptive load-balancing algorithm [9]. The "computational-space-decomposition" scheme is based on a novel idea that computational space shrinks where the workload density is high. The algorithm introduces a curvilinear coordinate transformation to partition workloads according to a uniform 3D-mesh topology in the curved computational space. This leads to curved partition boundaries in the physical Euclidean space. Subsequently simulated annealing is used to minimize load-imbalance and communication costs as a functional of the coordinate transformation. The multiresolution analysis based on wavelets allows compact representation of partition boundaries, leading to fast convergence of the minimization. The wavelet load balancer (WLB) achieves nearly perfect speed-up with negligible computational overhead.

Spacefilling-curve-based Adaptive Data Compression for Scalable I/O

For scalable input/output (I/O), we have designed a data compression algorithm, which uses octree indexing and sorts atoms accordingly on the resulting spacefilling curve [10]. By storing differences between successive atomic coordinates, the I/O requirement for the same error tolerance level reduces from $O(N\log N)$ to $O(N)$. An adaptive, variable-length encoding scheme is used to make the algorithm tolerant to outliers and optimized dynamically. The spacefilling-curve data compression (SCDC) algorithm achieves an order-of-magnitude improvement in the I/O performance for actual MD data with user-controlled error bounds.

Linear-scaling Quantum-Mechanical Algorithms

An atom consists of a nucleus and surrounding electrons, and quantum mechanics explicitly treats the electronic degrees-of-freedom. The DFT reduces the exponentially complex quantum problem to a self-consistent matrix eigenvalue problem, which can be solved with $O(N_{wf}^3)$ operations (N_{wf} is the number of independent wave functions, or electronic bands). The DFT can be formulated as a minimization of the energy, $E_{QM}(\mathbf{r}^N, \psi^{N_{wf}})$, with respect to electron wave functions, $\psi^{N_{wf}}(\mathbf{r}) = \{\psi_1(\mathbf{r}), \psi_2(\mathbf{r}), ..., \psi_{N_{wf}}(\mathbf{r})\}$, subject to orthonormalization constraints. Efficient parallel implementation of DFT is possible with real-space approaches based on higher-order finite differencing [11] and multigrid acceleration [12]. We include electron-ion interactions using norm-conserving pseudopotentials [13] and the exchange-correlation energy in a generalized gradient approximation [14]. For large systems ($N_{wf} > 1,000$), however, the $O(N_{wf}^3)$ orthonormalization becomes the bottleneck.

For scalable DFT calculations, linear-scaling algorithms are essential [15]. We have implemented [4] an $O(N_{wf})$ algorithm based on unconstrained minimization of a modified energy functional and a localized-basis approximation [16]. In the parallel linear-scaling DFT algorithm, the computation time scales as $O(N_{wf}/P)$ on P processors, whereas the communication time scales as $O((N_{wf}/P)^{2/3})$. This is in contrast to the $O(N_{wf}(N_{wf}/P)^{2/3})$ communication in the conventional parallel real-space DFT algorithm. Global communication for calculating overlap integrals of wave functions (which scales as $N_{wf}^2 \log P$ in the conventional DFT algorithm) is unnecessary.

Our parallel linear-scaling DFT (LSDFT) algorithm is highly scalable: For a 22,528-atom GaAs system on 1,024 Cray T3E processors, the parallel efficiency is 0.96 [4]. Figure 1 shows the performance of our scalable MD and QM algorithms on 1,024 T3E processors.

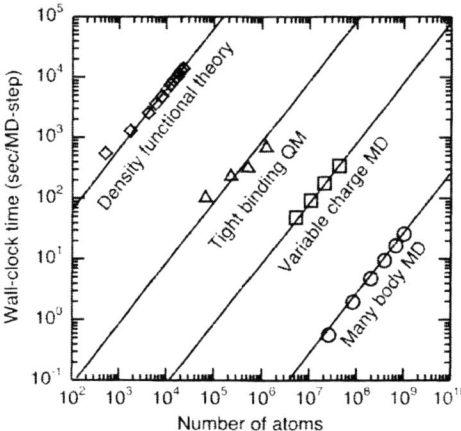

FIGURE 1. Computation time of MD and QM algorithms as a function of the number of atoms on 1,024 Cray T3E processors: (circles) classical MD based on many-body interatomic potentials; (squares) environment-dependant, variable-charge MD; (triangles) QM calculation based on the tight-binding method; and, (diamonds) self-consistent DFT-QM.

LARGE-SCALE ATOMISTIC SIMULATIONS OF NANOSTRUCTURED

The scalable simulation algorithms described in the previous section have been used to perform large-scale atomistic simulations of various nanosystems. In this section, we summarize some of our simulation results.

Sintering, Structure, and Mechanical Properties of Nanostructured Ceramics and Nanocomposites

Advanced structural ceramics are highly desirable materials for applications in extreme operating conditions. Light weight, elevated melting temperatures, high strengths, and wear- and corrosion-resistance make them very attractive for high-temperature and high-stress applications. The only serious drawback of ceramics is that they are brittle at low to moderately high temperatures. In recent years, a great deal of progress has been made in the synthesis of ceramics that are much more ductile than conventional coarse-grained materials [17]. These so called nanostructured materials are fabricated by *in-situ* consolidation of nanometer size clusters.

We have performed multimillion-atom MD simulations to investigate sintering, structure, and mechanical behavior of nanostructured Si_3N_4 [18-20], SiO_2 [21], and SiC [22].

Nanostructured Si_3N_4 is generated by consolidating a random cluster configuration [18-20]. Pair distribution functions and bond angle distributions reveal that interfacial regions in the consolidated nanophase Si_3N_4 are amorphous. Systems sintered at low pressures (1GPa) have percolating pores whose surface roughness exponents (0.46 and 0.86) are in excellent agreement with experiments. We also find that the dependence of elastic moduli on porosity and grain size in nanostructured Si_3N_4 can be understood in terms of a three-phase model for heterogeneous materials.

Nanocomposite materials often exhibit superior properties compared with conventional materials [23]. In recent years, large-scale MD simulations have played a central role in the investigation of dynamic fracture in such complex materials. For example, we are performing 1.5-billion-atom MD simulations to investigate fracture in a Si_3N_4 matrix reinforced with SiC fibers of linear dimension 0.3 μm. In order to simulate the effect of a glassy phase that lubricates the fiber-matrix interfaces, SiC fibers are coated with amorphous silica layer (Fig. 2). We are investigating the effect of interphase structure and residual stresses on the fracture toughness. Immersive and interactive visualization reveals a rich diversity of atomistic processes including fiber rupture, frictional pullout, and emission of molecular fragments, which must all be taken into account in the design of tough composites.

FIGURE 2. Atomistic model of fractured Si_3N_4 matrix reinforced with SiC fibers coated with amorphous silica. Small spheres represent silicon atoms, and large spheres represent nitrogen (green), carbon (magenta), and oxygen (cyan) atoms.

Structural Transformation in Semiconductor Nanocrystals

Despite numerous experimental studies, structural transformations in SiC and GaAs at high pressures are not well understood at the atomistic level. We have investigated the mechanisms of these transformations using MD simulations. In SiC, a reversible transformation between the four-fold coordinated zinc-blende structure and the six-fold coordinated rocksalt structure is found at a pressure of 100 GPa, in good agreement with experimental data [24]. We have found that the atomistic mechanism for the structural transformation is a cubic-to-monoclinic unit-cell transformation and a relative shift of Si and C sublattices in the [100] direction. This new transition path does not involve any bond breaking and it has a significantly lower activation energy compared with a previously proposed transformation mechanism.

We have also investigated pressure-induced structural transformations in GaAs nanocrystals of different sizes using MD simulations [25]. Semiconductor nanocrystals have numerous applications as optical devices and new synthetic paths to novel materials. It is found that the transformation from four-fold (zinc blende) to six-fold (rocksalt) coordination starts at the surfaces of nanocrystals and proceeds inwards with increasing pressure. Inequivalent nucleation of the rocksalt phase at different sites leads to an inhomogeneous deformation of the nanocrystal. For sufficiently large spherical nanocrystals, this gives rise to rocksalt structures of different orientations separated by grain boundaries (Fig. 3). The absence of such grain boundaries in a faceted nanocrystal of moderate size indicates sensitivity of the transformation to the initial nanocrystal shape. The pressure corresponding to the complete transformation increases with the nanocrystal radius and it approaches the bulk value for a spherical nanocrystal of ~ 5,000 atoms.

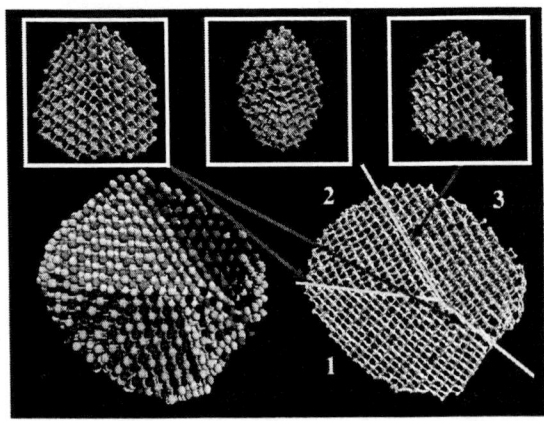

FIGURE 3. (Bottom left) Grain structure in a GaAs nanoparticle at a pressure of 22.5 GPa is color coded. Insets show grain boundary types.

Nanoindentation of Silicon Nitride

Nanoindentation testing is a unique probe of mechanical properties of materials. Typically, an atomic force microscope tip is modified to indent the surface of a very thin film. The resulting damage is used to rank the ability of the material to withstand plastic damage against that of other materials.

We have performed 10-million-atom MD simulations of nanoindentation of Si_3N_4 thin films [26]. The films were indented with a square-based pyramidal indentor to maximum depths of 8-9 nm. The hardness value calculated from the load-displacement curve at 300K (approximately 50 GPa) is reduced to approximately 90% of its value at 2000K. Figure 4 shows the local pressure distribution directly under the indenter for the fully loaded and fully unloaded configurations. These images have been used in conjunction with local bond-angle calculations to characterize a process of local amorphization under the indenter, which is arrested by either piling up of material along the indenter edges or by cracking under the indenter corners.

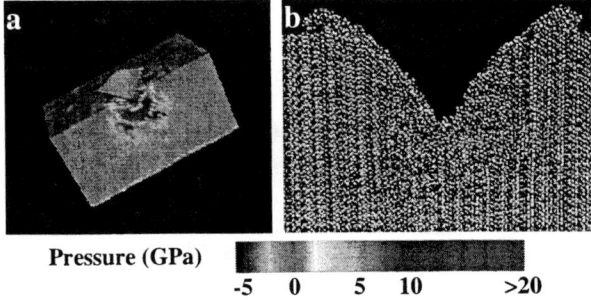

FIGURE 4. (a) A half-slice view of pressure in Si_3N_4 during nanoindentation. (b) An atomic view showing amorphization under the indentor and material pileup at the edges of the indentor. Red and yellow are silicon and nitrogen atoms, respectively.

Oxidation of Aluminum Nanoclusters

Oxidation plays a critical role in the performance and durability of various nanosystems. Oxidation of metallic nanoparticles offers an interesting possibility of synthesizing nanocomposites with both metallic and ceramic properties. We have performed the first successful MD simulation of oxidation of an Al nanoparticle (diameter 200Å, see Fig. 5) [27]. The MD simulations are based on an interaction scheme developed by Streitz and Mintmire, which can successfully describe a wide range of physical properties of both metallic and ceramic systems [28].

This scheme is capable of treating bond formation and bond breakage and changes in charge transfer as the atoms move and their local environments are altered [29,30].

The MD simulations provide detailed picture of the rapid evolution and culmination of the surface oxide thickness, local stresses, and atomic diffusivities. Clusters of OAl$_4$ coalesce to form a neutral, percolating tetrahedral network that impedes further intrusion of oxygen atoms into and of Al atoms out of the nanoparticle. As a result, a stable oxide scale is formed. Structural analysis reveals a 40Å thick amorphous oxide scale on the Al nanoparticle. The thickness and structure of the oxide scale are in accordance with experimental results.

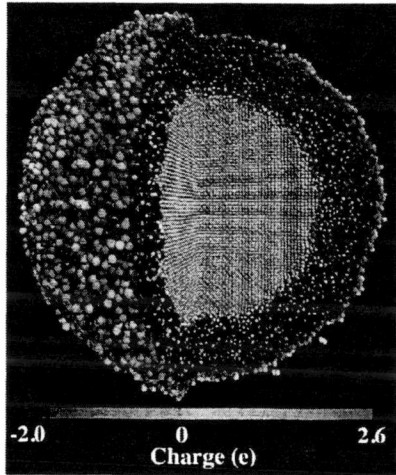

FIGURE 5. Snapshot of the Al nanocluster after 0.5 ns of simulation time. (A quarter of the system is cut out to show the aluminum/aluminum-oxide interface.) The larger spheres correspond to oxygen and smaller spheres to aluminum; color represents the charge on an atom.

10-million-atom DFT calculations, see Fig. 6. (Our latest benchmark on a 1,280-processor IBM SP3 including 8.1-billion-atom MD on silica and 140,000-atom DFT on GaAs indicates an even faster growth rate.)

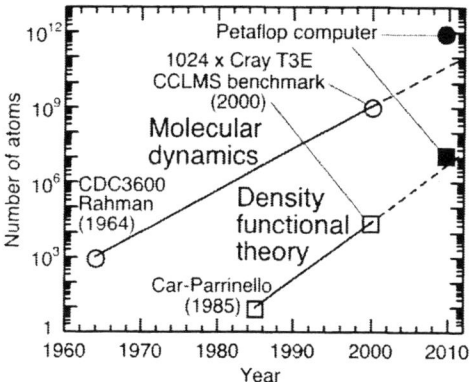

FIGURE 6. "Moore's law" in molecular dynamics.

In future, metacomputing on a Grid of geographically distributed supercomputers, mass storage, and virtual environment connected via high-speed networks will revolutionize computational research by enabling i) very large-scale computations that are beyond the power of a single supercomputer, and ii) collaborative, hybrid computations that integrate distributed, multiple expertise [33]. A multidisciplinary application that will soon require Grid-level computing is emerging at the forefront of computational science and engineering. We have recently developed such a multiscale simulation approach which seamlessly combines continuum mechanics based on the finite-element (FE) method, MD simulations to describe atomistic processes, and QM calculations based on the DFT to handle breakage and formation of atomic bonds [34].

CONCLUSION

Modern MD simulations of materials started in 1964, when Aneesur Rahman simulated 864 argon atoms on a CDC 3600 computer [31]. Assuming a simple exponential growth, the number of atoms that can be simulated in a classical MD simulation has doubled every 21 months to reach 1.02 billion atoms [4]. Similarly, the number of atoms in a DFT-based MD simulation (started by Roberto Car and Michele Parrinello in 1985 for 8 Si atoms [32]) has doubled every 16 months to reach 22,500 atoms [4]. Petaflop computers anticipated to be built in the next ten years will help maintain the growth rates in these "MD Moore's Laws" [4], and enable trillion-atom MD and

ACKNOWLEDGMENTS

This work is partially supported by AFOSR, DOE, NASA, NSF, and USC-LSU MURI. A few million-atom simulations were performed using in-house parallel computers at the Concurrent Computing Laboratory for Materials Simulations at Louisiana State University. 10 million to billion atom simulations were performed using parallel computers at Department of Defense's Major Shared Resource Centers under a DoD Challenge project.

REFERENCES

1. Pechenik, A., Kalia, R. K., and Vashishta, P., *Computer-Aided Design of High-Temperature Materials*, Oxford: Oxford Univ. Press, 1999.

2. Nakano, A., Kalia, R. K., and Vashishta, P., *Comput. Sci. Eng.* **1** (5), 39-47 (1999).

3. Kalia, R. K., Campbell, T. J., Chatterjee, A., Nakano, A., Vashishta, P., and Ogata, S., *Comput. Phys. Commun.* **128**, 245-259 (2000).

4. Shimojo, F., Campbell, T. J., Kalia, R. K., Nakano, A., Vashishta, P., Ogata, S., and Tsuruta, K., *Future Generation Comput. Sys.* **17**, 279-291 (2000).

5. Hoenberg, P., and Kohn, W., *Phys. Rev.* **136**, B864-B871 (1964).

6. Vashishta, P., Kalia, R. K., Nakano, A., Li, W., and Ebbsjö, I., in *Amorphous Insulators and Semiconductors*, edited by M. F. Thorpe and M. I. Mitkova, Dordrooht: Kluwer, 1996, pp. 151-213.

7. Greengard, L., and Rokhlin, V., *J. Comput. Phys.* **73**, 325-348 (1987).

8. Tuckeman, M. E., Berne, B. J., and Martyna, G. J., *J. Chem. Phys.* **97**, 1990-2001 (1992).

9. Nakano, A., *Concurrency: Practice and Experience* **11**, 343-353 (1999).

10. Omeltchenko, A., Campbell, T. J., Kalia, R. K., Liu, X., Nakano, A., and Vashishta, P., *Comput. Phys. Commun.* **131**, 78-85 (2000).

11. Chelikowsky, J. R., Saad, Y., Ögüt, S., Vasiliev, I., and Stathopoulos, A., *Phys. Stat. Sol. (b)* **217**, 173-195 (2000).

12. Fattebert, J.-L., and Bernholc, J., *Phys. Rev. B* **62**, 1713-1722 (2000).

13. Troullier, N., and Martins, J. L., *Phys. Rev. B* **43**, 1993-2006 (1991).

14. Perdew, J. P., Burke, K., and Ernzerhof, M., *Phys. Rev. Lett.* **77**, 3865-3868 (1996).

15. Goedecker, S., *Rev. Mod. Phys.* **71**, 1085-1123 (1999).

16. Mauri, F., and Galli, G., *Phys. Rev. B* **50**, 4316-4326 (1994).

17. Siegel, R. W., *Scientific American*, Dec. 1996, pp. 74-79.

18. Kalia, R. K., Nakano, A., Tsuruta, K., and Vashishta, P., *Phys. Rev. Lett.* **78**, 689-672 (1997).

19. Kalia, R. K., Nakano, A., Omeltchenko, A., Tsuruta, K., and Vashishta, P., *Phys. Rev. Lett.* **78**, 2144-2147 (1997).

20. Tsuruta, K., Nakano, A., Kalia, R. K., and Vashishta, P., *J. Am. Ceram. Soc.* **81**, 433-436 (1998).

21. Campbell, T. J., Kalia, R. K., Nakano, A., Shimojo, F., Ogata, S., and Vashishta, P., *Phys. Rev. Lett.* **82**, 4018-4021 (1999).

22. Chatterjee, A., Kalia, R. K., Loong, C.-K., Nakano, A., Omeltchenko, A., Tsuruta, K., Vashishta, P., Winterer, M., and Klein, S., *Appl. Phys. Lett.* **77**, 1132-1134 (2000).

23. Evans, A. G., *J. Am. Ceram. Soc.* **73**, 187-206 (1990).

24. Shimojo, F., Ebbsjö, I., Kalia, R. K., Nakano, A., Rino, J. P., and Vashishta, P., *Phys. Rev. Lett.* **84**, 3338-3341 (2000).

25. Kodiyalam, S., Kalia, R. K., Kikuchi, H., Nakano, A., Shimojo, F., and Vashishta, P., *Phys. Rev. Lett.* **86**, 55-58 (2001).

26. Walsh, P., Kalia, R. K., Nakano, A., Vashishta, P., and Saini, S., *Appl. Phys. Lett.* **77**, 4332-4334 (2000).

27. Campbell, T. J., Kalia, R. K., Nakano, A., Vashishta, P., Ogata, S., and Rodgers, S., *Phys. Rev. Lett.* **82**, 4866-4869 (1999).

28. Streitz, F. H., and Mintmire, J. W., *Phys. Rev. B* **50**, 11996-12003 (1994).

29. Ogata, S., Iyetomi, H., Tsuruta, K., Shimojo, F., Kalia, R. K., Nakano, A., and Vashishta, P., *J. Appl. Phys.* **86**, 3036-3041 (1999).

30. Ogata, S., Iyetomi, H., Tsuruta, K., Shimojo, F., Nakano, A., Kalia, R. K., and Vashishta, P., *J. Appl. Phys.* **88**, 6011-6015 (2000).

31. Rahman, A., *Phys. Rev.* **136**, A405-A411 (1964).

32. Car, R., and Parrinello, M., *Phys. Rev. Lett.* **55**, 2471-2474 (1985).

33. Foster, I., and Kesselman, C., *The Grid: Blueprint for a New Computing Infrastructure*, San Francisco: Morgan Kaufmann, 1999.

34. Ogata, S., Lidorikis, E., Shimojo, F., Nakano, A., Vashishta, P., and Kalia, R. K., *Comput. Phys. Commun.*, to be published.

PARALLEL SESSIONS

Artificial Intelligence

Momentum Reconstruction and Triggering Suggested for the ATLAS Detector

Gideon Dror[1+], Erez Etzion[2*]

1. Department of Computer Science, The Academic College of Tel-Aviv-Yaffo, Tel-Aviv 64044, Israel.
2. School of Physics and Astronomy, Raymond and Beverly Sackler Faculty of Exact Sciences, Tel-Aviv University, Tel-Aviv 69978, Israel.

Abstract. A neural network solution for a complicated experimental High Energy Physics problem is described. The method is used to reconstruct the momentum and charge of muons produced in collision of particles in the ATLAS detector. The information used for the reconstruction is limited to the output of the outer layer of the detector, after the muons went through strong and inhomogeneous magnetic field that have bent their trajectory. It is demonstrated that neural network solution is efficient in performing this task. It is shown that this mechanism can be efficient in rapid classification as required in triggering systems of the future particle accelerators. The parallel processing nature of the network makes it relevant for hardware realization in the ATLAS triggering system.

INTRODUCTION

The Large Hadron Collider (LHC) currently under construction is the largest particle accelerator ever built. It is scheduled to start operating in 5 years. The design luminosity is so high that a rate of 1 GHz of events is expected in its detectors. In such an environment it is important to distinguish between the relevant physics collision within a time frame of a few nano seconds. The goal of the trigger is to select 100 out of 1 billion events per second. One of the ways to select interesting events is to look for the signature of high transverse momentum (p_T) muons. Going through the ATLAS magnetic field the muons momentum could be derived from their curvature. The complicated inhomogeneous magnetic field in the forward area of the ATLAS detector makes it a very complicated problem to solve. However it is shown that a neural network (NN) implementation can be used for a fast and efficient triggering system. A standard feed forward network is trained to learn the characteristics of the detectors output, and to solve their complicated dynamics. The network successfully solves the inverse problem described below, and can efficiently be implemented in a hardware based triggering system.

EXPERIMENTAL SETTING

The goal is to use the momentum of muons produced in the interaction point of the ATLAS detector as a triggering signal. Muons with low angles with respect to the beam line cross the tracking device, shielding and calorimetry before they reach the Thin Gap Chambers (TGC) [1] located at the outer parts of the detector, 12-14 meters away from the interaction point. The muons emerging in the end-caps

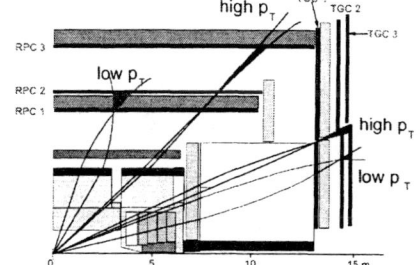

FIGURE 1. A side view of one ATLAS's quadrant. In black the muons trajectories. In brown the three TGCs layers.

[+] Gideon@server.mta.ac.il
[*] Erez@lep.tau.ac.il

undergo stochastic electric Coulomb scattering and their tracks are bent by highly inhomogeneous magnetic fields. The target is to deduce from the TGC hits the transverse momentum, p_T, with which the muon was created. As shown in Fig.1 the TGC coverage is limited to low polar angles corresponding to pseudorapidity $|\eta| > 1.05$.

THE NEURAL NETWORK

A standard feed forward NN was used to learn the mapping between the final muon track as detected by the TGC hits and its initial charge and momentum. Training was done with Lvenberg-Marquardt method on simulated events generated with the ATLAS simulation program, DICE[2],. The events studied were single muons in the area of $|\eta| > 1.05$ and $1 < p_T < 50$ GeV.

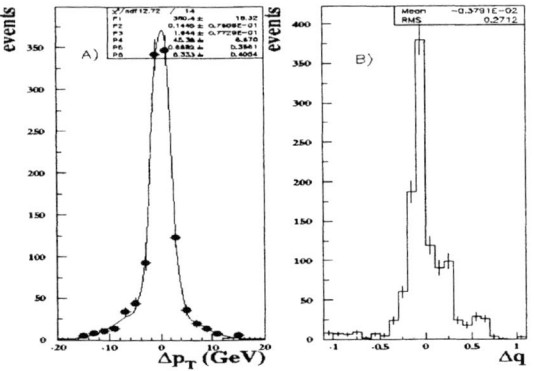

FIGURE 2. Differences between calculated and generated momentum, left - p_T, right - charge.

The NN is fed with 4 inputs, the slopes and intercepts of the projections of muon's tracks on xz an yz planes, calculated from LSQ linear fit. This preprocessing is required since while some of the hits in the raw data are missing, there are multiple hits in regions where TGC plates overlap. Earlier studies[3] used the Hough transform for preprocessing, however it turned out that the LSQ fit is sufficient. Four output linear neurons provide the value of the muon initial p_T, initial direction and its electric charge, q. Several NN architectures were examined. The chosen NN contains two hidden layers with seven sigmoidal transfer function each. Each network was trained using Levenberg-Marquardt method with 2500 events where the minimal number of events sufficient for reasonable solution is about 1400 events. Typically the training lasted several thousand epochs. Upon completion of the training phase each network was examined with a test set of 1829 events.

RESULTS

The difference between the NN estimates and the real p_T and charge as determined for the test set are shown in Fig. 2. The width of the two result among other from the real stochasticity distribution of the data due to the interaction point widths as well as the Coulomb scattering.

The relative error on p_T as a function of $|\eta|$ is shown in Fig. 3. The poor resolution for $|\eta| \approx 1.7$ results from the highly inhomogeneous magnetic field in the region between the end-cap and the barrel ATLAS toroid coils.

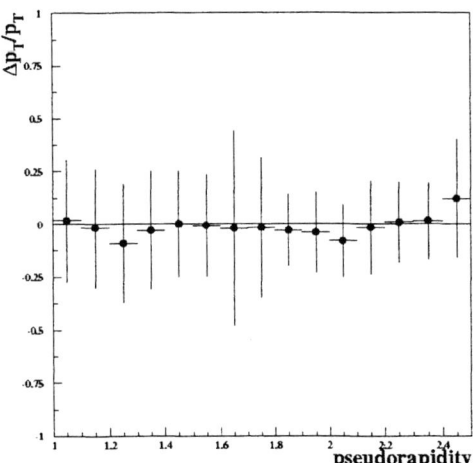

FIGURE 3. Mean relative error on p_T as a function of $|\eta|$. The error bars are the RMS of Gaussian fit of $\Delta P_T / P_T$.

The identification of the charge is more robust due to its discrete nature. In 98.5% of the events the charge was correctly identified, where naturally most of the misclassification occurs at high momentum.

CLASSIFICATION NETWORK

The speed of the network decision is crucial in triggering network. In our case the decision is simply based on the transverse momentum of the muon being bigger or smaller than a certain value (6 GeV). Only events with transverse momentum larger than 6 are stored for further analysis. The network described above yields about 94% correct classifications. As can be expected the misclassifications occur mainly near the threshold value.

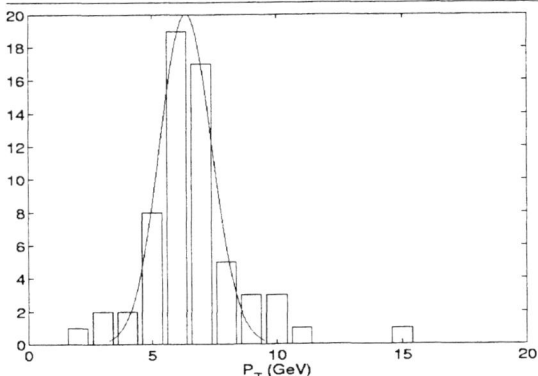

FIGURE 4. A Gaussin fit to the real p_T of the events which were misclassified by the tracking and selecting NN. The mean and RMS of the fit are 6.4 and 1.0 GeV respectively.

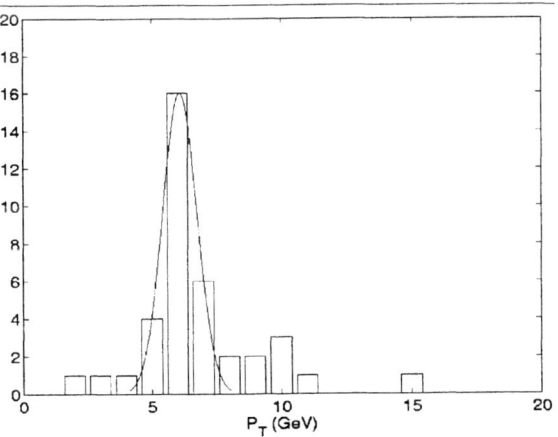

FIGURE 5. A Gaussian fit to the real p_T of the events that were misclassified by dedicated classification NN. The mean and RMS of the fit are 6.1 and 0.7 GeV respectively.

A more appropriate approach for triggering purposes is to directly train the network on the classification problem only. We used for this purpose a network with similar architecture, replacing the linear output neurons with a sigmoidal one.

We find that 96% of the events are correctly classified where again most misclassification occur near the threshold of transverse momentum around 6 GeV.

The p_T distributions of the misclassified events in the two methods are shown in Figs. 4 and 5. It turned out that most error are common to both networks and stem from misleading events produced in rare large-angle scattering. However, note that the width of the dedicated classification network is considerably narrower.

DISCUSSION

A feed forward NN capable of estimating the charge and momentum of muons in the ATLAS particle facility currently under construction is presented. With relatively simple architecture we were able to solve a complicated inverse problem. A similar NN can very efficiently be used in classification problem necessary for triggering purposes. Due to the parallel nature of the network, it opens a way for hardware realization of this problem. A. Chorti presented in the current workshop, a successful hardware implementation of this exact architecture [4].

ACKNOWLEDGMENTS

We wish to acknowledge the fruitful collaboration and the stimulating discussions with our colleagues Halina Abramowicz and David Horn from Tel-Aviv University. The research was partly supported by the Israeli Science Foundation.

REFERENCES

1. G. Bella et al., *Nucl. Inst. Meth* **A252**, 503 (1986); S. Dado et al., *Nucl. Inst. Meth* **A252**, 511 (1986); G. Mikenberg et al., *Nucl. Inst. Meth* **A265**, 23 (1988).

2. DICE Manual, *ATLAS note* **SOFT-NO-10** (1994).

3. G. Dror et al., *AIHENP99 proceedings*, edited by G. Athanasiu, D. Perret-Gallix, Elsevier North-Holland Publishers (1999).

4. A. Chorti, B. Granado, B. Denby, et. al. "An electronic system for simulation of neural networks with a microsecond real time constraint", these proceedings.

Neural network real time event selection for the DIRAC experiment

P. Kokkas, M. Steinacher, L. Tauscher, S. Vlachos*

Institute for Physics, University of Basel, Klingelbergstr. 82, CH-4056 Basel, Switzerland

Abstract. The neural network first level trigger for the DIRAC experiment at CERN is presented. Both the neural network algorithm used and its actual hardware implementation are described. The system uses the fast plastic scintillator information of the DIRAC spectrometer. In 210 ns it selects events with two particles having low relative momentum. Such events are selected with an efficiency of more than 0.94. The corresponding rate reduction for background events is a factor of 2.5.

INTRODUCTION

The DIRAC experiment (1) at CERN, is designed to measure the life-time of an atom consisting of two oppositely charged pions, henceforth called pionium. This is achieved by detecting the two charged pions emerging from the pionium's breakup.

Pioniums are produced when the CERN-PS 24 GeV/c proton beam hits a thin solid target. When they interact electromagnetically with matter, they break up into two oppositely charged pions with very small relative momentum (Q < 3 MeV/c) and, in particular, a very small angle between them (typically less than 0.3 mrad). The DIRAC detector is designed to measure the momenta of the two pions with highest accuracy. This is achieved with a double arm magnetic spectrometer.

The neural network model developed for the DIRAC first level trigger application uses a pair of hits downstream the DIRAC magnet (hits in the Ionization Hodoscope, IH) together with a hit in each spectrometer arm (Vertical Hodoscopes, VH hits). There are two separate Ionization Hodoscope detection planes each with 16 vertical slabs. Hence each IH plane is used independently in conjunction with the same downstream detectors, and the two event decisions are eventually combined together with a logical OR function. This trigger system is called DNA (Dirac Neural Atomic trigger).

DATA SAMPLES AND FORMAT USED

In what follows events that have to be selected by DNA are called 'good', and those to be rejected 'bad'. The definition of good events is the following:

- Q_x < 3 MeV/c (Q along the x, horizontal direction),
- Q_y < 10 MeV/c (Q along the y direction, direction of the magnetic field),
- Q_l < 30 MeV/c (longitudinal Q component).

Bad events are the ones not satisfying any of the above criteria. In total we have used 10000 events of each kind for the neural network training phase. An equal number of events has been used to estimate the DNA performance during development, before it was actually implemented.

The DNA trigger processes events with a maximum of 2 hits in each VH and a maximum of 5 hits in each IH layer. Events with more hits are rare and not evaluated by DNA (as combinatorics resolution is not feasible on-line). However for improved signal acceptance all such events are read-out for further off-line analysis.

Since there may be more than 2 hits in a IH layer, DNA uses only the pair of hits that are closer together. These are the ones that most probably correspond to the two charged pions of a decayed pionium. In case of a single hit in a IH layer, it is considered that both charged particles went through the same IH slab, and hence this slab is used twice.

The VHs can have one or two hits each. DNA forms all four possible combinations with a single hit in each VH. Hence four different input patterns are generated per IH layer per event. Each such pattern contains the same IH information. Four independent but identical neural net-

---

* e-mail: S.Vlachos@cern.ch

works evaluate further these input patterns (one per VH combination).

THE NEURAL NETWORK

For DNA we have used a feed-forward neural network model based on error back propagation (for a general description of neural network algorithms see for example (2) and (3)). It is well established that a feed-forward neural network with one hidden layer can represent any function of the input data. Consequently the DNA neural network has 18 input nodes, one hidden layer with two nodes and one node in the output layer. This model has been initially developed for a completely different application (4).

A simple but crucial choice we had to make was the selection of the appropriate transfer function for each neural network node. In order to avoid problems related with the special treatment of 0 in training algorithms, we have used instead a 'tanh' activation function (which has a sigmoid form from -1 to 1 for an input range from -1 to 1).

Using the data samples described in the previous section we have run the training algorithm for 2000 epochs (an epoch consists of presenting the whole training data sample once to the neural network, and obtain consequently one modification of all the weights).

The performance of the trained neural network is shown in Fig. 1. As all events with a neural network output above a given threshold are tagged as good and are accepted, the efficiency of the selection as a function of the threshold value can be established. Similarly one can establish the efficiency of selecting background. Fig. 1 shows these two efficiencies one against the other for various values of the neural network's threshold.

All four combinations for each of the two IH layers are evaluated in parallel by the same neural network developed above. Hence this network is used eight times in the DNA system. If any of these neural networks gives an answer (output value) greater than a common predetermined threshold, the event is accepted.

HARDWARE IMPLEMENTATION

The DNA application is based on a custom built hardware implementation of the neural network algorithm, originally developed for the CPLEAR experiment (5). The flexibility of this implementation allowed eventually the system to be incorporated in the L3 first-level trigger system (6) and now in DIRAC. A comparative study (7) has shown that more complicated algorithms or hardware architectures are in general less performant. A single card executes a feed-forward neural network with 1 single hidden layer with two nodes in 60 ns. This performance is obtained by using RAM memories with preloaded contents to perform the most computationally demanding part of the algorithm.

The complete DNA trigger consists of two identical chains, one for each IH detector plane. Each DNA chain is based on three separate electronics modules: The Interface and Decision card (IFD) (8, 9), the neural network (NN) cards (10) and a power-PC VME master CPU card (Motorola MVME–2302). The IFD card has been developed for the DNA trigger, the NN-cards are identical to the ones used earlier. Only the CPU-card is commercial (for ease of technical support and software development). The IFD card feeds four independent and identical NN cards (running the same NN algorithm), corresponding to the four combinatorial VH possibilities.

Since April 2000, the DNA trigger is operating in the DIRAC experiment as an integral part of the whole trigger system. Each event is fully classified in less than 210 ns.

EVENT SELECTION AND REAL-TIME PERFORMANCE

In order to study the DNA performance, data with DNA in flagging mode have been analysed. In that mode DNA only flags events as good or bad without rejecting any, so that events normally rejected by DNA can also be studied. In the following the threshold on the neural network output has been set to 0.0 (i.e. in the middle of the neural network output range).

The system's acceptance as a function of Q_{tot} is shown in Fig. 2. The value at very low Q_{tot} gives the DNA acceptance for good events. The overall DNA reduction rate can also be obtained from this figure by convoluting this distribution with the Q_{tot} distribution of two track events in DIRAC.

The performance of the DNA system can be summarized in the following two numbers:

- Acceptance of low Q_{tot} pion pairs: 0.94, derived from Fig. 2,

- Background rate reduction: 2.5, measured during data-taking.

Note here that the above numbers are higher than the ones concerning the neural network performance as per event up to four possible hit combinations are used and if any of them gives a high neural network output (above threshold) the event is accepted.

CONCLUSIONS

The DNA trigger is successfully used in the DIRAC experiment since it was first installed. It reduces the event rate by a factor of 2.5 while keeping more than 0.94 of low relative momentum pionic pairs. The hardware implementation based on custom built electronics allows a decision time of only 210 ns. No particular bias is introduced by the DNA event selection.

REFERENCES

1. B. Adeva et. al. Lifetime measurement of $\pi^+\pi^-$ atoms to test low energy QCD predictions, CERN/SPSLC 95–1 (1995).
2. J. Hertz et al., Introduction to the theory of neural computation Addison-Wesley, New York 1991.
3. C. Bishop, Neural networks for pattern recognition, Claredon press, Oxford 1995.
4. G. Athanasiu et. al., Nucl. Instr. and Meth. A324, 320 (1993).
5. F. Leimgruber et. al., Nucl. Instr. and Meth. A365, 198 (1995).
6. D. Haas et. al. Nucl. Instr. Meth. A401, 19 (1999).
7. S. Vlachos et. al., Nucl. Instr. and Meth. A385, 361 (1997).
8. S. Vlachos, 'A global overview of the Basel neural network project after 10 years of experience', 'Sixth International Workshop on Software Engineering, Artificial Intelligence Neural Nets, Genetic Algorithms, Symbolic Algebra, Automatic Calculation (AIHENP99)', 12–16 April 1999, Heraclion (Greece), proceedings to appear.
9. M. Steinacher, 'Interface and Decision Card; The design', Revision 2.0, Internal Report, University of Basel, 9–6–1999.
10. M. Steinacher, 'Hardware implementation of a fast neural network', 'Sixth International Workshop on Software Engineering, Artificial Intelligence Neural Nets, Genetic Algorithms, Symbolic Algebra, Automatic Calculation (AIHENP99)', 12–16 April 1999, Heraclion (Greece), proceedings to appear.

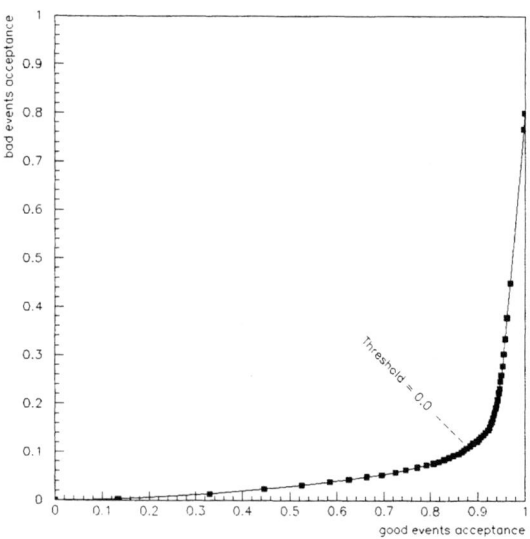

FIGURE 1. The relation between the signal acceptance (good events) and background acceptance (bad events). Each point corresponds to a different neural network threshold. The standard DNA working point (threshold = 0.0) is also indicated on the figure for completeness.

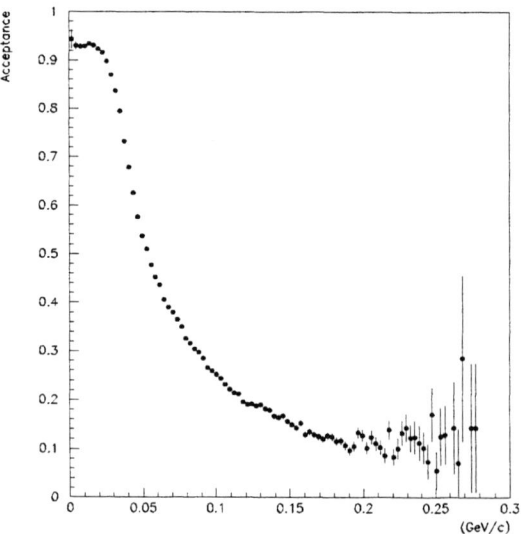

FIGURE 2. The efficiency of DNA as a function of Q_{tot}.

Intelligent preprocessing for neural networks in the H1 experiment

JC.Prévotet[1], B.Denby[1], P.Garda[1], B.Granado[1],
W.Fröchtenicht[2], G.Grindhammer[2], L.Janauschek[2], C.Kiesling[2], T.Kobler[2],
B.Koblitz[2], S.Schmidt[2], B.Tzamariudaki[2], S.Udluft[2]

[1] *Laboratoire des Instruments et Systèmes, Université Pierre et Marie Curie, Paris VI, France*
[2] *Max Planck Institut für Physik München, Germany*

Abstract. After the upgrade of the HERA machine at DESY in 2001, an increase in the luminosity of a factor 5 is expected. Since the data output rate of the L2 trigger should be kept at the pre-upgrade level, a smarter way of preprocessing data has been developed, extracting the most physically relevant information in order to optimize the neural networks. We describe here the new neural preprocessor DDB2 (Data Distributed Board) and focus especially on the algorithmic principles. A general overview of the hardware implementations of such algorithms is then discussed, notably the use of the current, fast FPGA technology to combine parallelism and speed.

INTRODUCTION

The L2 neural network trigger in the H1 experiment has shown its effectiveness since 1995, producing interesting physics results. Some of them are described elsewhere (1). The next technical improvements of the Hera collider will lead to better performances forcing the trigger selectivity to improve. For this purpose, a more intelligent pre-processing has been envisaged to keep the data flow workable but reduce the dimensionality of the problem by exploiting more correlations between data. The information is built on properties of particles emitted during a collision (momentum, energy, type of particle) collected from all sub-detectors in the H1 experiment. Using such "intelligent" input variables, the efficiency of the neural networks is consequently improved, resulting in a better discrimination between interesting events and background. The pre-processing is performed in four steps. The first is the clustering in which all signals within one sub-detector, localized in the same area are associated in order to constitute one entity. Clusters found in all subdetectors are then combined into objects according to their angular coordinates before being sorted according to particular criteria. The final step is called post-processing and is used to execute straightforward calculations and generate variables to make them directly usable for the neural networks. All these algorithms have to be implemented in hardware since the process must be executed in real time ($8\mu s$). We summarize here the hardware specifications and describe the DDB2 architecture.

ALGORITHMIC PRINCIPLES

Clustering

The basic idea of the clustering is to arrange signals within a specific subdetector from their position within a θ, ϕ plane. Due to the nature of the subdetectors, data may represent energies coded in 8 bits or have single bit fields, leading to two different kinds of processing.

Energy Clustering

LAR clustering: Clustering in the liquid argon (LAR) calorimeter is performed separately in the electromagnetic (EM) and the hadronic (HA) parts of the calorimeter. Data are coded in 16 bits (8 address bits + 8 data bits) representing respectively their position in the (θ, ϕ) plane and their energy value (Fig.1). The incoming data are picked out by energy and the cluster having the largest one is taken as a center. The positions of its neighbors are then determined via LUTs (Look Up Tables) and their energies are summed. Up to 16 neighbors are tolerated although 9 or 10 are sufficient in principle. As output from this algorithm, parameters such as the energy of the cluster's center (E_{cent}), the summed energies of the adjacent neighbors (E_{ring}), the total energy (E_{tot}), the number of non-zero cells in the neighborhood (N_{hit}) and the angular coordinates of the cluster's center (θ, ϕ) are obtained.

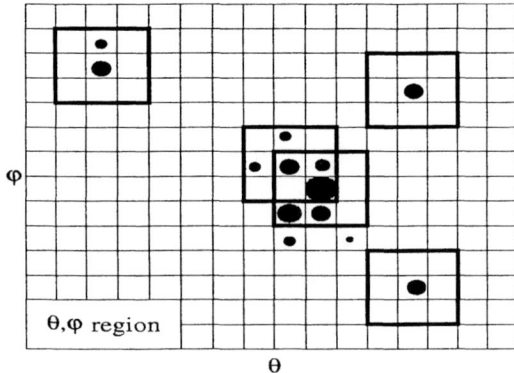

FIGURE 1. Clustering algorithm for the calorimeter information

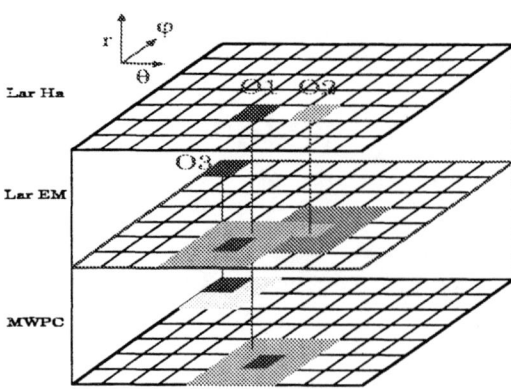

FIGURE 3. Example of the matching algorithm in the first 3 layers. Here are 3 objects found.

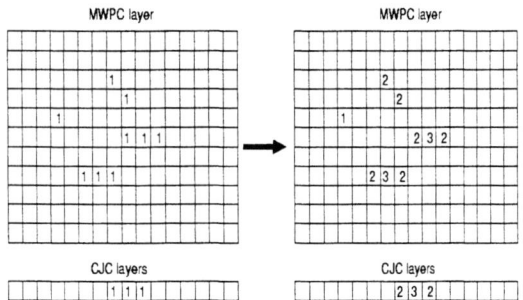

FIGURE 2. Preclustering algorithm for bit fields: MWPC above, and Drift chambers below. This step is necessary before applying the clustering algorithm.

Bit field clustering

MWPC clustering: Data coming from the multiwire proportional chambers (MWPC) are bit fields distributed on a 16x16 grid. Because of the binary nature of data, a preclustering step is necessary before applying the clustering algorithm itself. All non zero bits in the map are summed with their corresponding neighbors and the map is updated. See Fig.2.

Similar clustering principles as in the LAR are then performed starting with the largest cluster. Parameters such as P_{tot}, P_{cent}, P_{ring}, N_{hit}, θ, ϕ are also calculated here.

CJC clustering: Data coming from the central jet chamber (CJC) provide only the ϕ information, delivering a one dimensional array of 45 bits. Four arrays are provided, for high momentum negative tracks, high momentum positive tracks, low momentum positive tracks, low momentum negative tracks. The clustering procedure is simpler than for the other sub-detectors, due to the single dimensionality. Drift chambers parameters such as D_{tot}, D_{cent}, D_{ring}, n_{hits}, θ, ϕ are provided as outputs.

The matching algorithms

Two matching algorithms work on data coming from subdetectors belonging to two different angular ranges. The first one matches the angular coordinates of clusters coming from the HA and EM part of the calorimeter, MWPC, CJC, muon, FTT *(fast track trigger)*. The second processes the data coming from another angular region of the detector : The SpaCal trigger chambers, the muon chamber and the backward silicon tracker (BST). These algorithms consist in detecting a cluster with the maximum energy in a seed layer and compare its coordinates (θ, ϕ) with the coordinates of all available clusters in the other sub-detectors. See Fig.3. A match is defined by ($|\Delta\theta|, |\Delta\phi|$ <n where n can be set to a value in the range 0-16). The number of clusters within a layer is configurable but limited to 16 and each layer is transformed according to a universal (θ, ϕ) grid in order to have the same basis for coordinate comparisons. Clusters matched are removed to avoid double-counting, except those belonging to the CJC layers which can be used several times since they only contain phi information. The outputs of these algorithms consist of maximum 16 objects, constituted by the properties of the clusters that form it.

Sorting

All objects are sorted according to selectable properties. Three sorting procedures are executed in parallel, providing three different lists of sorted objects. If the sorting key is not unique, a second criterion is taken. Quantities like energies and angular coordinates can be sorted.

The second role of this algorithm is to modify the object's coordinates, redetermining the angular information relative to the biggest object in the lists.

Post-processing

This algorithm constitutes the last step of the preprocessing chain in which many types of calculations can be made in order to optimize the data input for the neural networks. At present, calculation of straightforward quantities such as cluster counting, angular differences between leading clusters are performed. The addition of new features is still under investigation.

HARDWARE CONSIDERATIONS

The different preprocessing algorithms are destined to be implemented on programmable boards with an execution time constraint of $8\mu s$.

L2 Trigger Hardware organization

The same configuration has been kept as in the previous preprocessing. More details are given in (2). Data coming from the sub-detectors are collected in receiver cards and sent via a 8 by 16 bits L2 bus to the DDB boards in which the preprocessing is performed. DDB boards reside in a VME crate and are connected to their corresponding CNAPS board which implement the neural networks. All networks are 3 layer feed forward MLPs and are trained individually for specific physics processes. The CNAPS boards run at 20.8 MHz and deliver a result in $8\mu s$. Further technical details about the L2 neural network trigger are given elsewhere (3). Five new DDB2 boards and their associated CNAPS boards are foreseen to be concurrently added to the previous existing ones, offering two kinds of preprocessing.

The DDB2 (Data Distribution Board)

An overview of the board is shown in Fig.4. Data coming from all sub-detectors arrive in parallel on 8 sub-buses constituting the L2 bus and are directly provided to the corresponding clustering modules in which they are sorted. A preclustering step is previously executed in the case of bit field data in order to find a maximum before actual clustering. The addresses are then sent to their associated matching modules to determine the vicinity within all layers, while the parameters related to a specific cluster are stored in memories.

After the matching step, addresses of matched clusters constituting an object are provided and used as pointers to the RAMs in which the parameters have been previously stored.

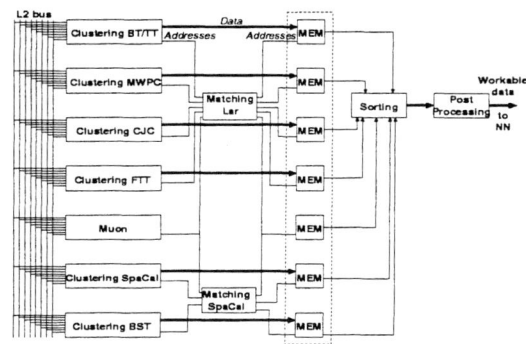

FIGURE 4. Description of the DDB2 board

Parameters are then accessible and are sent to the sorting machine for list generation before being processed in the last post-processing unit and sent to the neural network.

All modules have made intensive use of pipeline structures and parallel computation techniques in order to meet the timing constraint requirement. Moreover, the need of a large amount of small memories used as LUTs has opened the way to use specific FPGA circuits. The Xilinx Virtex family was chosen to implement such designs providing more flexibility and lower development costs than standard ASICs.

CONCLUSION

The DDB2 preprocessor represents an important innovation in the preprocessing step of the L2 neural network trigger because it takes advantage of more physically relevant information. The design have already been simulated and significant improvements have been noticed. More details about the DDB2 can be obtained in (4). Concerning the hardware, FPGA circuits present an effective solution in terms of speed and flexibility and seem appropriate to process all algorithms in $8\mu s$.

REFERENCES

1. Janauschek, L. et al., in *New Computing Techniques in Physics Research VI, Greece*.
2. Kiesling, C. et al., in *these proceedings*.
3. Köhne, J., *Nuclear Instruments and Methods in Physics research*.
4. Schmidt, S., *Untersuchungen zur Verbesserung des neuronalen Triggers beim H1 experiment*, Master's thesis, Fakultät für Physik der Technischen Universität München, unpublished, 2000.

An electronic system for simulation of neural networks with a micro-second real time constraint

Arsenia Chorti, Bertrand Granado, Bruce Denby and Patrick Garda

Laboratoire des Instruments et Systèmes
Université Pierre et Marie Curie
Paris France

Abstract. Neural networks implemented in hardware can perform pattern recognition very quickly, and as such have been used to advantage in the triggering systems of certain high energy physics experiments. Typically, time constants of the order of a few microseconds are required. In this paper, we present a new system, MAHARADJA, for evaluating MLP and RBF neural network paradigms in real time. The system is tested on a possible ATLAS muon triggering application suggested by the Tel Aviv ATLAS group, consisting of a 4-8-8-4 MLP which must be evaluated in 10 microseconds. The inputs to the net are dx/dz, x(z=0), dy/dz, and y(z=0), whereas the outputs give pt, tan(phi), sin(theta), and q, the charge. With a 10 MHz clock, MAHARADJA calculates the result in 6.8 microseconds; at 20 MHz, which is readily attainable, this would be reduced to only 3.4 microseconds. The system can also handle RBF networks with 3 different distance metrics (Euclidean, Manhattan and Mahalanobis), and can simulate any MLP of 10 hidden layers or less. The electronic implementation is with FPGA's, which can be optimized for a specific neural network because the number of processing elements can be modified.

INTRODUCTION

Neural networks implemented in hardware can perform pattern recognition very quickly, and as such have been used to advantage in the triggering systems of certain high energy physics experiments. But the time constraint of such implementation is about few microseconds which is a very difficult. In our laboratory we have developped, MAHARADJA, an electronic system to simulate neural networks with real time simulation constraints. In this paper, we present the evaluation of this system to a possible ATLAS muon MLP triggering application suggested by the Tel Aviv ATLAS group with a time constraint of 10 microseconds.

MAHARADJA

MAHARADJA is realized to simulate two kind of neural networks models:

- Radial Basis Function Neural Networks
- Multi-Layer Perceptrons

In this article we only discuss the Multi-Layer Perceptrons implementation, the Radial Basis Function implementation is described in (1).

Our goal here is to know if our system can simulate High Energy Physics MLP with very difficult constraint. To investigate this question we benchmark MAHARADJA with a real network defined by Tel Aviv ATLAS group.

SIMULATED NETWORK

The simulated network is used in the L2 trigger and has 4 layers:

- a 4 neuron input layer and the inputs are dx/dz, x(z=0), dy/dz, and y(z=0)
- two 8 neuron hidden layers
- a 4 neuron output layer containing pt, tan(phi), sin(theta), and q, the charge

The required latency of this MLP is $10\mu s$.

MAHARADJA DESCRIPTION

Our system is based on four principal components:

- **A sequential processor**: this processor can execute sequential part of the algorithm and manage the system.

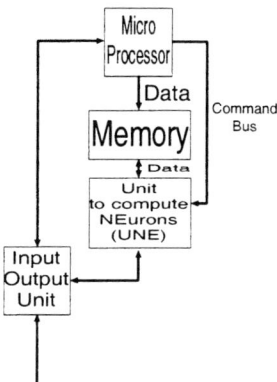

FIGURE 1. Architecture of MAHARADJA

- **A Unit to compute NEurons (UNE)**: this unit accelerate the neuron computations to obtain very fast simulation.

- **An Input-Output unit**: this unit can provide a high input bandwidth to the UNE.

- **A shared memory**: this memory is shared between the UNE and the processor. It contains the neural network parameters, such as weights and size of layers.

ARCHITECTURE OF *UNE*

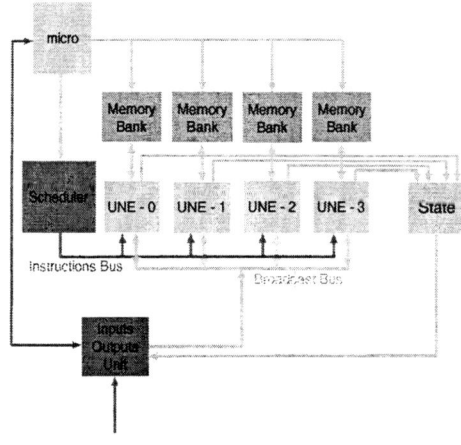

FIGURE 2. *UNE* unit organization

The *UNE* unit has two parts:

- four processors to compute post-synaptic potentials
- a component to compute the neuronal states

The processor of the *UNE* unit of MAHARADJA is organized in a SIMD [1] fashion. The interconnections are made by a 16-bit broadcast bus connected to the INPUT-OUTPUT unit.

A *Scheduler*, shown on Fig. 2, controls this unit by a 10-bit command bus.

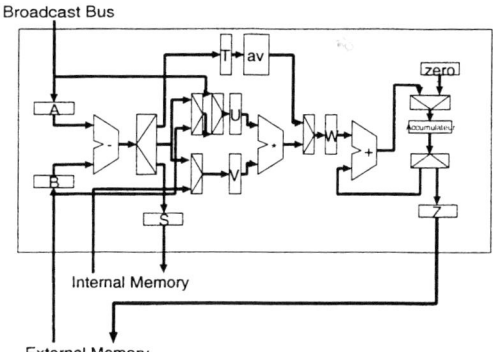

FIGURE 3. a processor of the *UNE* unit - A, B, S, T, U, V, W and U are registers that synchronize computation. - AV is the absolute value unit and ACCU the accumulation register.

Each processor of *UNE* unit compute one or more post-synaptic potentials as shown in Fig. 3.

The architecture of a *UNE* processor has

- a 16-bit subtractor to compute the first step of a distance computation in Radial Basis Functions.

- a 16-bit multiplier to compute the second step of Euclidean or Mahalanobis distance in Radial Basis Function or to compute the first step of Multi-Layer Perceptrons.

- a 16-bit absolute value unit to compute the second step of the Manhattan distance in Radial Basis Functions.

- a 32-bit adder to accumulate the result provided by the multiplier or the absolute value unit. This is the third step in computing distance in Radial Basis Functions or in computing Multi-Layer Perceptrons.

With these 4 operators, it is possible to compute all the post-synaptic potential for Radial Basis Functions with Manhattan, Euclidean or Mahalanobis distances and for Multi-Layer Perceptrons.

Behind the processor of the *UNE* unit there is a component to compute the neuron states. This computation is realized with a Lookup Table store in a memory.

---

[1] Single Instruction stream Multiple Data stream

UNE CONTROL

The *scheduler* can place the *UNE* unit in a functional mode. This mode can be:

- RBF with 3 different metrics:
  - Manhattan distance
  - Euclidean distance
  - Mahalanobis distance
- MLP

This gives 4 modes. When a mode is choosen, it is impossible to switch to another in a dynamic way. To realize this, one must reinitialize the system.

The management of the *UNE* unit is realized by request. The beginning and the end of the computation is made by control signals, a *begin* signal and an *end* signal.

MEMORY

With each processor of the *UNE* unit, there is 256-KB of associated memory to store the simulated neural network parameters such as post-synaptic weights or sizes of layers.

INPUT-OUTPUT UNIT

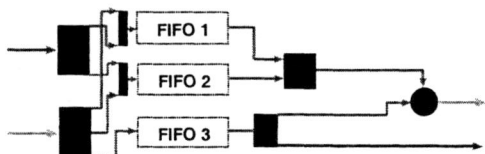

FIGURE 4. The input-output unit

This unit realizes the connection between MAHARADJA and the external world, for example the converter of the analog signals in the calorimeter of a collider. To obtain a high input bandwith we use a 3 FIFO structure, shown in Fig. 4. With such a structure the system makes a pipeline beetween the computation in the *UNE* unit and the acquisition of new data in the *Input-Output* unit.

In Fig. 5 we can see how the *Input-Unit* calculation for the HEP MLP is simulated. As we can see, the 2 fifo [2] (FIFO1 and FIFO2) are used to acquire new data from the external world. When the FIFO1 is used to store new

---

[2] first-in first-out

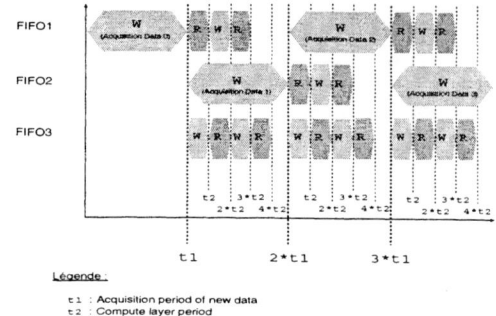

FIGURE 5. Use of the 3 FIFOS

data, FIFO2 is used for the computation and vice-versa. The third fifo (FIFO3) is used only for the computation and to store intermediate and final results.

IMPLEMENTATION

MAHARADJA is implemented with Altera APEX20K FPGA. We can with such an implementation modify some parameters like the number of processors to accelerate neural network computation.

TIMING ANALYSIS

We have carried out a time analysis of the MAHARADJA system, based on a prediction and evaluation methodology that we developed in our laboratory (2, 3). We first extract an analytical model of the evaluated system. This model is shown in table 1.

The variables in Table 1 are:

- nbp : Number of post-synaptic potentials (Terminal layer of the connection)

- nbe : Number of neurons (Initial layer of the connection)

- N : Neural Nework layer number

Now we can use the analytical model to predict the simulation time of the HEP neural network. This time is given in Table 2. Note that MAHARADJA can simulate this MLP with a simulation 1.5 times less than the desired $10\mu s$ if we use a 10 MHz clock and 4 processors in *UNE* unit, and it takes 6.5 times less if we use a 20 MHz clock and 8 processors in *UNE* unit.

Table 1. Time analysis of MAHARADJA

Multi Layer Perceptrons
$\sum_{i=1}^{N} \lceil \frac{nbe_i}{4} \rceil * (nbp_i + 2) + nbe_i + 1$

Manhattan Distance
$\lceil \frac{nbp}{4} \rceil * (nbe + 4) + \lceil \frac{nbe_s}{4} \rceil * (nbp_s + 2) + nbe_s + 1$

Euclidean Distance
$\lceil \frac{nbp}{4} \rceil * (nbe + 4) + \lceil \frac{nbe_s}{4} \rceil * (nbp_s + 2) + nbe_s + 1$

Mahalanobis Distance
$\lceil \frac{nbp}{4} \rceil * (nbe^2 + 6 * nbe + 2) + \lceil \frac{nbe_s}{4} \rceil * (nbp_s + 2) + nbe_s + 1$

Table 2. Comparison of Simulation Time for an MLP Predicted by the Analytical Model.

System	Frequence (MHz)	Time (μs)
MAHARADJA 4 UNE	10	6.5
MAHARADJA 8 UNE	10	3.2
MAHARADJA 8 UNE	20	1.6

CONCLUSION

In this article we propose an electronic system, MAHARADJA, which can calculate the result of a HEP MLP in 6.5 microseconds at 10 MHz, or at 20 MHz, 3.4 microseconds. The system can also handle RBF networks with 3 different distance metrics (Euclidean, Manhattan and Mahalanobis), and can simulate any MLP of 10 hidden layers or less. The electronic implementation is with FPGA's, which can be optimized for a specific neural network because the number of processing elements can be modified.

REFERENCES

1. Granado, B., *Architecture des systèmes électroniques pour les réseaux de neurones - Conception d'une rétine connexionniste*, Ph.D. thesis, Université Paris XI, November 1998.

2. Granado, B., and Garda, P., in *Proceedings of ICANN'96*, Juillet 1996.

3. Granado, B., and Garda, P., in *Proceedings of IWANN'97*, Lanzarote - Canary Islands, Spain, June 1997.

Selection of W-Pair-Production in DELPHI with Feed-Forward Neural Networks

K.-H. Becks, P. Buschmann, J. Drees, U. Müller*, H. Wahlen

Fachbereich Physik, Bergische Universität-Gesamthochschule, D-42097 Wuppertal and DELPHI collaboration

Abstract. Since 1998 feed-forward networks have been applied for the separation of hadronic WW-decays from background processes measured by the DELPHI collaboration at different center-of-mass energies of the Large Electron Positron collider at CERN. Prior to the publication of the 189 GeV results (1) intensive studies of systematic effects and uncertainties were performed. The methods and results will be discussed and compared to standard selection procedures.

INTRODUCTION

In 1996 the energy of the Large Electron Positron collider (LEP) at CERN[1] crossed the threshold energy of 161 GeV for W-pair-production $e^+e^- \to W^+W^-$. The determination of the W-production cross-section and the direct measurement of the mass of the W-boson allows new tests of the standard model and cross-checks with earlier electroweak measurements.

In 1998 feed-forward neural networks were applied in the selection of hadronic WW-decays $W^+W^- \to q\bar{q}q\bar{q}$ measured by the DELPHI[2] detector at the center-of-mass energy of 189 GeV. Due to convincing results this analysis tool has also been used at all higher center-of-mass energies.

SIGNAL AND BACKGROUND PROCESSES

Standard QCD-events $e^+e^- \to Z^0/\gamma^* \to q\bar{q}$ are the dominant background to W-pair-production. The second important background is the hadronic decay of the Z-pair-production $e^+e^- \to ZZ \to q\bar{q}q\bar{q}$ which ends in the same final state with the similar topology like the signal. It is therefore quite important to separate signal from background with high efficiency and purity.

The hadronic WW-decay shows a 4-jet event topology. The former selection process of the "W working group" of the DELPHI collaboration was based upon linear cuts on four variables: the effective center-of-mass energy, the number of jets, their total track multiplicity and the combined variable $D = \frac{E_{min} \cdot \Theta_{min}}{E_{max} \cdot (E_{max} - E_{min})}$, which consists of the minimum and the maximum jet energy and the minimum angle between two jets (3, 4).

FEED-FORWARD ANALYSIS

The analysis starts with a loose preselection against non-4-jet events and events with reduced center-of-mass energy due to initial state photon radiation.

Then a feed-forward network based upon the JETNET package (2) with a 13 - 7 - 1 architecture is used. One hidden layer with 7 nodes and one single output follow the 13 inputs which consist of physical observables which describe the event or jets of WW-decays or background events. All variables have discriminating power known from earlier studies with self-organizing maps (5), the old standard or other high-energy physics analyses.

The training was performed over 1000 cycles with samples consisting of 3500 DELPHI WW- and QCD-Monte-Carlo events each generated with ARIADNE (7) or PYTHIA / JETSET (6). The influence of the DELPHI detector (described in (8)) was simulated with the DELSIM or the FASTSIM package (9).

Tests with an additional ZZ-Monte-Carlo training sample and 3 output nodes gave compatible results but required a higher CPU time. So the easier and faster network was chosen.

After finishing the training, the network output was calculated for independent samples of simulated *WW*, *QCD* and *ZZ* events and for real data.

---

* corresponding author: mueller@whep.uni-wuppertal.de
[1] European Organization for Nuclear Research
[2] **D**etector with **L**epton, **P**hoton and **H**adron **I**dentification

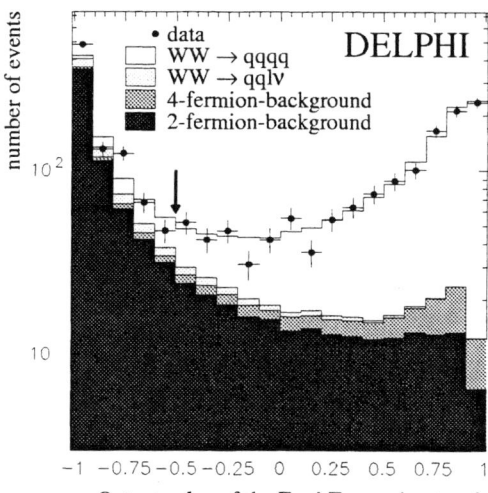

FIGURE 1. neural net output with data-MC agreement

SELECTION QUALITY

Figure 1 shows the output distribution of the neural network for data and simulated events. The points show the data and the histograms the predicted distributions for the signal and the different backgrounds. Events to the right of the vertical arrow were accepted for the event sample. The data-MC agreement is quite good.

Table 1 shows characteristic numbers of the selection quality at 189 GeV for the neural net compared to the standard linear cut analysis. The product of efficiency and purity is a good indicator for the selction quality and the expected statistical error. The neural net analysis shows a clear improvement.

The good data-MC agreement and the improvement in the selection quality were also found at all higher LEP energies. So this neural net was chosen for the official DELPHI cross section analysis in the hadronic decay channel.

SYSTEMATIC STUDIES

This choice made detailed studies of the systematic uncertainties necessary. Two different kinds of tests were performed to determine the systematic errors of the signal efficiency and the remaining background.

First different studies of the network stability were made:

- tests of different network architectures: one or two hidden layers with more or less hidden nodes ;
- variations of important network parameters like learning rate and momentum term around the working point ($\eta = 0.0025^{+0.015}_{-0.0015}$, $\alpha = 0.56 \pm 0.3$) ;
- use of different numbers of training events and different training samples .

All these tests gave compatible results within the statistical uncertainties and so no contribution to the systematic errors were assumed.

On the other hand studies were performed with techniques known from the old or other high-energy analyses. Here the neural net was only used as a mathematical function with a fixed training, fixed weights and always the same cut in the output distribution:

- comparison of Monte-Carlo generators with different hadronisation models and different MC parameter settings ;
- data-MC agreement using the technique [3] of mixed Lorentz-boosted Z^0 for data taken at the Z-resonance at a center-of-mass energy of 91.2 GeV ;
- smearing of input variables taking detector resolution into account ;
- influence of final state interactions on signal efficiency (Bose-Einstein correlation and color reconnection) .

Each method gave a systematic effect on the selection efficiency and the background. Afterwards the different systematics were combined, trying to take into account correlations between the methods.

---

[3] two independent hadronic Z decays were transformed into a pseudo W pair event by applying an appropriate boost to the particles of each Z event

Table 1. selection results of neural net compared to linear cut analysis

		NN	cuts
signal efficiency	[%]	88.74	85.58
remaining bg	[pb]	1.886	2.228
selection purity	[%]	77.84	74.14
efficiency × purity	[%]	69.08	63.45
selected events		1298	1342

FINAL RESULTS

Table 2 shows the final cross section numbers for 189 GeV and all higher center-of-mass energies. All results were determined from a binned maximum likelihood fit to the distribution of the neural net output variable above a cut value taking into account the expected background in each bin. The fit also gave the final systematic error for 189 GeV using the uncertainties of efficiency and background as well as other analysis-independent errors (e.g. of the luminosity and the beam energy). The result is:

$$\sigma_{W^+W^-\to q\bar{q}q\bar{q}} = 7.36 \pm 0.26\,(stat) \pm 0.10\,(syst)\,pb.$$

The result from the old linear cut analysis was also determined as a comparison:

$$\sigma_{W^+W^-\to q\bar{q}q\bar{q}} = 7.56 \pm 0.28\,(stat)\,pb.$$

Here the systematic error can be expected from the previous years to be compatible to the neural net analysis.

The results agree within the uncertainties and the systematic errors were found to be compatible. But the statistical error of the neural network is lower due to the better selection quality.

SUMMARY

A feed-forward neural network is successfully used in the DELPHI collaboration to select hadronic WW-candidates and determine the production cross section. The selection quality is clearly improved compared to the previous standard analysis using linear cuts. For this reason the statistical error could be reduced using the neural network. The systematic error was determined performing studies of the network stability and using the neural net as a fixed mathematical function in well known techniques. Finally the systematic error was found to be compatible compared to the standard analysis.

REFERENCES

1. DELPHI Collaboration, *Physics Letters B* **479**, 89 (2000).
2. Lønnblad, L., Peterson, C., Pi, H. and Røgnvaldsson, T., JETNET *3.1 - A Neural Network program for jet discrimination and other High Energy Physics triggering situations* Department of Theoretical Physics, University of Lund, Sweden (1994).
3. DELPHI Collaboration, *Physics Letters B* **397**, 158-170 (1997).
4. DELPHI Collaboration, *E.Phys.J.C* **2**, 581 (1998).
5. Becks, K.-H., Drees, J., Flagmeyer, U. and Müller, U., *Nucl.Instr.Meth. A* **426**, 599-604 (1999).
6. Sjøstrand, T., PYTHIA 5.7 *and* JETSET 7.4 *Physics and Manual* CERN-TH.7112/93, (1993).
7. Lønnblad, L., *Comp. Phys. Comm.* **71**, 15 (1992).
8. DELPHI Collaboration, *Nucl.Instr.Meth. A* **378**, 57 (1996).
9. DELPHI Collaboration, *DELPHI Note* 89-67 and 89-68 *PROG* 142 and 143 (1989).

Table 2. cross section results with statistical errors for different center-of-mass energies (ECM)

ECM [*GeV*]	xsec [*pb*]
188.6	7.36 ± 0.26
191.6	7.86 ± 0.65
195.6	8.23 ± 0.39
199.5	7.90 ± 0.36
201.6	7.98 ± 0.53
204.9	8.33 ± 0.40
206.8	7.74 ± 0.37

Use of Neural Networks in a Search for Single Top Quark Production at DØ

Lev Dudko, for the DØ Collaboration

Institute of Nuclear Physics, Moscow State University

Abstract. We present a search for electroweak production of single top quarks in the DØ detector at the Tevatron collider. After initial selections, the signal forms less than one percent of the background, and requires a powerful analysis tool to separate it from background. For this purpose, we employ the neural network package MLPfit, and train it on Monte Carlo (MC) models of two processes for signal, and on data and MC models of five processes for background. Based on an analysis of singularities in the Feynman diagrams for single-top production, we choose an optimal set of kinematic variables as inputs to the networks. We use separate networks for each signal-background pair. For the dominant backgrounds, we use sequential nets.

INTRODUCTION

After the discovery of the top quark in $t\bar{t}$ pair production via the strong interaction (1), the next natural step is to test its properties and the parameters of top physics. One of the most interesting way to do that is to examine the electroweak production of single top quarks (2). This production mode can provide directly the magnitude of the CKM element V_{tb}, the width of the top quark, and certain spin effects resulting from the (V–A) vertex. Beyond the standard model, we can test the Wtb couplings for anomalous behavior, and search for FCNC couplings or W' production. The rate for single top quark production is sufficiently large to pose a significant background to the search for the Higgs boson and other new physics processes. Single top quarks can be produced at the Tevatron in two main processes, via s-channel and t-channel mechanisms (Fig. 1), with cross sections of:

$$\sigma_{\text{NLO}}(p\bar{p} \to t\bar{b}(\bar{t}b)X) = 0.73 \pm 0.04 \text{ pb} \quad (3)$$
$$\sigma_{\text{NLO}}(p\bar{p} \to tq\bar{b}(\bar{t}qb)X) = 1.70 \pm 0.19 \text{ pb} \quad (4)$$

We perform our search in the leptonic decay channels of the W boson from top quark decay, with $W \to e\nu_e$, $\mu\nu_\mu$, and $W \to \tau\nu_\tau \to e\nu_e\bar{\nu}_\tau\nu_\tau$, $\mu\nu_\mu\bar{\nu}_\tau\nu_\tau$. This signal is mimicked by several background processes (2):

$\sigma_{\text{LO}}(p\bar{p} \to Wb\bar{b}, Wc\bar{c}, Wcs, Ws\bar{s}) = 35$ pb

$\sigma_{\text{LO}}(p\bar{p} \to Wjj, j = u,d,g) = 966$ pb

$\sigma_{\text{NLO}}(p\bar{p} \to WW, WZ) = 13$ pb

$\sigma_{\text{NNLO-NNLL}}(p\bar{p} \to t\bar{t}) = 6.5$ pb

QCD multijet events with misidentified leptons

Background rates are huge relative to signal, and different backgrounds have essentially different kinematics. These are the reasons to use the neural network (NN) technique in the analysis. After several attempts to improve the signal-to-background ratio, we have developed the NN strategy described below, and train each signal process versus each kind of background.

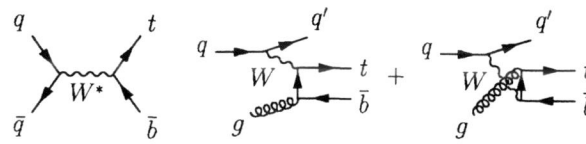

FIGURE 1. Representative Feynman diagrams for single top quark production.

APPLYING THE NN TECHNIQUE

Training the Neural Networks.

The first step in applying the NN technique for separating signal from background involves making a proper choice of input variables. A useful observation that helps to choose an optimal set of input variables is that the rate for a scattering process is greatest in the regions of phase space near singularities of the corresponding matrix element. If such singularities occur in different places for signal and background, then the dependence of the corresponding variables should differ markedly between signal and background. In general, there are two types of singularities: s-channel, $M^2_{f1,f2} = (p_{f1} + p_{f2})^2$, and t-channel $\hat{t}_{i,f} = (p_f - p_i)^2$. Here p_f is the four momentum of the final particle and p_i of the initial particle (parton). The position of the singularity in each of these variables is determined by the denominator of the corresponding Feynman propagator. We use this idea to choose an optimal set of kinematic variables (5).

Before implementing the final NN analysis, we choose to restrict all event samples by eliminating regions of phase space where background is large and expected signal small. In these regions, the kinematic characteristics of background and signal differ greatly, and such events can therefore be easily separated, and the background removed at essentially any stage of the analysis. (In our specific analysis, the excluded regions also correspond to regions where the background tends to be poorly modeled.) The background that remains after such preprocessing, being closer in character to signal, is therefore more difficult to distinguish from signal. However, with sufficient acceptance, the small loss of signal from such preprocessing is not debilitating, and the NN can then be trained using the most sensitive variables for separating signal and backround. An example of the NN outputs for such preprocessed and not preprocessed events is given in Fig. 2, and shows the impact of the preprocessing. For the training of our networks we use

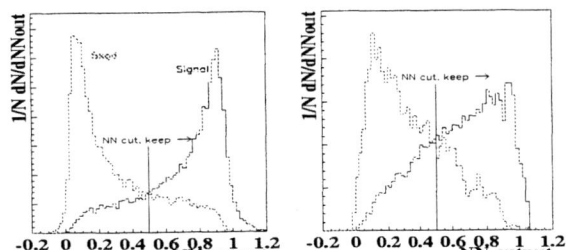

FIGURE 2. NN outputs for one signal and one background after looser (left plot) and tighter (right plot) preprocessing selection.

the MLPfit package (6), with our own code interface and a Hybrid Linear-BFGS training method. We obtain best results using the structure of the NN shown in Fig. 3. The numbers in the boxes correspond to the numbers of nodes on the input, hidden and output layers. We use a structure of three subnetworks for each of the Wjj and Wbb backgrounds. In this structure, we combine two different kinds of input information to improve the extraction of signal from these backgrounds. In addition to kinematic information, we use muon-tagging information for b-jets. In the first subnetwork (Wjj NN1), we combine the muon tagging information and kinematic information. In the second subnetwork (Wjj NN2), we use only kinematic information. We then combine the outputs of the first and second subnetworks into a third one. As shown in Fig. 4, this mix of b-tag information and kinematics leaves more signal and less background than is possible in the simpler structure: e.g., a cut on

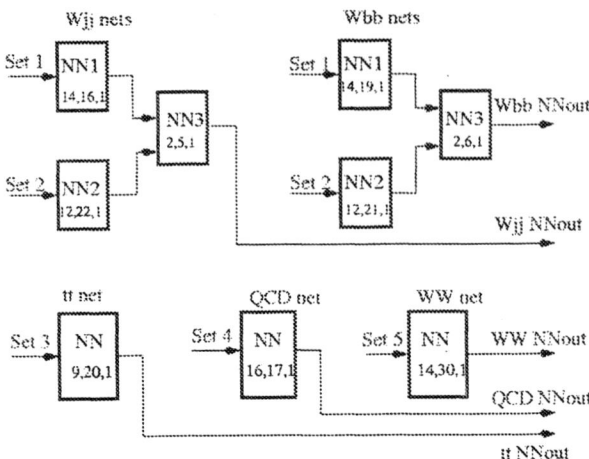

FIGURE 3. Neural network architecture.

only the first or on the second subnetwork (horizontal or vertical line on the plot) in comparison to that on the NN3 output. For the $t\bar{t}$, WW, and QCD backgrounds,

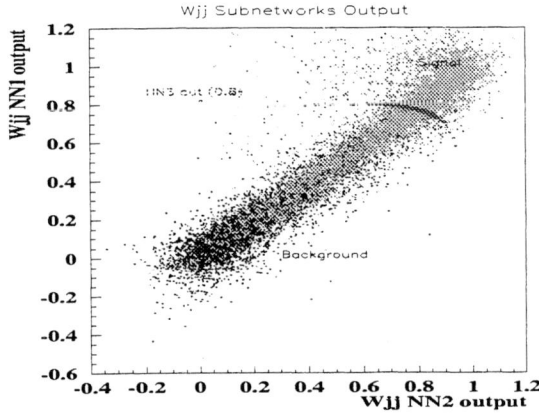

FIGURE 4. Output of the first Wjj subnetwork, versus that of the second. The cut on the third subnetwork is shown.

we train one network for each to separate each from signal. The outputs of each of these networks for signal and background are shown in Fig. 5, with the distributions normalized to unity. To perform the analysis, we pass all data, background and signal samples, through all networks in parallel, and cut on the output of each of the networks.

Check of the NN

To check the stability of the NN analysis, we perform two tests of the networks. In the first step, during training, we divide the input sample (signal and background) into two parts. We train the NN on one of these parts, and at every cycle of training we check for improvement of the training using the test sample, as shown in Fig. 6. This check is important to prevent "over-fitting" the NN. The second test is performed after training. We check

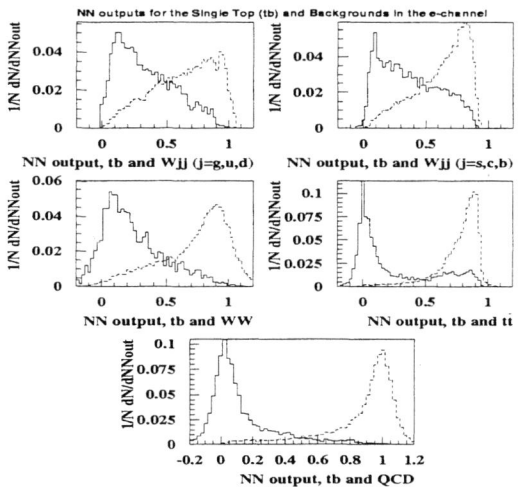

FIGURE 5. NN outputs of trained signal/background pairs. The signal peak lies near one, and the background peak near zero.

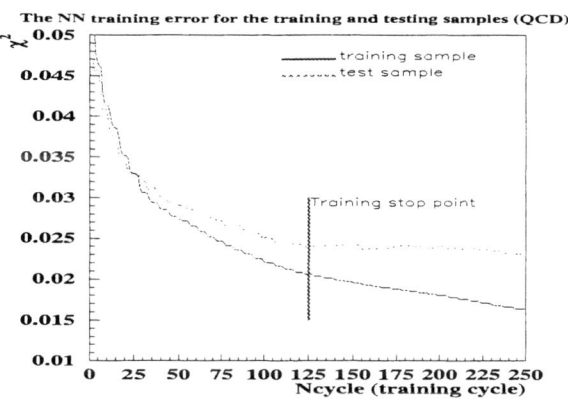

FIGURE 6. Prevention of the over-fitting problem. χ^2 is the error function.

the overall consistency of response of the NNs and our modeling of signal and background on the preliminary data sample, collected during Run 1. Figure 7 shows the NN outputs for the sum of signal and background models (central curve with gray error band) and data (triangles). Based on this agreement between the model and data, we conclude that our model describes signal and background, and that the response of the NN is reliable (e.g., there is no over-fitting of the NN).

CONCLUSION

After performing these checks, the NNs are ready for the final implementation of cutoffs, after which we can

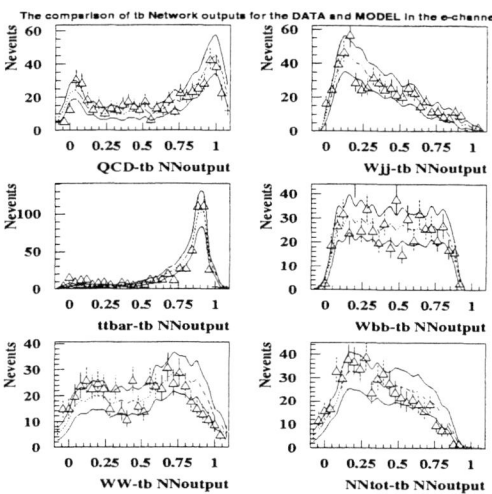

FIGURE 7. Comparison of the NN outputs for the complete signal+background model and DØ Run 1 data.

count the expected and observed numbers of events. For our chosen set of cutoffs on the 5 NN outputs, we obtain the following final overall NN efficiencies:

Signal Eff. = Eff.Preprocessing.Cuts × Eff.NN.Cuts = 78% × 21% = 16%

Background Eff. = Eff.Preprocessing.Cuts × Eff.NN.Cuts = 27% × 2.7% = 0.73%

We have applied this analysis technique to a search of DØ data for single top quark production, and the results are being prepared for publication.

REFERENCES

1. F. Abe *et al.* (CDF Collaboration), Phys. Rev. Lett. **74**, 2626 (1995); S. Abachi *et al.* (DØ Collaboration), Phys. Rev. Lett. **74**, 2632 (1995).

2. B. Abbot *et al.* (DØ Collaboration), to appear in Phys. Rev. D, hep-ex/0008024; P. Savard, to appear in the proceedings of the Meeting of the Division of Particles and Fields of the American Physical Society, Columbus, OH, August 2000

3. M.C. Smith and S. Willenbrock, Phys. Rev. D **54**, 6696 (1996)

4. T. Stelzer, Z. Sullivan, and S. Willenbrock, Phys. Rev. D **56**, 5919 (1997)

5. E. Boos, L. Dudko and T. Ohl, Eur. Phys. J. C **11**, 473-484 (1999); L. Dudko in the Proceedings of AIHENP 99; E. Boos, L. Dudko, DØNote 3612

6. MLPfit v1.40, J. Schwindling, B. Mansoulie, http://home.cern.ch/~schwind/MLPfit.html

A Hybrid Training Method for Neural Energy Estimation in Calorimetry

P.V.M. da Silva[a], J.M. Seixas[a] and J. Seixas[b]

[a]*Signal Processing Laboratory (LPS)*
COPPE/EE/Federal University of Rio de Janeiro
C.P. 68504, Rio de Janeiro 21945-970, RJ - Brazil
[b]*Superior Technical Institute, Lisbon, Portugal*

Abstract. A neural mapping is developed to improve the overall performance of Tilecal, which is the hadronic calorimeter of the ATLAS detector. Feeding the input nodes of a multilayer feedforward neural network with the energy values sampled by the calorimeter cells in beam tests, it is shown that the original energy scale of pion beams is reconstructed over a wide energy range and linearity is significantly improved. As it happens for classical methods, a compromise between nonlinearity correction and the optimization of the energy resolution of the detector has to be accomplished. A hybrid training method for the neural mapping is proposed to achieve this design goal. Using the backpropagation algorithm, the method intercalates an epoch of training steps, for which the neural mapping mainly focus on linearity correction, with another block of training steps, in which the original energy resolution obtained by linearly combining the calorimeter cells becomes the main target.

INTRODUCTION

For the LHC experiment, the ATLAS detector is currently being developed at CERN. Among its subdetectors, calorimeters play an important role, as they measure accurately the energy of the incoming particles and provide a detailed energy deposition profile.

The hadronic calorimetry in ATLAS is performed by the scintillating tile calorimeter (Tilecal), which comprises three sections: one barrel and two extended barrels (1). Each section is made of 64 modules and the whole calorimeter system has more than ten thousand readout channels.

In a practical calorimeter design, nonlinearities typically arise (6). Therefore, a sort of nonlinearity compensation scheme has to be provided and multiparameter functions are usually employed for this purpose. However, very often an improvement on linearity is followed by a deterioration of the energy resolution of the detector, so that a compromise between these two design parameters has to be accomplished. Moreover, the multiparameter function approach typically results in energy dependent parameters, which makes the compensation algorithm rather complex.

The approximation of a mathematical function using examples of its action on experimental data points appears in many different areas of basic science and engineering. Neural networks have proved to be valuable in many of such problems, as their nonlinear processing capability usually provides a good solution even for complex mapping problems. In particular, feature-based networks, like multilayer networks trained by means of backpropagation (2), can implement an input/output mapping as a result of the supervised learning process, which modifies the synaptic weights of the network. During the training phase, the mapping is continuously modified according to the patterns fed into the input nodes of the network and their respective target values, which represent the knowledge of the action of the mapping on such input patterns (3).

This work aims at developing a neural mapping for estimating the actual energy being deposited in the Tilecal. Recent studies (5) shown that a multilayer neural network tends to correct effectively the linearity response of the Tilecal for a wide energy range, but the energy resolution suffers from deterioration, as for the classical approach. Thus, for achieving both good linearity and high energy resolution, a hybrid training method for the multilayer neural mapping is proposed in this paper.

THE EXPERIMENTAL SETUP

The Tilecal is a three-layer segmented hadronic calorimeter. A calorimeter module (see Figure 1) consists of iron (to absorb the energy of the interacting par-

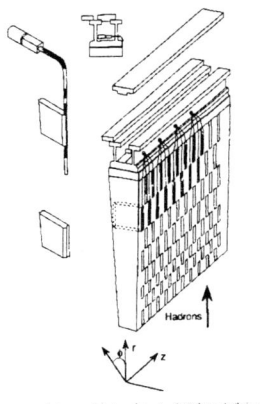

FIGURE 1. A calorimeter module.

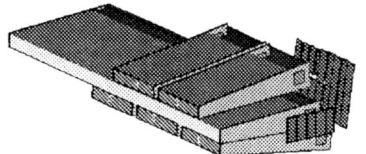

FIGURE 2. Experimental setup.

ticles), scintillating tiles (for sampling the energy being deposited in the detector), and optical fibers (to carry the light produced in the tiles towards the outer part of the detector). Groups of fibers are optically coupled to a photomultiplier, defining the detector segmentations.

The development of the neural estimator used experimental pion data from a last generation (Module 0) prototype. The experimental arrangement is shown on Figure 2. In these tests, one half (right side) of the Module 0 prototype of the barrel section was instrumented, producing 46 readout channels. To compensate for the unavoidable energy leakage, which results from the use of such reduced (single module) calorimeter system in the tests, five modules from a previous generation of prototypes were placed around the Module 0 (two above and three below), representing additional 200 channels to be readout. These five modules also simulate the information of neighboring modules in the actual arrangement for LHC operation. For all data used in this paper, the pion beam was aligned to enter at the center of the instrumented part of Module 0 prototype.

The performance of the Tilecal prototype and its neural estimator for pions was evaluated from tests for which ten nominal beam energies were selected, ranging from 10 to 400 GeV. On average, 6000 events were accumulated for each nominal energy value.

The Tilecal provides very good calorimeter performance. If its original response is evaluated by linearly adding the raw data (digitized energy values from analog-to-digital converters with pedestal subtraction) from calorimeter cells for each event (without any attempt for corrections that have been shown possible to be made (1)), some nonlinearities can be observed. As expected, Gaussian distributions are obtained, although some energy leakage effect can be noted (above 100 GeV) in the form of lower-energy tails.

For evaluating linearity, a least squares straight line fitting was performed on the data points and the maximum relative deviation with respect to the fitting was measured.

To compute the energy resolution, a model based on a linear combination of scaling and constant factors was employed (6):

$$\frac{\sigma}{E} = \frac{a}{\sqrt{E}} + b \quad (1)$$

The scaling factor a dominates the energy resolution at higher energy values, where the constant term b becomes dominant. From raw experimental data (no corrections), a maximum non-linearity of 8.5% and an energy resolution of $\frac{39.6\%}{\sqrt{E}} + 5.0\%$ were measured (see Figure 3).

NEURAL ESTIMATION

The feedforward neural network used to perform the desired energy estimation was fed with the energy sampled by each calorimeter cell (246 input nodes). The network had 20 neurons in the hidden layer. For all such neurons, a sigmoidal activation function was used. A single linear neuron in the output layer then provided the desired linear reconstruction of the energy scale.

First, the network was fully trained with backpropagation method. The experimental data sets were equally split into two for building the required training and testing sets. The input data were normalized by a constant factor, which was obtained (mean value plus one sigma) from a distribution of all cell energies for the various nominal beam energies. As neuronal saturation has been observed to play an important role for the energy resolution of the neural estimator (4), the temperature factor of the sigmoids were controlled to guarantee proper operation of the hidden neurons. Normalized nominal beam energies defined the target values of the training phase.

The nonlinearity of the original response was pretty well compensated for. However, the energy resolution was worse, and the model from equation (1) did not describe the behavior of the estimator. As was mentioned, this is in agreement with recent studies on the development of the neural mapping for the Tilecal (5).

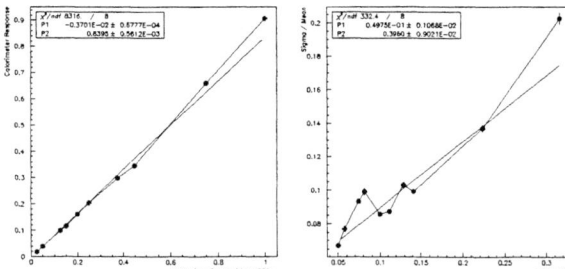

FIGURE 3. Linearity (left) and energy resolution (right) from the original calorimeter response (linear combination of cells).

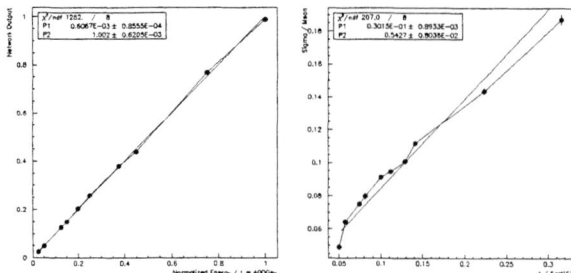

FIGURE 4. Linearity (left) and energy resolution (right) from the neural estimator and hybrid training.

In order to improve the performance of the neural estimator, it would be desirable to retain the proven ability of the estimator for nonlinearity compensation, but introduce a way to reduce the deterioration on energy resolution, recovering, at least, the original response that is obtained when a linear combination of cells is performed.

To achieve this goal, a hybrid method was developed for the training phase of the network. Thus, using backpropagation, the neural network was trained mixing a bunch of training steps, in which the target values were set as nominal beam energies (main focus on nonlinearity compensation), with a set of training steps, for which the network tries to follow the original calorimeter response (focusing on achieving a good energy resolution), and the target values being the total energy obtained from the linear combination of calorimeter cells.

The determination of the number of cycles of each training phase of the proposed method deserved some attention. It should be observed how long it would take for the network to learn the original calorimeter response and switch from one phase to another. Thus, 30000 training steps were used in the phase the network follows the nominal beam energy targets, and 7500 training steps were required to learn the original calorimeter response.

The performance for this training approach can be evaluated from Figure 4, representing a linearity better than 2.8% and an energy resolution of $\frac{54.3\%}{\sqrt{E}} + 3.0\%$. It can be depicted from this figure that the behavior of the energy resolution of the estimator tends to align with the model of equation (1), in this case.

CONCLUSIONS

A neural mapping for compensating for nonlinearities that may arise from raw data analysis over the experimental response of the Tilecal was presented. In order to achieve a significant improvement on the linearity of the original calorimeter response and keep the original good energy resolution of the detector, a hybrid training method for the neural estimator was developed using the standard backpropagation algorithm. The method successfully learns the original calorimeter response as a way to keep in track the energy resolution figure, while correcting for the observed nonlinearities.

As the neural estimator generally exhibits a different behavior for low and high energy ranges, a multiple neural network estimator is being also developed. Combining networks that would be able to learn in more detail the energy deposition profiles for different energy ranges may improve the overall estimation performance.

ACKNOWLEDGEMENTS

We are thankful to CNPq and FAPERJ (Brazil), and ICCTI (Portugal) for the support of this work. We also thank the Tilecal collaboration at CERN, for providing the experimental data used and contributing with valuable discussions to this work.

REFERENCES

1. Ariztizabal, F. e. a., *Nuclear Instruments and Methods* **A(349)**, (1994) 384–397.
2. Haykin, S., *Neural Networks: a Comprehensive Foundation*, second edn., Prentice-Hall, 1999.
3. Hecht-Nielsen, R., *Neural Computation*, Addison-Wesley, 1990.
4. Seixas, J., in *Symposium on Scientific Applications on Neural Nets*, Bad Honnef, Germany, 1998 pp. 1–16.
5. Seixas, J. M., da Silva, P. V. M., and Calôba, L. P., in *IEEE Computer Society Edition for the Brazilian Symposium on Neural Networks*, 1998 pp. 204–209.
6. Wigmans, R., *Rev. Nucl. Sci.* **41**, (1991) 133–148.

Principal Component Analysis for Neural Electron/Jet Discrimination in Highly Segmented Calorimeters

M. R. Vassali and J. M. Seixas

Signal Processing Lab. - COPPE/EE/UFRJ
Federal University of Rio de Janeiro
CP 68504, Rio de Janeiro 21945-970, RJ, Brazil
{vassali, seixas}@lps.ufrj.br

Abstract. A neural electron/jet discriminator based on calorimetry is developed for the second-level trigger system of the ATLAS detector. As preprocessing of the calorimeter information, a principal component analysis is performed on each segment of the two sections (electromagnetic and hadronic) of the calorimeter system, in order to reduce significantly the dimension of the input data space and fully explore the detailed energy deposition profile, which is provided by the highly-segmented calorimeter system. It is shown that projecting calorimeter data onto 33 segmented principal components, the discrimination efficiency of the neural classifier reaches 98.9% for electrons (with only 1% of false alarm probability). Furthermore, restricting data projection onto only 9 components, an electron efficiency of 99.1% is achieved (with 3% of false alarm), which confirms that a fast triggering system may be designed using few components.

INTRODUCTION

By 2005, the LHC will be colliding protons of 7 TeV at periods of 25 ns. This collision frequency will produce an enormous amount of data that shall be analyzed by an online multistage triggering system, in order to reject the huge background noise of the experiment. One of the detectors to be placed around a collision point of the LHC is ATLAS, for which the trigger system is being designed in three levels of analysis. For the triggering system, calorimeters play an important role. Both electromagnetic (e.m.) and hadronic sections are highly-segmented into cells, in order to provide a detailed energy deposition profile for incoming particles. The second-level (L2) trigger system will have access to full calorimeter granularity in the restricted areas of the detector for which significant signal can be observed. These are the regions of interest (ROIs), which are identified by the first-level (L1) system (3).

In this paper, an electron/jet discriminator for the L2 system is developed using neural processing over calorimeter information. In these simulations, a ROI for the e.m. calorimeter comprised three segments, with each one formed by a matrix of 12x12 cells, while the hadronic section comprised 4 segments formed by 4x4 cells each. Figure 1 illustrates such arrangement. More recent updates on the calorimeter system show different granularity for the e.m. segments and only three segments for the hadronic section. However, as from recent studies, these changes proved not to be very significant in terms of the efficiency achieved by the neural discriminator (4).

As depicted in Figure 1, a ROI in the calorimeter was described by 496 cells, which makes the discrimination between electrons and jets a very complex task. As speed is one of the main issues of the L2 system, the reduction of this high dimensionality of the input data space is very attractive. Principal component analysis (2) has extensively been used for data compaction in a variety of applications. However, as the calorimeter system provides a highly-segmented readout of particle interactions, in this paper principal component analysis is performed on each calorimeter segment. This alternative approach avoids extracting principal components from grouped calorimeter data, which is typically performed, as some of the outermost cells of each calorimeter segment may provide important discriminating information in spite of their poor signal to noise ratio. By extracting segmented principal components, one expects an improvement in the overall performance of the discriminator, when original data are projected onto just a few components. This would allow a fast and highly efficient triggering system to be designed.

PRINCIPAL COMPONENT ANALYSIS

The principal component analysis (PCA) is based on the Karhunen-Loève series. PCA aims at finding a set of M orthogonal vectors in input data space that account for

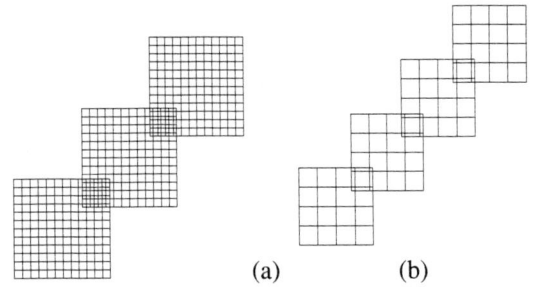

FIGURE 1. The electromagnetic (a) and hadronic (b) sections.

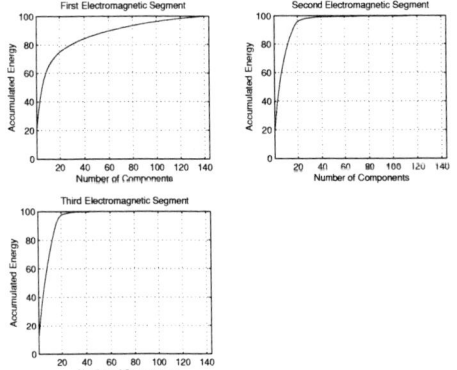

FIGURE 2. Cumulative energy of the first principal components of the electromagnetic segments.

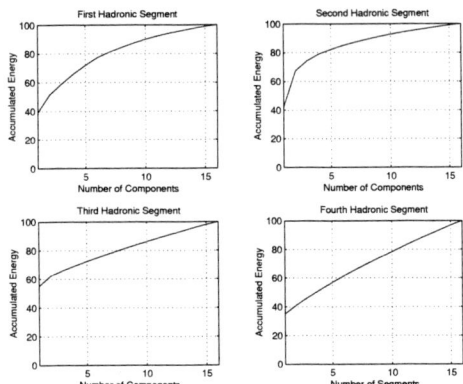

FIGURE 3. Cumulative energy of the first principal components of the hadronic segments.

as much as possible of the data's variance. Thus, the differences in classes of events originally present in the data set may still be identified by projecting data of the original N-dimensional input space onto the M-dimensional subspace spanned by these vectors. This would perform a dimensionality reduction with the preservation of the information spread around the full input space, as normally $M \ll N$.

The extraction of these components can be made by classical or neural methods. In the classical method, the components are the eigenvectors of the autocorrelation matrix of the random process under consideration, sorted by the decreasing order of the corresponding eigenvalues. Alternatively, a neural network can be used, and the training phase may be supervised or non-supervised (2).

Figures 2 and 3 show the cumulative energy (in percentage of the total energy in the segment) of the first principal components, when data compaction is performed on each segment. It can be clearly seen that the main calorimeter information is preserved in just a few components, which allows a significant compaction ratio to be achieved.

NEURAL DISCRIMINATION

For the classifier design, a three layer fully-connected feedforward neural network was trained (using the back-propagation method (2)) to separate electrons from jets. The hyperbolic tangent was the activation function for all neurons. The number of input nodes of the final classifier was dependent on the number of principal components selected. The single hidden layer was composed of 33 neurons, whereas the output layer had just one neuron.

The whole L2 simulation data set contained 2528 electrons and 685 jets. This data sample was split into two to form the training and testing sets. Data normalization for e.m. segments was performed by dividing the energy samples of each cell by the total energy deposited in the full e.m. calorimeter. For the hadronic section, there was no need to apply any normalization scheme.

Considering only 33 components, the neural discriminator achieved an electron efficiency of 98.9% for the testing set, with less than 1% of false alarm probability (jets misclassified as electrons). Table 1 shows how these 33 components were distributed among the calorimeter segments. It confirms that the extraction of segmented principal components was able to reveal the important information of the hadronic and the two outer e.m. segments, from which almost all of such important components were extracted. The high performance of the classifier can be observed from Figure 4, which shows the particle distributions of the network output. Target values were arbitrarily chosen as -1 for electrons and 1 for jets.

Another attractive result was found by selecting just 9 components[1]. Obviously, this further compaction in-

---

[1] These 9 components were chosen according to a relevance study (1) that showed the main components for the neural discrimination process.

Table 1. The distribution of the 33 components among calorimeter segments.

Segment	Electromagnetic Components	Hadronic Components
First	2	7
Second	4	7
Third	4	7
Fourth	-	2

Table 2. Comparison to other methods of discrimination.

Method used for Discrimination	Electron Efficiency (%)	Misclassified Jets (%)
Neural Network using 33 Non-segmented principal components*	72.4	39
Matched Filters	76.0	32
Neural Network directly on energy cells	97.9	10
Neural Network using 9 components	99.1	3
Neural Network with topological mapping (Rp) and rings(6)	98.7	2
Neural Network using 33 components	98.9	1

* These components were extracted from grouped data.

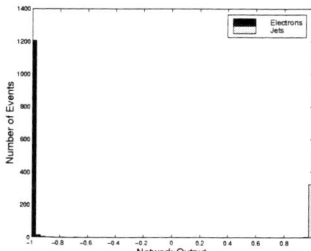

FIGURE 4. Output of the neural discriminator.

creases the false alarm probability, which becomes 3%, for the same level of electron efficiency. However, this slight deterioration in discrimination performance may be worth the considerable increase in processing speed such reduced set of components provides.

These results may be compared to others obtained from different design approaches using the same data set (see Table 2). It is shown that segmented principal component analysis outperforms a variety of methods, among which neural processing on non-segmented components, full input data space and topological data description in terms of concentric ring sums. It also outperforms classical analysis from matched filters (5).

CONCLUSIONS

It can be seen from Table 2 that the extraction of principal components for each segment of the calorimeter system performs much better than the non-segmented extraction of components. Projecting data onto 33 segmented principal components proved to be even better than a direct neural processing over the whole group of calorimeter cells, when data compaction of any kind is not realized. Thus, the segmented principal component analysis not only compacts efficiently the original data input space, enabling to reduce the processing time for the online operation, but also improves the neural discrimination efficiency. It should also be noted that the segmented principal component approach may be combined with topological methods, for improving the overall performance. This approach is presently under investigation.

Envisaging an even faster processing speed, a more compact design using just 9 components proved to be feasible, as the discrimination performance was only slightly deteriorated. An implementation of such compact triggering system is being attempted using digital signal processing (DSP) technology.

ACKNOWLEDGEMENTS

We are thankful to CAPES, CNPq, FAPERJ and FUJB (Brazil) for the support to this work. We are also thankful to the ATLAS collaboration at CERN, for making available the data used in this work.

REFERENCES

1. Gruber, A., and Moeck, J., *Performance of Backpropagation Networks in the Second-Level Trigger of the H1-Experiment*, 1993. Oberammergau, Germany.
2. Haykin, S., *Neural Networks: a Comprehensive Foundation*, second edn., Prentice-Hall, 1999.
3. ATLAS/Level-1 Trigger Group, Atlas level-1 trigger technical design report, CERN/LHCC/98-14, June 1998. TDR 12.
4. Rabello, A., and Seixas, J. M., *Ring Mapping for Neural Electron/Jet Separation using Multilayered and Multisegmented Calorimeters*, in Portuguese, 2000. XXI Brazilian Meeting on Fields and Particle Physics.
5. Trees, H. L. V., *Detection, Estimation and Modulation Theory, part I*, John Wiley & Sons, 1968.
6. Vassali, M. R., and Seixas, J. M., *A Topological Mapping for a Particle Neural Discriminator based on Segmented Calorimeters*, in Portuguese, 2000. XXI Brazilian Meeting on Fields and Particle Physics.

Particle Identification at DØ using Neural Networks

D. Chakraborty

Department of Physics and Astronomy, State University of New York at Stony Brook, Stony Brook, NY 11794-3800
(for the DØ collaboration)

Abstract. We have investigated the possibility of employing neural networks for identification of electrons and τ leptons with the DØ detector in the upcoming run of the Tevatron collider at Fermilab. Preliminary results based on Monte Carlo simulations indicate that for any acceptable level of signal efficiency, neural networks consistently outperform covariance matrices so far employed for the same purpose. Using a subset of variables used by a covariance matrix, a properly trained neural network offers 2 times better background rejection for taus, and 10 times for electrons, at 90% signal efficiency. Similar enhancements can be expected for identification of other objects (such as muons, b or c jets, quark vs gluon jets, neutrinos etc).

INTRODUCTION

Effective particle identification is a central challenge in Experimental High Energy Physics. Most of the interesting processes are buried in enormous backgrounds. "Smart" algorithms are needed to identify particles that feature in signal but not in background. Here we address the identification of electrons and τ leptons which appear in the final states of many important processes at the Tevatron collider experiments at Fermilab. During Run 1 (1992-96), the Tevatron delivered ~ 125 pb^{-1} of proton-antiproton collisions at a center-of-mass energy of 1.8 TeV to each of its two detector experiments, DØ and CDF. After an extensive upgrade to both the collider and the detectors, collisions will resume in the spring of 2001 (Run 2). The collider upgrade will result in a higher center-of-mass energy (2.0 TeV), and an integrated luminosity of $2-20$ fb^{-1} per experiment. The upgraded detectors will have superior momentum and position resolution.

DØ is a general-purpose detector (1). In Run 2, it will have a central cylindrical tracking volume consisting of silicon and scintillating fiber trackers immersed in a coaxial magnetic field of 2 Tesla where the trajectory of a charged particle can be mapped with a precision of a few microns, and the magnetic field will allow momentum measurement from the curvature of the track. Surrounding the tracker and the solenoidal magnet are preshower detectors and calorimeters providing an almost full solid angle coverage of the interaction point. The active medium of the calorimeter (preshower detector) is liquid argon (plastic scintillator) while the passive absorbers are made of depleted uranium or steel (lead). With the exception of muons and neutrinos, all particles leaving the tracking volume deposit all of their energy in the calorimeter. Momenta of the muons, which penetrate the calorimeter, are measured by tracking them before and after their passage through a thick layer of magnetized iron. The presence of any neutrino in an event can only be inferred from the imbalance in the transverse component of the total momentum. In Run 1, covariance matrices were employed to identify electrons and τs using several measurements obtained from different parts of the detector. Here we examine the merits of using artificial neural networks as an alternative to covariance matrices in the context of Run 2.

THE COVARIANCE MATRIX APPROACH

For a set of characteristic variables $\{x_1, x_2, \ldots x_n\}$, the $\{i,j\}$th element of the covariance matrix is given by

$$\sigma_{i,j} = \langle x_i x_j \rangle - \langle x_i \rangle \langle x_j \rangle, \quad (1)$$

determined from a training sample of signal or background, either simulated or obtained from test-beam. For a given candidate with corresponding characteristics $\{X_1, X_2, \ldots X_n\}$, we define

$$\chi^2 = \sum_{i,j} X_i H_{i,j} X_j, \quad (2)$$

where H, also called "H-matrix", is the inverse of the covariance matrix:

$$H_{i,j} = (\sigma^{-1})_{i,j}. \quad (3)$$

For a suitably chosen set of characteristics, the χ^2 can be used as a multivariate discriminant to separate signal from background. If the matrix is trained on a signal sample, then, on average, χ^2s for signal will be smaller than those for background. This approach involves a simple linear transformation which is optimal if the variables follow a multivariate Gaussian distribution.

THE NEURAL NETWORK APPROACH

For the problems in our hands, however, the measured characteristics have asymmetric distributions and bear non-linear correlations. Neural networks, which can handle such features, are therefore expected to offer superior separation between signal and background. Specifically, we have tested the MLPfit neural network (2). We used a single layer of hidden variables or neurons. The jth neuron performs a linear combination of all the inputs,

$$h_j = c_j + \sum_i w_{i,j} x_i, \qquad (4)$$

and delivers the output after a sigmoidal transformation

$$s(x) = (1 + e^{-x})^{-1}. \qquad (5)$$

The output of the network is a linear combination of the outputs of the neurons:

$$y = k + \sum_j p_j s(h_j). \qquad (6)$$

The network determines the constants $c_i, w_{i,j}, p_j$, and k, by fitting the functions to multidimensional distributions of input variables from samples of signal and background. We opted for a hybrid-linear Broyden, Fletcher, Goldfarb, Shanno (BFGS) method for minimizing the error function (3).

IDENTIFICATION OF ELECTRONS

We focus on identifying high-energy central electrons. Specifically, we require $|\eta| < 1.0$ and $E > 10$ GeV where $\eta \equiv -\ln(\tan\frac{\theta}{2})$ is the pseudorapidity with θ being the polar angle with respect to the beam axis, and E the energy of the electron. Our training samples consist of 2600 Monte Carlo single electrons, uniformly distributed in 10 GeV $< E < 100$ GeV, for signal, and 1600 QCD dijet events with at least one jet hadronizing into a final state with π^0 fraction greater than 65% for background (hadronic jets with smaller π^0 fraction are not likely to mimic an electron). Similar samples were used for testing.

The energy deposited by an electron is contained in a short and narrow volume in the calorimeter. The DØ calorimeter is finely segmented in transverse as well as longitudinal direction. The 4 innermost layers of

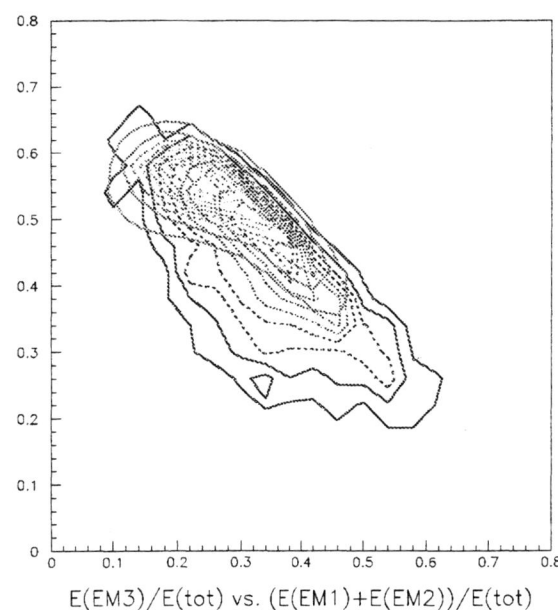

FIGURE 1. Correlation between energy fractions in EM layer 3 and the sum of layers 1 and 2 for signal (light contours) and background (dark contours).

the calorimeter, EM1, ...EM4, and the central preshower detector (CPS) contain most of an electron's energy, while the shower widths in the azimuthal angle (ϕ) and pseudorapidity (η) give a measure of the lateral spread. The following 7 variables are given to the network: $E_{\text{EM1+EM2}}/E$, E_{EM3}/E, E_{EM4}/E, E_{CPS}/E, $E/100$, σ_ϕ, and σ_η. Figure 1 shows the non-linear correlation between two of these variables.

After some optimization, we adopted a network with 16 neurons in the hidden layer and trained it over 400 epochs. The covariance matrix, or H-matrix, had a much more detailed picture consisting of 41 variables. Figure 2 shows the comparison between the H-matrix and the neural network results. We find that for any acceptable level of signal efficiency, the neural network affords us a level of background suppression 5 to 10 times better than that of the H-matrix.

IDENTIFICATION OF τ LEPTONS

The preselection of τ lepton candidates are similar to those of electrons: $|\eta| < 1.0$ and $E_T > 10$ GeV where E_T is the transverse component of the momentum. Our training samples consist of about 1150 hadronically decaying τs chosen from 2000 Monte Carlo $Z \to \tau\tau$ events based on the above criteria for signal, and about the same number of QCD jets for background. Similar samples are used for testing.

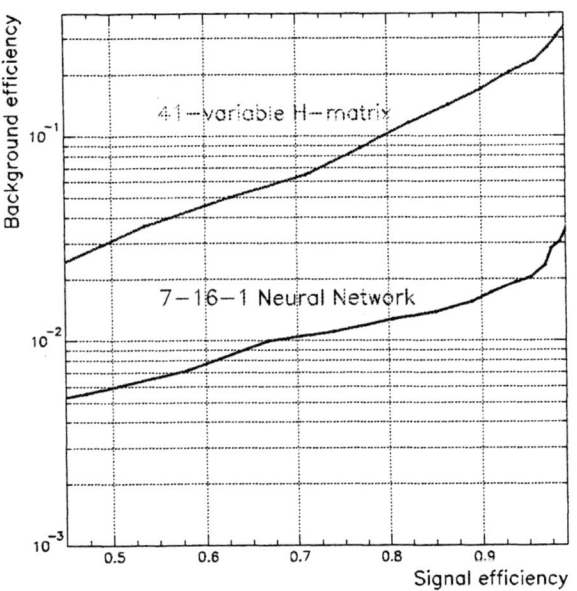

FIGURE 2. Fraction of background retained as a function of electron signal efficiency: for *H*-matrix (upper curve) and MLPfit neural network (lower curve).

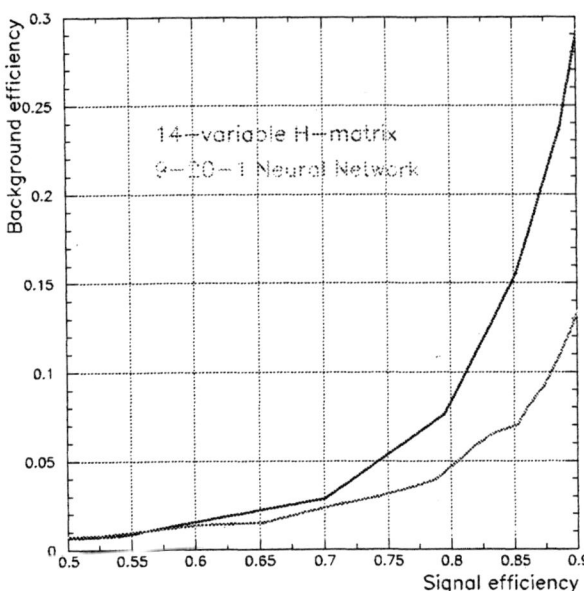

FIGURE 3. Fraction of background retained as a function of τ signal efficiency: for *H*-matrix (upper curve) and MLPfit neural network (lower curve).

Typically, the energy profile of a τ is narrower than that of a QCD jet although not as narrow as that of an electron. Also, on average, a τ has fewer associated tracks than a QCD jet. The following 9 variables are given to the tau neural network: E_{EM3}/E, E_{EM4}/E, E_{FH}/E, RMS(cal), E_{hot2}/E, $\log_{10} E - 1$, $\log_{10}(N_{0.4} + 1)$, $(N_{0.4} - N_{0.2})/(N_{0.4} + 1)$, and RMS(trk), where FH refers to the first calorimeter layer after the four EM layers, and N_R is the number of tracks within a cone of radius R centered on the line joining the primary interaction vertex and the calorimeter cluster centroid. RMS(cal) is the root-mean-squared width of the calorimeter cluster in the $[\eta, \phi]$ space, while RMS(trk) is the momentum-weighted root-mean-squared width determined from the associated tracks with respect to the calorimeter cluster centroid. E_{hot2}/E is the fraction of the total energy in the two most energetic towers in the calorimeter cluster. This variable is the most powerful discriminant for τs. We adopted a network with 20 neurons in the hidden layer and trained the network over 500 epochs. The covariance matrix, or *H*-matrix, had 14 variables including all of the above except RMS(cal). Instead of RMS(cal), it had two other variables to quantify the lateral spread of the cluster. Figure 3 shows the comparison between the performances of the *H*-matrix and the neural network. For large signal efficiencies, the neural network admits only half as much background as the *H*-matrix, while performing comparably for smaller efficiencies.

CONCLUSION

Our studies indicate that the neural network consistently offers better background rejection for identification of electrons and τ leptons than the covariance matrix method. Preliminary studies suggest that further improvements can be expected from careful optimization. The findings also make a strong case for adopting neural networks for identification of other objects, such as muons, *b* and *c* jets, neutrinos, and separation of quark- and gluon-initiated jets.

REFERENCES

1. DØ Collaboration, S. Abachi et al. Nucl. Instr. and Methods A**338**, 185 (1994).
2. `http://www.cern.ch/~schwind/MLPfit.html`
3. P.E. Gill, W. Murray, M.H. Wright, "Practical Optimization" Academic Press, Inc., London (1981).

Vertex Reconstructing Neural Network at the ZEUS Central Tracking Detector

Gideon Dror[1+], Erez Etzion[2*]

1. Department of Computer Science, The Academic College of Tel-Aviv-Yaffo, Tel-Aviv 64044, Israel.
2. School of Physics and Astronomy, Raymond and Beverly Sackler Faculty of Exact Sciences, Tel-Aviv University, Tel-Aviv 69978, Israel.

Abstract. An unconventional solution for finding the location of event creation is presented. It is based on two feed-forward neural networks with fixed architecture, whose parameters are chosen so as to reach a high accuracy. The interaction point location is a parameter that can be used to select events of interest from the very high rate of events created at the current experiments in High Energy Physics. The system suggested here is tested on simulated data sets of the ZEUS Central Tracking Detector, and is shown to perform better than conventional algorithms.

INTRODUCTION

The z coordinate of the interaction vertex position in collider experiments can be used for a prompt decision whether to record the event or reject it. Here we explore an unconventional neural network (NN) approach to vertex finding tested with simulation of data collected by the ZEUS[1] detector at HERA. A collision of electron and proton occurs at HERA every 96 nsec. A very small rate of signal is selected from the background by a three-level triggering system that performs filtering through software and hardware. The challenge is to find an efficient way to extract the location of the interaction vertex along the z-axis, determined by the direction of the colliding protons. In the following we use simulated Central Tracking Data (CTD)[2] events and compare our results with those obtained by the second level trigger. Such a NN calculation may be very fast when implemented on dedicated hardware.

The basic idea is to use a network with fixed architecture that is inspired by the way our brain processes visual information. It is similar to the orientation selective NN employed by [3] which was used to select linear tracks. The present network is based on our identification method. Previous successful attempt[4] required a large network size for hardware realization. Here we show that a much smaller network can be used without any precision loss.

ZEUS CTD

The ZEUS CTD, placed in a magnetic field of 1.43T, surrounds the interaction point and is designed to measure direction, charge and momentum of charged particles. It has a cylindrical shape around the z-axis. 4608 signal wires are organized in 9 super-layers; each consists of 8 wire layers. Wires in odd super-layers run parallel to the z-axis while 5^0 tilt wires in even ones. The three inner layers provide also a z position by comparing the pulse time of arrival at both ends of the chamber.

[+] Gideon@server.mta.ac.il
[*] Erez@lep.tau.ac.il

THE INPUT DATA

The input consist of the (*x,y*) positions of hits in the axial super-layers (1,3,5,7,9) and the *z* timing for hits in the super-layers 1,3,5.

Fig. 1 shows an example of an event projected onto *z*=0 plane. Several slightly curved tracks (arcs) starting at the origin can be seen, each of which is made of 30-40 data points.

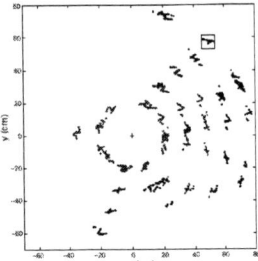

FIGURE 1. A typical event projected into the *z*=0 plane. The dots represent hits in the CTD. Red dots contain *z* information.

Since the tracks are expected to form a helix shape with small curvature, one expects a linear dependence of the z coordinates of the hits on their radial position. It was demonstrated that trying to fit the hits to a straight-line resulted in a considerable scatter of the data.

THE NETWORK

The network is based on step-wise changes in the representation of the data, moving from the input points, to local line segments to local arcs and finally to global arcs. Two parallel calculations deal separately with the two problems: *xy* and *z*-coordinates. The first NN, which handles the *xy* information, is responsible for constructing the arcs that correctly identify some of the particle tracks in the event. The second process uses the information to evaluate the *z* location of the point where all tracks meet.

The arc identification NN processes information in the fashion akin to the way visual information is processed by the primary visual system [5]. The input layer is made of a large number of neurons. The total area to cover in *xy* is 5000 cm^2 and we initially covered it with 100000 input neurons[4].

Neurons of the first layer are line segment detectors labeled by (*XY*α) – the coordinates of the segment center and its orientation. The activation of the first layer is given by $V_{XY\alpha} = g(\sum_{XY} J_{XY\alpha,xy} V_{xy} - \vartheta_1)$

where *J* is 1 when the distance between *xy* and *XY* < 0.5, -1 when 0.5<distance<2, and zero elsewhere. Fig. 2 (Left) represents the output of this layer.

Neurons of the second layer transform the representation of local line segments to local arc segments. They are defined by (κ,θ,i) their curvature κ=1/R, θ, the slope at the origin and i which relates each arc segment to the super-layer it belongs to. The mapping between the second and third layers is done in the following way: For a given local arc segment, we take the arc segment which is closest to it.

The neurons belonging to the last layer are global arc detectors, and detect projected tracks on to the *z*=0 plane. These neurons are denoted by (κ,θ) and are connected to the second layer in a simple fashion, $V_{\kappa\vartheta} = g(\sum_{\kappa'\vartheta'i} \delta_{\kappa,\kappa'} \delta_{\vartheta,\vartheta'} V_{\kappa',\vartheta',i} - \vartheta_3)$. Fig. 2 (Right) represents the activity of the third layer neurons. Each active neuron is equivalent to an arc in the figure.

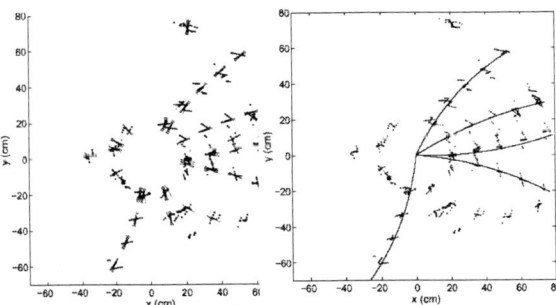

FIGURE 2. Representation of the activity of (Left) first layer neurons *XY*α for Fig. 1 input. At some of the points several line segments with different directions occur due to the low threshold (θ_1). On the right last layer neurons plotting the appropriate arcs in κθ plane.

The input to vertex *z* location finding is the mean value of points within the receptive field of the first NN input neurons.

The first layer neurons compute the mean value of the *z* coordinate of the first layer neurons in their receptive field, averaging over all the neurons within the section. This is similarly propagated to the second layer neurons.

The third layer neurons evaluate the *z* value of the arcs origin by simple extrapolation. The final *z* estimate of the vertex is calculated by averaging the output of all active third layer neurons.

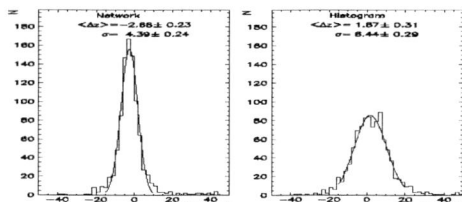

FIGURE 3. A distribution of the error in the estimation of the z values, the NN estimate on the left and the conventional histogram method on the right.

Fig. 3 demonstrates how the NN z estimate is more precise than the conventional histogram method [6].

RESULTS

The results of the network were tested with 992 simulated HERA events, comparing the results to the actual location of the vertex z as well as to the histogram method. There is a large overlap between events with no estimation. However, the resolution of the NN method ($\sigma \sim 4.5$ cm) is by almost a factor of two better than by the conventional method ($\sigma \sim 8.4$ cm). ZEUS full off-line reconstruction leads to a resolution of 0.1 cm. Taking the number of output neurons to represent the number of track candidates passing through the vertex, we compared it to the histogram method and to the number of tracks in the full off-line reconstruction. Both methods were comparable and both tend to slightly underestimate the number of tracks from the vertex.

More than 300 NN differing in the number of neurons and their configurations, were examined. Fig 4. shows that we could decrease the number of input neurons as well as the resolution in the second layer without significantly affecting the performance of the network. Lateral connection between 1st layer neurons enabled us to reduce the threshold and by that to reduce the network size.

DISCUSSION

We have described a feed-forward NN that performs a task of pattern identification by threshold and data subsets selection. The network uses a fixed architecture; this allows a hardware implementation, which may be crucial for fast triggering purposes. The pattern type we were after motivated the fixed architecture and the present synaptic connections. The obtained results are better than conventional methods, which enable the opportunity of new NN implementation in triggering devices of HEP experiments.

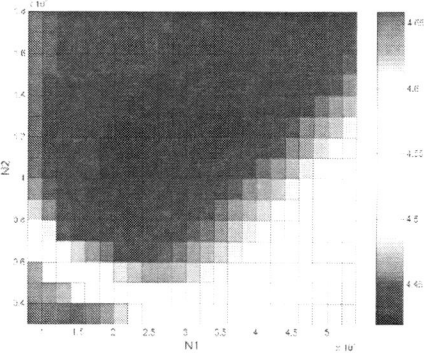

FIGURE 4. The network output width as a function of the number of neurons in the input layer (N1 axis) and second layer (N2 axis). (Blue- smaller σ, Red – Larger σ)

ACKNOWLEDGMENTS

The study has been partly pursued in the framework of the ZEUS Collaboration and we would like to acknowledge the help of our colleagues Halina Abramowicz from ZEUS in providing the necessary data sample and for stimulating discussions. We wish to thank David Horn from Tel-Aviv University for the fruitful collaboration. This research was partly supported by the Israeli National Science Foundation.

REFERENCES

1. ZESU Collab., The ZEUS Detector, Status Report 1993, DESY 1993; M. Derrick et al.,. *Phys. Lett.* **B293**, 465 (1992).

2. B. Foster et al., *Nucl. Inst. Meth.* **A338**, 254 (1994).

3. H. Abramowicz, D. Horn, U. Naftaly and C. Sahar-Pikielny, *Nucl. Inst. Meth.* **A378**, 305 (1996); *Advanced in Neural Information Processing Systems 9*, edited by M.C. Mozer, M.J Jordan and T. Petsche, MIT Press, 1997, pp. 925.

4. G. Dror et al., *AIHENP99 proceedings*, edited by G. Athanasiu, D. Perret-Gallix, Elsevier North-Holland Editors, (1999).

5. D.H. Hubel ,T.N Wiesel, *J. Physiol.*, **195**, 215 (1968).

6. Quadt, Master of Science Thesis, University of Oxford (1997).

Neural Networks for Higgs Physics

Silvia Tentindo-Repond[b], Pushpalatha C. Bhat[a], Harrison B. Prosper[b]

[a]*Fermi National Accelerator Laboratory*, [b]*Department of Physics, Florida State University, Tallahassee, FL 32306*

Abstract. The main application of neural networks (NN) in Higgs physics so far has been to optimize the signal over background ratio. The positive result obtained imply that the use of NN will lead to a big reduction in the integrated luminosity required for the discovery of the Higgs in RunII. Neural Networks have also been recently used in Higgs physics to set up tagging algorithms to identify the heavy flavor content of jets. Whereas in the previous studies the NN b-tagging methods used are channel-independent, a channel-dependent method has been used in the present work. The signal $p\bar{p} \to WH \to \ell\nu b\bar{b}$ has been studied against the dominant background $p\bar{p} \to Wb\bar{b}$, in an attempt to improve the signal over background ratio by trying to push the invariant mass of the background events further away from the signal. This result would get the equivalent effect of an improved mass resolution.

INTRODUCTION

Higgs Physics

The Higgs boson is the last missing particle in the Standard Model (SM) that would complete the description of almost all phenomena in elementary particle physics in a simple way, and would explain the fundamental mechanism that governs the electro-weak symmetry breaking, the process that provides mass to all fermions.

The SM parameters, the W boson and the top quark masses, are related to the Higgs boson mass. Experimental high precision electroweak measurements constrain the Higgs mass to $M_{H_{SM}} = 107^{+67}_{-45}$ GeV/c^2 and give a 95% confidence level upper limit of 225 GeV/c^2 (1). Some theories like the Minimal Supersymmetric SM (MSSM) predict an upper limit of 130 GeV/c^2.

The low mass of the Higgs boson motivates its search at the Tevatron during Run II. Typical predicted cross sections, for example, for a Higgs mass of 100 GeV/c^2 are of the order of 1 pb; cross section times branching ratios for the processes studied here are 85 fb (WH) and 3500 fb for the background (W bb). Typical luminosities required at the Tevatron for a signal exclusion at 95% confidence level, or for 3σ or 5σ discovery combining both CDF and D0 data, and all the decay channels, are of the order of respectively : 10, 20 and 40 fb^{-1}, for a 130 GeV/c^2 Higgs boson.

Dominant decay channels, and detection challenges (b-tagging)

The dominant decay channel for a low mass Standard Model Higgs, up to 130 GeV/c^2, is the $b\bar{b}$ channel, with a branching ratio of about 85%. The $c\bar{c}$ channel and $\tau\tau$ channels are one order of magnitude lower. At higher energies the WW and $t\bar{t}$ decay channels prevail, imposing different detection constraints. For a low mass SM Higgs it is essential to have a good b tagging method.

b-tagging algorithms that enhance the efficiency for detection of the signal $WH \to \ell\nu b\bar{b}$ do nothing to reduce the dominant background $Wb\bar{b}$. A method that discriminates b-tagged jets resulting from Higgs decay, from b-tagged jets not coming from Higgs, is needed. Neural networks can be trained to do this task.

MULTIVARIATE VERSUS TRADITIONAL METHODS IN HIGGS PHYSICS

One of the most important goals of multivariate methods is to improve the ratio of signal over background. This task becomes particularly critical when the signal is very small, and when its properties are very similar to the properties of the background.

In Higgs physics, the first application of multivariate methods has been to maximize the chance to discover the Higgs boson, and to reduce the required luminosity for a 3σ or 5σ discovery (2, 3).

Studies show that in RunII, for a given integrated luminosity, the NN analysis, as compared to the traditional analysis, provides a much higher discovery reach in mass. Typically, for a Higgs mass of 130 GeV/c^2, a factor 2 better in integrated luminosity is reachable.

Another important application of NN in Higgs physics has been to improve the b-tagging efficiency. Along with tagging leptons it will be possible in RunII to do b-tagging by using b lifetime variables. By traditional methods, the efficiency for b-tagging through secondary vertex and impact parameter reconstruction is estimated to be about 45% per single b particle; if two b jets are tagged, this efficiency becomes critically low, and improved methods of tagging are needed. Besides, the method of traditional b-tagging will not be able to discriminate the b component from the c component (about 10% of heavy flavor jets are c-jets for Higgs up to masses of 130 GeV/c^2). A Neural network analysis has been used (9) to combine lifetime variables (tracks consistent with secondary vertex, impact parameter), and kinematic variables (mass, fragmentation). This tagging method can potentially outperform existing tagging algorithms.

Another potential field of application of NN in Higgs physics could be to improve the Higgs invariant mass resolution (P.C.Bhat, APS '99). The di-jet invariant mass variable has proven to be a critical one to discriminate signal from background in Higgs physics. The assumed mass resolution in the recent Run II Susy/Higgs Workshop (5) is 10%.

The mass resolution gets contributions from different processes, at the parton level, particle level and detector level. Improving the mass resolution will be a very challenging task, and the appropriate methods have yet to be worked out to reach the expected resolution. A method that could reduce the background, and at the same time increase the signal-background separation in the mass spectrum, for example, by pushing the background away from the signal, would reach the same goal as to increasing the mass resolution. This is the new application that is attempted here using NN.

CHANNEL-DEPENDENT B-TAGGING

We remark that the methods developed so far (2, 9) for b-tagging with NN are channel-independent, in the sense they are applicable independently of the decay channel studied. In the following we want to investigate the possibility to develop an NN method optimized to a given decay channel. We will call this " channel-dependent b-tagging ", the aim of which is to reduce the background.

Simulation

Here we consider the signal $p\bar{p} \to WH \to \ell\nu b\bar{b}$, and its main background $p\bar{p} \to Wb\bar{b}$.

The WH events were simulated using the PYTHIA program (6) for Higgs boson masses of M_H = 100, 120 and 130 GeV/c^2. The $Wb\bar{b}$ sample was generated using CompHEP (7), with fragmentation done using PYTHIA. To model the expected response of the CDF/DØ Run II detectors we used the SHW program (8).

An initial loose set of cuts on the variables is imposed, based on knowledge of the detector properties, trigger thresholds and kinematic properties of the events of interest:

- the transverse momentum of the isolated lepton $P_T^\ell > 15$ GeV/c
- the pseudo-rapidity of the lepton $|\eta_\ell| < 2$
- the missing transverse energy in the event $\not{E}_T > 20$ GeV
- two or more jets in the event with $E_T^{jet} > 10$ GeV and $|\eta_{jet}| < 2$.

NN Analysis

The feed-forward NN package used is MLPfit (version 1.40) (4). The NN configuration chosen here is 3-6-1, 3 input nodes, 6 hidden nodes and one output. One sample (the WH signal data) is used to train the NN, for b-tagged jets (NN output set to 1) and for non b-tagged jets (NN output set to 0).

The set of variables that has been chosen as input to the NN is the following:

- E_T^b – transverse energy of the b-tagged jets.
- N_{trk} – charged track multiplicity in the jets.
- *Width* – width of jets.

In Fig. 1 are plotted the jet variables used for training the NN, for signal and for background.

The NN function is then used to discriminate b-tagged jets coming from the signal $p\bar{p} \to WH \to \ell\nu b\bar{b}$ from b-tagged jets coming from the background. The hope is that the b-jets in the signal are identified with higher efficiency than those in the background.

In Fig. 2 is plotted the NN output for the most energetic jet : 916 out of 1000 of the most energetic jets are b-jets. (For the second most energetic jet, 787/1000 are b-jets).

These tagged b jets are channel dependent b-tags. The NN output function if applied to the background sample,

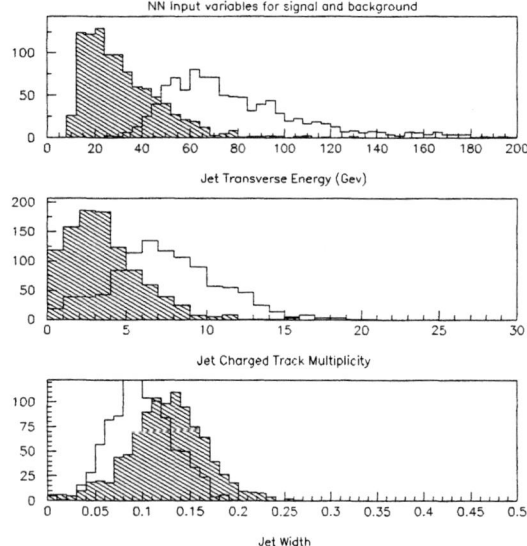

FIGURE 1. Jet variables used as input for training NN in channel-dependent b-tagging studies for signal (unshaded) and background (shaded)

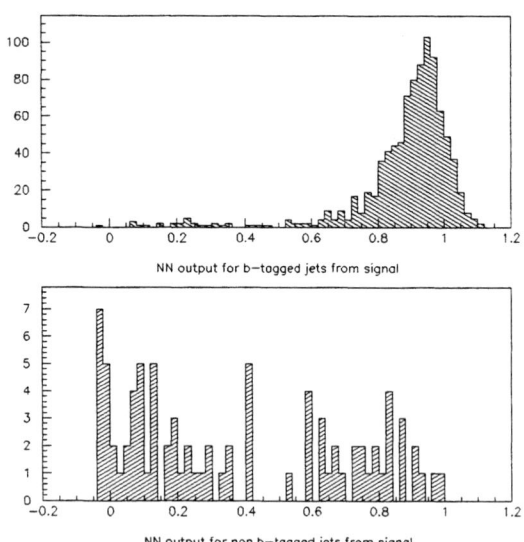

FIGURE 2. NN output distribution for most energetic b-tagged jet and most energetic non b-tagged jet

peaks at 0. The invariant mass plot can be used to see the discrimination power of the NN output. By imposing a N-N cut at 0.4 to jet1 and jet2, 65% (M_{Higgs} = 100 GeV/c^2), 74% (120GeV/c^2) and 58% (130GeV/c^2) signals survive, whereas the background is reduced to 24%.

These studies have just begun, and the method has not been optimized yet. The result of this NN method brings no mass resolution improvement since that is not its purpose. It tries to reduce the background by pushing the remaining background events lower in the invariant mass spectrum.

SUMMARY

Neural networks methods in Higgs physics have been used to maximize discovery potential. They have been used as a b-tagging algorithm and specifically as a discriminator of charm and bottom content in heavy flavor jets. Channel - dependent b-tagging is being studied and has been applied to the channel $p\bar{p} \to WH \to \ell\nu b\bar{b}$ against the dominant background $p\bar{p} \to Wb\bar{b}$, and better separation between signal and background has been obtained. We plan to continue the systematic study of these NN methods, as well as those that may improve the di-jet invariant mass resolution.

REFERENCES

1. LEP Electroweak Group, http://www.cern.ch/LEPEWWG/plots/summer99.
2. Bhat, P. C., Gilmartin, R., Prosper, H. B., *Phys.Rev.D, 62, 074022(2000)*.
3. Bhat, P. C., these proceedings.
4. MLPfit v1.40, Schwindling, J., Mansoulie, B. http://home.cern.ch/ schwind/MLPfit.htlm
5. Run II Higgs Working Group of the Run II SUSY/Higgs workshop, http://fnth37.fnal.gov/higgs.html; T. Han, A. S. Turcot and R. Zhang, *Phys. Rev.* D **59**, 093001 (1999).
6. PYTHIA, T. Sjöstrand, *Comp. Phys. Comm.* **82**, 74 (1994).
7. CompHEP, A. S. Belyaev, A. V. Gladyshev and A. V. Semenov, e-print hep-ph/9712303; E.E. Boos *et al.*, e-print hep-ph/9503280.
8. SHW v2.0, Conway, J. - available at http://www.physics.rutgers.edu/~jconway/soft/shw/shw.html (unpublished).
9. Wolinski, D., Amidei, D., Demina, R.- CDF/ANAL/EXOTIC/CDFR/5051 (Sept 1999).

Top Quark Mass Measurements Using Neural Networks

Suman B. Beri[a], Pushpalatha C. Bhat[b], Rajwant Kaur[a], Harrison B. Prosper[c]

[a]*Department of Physics, Panjab University, Chandigarh, India*
[b]*Fermi National Accelerator Laboratory, Batavia, Illinois, USA* *
[c]*Department of Physics, Florida State University, Tallahassee, Florida, USA*

Abstract. A major goal of high energy physicists over the next few years is to reduce the uncertainty in our knowledge of the top quark mass. Neural network based methods may play a useful role in this regard, as is borne out by the preliminary results reported here.

INTRODUCTION

Over the next few years, high energy physicists hope to effect a substantial reduction in the uncertainty in our knowledge of the top quark mass (1). The success of this endeavor is contingent on the Tevatron delivering substantially more data in Run II than was collected during Run I, and, on our finding ways to reduce significantly systematic errors. As evidenced by the presentations at this workshop, neural networks (2) are extraordinarily powerful data analysis tools. We are, therefore, motivated to explore their use in the measurement of mass, with the goal of discovering methods with intrinsically smaller systematic errors. We report some preliminary results.

To provide some context we define a few pertinent concepts. A *pattern* is characterized by a d-dimensional vector $\mathbf{x}$ of real-valued numbers called *features*. Usually, one has a sample of N such *feature vectors* $\mathbf{x}_1, \ldots, \mathbf{x}_N$, each unambiguously classified into one of K classes. The feature vectors of class k inhabit a d-dimensional *feature space*, and are sampled from a probability density function $f(\mathbf{x}|k)$. An important goal of a high energy physics analysis is to assign an event of unknown identity (characterized by a vector $\mathbf{x}$) to its proper class. This can be done using the *Bayes decision rule*: Compute the (Bayesian posterior) probability $P(k|\mathbf{x})$ of $\mathbf{x}$ belonging to class k and assign $\mathbf{x}$ to the class with the highest probability. This rule minimizes the probability of making mistakes.

About a decade ago, Ruck (3), Wan (4), Blum and Li (5) showed that the outputs of a feed-forward neural network directly approximate Bayesian posterior probabilities. It is this insight that forms the basis of the methods we are studying. Although the Bayesian connection was proven in the context of neural networks the result, in fact, is much more general and pertains to *any* function able to model probability densities. Typically, such a function is obtained by minimizing an error function that has a number of adjustable parameters. It is the nature of the error function that is minimized that yields the Bayesian interpretation rather than the specific functional form that is used to model the densities. Neural networks are of interest because they are universal approximators (6) and can, therefore, in principle, model any smooth probability density.

Let $F(m, \mathbf{x}, \omega)$ be a function that is able to model, to arbitrarily high accuracy, a given probability density. In the present context F is a feed-forward neural network. The quantity m is an integer in the range 1 to M, corresponding to a network with M outputs. The symbol ω denotes the set of parameters that are to be adjusted so as to minimize the error function,

$$E_N = \frac{1}{N} \sum_{n=1}^{N} \sum_{m=1}^{M} [D(m, \mathbf{x}_n) - F(m, \mathbf{x}_n, \omega)]^2 w(\mathbf{x}_n), \quad (1)$$

where D is a known function and $w(\mathbf{x})$ is an (optional) pattern dependent weight function. We choose the function D to be

$$D(m, \mathbf{x}_n) = d_{m,k} \text{ if pattern } \mathbf{x}_n \text{ belongs to class } k, \quad (2)$$

where $d_{m,k}$ is a matrix of known constants. In the limit $N \to \infty$, one obtains

$$F(m, \mathbf{x}, \omega) = \sum_{k=1}^{K} d_{m,k} P(k|\mathbf{x}), \quad (3)$$

where the probabilities $P(k|\mathbf{x})$ are given by Bayes' theorem

$$P(k|\mathbf{x}) = \frac{w_k(\mathbf{x}) f(\mathbf{x}|k) P(k)}{\sum_k w_k(\mathbf{x}) f(\mathbf{x}|k) P(k)}, \quad (4)$$

---

* Operated by Universities Research Association under contract to the U.S. Department of Energy.

and where $P(k)$ is the prior probability of the kth class. We emphasize that this result holds in the limit of an infinite sample of patterns and arbitrarily accurate functional approximation. In practice, of course, neither condition holds. However, experience suggests that practical neural networks can often provide good approximations to Eq. (3), provided that the sample of patterns is large enough.

A PRELIMINARY STUDY

We consider the reaction $p\bar{p} \to t\bar{t} \to e\mu + X$, in which one top quark yields a b quark and an electron (e) and the other a b quark and a muon (μ). Although the branching ratio for the $e\mu$ channel is small, $\sim 2.4\%$ to be compared with $\sim 15\%$ for each single lepton channel, this channel is interesting because of its relatively smaller backgrounds, principally from the reactions $p\bar{p} \to Z \to \tau\tau \to e\mu$ and $p\bar{p} \to WW \to e\mu$. For this study, the $t\bar{t}$, Z and WW events were simulated using the PYTHIA program (10) and the detector was simulated using SHW (11).

The preferred way to measure the mass of a particle created in a high energy event is to perform a complete kinematic reconstruction of the event. However, for the $e\mu$ channel, as is true of the other di-lepton channels, there is insufficient information available to effect a complete kinematic reconstruction, although it is possible to perform a partial reconstruction (7). An alternative strategy is to use mass-dependent quantities that are not necessarily derived from kinematic fits (8, 9). In this study we use the variable $x(l,b) \equiv \sqrt{2l \cdot b + m_W^2}$, where m_W is the W boson mass, and l and b are lepton and b-jet four vectors. This variable, as pointed out in Ref. (8), has the virtue of being insensitive to jet systematic uncertainties, and has a distribution whose mean is an almost linear function of the top quark mass. Moreover, if the b-jet and the lepton l come from the same top quark the distribution of x ends at the top quark mass, as shown in Fig. 1a. But to avail oneself of this useful feature requires an efficient algorithm for correctly pairing the leptons and the b-jets.

In the present study we did not attempt to do this. Instead, we considered the four pairings of the two leptons and the two b-jets. As shown in Figs. 1b-e, the effect of the mispairing is to broaden the x distributions beyond the top quark mass. However, these variables are still mass-dependent and can therefore be used to measure the top mass. We associate each class label k with a top quark mass m_k; $P(k|\mathbf{x})$ may then be interpreted as the probability that the top quark mass is m_k given the measured vector $\mathbf{x} = [x(e, j_1), x(\mu, j_2), x(\mu, j_1), x(e, j_2)]$. If, furthermore, we take $M = 1$ and set $d_{1,k} = m_k$, Eq. (3) becomes the mean of the posterior probability of the top quark

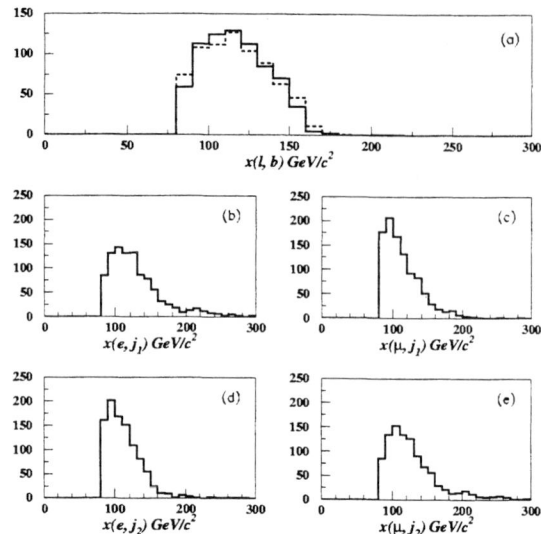

FIGURE 1. (a) The distribution of x, after event reconstruction (solid line), for correctly paired lepton and b-jet, compared with the distribution at the parton level (dashed line). We note how closely the two distributions agree. (b) to (e) The distribution of x, after reconstruction, for all pairings of leptons and jets. The distributions are for a top quark mass of 170 GeV/c^2.

mass, which, from a Bayesian viewpoint, is a natural estimate of that quantity, and, moreover, is a quantity that can be directly estimated using a neural network, as noted above.

To test this idea, we estimated the posterior mean using a (4,n,1) neural network (with n, the number of hidden nodes, typically equal to 5). During training we set the target value equal to the top quark mass m_k associated with the feature vector $\mathbf{x}$ (where $m_1 = 100, m_2 = 110, \ldots, m_{15} = 240$ GeV/c^2). Three hundred feature vectors were used for each top quark mass and the training was done using the program MLPfit (12) and its default minimization algorithm. To avoid over-training the network was validated on an independent sample.

Figure 2 shows the mean of the network output distribution (that is, the mean of the distribution of posterior mean) as a function of the top quark mass over the range 100 to 240 GeV/c^2. The points define a curve that is linear within the range 140 to 190 GeV/c^2 and non-linear outside. Its sigmoidal shape is easy to understand. At the lower end, the truncation at 100 GeV/c^2 and the finite width of the posterior distribution forces the posterior mean to be on average greater than 100 GeV/c^2. Likewise, at the upper end, the posterior mean is forced to be on average less than 240 GeV/c^2.

For a single measurement the method is clearly biased.

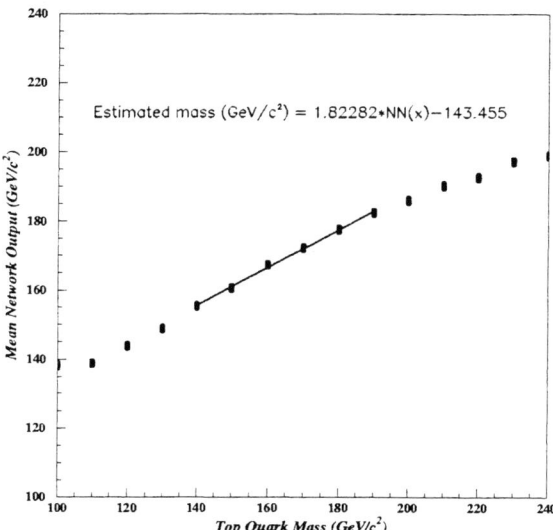

FIGURE 2. The mean of the distribution of network output as a function of the top quark mass. The solid line is the best linear fit to the specified points while the expression shown is the linear correction function derived from the fit.

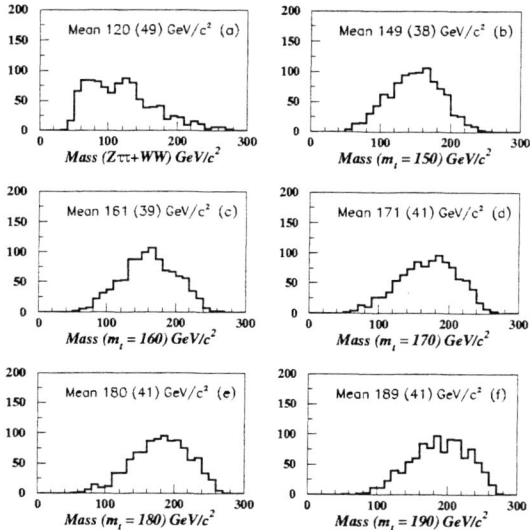

FIGURE 3. Distributions of estimated top quark mass. (a) $Z \to \tau\tau$ and WW backgrounds. (b) to (f) Signal distributions for different top quark masses. The numbers in parentheses are the estimated uncertainties for a *single* unbiased measurement, that is, for a mass measurement based on a single $e\mu$ event.

However, we argue that the relevant issue is not the bias for a sample of size one but rather the magnitude of the root mean square difference between the true mass and the measured mass for a sample whose size is likely to be much greater than one, and perhaps as large as 100. However, in this preliminary study we assume that a lack of bias for a single measurement is desirable. The simplest way to achieve this is to apply a linear correction, shown in Fig. 2, to the network output (that is, the posterior mean). When such a correction is applied one obtains the distributions shown in Fig. 3, which demonstrates the promise of the approach we have outlined.

This work was supported in part by the U.S. Department of Energy and the Department of Science and Technology of India.

REFERENCES

1. CDF and DØ Collaborations, Demortier, L. *et al.*, "The Top Averaging Group," *Fermilab-TM-2084* (1999); Bhat, P.C., Prosper, H.B., and Snyder, S.S., *Int. J. Mod. Phys.* A **13**, 5113 (1998).
2. Bishop, C.M., *Neural Networks for Pattern Recognition*, Clarendon Press, Oxford, 1998; DØ collaboration, Bhat, P.C, in *Proceedings of the 10th Topical Workshop on Proton-Antiproton Collider Physics*, AIP, Woodbury, New York, 1995, p. 308.
3. Ruck, D.W. *et al.*, *IEEE Trans. Neural Networks* **1** (4), 296 (1990).
4. Wan, E.A., *IEEE Trans. Neural Networks* **1** (4), 303 (1990).
5. Blum, E.K., and Li, L.K., *Neural Networks* **4**, 511 (1991).
6. Hornik, K., Stinchcombe, M., and White, H., *Neural Networks* **2**, 359 (1989); Hornik, K., *Neural Networks* **6**, 1069 (1993);
7. DØ Collaboration, Abachi, S. *et al.*, *Phys. Rev. Lett.* **80**, 2063 (1998); Kondo, K., *J. Phys. Soc. Jpn.* **57**, 4126 (1988); *ibid.*, *J. Phys. Soc. Jpn.* **60**, 836 (1991); *ibid.*, *J. Phys. Soc. Jpn.* **62**, 1177 (1993); Dalitz, R.H., and Goldstein, G.R., *Phys. Rev.* D **45**, 1531 (1992); *ibid.*, *Phys. Lett.* B **287**, 225 (1992);
8. Prosper, H.B., *Phys. Lett.* B **335**, 515 (1994).
9. CDF Collaboration, Abe, F. *et al.*, *Phys. Rev. Lett.* **80**, 2779 (1998).
10. PYTHIA v6.143, Sjöstrand, T., *Comp. Phys. Comm.* **82**, 74 (1994). Available at http://www.thep.lu.se/tf2/staff/torbjorn/Pythia.html
11. SHW v2.3, Conway, J., available at http://www.physics.rutgers.edu/~jconway/soft/shw/shw.html
12. MLPfit v1.40, Schwindling, J., and Mansoulié, B., available at http://schwind.home.cern.ch/schwind/MLPfit.html

Clopper-Pearson Bounds from HEP Data Cuts

B. A. Berg *

Department of Physics, The Florida State University, Tallahassee, FL 32306, USA.
E-mail: berg@hep.fsu.edu

Abstract. For the measurement of N_s signals in N events rigorous confidence bounds on the true signal probability p_{exact} were established in a classical paper by Clopper and Pearson [Biometrica 26, 404 (1934)]. Here, their bounds are generalized to the HEP situation where cuts on the data tag signals with probability P_s and background data with likelihood $P_b < P_s$. In particular, the method may be of interest in connection with the statistical analysis part of the ongoing Higgs search at the LEP experiments. The relevant Fortran code is available on the web.

INTRODUCTION

The general theory of confidence bounds (or fiducial intervals) was developed by Fisher (1), Neyman and Pearson (2). We consider a particular problem which is of interests when cuts are used to analyze high energy physics data. Typically, a neural network or some other method of performing the cuts results in probabilities (efficiencies) to tag signals more likely than background events. For instance by means of Monte Carlo (MC) simulations, these probabilities can normally be calculated. Let P_s be the probability to tag a signal and P_b be the likelihood to tag a background event, $0 < P_b < P_s < 1$. Out of a total number of N data one gets in this way

$$N^Y \quad \text{tagged data.} \quad (1)$$

It is easy to find from this the mean expectation for the signal probability. Assume that there are N_s signals and N_b background events in the data. Then we have

$$N = N_s + N_b \quad \text{and} \quad N^Y = P_s N_s + P_b N_b$$

and these two equations solve for

$$p_{\text{mean}} = \frac{N_s}{N} = \frac{N^Y - P_b N}{N(P_s - P_b)}. \quad (2)$$

The question is, what are the implied confidence limits on the signal probability?

The special case $P_s = 1$ and $P_b = 0$ (sure signal detection) has been treated by Clopper and Pearson (3) in 1934. After briefly reviewing their approach in the next section, we derive and illustrate the general case in the subsequent section. This is, in part, based on Ref.(4). In particular, the method is valid when the number of tagged data is small and returns the probability P_0 for the case that there is no signal, i.e. that the exact signal probability is $p_{\text{exact}} = 0$. This is of interest for the statistical analysis of the ongoing Higgs search at LEP (5). Discovery of the Higgs particle on the 5σ level would mean $P_0 \leq 0.287\, 10^{-6}$. Conclusions follow in the final section.

The Fortran code which produces the illustrations of this paper is available on the web. Start at the author's homepage (6) www.hep.fsu.edu/~berg and follow the research and from there the Clopper-Pearson hyperlink. Its use is explained in Ref.(7).

THE CLOPPER–PEARSON CONFIDENCE LIMITS

Let p be the likelihood that a data point is a signal. For N measurements the probability to observe k signals is given by the binomial coefficient

$$b_N(k, p) = \binom{N}{k} p^k q^{N-k} = \frac{N!}{k!(N-k)!} p^k q^{N-k} \quad (3)$$

with $q = p - 1$. The probability to observe $k \geq N_s$ signals is given by

$$P_{k \geq N_s}(p) = \sum_{k=N_s}^{N} b_N(k, p) \quad (4)$$

and the probability to observe $k \leq N_s$ signals by

$$P_{k \leq N_s}(p) = \sum_{k=0}^{N} b_N(k, p). \quad (5)$$

For $N = 26$ and $N_s = 10$ the functional forms of $P_{k \geq N_s}(p)$ and $P_{k \leq N_s}(p)$ are depicted in figure 1.

---

* In part supported by the DOE Grant DE-FG02-97ER41022.

Assume that N_s signals are found in N measurements and that a probability $Q^c < 0.5$ (typical values are $Q^c = 0.16$ or $Q^c = 0.025$) is given. We can solve equation (4) for $P_{k \geq N_s}(p_-) = Q^c$ and p_- is a lower bound on the true signal probability p_{exact}, such that the likelihood to find $k \geq N_s$ signals in N measurement is smaller than Q^c for every $p_{\text{exact}} < p_-$. Correspondingly, we can solve equation (5) for $P_{k \leq N_s}(p_+) = Q^c$ and p_+ is an upper bound on the true signal probability p_{exact}, such that the likelihood to find $k \leq N_s$ signals in N measurement is smaller than Q^c for every $p_{\text{exact}} > p_+$. Together, this combines into the Clopper–Pearson bounds: The probability to find the true signal probability in the range

$$p_- \leq p_{\text{exact}} \leq p_+ \quad \text{is larger or equal to} \quad P^c = 1 - 2Q^c. \tag{6}$$

In more details the meaning of the inequality is discussed in (4). For $P^c = 0.68$ ($Q^c = 0.16$) the $p^\pm$ values are indicated in figure 1. Approximately, this range corresponds to the confidence of a 1σ error bar. Similarly the confidence range corresponding to a 2σ error bar, etc., can be found.

CONFIDENCE LIMITS FROM HEP DATA CUTS

We are interested in the situation where signal and background data can no longer be distinguished unambiguously. Instead, a neural network or other device yields statistical information by tagging signals with efficiency P_s and background data with probability P_b, as discussed in the introduction.

Applying the cuts to all N data results in N^Y tagged data ($0 \leq N^Y \leq N$), composed of $N^Y = N_s^Y + N_b^Y$, where N_s^Y is the number of tagged signals and N_b^Y is the number of tagged background data. Of course, the values for N_s^Y and N_b^Y are not known. Our task is to determine confidence limits for the signal probability p from the sole knowledge of N^Y. We proceed by writing down the probability density of N^Y for given p and, subsequently, generalizing the Clopper-Pearson method.

First, assume fixed N_s. The probability densities of N_s^Y and N_b^Y are binomial and thus the probability density for N^Y is given by the convolution

$$P^Y(N^Y | N_s) = \sum_{N_s^Y + N_b^Y = N^Y} b_{N_s}(N_s^Y, P_s) b_{N_b}(N_b^Y, P_b). \tag{7}$$

Proof: For a signal event the probability to be tagged is P_s, so $b_{N_s}(N_s^Y, P_s)$ is the probability to tag N_s^Y out of the N_s signals. Similarly, the probability for a background event to become tagged is P_b and $b_{N_b}(N_b^Y, P_b)$ is the probability to tag N_b^Y of the $N_b = N - N_s$ background events. As these two probabilities are independent, the likelihood that precisely N_s^Y of the signals and N_b^Y of the background events are tagged becomes the product $b_{N_s}(N_s^Y, P_s) b_{N_b}(N_b^Y, P_b)$. Summing over all possibilities which add up to $N_s^Y + N_b^Y = N^Y$ gives the result.

Summing over N_s in (7) removes the constraint of fixed N_s and, with N, p fixed, the probability to tag k events becomes

$$b_N^Y(k, p) = \sum_{N_s=0}^{N} b_N(N_s, p) P^Y(k | N_s). \tag{8}$$

Fourier transformation of the convolution (7) allows for an efficient numerical calculation of the $P(k|N_s)$ coefficients. In analogy with equations (4) and (5) we find the probabilities to tag $k \geq N^Y$ and $k \leq N^Y$ events to be

$$P_{k \geq N^Y}^Y(p) = \sum_{k=N^Y}^{N} b_N^Y(k, p) \quad \text{and} \quad P_{k \leq N^Y}^Y(p) = \sum_{k=0}^{N^Y} b_N^Y(k, p). \tag{9}$$

For $N = 35$, $N^Y = 12$, $P_s = 0.8$ and $P_b = 0.05$ the functions $P_{k \geq N^Y}^Y(p)$ and $P_{k \leq N^Y}^Y(p)$ are depicted in figure 2. The 68% confidence range (6) is also indicated in figure 2, where the bound values $p_\pm$ are now defined as solutions of the equations

$$Q^c = P_{k \geq N^Y}^Y(p_-) \quad \text{and} \quad Q^c = P_{k \leq N^Y}^Y(p_+). \tag{10}$$

The range $[p_-, p_+]$, obtained with $Q^c = 0.16$, guarantees the standard one error bar confidence probability of 68% for every true signal probability p_{exact}. For almost all values the actual confidence will be better. However, the bounds cannot be improved without violating the requested confidence probability for the case that p_{exact} happens to agree with either p_- or p_+. In the same way, bounds calculated with $Q^c = 0.023$ ensure the standard two error bar confidence level of 95.4% or better, and so on.

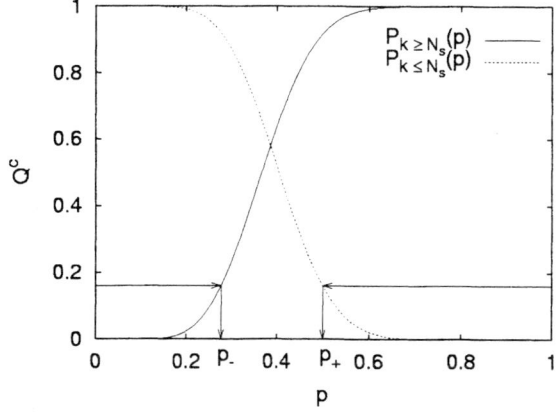

FIGURE 1. The probability functions $P_{k \geq N_s}(p)$ and $P_{k \geq N_s}(p)$ for $N = 26$ and $N_s = 10$ together with 68% confidence bounds.

Data Sets with Few Signals

As outlined, data sets with few signals are of of particular interest in high energy physics. Let us replace $N^Y = 12$ of figure 2 by $N^Y = 3$. The resulting graph is depicted in figure 3. From the $P^Y_{k>N^Y}(p)$ curve we read off the finite probability $P_0 = 0.254$ for the likelihood that the true signal probability $p_{\text{exact}} = 0$ generates $k \geq N_y$ tags. Due to this probability the lower 68% confidence bound p_- disappears, whereas the upper p_+ bound does still exist. Let us note that for the data of figure 2 we have $P_0 = 0.69\,10^{-7}$, i.e. there $p_{\text{exact}} = 0$ is ruled out on the 5σ level.

CONCLUSIONS

We have calculated confidence limits for an unknown true signal likelihood p_{exact}. The input used are the efficiency P_s for tagging signals, the probability P_b for tagging background events, the number N^Y of tagged data and the total number of data N. In particular, the method allows to deal with the situation where only few signals occur and yields then a finite probability for the likelihood that $k \geq N^Y$ tags are observed if the true signal probability is $p_{\text{exact}} = 0$. In real life the probabilities P_s and P_b are most likely estimators by themselves, i.e. quantities with error bars. This causes no major problem, one just has to apply our confidence calculations to an appropriate sample and to average over the results.

REFERENCES

1. R.A. Fisher, Proc. Camb. Phil. Soc. 26, (1930) 528; Proc. Roy. Soc. A139, (1933) 343.
2. J. Neyman and E.S. Pearson, Phil. Trans. Roy. Soc. A231, (1933) 289.
3. C.J. Clopper and E.S. Pearson, Biometrika 26, (1934) 404. For a textbook treatment see S. Brandt, *Statistical and Computational Methods in Data Analysis* (North-Holland, 1983).
4. B.A. Berg and J. Riedler, Comp. Phys. Commun. 107, (1997) 39.
5. http://alephwww.cern.ch/ALPUB/seminar/wds/
6. The address of the authors homepage and its tree structure are expected to be stable, whereas the absolute address where the programs are located is likely to change.
7. B.A. Berg, hep-ex/0010061.

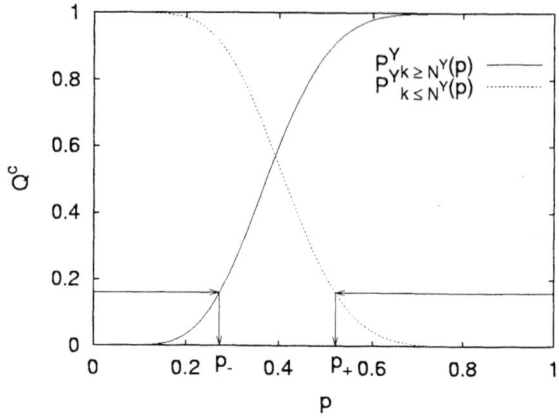

FIGURE 2. The probability functions $P^Y_{k \geq N_Y}(p)$ to find $k \geq N^Y$ tags in N events and $P^Y_{k \leq N^Y}(p)$ to find $k \leq N^Y$ tags in N events (9) are depicted for $N = 35$, $N^Y = 12$, $P_s = 0.8$ and $P_b = 0.05$. Symmetric 68% confidence bounds, found for $p_- = 0.274$ and $p_+ = 0.521$, are also indicated.

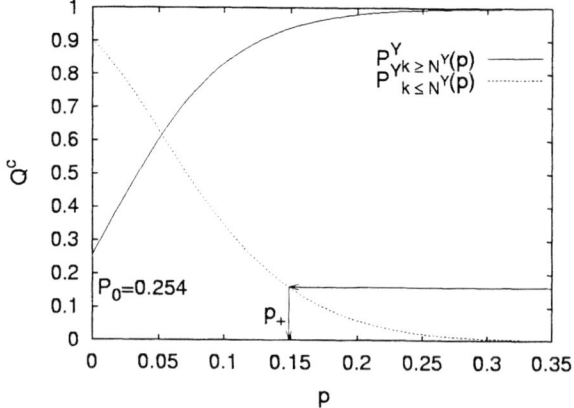

FIGURE 3. The same functions as in figure 2 are depicted, but for $N^Y = 3$ instead of $N^Y = 12$ tags. The lower 68% confidence bound does not exist anymore, instead the likelihood for $p_{\text{exact}} = 0$ has become $P_0 = 0.254$. The upper 68% confidence bound is found at $p_+ = 0.149$.

Experimenting with rule induction algorithms in HEP data analysis

N. Stepanov

Helsinki Institute of Physics, Helsinki, Finland *

Abstract. The simple prototype rule induction system for the HEP analysis applications is presented. The feasibility of the approach is demonstrated on the benchmark classification task taken from the LHC physics.

INTRODUCTION AND MOTIVATION

It is a common practise now to use the machine learning methods in high energy physics (HEP) multi-variable data analysis. In particular, neural network systems (NN) are proven to be useful in numerous classification, prediction and regression applications. NN approach is certainly very attractive but has its doubters. Most of them are some how connected with the fact that NN is a black box system. The black box develops its own concept representation that to be used for concept recognition purposes, however, this internal description cannot be easily interpreted by the user and provides neither insight nor explanation of the recognition process. There are knowledge-oriented methods capable of developing descriptions understandable to the human being directly manipulating symbolic structures, e.g. rule induction algorithms (RIA). The purpose of this paper is to present a simple framework for a rule induction chain and to demonstrate the usefulness of the approach on the benchmark classification task taken from the LHC physics.

The rules we are going to use incorporate an "attributional" logic which is roughly equivalent to the zero-order logic (1): **rule := if<condition> then predict <class>**, where **<condition>** is a conjunct of attribute tests. In this scheme the concept to learn (e.g. "Higgs event") is characterised by values of some predefined set of attributes, such as "event E_t^{miss}" or "number of the isolated leptons in event" and each concept example is a vector of attribute values of the finite fixed length. The goal of our rule induction chain is to represent the target concept by the **decision list**, i.e. ordered rule set requiring the rule evaluation in a sequential order. Three attribute types are introduced: **symbolic** attribute (takes values from the finite set) with the only allowed test on the attribute value: $attr\_val = Val$; **integer** attribute (takes values from N) with possible tests: $attr\_val = Val; Val_1 < attr\_val \leq Val_2; Val_1 < attr\_val$ and **real** attribute (takes values from R) with tests: $attr\_val = Val; Val_1 < attr\_val < Val_2; Val_1 < attr\_val$.

RULE INDUCTION CHAIN

The rule induction system we are using contains three basic modules.

• Rule engine. It is the main module responsible for the raw rule set generation given the data. It could be any RIA discovered so far. At the moment we implemented and tested just two RIA's representing two main algorithm families: 'divide-and-conquer' and 'separate-and-conquer' (2). Our first RIA is inspired by the Quinlan's C4.5rule (3). Strictly speaking, it's not the RIA but the top down decision tree induction algorithm (TDIDT). However, any tree can be easily converted to the equivalent rule set. The second one is really RIA using the top down rule induction technique invented primarily by Michalski for his AQ family (4). We tested both RIA's on the same task with similar results - both demonstrate about the same performance and have the same problems, in particular, both need intensive pruning.

• Pruning module. Any rule set generated has to be pruned to handle noisy data and to control the overfitting. There are two pitfalls of the general post-pruning algorithms. The first one is that they have bad time complexity, e.g. Reduced Error Pruning (REP) has a complexity $O(n^4)$ (5) while the rule generation itself is only $O(n^2 \times \log n)$. Second, the results are not guaranteed because the greedy hill-climbing strategy is used and some potentially useful rules can be lost. In our case these pitfalls become of particular importance since we are experimenting with the interactive system. Much more simple two-step post-pruning scheme is adopted here. First,

---

* on leave from ITEP, Moscow, Russia

each rule is simplified by the dropping and modifying the "overfitted" tests from the rule body. The entire rule can be also dropped if: it contains just the "bad" tests or it becomes equivalent to some previously simplified rule (or its trivial specification). Second, simplified rule set is tested using another "pruning" data sample. Some new rules have now bad purity and thus are dropped. The resulting rule set is ordered in descending order according to the rule coverage. This simple scheme is sufficient for our purposes. The price is the loss of generality and new subjective bias introduced. This is because the "overfitness" criteria is task specific.

- Interactive fine-tuning module. This module helps user to update the generated decision list iteratively, monitoring the performance in each step. It is allowed both to drop rule or to modify rule. The modification includes the possibilities to drop or modify the existing test and to add the new one. The target concept can be now more complicated than the target used at the rule induction stage (see example below).

CASE STUDY

The test target was to optimise the b-tagging performance for the particular channel: $pp \to t\bar{t}(H \to b\bar{b}) \to l, 4b, jj, E_t^{miss}$. As it was shown recently (6) this channel allows about the only opportunity to observe the SM Higgs via $b\bar{b}$ decay up to $M_H \sim 130$ GeV and to discover the lightest Higgs of the MSSM almost everywhere in $M_A, \tan\beta$ parameter space. However, it requires the ultimate detector performance, in particular, to suppress the backgrounds, four b-jets have to be identified per event. The main background is the $pp \to t\bar{t} + X$ itself.

Simulation details

The $t\bar{t}H$ ($M_H = 120$ GeV) and $t\bar{t} + X$ events were generated by PYTHIA (7). Only events with at least 6 jets with $E_t > 15$ GeV and isolated lepton with $P_t > 15$ GeV in η range $|\eta| < 2.4$ were selected. To learn the classifier, three jet classes were introduced: b-jets, c-jets and qg-jets. The jet classification was done using the parton level information. To avoid ambiguities, each heavy quark was allowed to classify the only closest jet.

To obtain the realistic b-tagging performance estimates, the detector response (in our case CMS detector) has to be simulated properly. The compromise approach was used. We applied the tuned fast MC for the Calorimeter system (8) to reconstruct jets but the detailed GEANT simulation for the Tracker and dedicated CMS pattern recognition packages for the track and vertex reconstruction (9, 10, 11).

Jet example structure

To b-tag jet one usually searches for high quality tracks with large transverse impact parameter (TIP) and (or) secondary vertices inside the jet cone. The proper quantities are track signed TIP significance defined as $SIG = sign(\cos(\vec{TIP} * \vec{P_t^{jet}})) \times TIP/err(TIP)$ and similarly defined secondary vertex significance SIV. Only "good" tracks satisfying the minimal quality criteria: $SIG > 2; P_t > 0.7\,\text{GeV}; N_{hit} > 5; N_{pixelhit} > 1; \chi^2/ndf < 5$ and $TIP < 0.2$ cm were used for the analysis. In total, each example (jet) was characterised by 9 (correlated) attributes: 1) **integer jet** E_t (1: $E_t < 30$ GeV; 2: 30 - 60; 3: 60 - 120; 4: 120 - 200 and 5: $E_t > 200$ GeV); 2) **integer jet** η (1: $|\eta| < 0.7$ (barrel); 2: 0.7 - 1.2; 3: 1.2-1.6; 4: 1.6-2.0; 5: 2.0-2.4 ($\sim$ maximal CMS Tracker acceptance); 3) **integer NTR** (number of "good" tracks in a cone 0.4); 4) **real TSS** (total significance of associated tracks ΣSIG_i); 5) **real TPT** (ΣP_t^i); 6) **real MST** (effective mass of associated tracks); 7) **integer NSV** (number of secondary vertices associated with a jet); 8) **real TSV** (ΣSIV_i); 9) **real MSV** (effective mass of all tracks trapped in associated vertices)

Each jet example was also labeled with jet class ID.

Results

To estimate the system performance we introduce six parameters. First three monitor the overall b-tagging performance: **BE** (probability to b-tag b-jet); **QGI** (probability to b-tag qg-jet) and **CI** (the same for c-jets). At the rule generation and pruning stage the system was learned just to recognise b-jets. Three more parameters: **ES4** (fraction of "good" signal events 4-b-tagged); **EGL** (fraction of signal events 4 b-tagged) and **TTI** (fraction of $t\bar{t} + X$ events 4 b-tagged) were used at the interactive stage to tune the b-tagging rules for the particular $t\bar{t}H$ channel. Two different parameters for the signal efficiency monitoring were required since just ~ 0.53 selected $t\bar{t}H$ events with at least 6 jets actually contain at least 4 jets matched with b-quarks.

We made 3-fold cross-validation tests randomly dividing all simulated data on three subsets: 1000 ($t\bar{t}H$) + 1000 ($t\bar{t} + X$) events in the rule generation set, 2000 + 3000 in pruning and tuning one and 2000 + 3000 in the testing one. The best decision list generated was tested using the whole simulated event set. As a "straw man" we used a simple b-tagging rule learned by hand: a jet is b-tagged provided there are at least two good tracks with $SIG > 2$ and the reconstructed vertex in the jet cone (i.e. $NTR > 1 \vee NSV > 0$). However, it looks not so easy to beat the straw man performance: **BE = 0.569; QGI = 0.0194; CI = 0.1436; ES4 = 0.1446; EGL = 0.089; TTI = 0.00392**. Decision list performance (averaged over 3 tests) is indeed better: **BE = 0.623; QGI = 0.0198; CI = 0.163; ES4 = 0.2174; EGL = 0.134; TTI = 0.0032**.

The best decision list generated contains 12 rules: $NTR > 1 \vee TPT > 4 \vee TSV > 3$; $NTR > 2 \vee TPT > 8 \vee TSS > 8$; $NTR > 1 \vee TSV > 2 \vee MSS > 1$; $NTR > 1 \vee TPT > 15 \vee TSS > 20$; $NTR > 1 \vee TPT > 9 \vee TSS > 9 \vee E_t \leq 3 \vee |\eta| \leq 2$; $NTR > 1 \vee TPT > 5 \vee TSS > 7 \vee NSV > 0 \vee E_t \leq 2$; $NTR > 0 \vee TPT > 1 \vee TSV > 6 \vee NSV > 1 \vee E_t \leq 2$; $NTR > 1 \vee TPT > 5 \vee TSS > 7 \vee E_t = 1$; $NTR > 1 \vee TPT > 4 \vee MSS > 0.5 \vee E_t = 1 \vee |\eta| \leq 3$; $NTR = 1 \vee TSV > 6 \vee 0.8 < MSV < 3 \vee TPT > 3 \vee E_t \leq 3 \vee |\eta| \leq 2$; $NTR = 1 \vee TSV > 8 \vee TPT > 5 \vee TSS > 6 \vee E_t \leq 3 \vee |\eta| \leq 2$; $NTR = 1 \vee TSV > 5 \vee TPT > 2 \vee TSS > 6 \vee E_t \leq 4 \vee |\eta| \leq 2 \vee 1 < MSV < 5$

The structure of this decision list looks quite logical and understandable - first four general rules in the list cover most of examples, the rest suits to catch the low E_t and (or) small η exceptions. The truncated list containing just first 4 rules demonstrates excellent purity parameters and outperform the straw man in the efficiency: **BE = 0.584; QGI = 0.0160; CI = 0.141; ES4 = 0.1588; EGL = 0.0975; TTI = 0.00143**. The full list allows one to gain in efficiency by some impurity expenses: **BE = 0.628; QGI = 0.0204; CI = 0.167; ES4 = 0.22; EGL = 0.136; TTI = 0.00356**.

Unfortunately, the present $t\bar{t}$ statistics accumulated does not allow to estimate accurately the signal significance improvement, but preliminary rough estimates indicate that we are close to the level of 5σ signal significance for $L_{int} = 30 fb^{-1}$ which is much better than all previous attempts. In near future we are planning to increase the background statistics in ten times. This will allow to monitor directly the $t\bar{t}H$ signal significance at the interactive stage.

SUMMARY

We presented the prototype framework for the rule induction system and demonstrated the feasibility of the approach on the challenging classification problem, namely, b-jet identification for the particular $pp \rightarrow t\bar{t}H$ physics channel. The rules set generated is simple but it easily outperforms the standard hand-written b-tagging algorithms developed for CMS Tracker so far (12). At the moment the system relies too heavily on the human expert help at the pruning and interactive stage of the algorithm thus creating additional subjective bias. But our goal was to develop the user assistant interactive system rather than the autonomous discovery system. However, we are planning to experiment with algorithms incorporating the background knowledge in more general ways (see e.g (13)).

The physics is simple and intrinsically logical, it is the natural room for the symbolic learning algorithms.

REFERENCES

1. Michalski, R. S., *Discovering Classification Rules Using Variable-Valued Logic System CL1*. Proceedings of the 3rd International Conference on Artificial Intelligence, IJCAI, 1973, pp.162-172

2. Kubat, M, Bratko, I, Michalski R, *A Review of Machine Learning Methods*. in Machine Learning and Data Mining: Methods and Applications, Edited by R.S.Michalski, I.Bratko and M.Kubat, John Willey & Sons Ltd, 1996

3. Quinlan, J. R., *C4.5: Programs for Machine Learning*. Morgan Kaufman, San Mateo, CA, 1993

4. Michalski, R. S., *On the quasi-minimal solution of the covering problem*. In Proceedings of the 5th International Symposium on Information Processing (FCIP-69), Vol.A3, pp. 125-128, 1969.

5. Cohen, W. W., *Efficient pruning methods for separate-and-conquer rule learning systems*. In Proceedings of the 13th International Joint Conference on Artificial Intelligence, pp. 988-994, 1993

6. *ATLAS Detector and Physics Performance Technical Design Report*. CERN-LHCC-99-15, ATLAS TDR 15, 1999

7. Sjöstrand, T. http://www.thep.lu.se/ torbjorn/Pythia.html

8. Abdullin, S., Khanov, A., Stepanov, N. *CMSJET* CMS TN/94-180, (last update 1999)

9. *CMSIM. CMS Simulation and Reconstruction Package*. http://cmsdoc.cern.ch/cmsim/cmsim.html

10. Caner, A. et al. *Track and Vertex Finding Performance with the CMS Inner Tracker*. Nucl. Inst. Meth. A435, pp. 118-143, 1999

11. Stepanov, N., Khanov, A. *Vertex finding with deformable templates at LHC*. Nucl. Inst. Meth. A389, pp. 177-182, 1997

12. *The Tracker Project Technical Design Report*. CERN-LHCC-98-6, CMS TDR 5, 1998

13. Clark, P. and Matwin, S. *Learning Domain Theories using Abstract Background Knowledge*. Tech.Report TR-92-95, Dept CS, Univ. Ottawa, Canada, 1992

Looking for Instanton-induced Processes at HERA Using a Multivariate Technique Based on Range Searching

T. Carli[1] and B. Koblitz[1,2]

[1] *Universität Hamburg and* [2] *MPI für Physik, Föhringer Ring 6, D-80805 München*

Abstract. We present a method to discriminate instanton-induced processes from standard DIS background based on Range Searching. This method offers fast and automatic scanning of a large number of variables for a combination of variables giving high signal to background ratio and the smallest theoretical and experimental uncertainties.

INSTANTON-INDUCED PROCESSES

Instantons (1) are a fundamental non-perturbative aspect of QCD, inducing hard processes that are absent in perturbation theory. The expected cross section as calculated in "instanton-perturbation-theory" is sufficiently large (2, 3, 4) to make an experimental discovery possible (5, 6). For a more detailed introduction to instantons (*I*) see e.g. (7).

We study the prospect of a search for *I*-induced events modelled by the Monte Carlo Generator QCDINS (8) which generates *I*-induced events in deep-inelastic *ep*-scattering where a quark emerging from a $q\bar{q}$-splitting of the exchanged photon fuses with a gluon emitted from the proton. In the *I*-induced process $q\bar{q}$-pairs of each of the three light quark flavours and on average 2-3 gluons are produced. In the hadronic CMS they form a band (of about two units in pseudo-rapidity) of particles with high transverse energy which are homogeneously distributed in azimuth. Since in every event a pair of strange quarks is produced, in this band an increased number of kaons compared to standard DIS events is expected. Finally, the quark out of the split photon not participating in the instanton subprocess forms a hard jet.

The predicted cross section $\sigma^{(I)}_{HERA} = 29.2^{+9.9}_{-8.1}$ pb (4, 7) in a kinematic region where "instanton-perturbation-theory" ($x_B > 10^{-3}, 0.1 < y < 0.9, Q^2 > 113\,\text{GeV}^2$) is applicable, is two orders of magnitude smaller than the DIS cross section $\sigma_{DIS} \approx 3000$ pb. Therefore the highest possible signal to background ratio has to be achieved by exploiting observables characterising *I*-induced processes. To find these observables a large number of promising event variables have to be investigated and the sensitivity to systematic details in the modelling of the hadronic final state has to be tested. This requires a sophisticated and fast discrimination method to find the appropriate combination of event variables.

RANGE SEARCHING

Events can be classified as signal or background by estimating the probability density ρ of both these classes at the point of the event in the event-variable phase space, employing a Monte Carlo (MC) generator to sample the densities. In the case of neural networks (NN) this is done by fitting the probability-density with the adjusted weights of the neurons. To circumvent this time consuming procedure the density at each point can be directly estimated by counting the number of background and signal events in a surrounding box V. Given the ratio

$$\ell := \frac{\rho(I)}{\rho(DIS)} = \frac{\#I(V)}{\#DIS(V)}$$

the probability of an event to be a signal event is $D = \ell/(1+\ell)$. Compared to NN's this method also has the advantage of not extrapolating into phase space regions where there are no sample events available. Thus signals from data events outside the region covered by the MC simulation can be avoided. This is not the case for NN's which extrapolate into regions where there is no test data. Counting the number of events in the vicinity of a certain point is a problem known as Range Searching.

Range Searching algorithms have been developed which allow a search time $\sim \log(n)$, where n is the number of points that have to be searched (9). We employ an algorithm (11) suitable especially for a large number of events and dimensions (i.e. observables). The MC events are successively filled into the nodes of two binary trees, one for the signal events and one for the background events, where the criterion by which the decision is taken

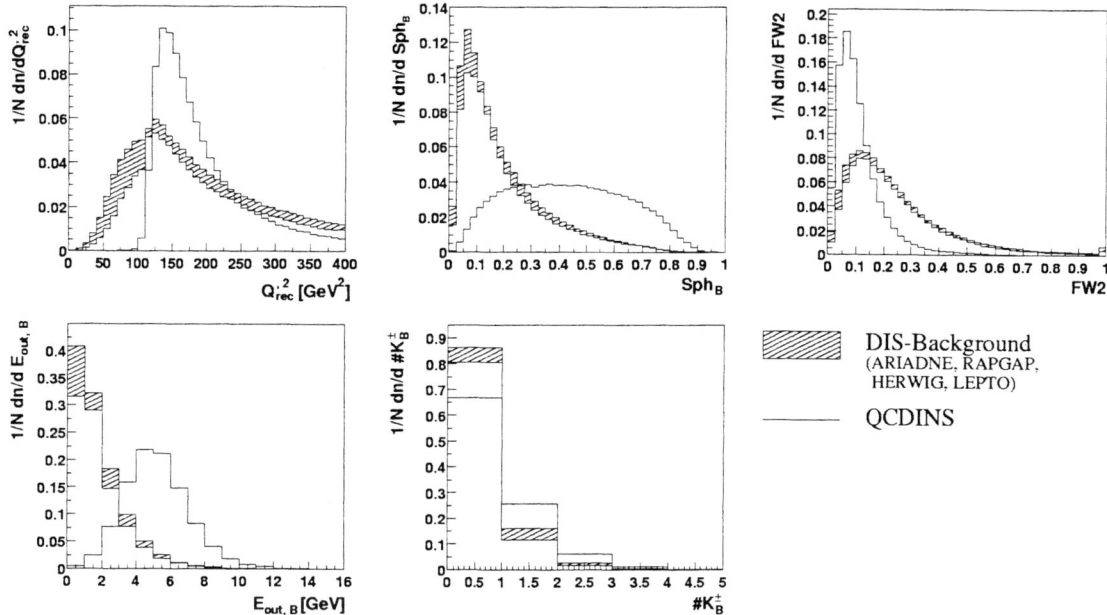

FIGURE 1. The characteristic event variables providing good instanton separation with small systematic uncertainties. Shown is the reconstructed virtuality of the quark entering the I-subprocess Q'^2_{rec}, the sphericity of the particles in the I-Band in their rest system, the second Fox-Wolfram moment of these particles and the event shape variable $E_{out,B}$ which is the projection of the particle transverse energy onto the axis, that makes this quantity maximal (see (6)). Finally the number of charged kaons in the I-Band is shown.

to descend to the left or right of a node is given by the value of one of the event variables. While descending the tree this variable cycles through all the ones considered. After filling, the position of every event in the tree is given by its coordinates in the event variable space. Classification of an event is done by searching in the trees for all background and signal events in the box V. This is done in the same manner as filling the tree.

RESULTS

Starting with 35 variables based on the hadronic final state the best 12 were chosen by calculating the discriminant with all 2-combinations (pairs) of the initial variables and taking those variables which provide a high separation power $S = \varepsilon(I)/\varepsilon(DIS)$ demanding an efficiency for instantons of $\varepsilon(I) = 10\%$. The number of considered variables is further reduced by calculating all 5-combinations and selecting those with highest separation power and a small systematic variation of the background. The systematic uncertainty was obtained by using four standard DIS-MC simulators which were tuned to data on representative hadronic final state quantities, in the range $Q^2 > 100\,\text{GeV}^2$ at HERA (10). The variables forming the best combination is shown in Figure 1. The separation power for $\varepsilon(I) = 10\%$ is $S = 126$. In Figure 2 the shape normalised discriminant D is shown for the I-induced and the background events, as well as the distribution for $D > 0.9$ normalised to a luminosity of $100\,\text{pb}^{-1}$ which is comparable to that already collected by each of the HERA experiments H1 and ZEUS. An event sample can be isolated where half of the events are instantons while the I-efficiency is still 10%.

For a search method to be reliable and easy to apply it is important to have as few free parameters as possible. In the case of Range Searching these are the number of events in the search trees, the size of the neighbourhood V and the minimum number of events in this neighbourhood to classify an event. To reduce the number of parameters for the box size the ratios of the box edge lengths were fixed by defining a box which contains most of the events and letting V be a scaled version of this large box. The projections onto these box edges are shown in Figure 1. The variation of the result depending on the size of V is shown in Figure 3. Clearly the separation increases for smaller boxes with the number of events that populate the search trees, while for larger boxes this difference vanishes. The plateau is increasing in width with the number of tree events and reaches nearly an order of magnitude in size, thus allowing to use only an approximate size parameter and reducing the need for fine tuning, if enough MC statistics is available.

In addition a comparison with a single hidden layer feed forward NN was done. The network performing

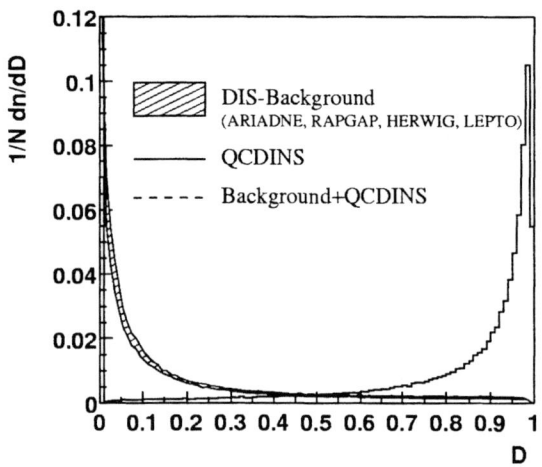
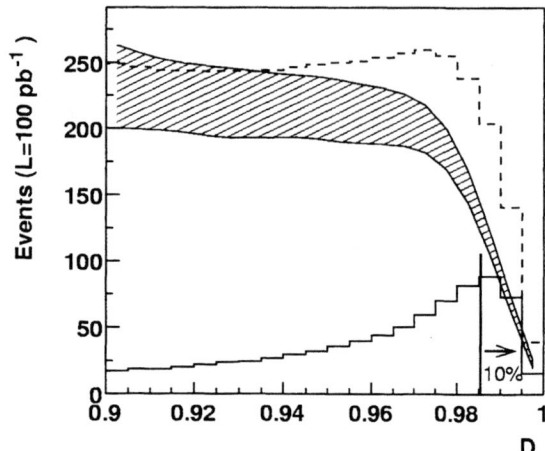

FIGURE 2. To the left, the shape normalised discriminant D for the instanton events using QCDINS and for standard DIS events using four MC simulators is shown. The second plot shows a zoom into the rightmost part and is now normalised to a luminosity of $L = 100\,\text{pb}^{-1}$. At $\varepsilon(I) = 10\%$ 178 I-induced events are expected.

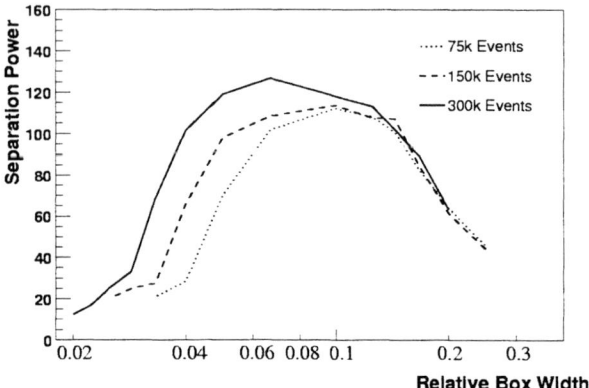

FIGURE 3. The separation power S at $\varepsilon(I) = 10\%$ for different box sizes and different numbers of events in the search trees.

best had 100 hidden nodes and was trained with the same input data. It reached a separation of $S = 116$ at an I-efficiency of 10%, being slightly worse than the Range Search method. Training the net was rather time consuming [1] and a lot of human intervention had to be done to adjust the input scales and training parameters.

CONCLUSIONS

The multivariate discrimination method based on Range Searching performs at least as good as a NN's when applied to the search for instantons at HERA. It is much less time consuming and can be easily used to automatically scan a large number of appropriate variables. The short processing time allows extensive searches for the best discriminating variables taking systematic effects into account. In a region where I-perturbation theory can be safely employed this novel discrimination method results in an 50% I-enriched data sample while the I-efficiency is still 10%.

REFERENCES

1. A. Belavin et al., *Phys. Lett.* B **59**, 85 (1975); G. 't Hooft, *Phys. Rev.* D **14**, 3432 (1976).
2. S. Moch, A. Ringwald and F. Schrempp, *Nucl. Phys.* B **507**, 134 (1997).
3. A. Ringwald and F. Schrempp, *Phys. Lett.* B**438**, 217 ('98).
4. A. Ringwald and F. Schrempp, *Phys. Lett.* B**459**, 249 ('99).
5. A. Ringwald and F. Schrempp, hep-ph/9411217, in *Quarks '94*, eds. D. Yu. Grigoriev et al. (World Scientific, 1995).
6. T. Carli, J. Gerigk, A. Ringwald and F. Schrempp, hep-ph/9906441, in *Monte Carlo Generators for HERA Physics*, eds. A. T. Doyle et al., DESY-PROC-1999-02.
7. A. Ringwald and F. Schrempp, hep-ph/9909338, in *New Trends in HERA Physics 1999*, eds. G. Grindhammer et al. (Springer, 2000).
8. A. Ringwald and F. Schrempp, hep-ph/9911516, in *Comput. Phys. Commun.* 132, 267 (2000)
9. R. Sedgewick, *Algorithms in C++*, Addison Wesley, 1992, Chapter 26
10. N. Brook et al., in *Monte Carlo Generators for HERA Physics*, eds. A. T. Doyle et al., DESY-PROC-1999-02.
11. Source available at http://www.desy.de/~koblitz

---

[1] 4h compared to 20 min for the Range Search method on a Linux PC

A Self-Organizing Neural Network for Job Scheduling in Distributed Systems

Harvey B. Newman, Iosif C. Legrand

Charles C. Lauritsen Laboratory of High Energy Physics
California Institute of Technology, Pasadena, CA 91125, USA

Abstract. The aim of this work is to describe a possible approach for the optimization of the job scheduling in large distributed systems, based on a self-organizing Neural Network. This dynamic scheduling system should be seen as adaptive middle layer software, aware of current available resources and making the scheduling decisions using the "past experience. It aims to optimize job specific parameters as well as the resource utilization. The scheduling system is able to dynamically learn and cluster information in a large dimensional parameter space and at the same time to explore new regions in the parameters space. This self-organizing scheduling system may offer a possible solution to provide an effective use of resources for the off-line data processing jobs for future HEP experiments.

INTRODUCTION

Finding and optimizing efficient job scheduling policies in large distributed systems, which evolve dynamically, is a challenging task. It requires the analysis of a large number of parameters describing the jobs and the time dependent state of the system. This study uses the MONARC Simulation system [1] to develop an approach to the job-scheduling task in distributed architectures, based on self-organizing neural networks [2]. The use of these networks enables the scheduling system to dynamically learn and cluster information in a high-dimensional parameter space. We have applied this approach to the problem of distributing offline data processing jobs among the worldwide-distributed regional centers in the LHC computing model. These jobs need random access to very large amounts of data, which are assumed to be organized and managed by distributed federations of OODB systems. Such a scheduling system may also help manage the way data are distributed among regional centers as a function of time, making it capable of providing useful information for the establishment and execution of data replication policies.

THE JOB SCHEDULING

A large number of parameters, most of them time dependent, must be used for the job scheduling in large distributed systems. The problem is even more difficult when not all of these parameters are correctly identified, or when the knowledge about the state of distributed system is incomplete or/and known with a certain delay in the past.

A "classical" model

A "classical" scheme to perform job scheduling is based on a set of rules using (part of) the parameters and a list of empirical constraints based on experience. It can be implemented as a long set of hard coded comparisons to achieve a scheduling decision for each job. In general, it can be represented as a function, which may depend on large numbers of parameters describing the state of the systems and the jobs. After a job is executed based on this decision, a performance evaluation can be done to quantify it.

A self organizing model

This approach is based on using the "past experience" from jobs that have been executed to create a dynamic decision making scheme. A competitive learning algorithm is used to "cluster" correlated information in the multi-dimensional input space defined by the parameters describing the systems, the jobs, the decisions and the results.

In the area of competitive learning quite a large number of modes exist which may have similar goals but differ considerably in the way they work or the implementation is done to solve certain problems. In our case, a feature mapping architecture able to identify correlation and cluster data distributed in a high-dimensional input space is done using, a growing self organizing network [2]. The incremental network models do not have a predefined structure (as in the case of self-organizing maps) and the addition and removal of neurons from the structure are done as part of the learning procedure. Known also as a neural gas, due to the fact that it does not have a static topological structure, such structure can offer an effective approach for feature extraction in multi dimensional space. The aim of the learning process is to cluster the input data into a set of partitions such that the intra-cluster variance remains small compared with the inter-cluster variance and to estimate the probability density function. This clustering scheme seems possible as we expect a strong correlation between the parameters involved. The strategies in modifying the network topology during the learning process (adding, deleting and clustering neurons) are to improve the pattern identification in the data and at the same time to try to increase the entire entropy. The processes of clustering neurons during the learning may be compared with creating molecules in a gas.

A very simple toy example

This over-simplified example aims to schematically present the way a self-organizing network can identify correlations and may be used to cluster the information. We assume that the time to execute a job in the local farm having a certain load (α) is:

$$T_1 = T_0(1 + f(\alpha)) \qquad (1)$$

Where T_0 is the theoretical time to perform the job and $f(\alpha)$ describes the effect of the farm load in the job execution time. If the job is executed on a remote site, an extra factor ($\beta>1$) is introduced increasing the turnaround time:

$$T_r = T_0\beta(1 + f(\alpha_r)) \qquad (2)$$

The ratio between the assumed execution time and the theoretical one (T_0) for a logarithmic load function is presented in Fig. 1.

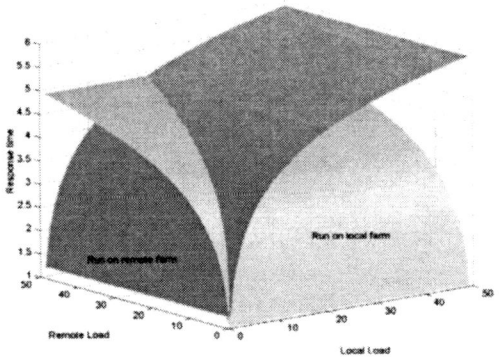

Figure 1. A simple example of the execution time for jobs executed on local or remote systems having different load factors.

The way a competitive learning algorithm for a neural gas like system performs for this problem is shown in Fig. 2. "Neurons" connected by lines are relatively close in the parameter space.

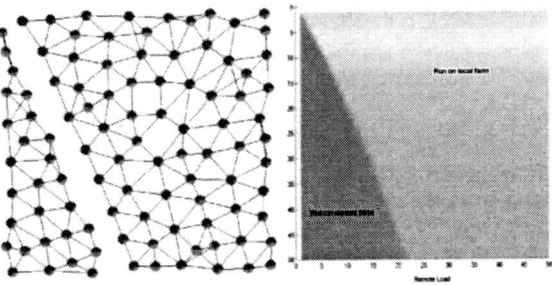

Figure 2. The self-organizing clusters compared with the projected decision types to obtain the best response time.

The Scheduling Decision

The decision for future jobs should be based on identifying the clusters in the total parameter space, which are close to the hyper plane defined in this space by the subset of parameters describing the job and the state of the system (parameters known before the job is submitted). In this way the decision can be done evaluating the typical performances of this list of close clusters and choose a decision set which meets the expected / available performance / resources and cost.

However, the self-organizing network has, in the beginning, only a limited "knowledge" in the parameter space and exploring efficiently other regions is a quite difficult task.

In this approach for scheduling, the difficult part is not the learning from previous experience, but making decisions when the system does not know (never tried) all possible options for a certain state of the system [3]. Even more difficult is to bring the system into a certain load state, which may require a long sequence of successive decisions taken to achieve it (like in a strategic game problem). The result of each decision is seen only after the job is finished, which also adds to the complexity of quantifying the effect of each decision. For this reason a relatively short history of the previous decisions taken by the system is also used in the learning process. We assume a relatively long sequence of decision (a policy) can be described with a set of a few points decision history, which partially overlap. This means that we can build a trajectory in the decision space by small segments that partially overlap.

An Example

We consider three Regional Centers ("caltech", "kek", "cern") connected. For two of them ("caltech" and "kek") during a certain period of time, every day, the number of data processing jobs submitted for execution exceed the locally available processing power while the third one ("cern") does not have any activity. At Day=0 (initial condition) a "classical" scheduling algorithm is used and all the jobs are submitted only to the local regional center. As the rate of sending jobs into the local regional center during a certain period of time exceed the local available processing power, many jobs are put into an execution queue. This job queuing makes the mean value for the turnaround time to be quite modest. In the case a job is exported to another regional center, we considered that the job works with data from its original Regional Center. Due to bandwidth limitation and much higher round trip time (RTT) the execution time on a remote site is longer than running it locally when the processing power is not overloaded. Therefore there is a penalty when jobs are executed on a remote site.

The decision making scheme together with the self organizing networks are "learning" what happens if jobs are exported, and try to provide a better mean return time for all jobs in each Regional Center After several days in which the usage pattern was similar as on the first day, the self-organizing scheduler has investigated new possible options and tried to optimize the turnaround time for jobs. The improvement in the mean turnaround time for both active regional centers during this learning procedure is presented in Fig. 3.

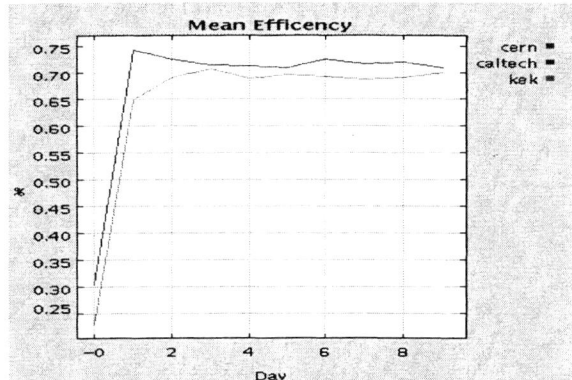

Figure 3. The improvement of the mean job efficiency (CPU time/ total time, during several days of learning.

SUMMARY

Using such a scheduling approach on quite simple problems in a realistic environment provided by the MONARC simulation tool seems to offer a possible solution for an efficient job distribution between Regional Centers. The difficult part in this approach is not the learning part but the decision making scheme which needs to effectively explore the "unknown" parts of the parameter space. Compared with a "classical" model we expect that such an approach may offer a better way to analyze the possible options and can evolve and improve itself dynamically.

REFERENCES

1. I.C. Legrand, H.B. Newman, "The Monarc Toolset for Simulating Large Network-Distributed processing Systems", Proc. of the 2000 Winter Simulation Conference. And MONARC Simulation Tool http://www.cern.ch/ MONARC/sim_tool/

2. B.Fritzke, "A self-organizing network that can follow non-stationary distributions, " Proc. of the International Conference on Artificial Neural Networks '97, Springer, 1997, pp. 613-618.

3. M. Norgaard, O.Ravan, N,K, Poulsen and L.K. Hansen, Neural Networks for Modelling and Control of Dynamic Systems, Springer, 2000.

Innovative Software Algorithms & Tools

Tailorable Software Architectures in the Accelerator Control System Environment

Igor Mejuev[*], Akira Kumagai[*] and Eiichi Kadokura[¶]

[*]PFU Limited, 658-1 Tsuruma, Machida, Tokyo 194-8510, Japan
[¶]High Energy Accelerator Research Organization, Oho 1-1, Tsukuba, Ibaraki 305-0801, Japan

Abstract. Tailoring is further evolution of an application after deployment in order to adapt it to requirements that were not accounted for in the original design. End-user tailorability has been extensively researched in applied computer science from HCI and software engineering perspectives. Tailorability allows coping with flexibility requirements, decreasing maintenance and development costs of software products. In general, dynamic or diverse software requirements constitute the need for implementing end-user tailorability in computer systems. In accelerator physics research the factor of dynamic requirements is especially important, due to frequent software and hardware modifications resulting in correspondingly high upgrade and maintenance costs. In this work we introduce the results of feasibility study on implementing end-user tailorability in the software for accelerator control system, considering the design and implementation of distributed monitoring application for 12 GeV KEK Proton Synchrotron as an example. The software prototypes used in this work are based on a generic tailoring platform (VEDICI), which allows decoupling of tailoring interfaces and runtime components. While representing a reusable application-independent framework, VEDICI can be potentially applied for tailoring of arbitrary compositional Web-based applications.

INTRODUCTION

End-user programming is considered as a promising approach capable of reducing maintenance cost and achieving greater flexibility of software products. In this context the architecture of end-user tailorable systems allows modifying the application within the context of its use and can be considered as a methodology of practical implementation of end-user programming paradigm. The application of runtime tailoring can solve the contradiction between dynamicity of requirements and inherent complexity of software present in particular application domains.

An accelerator control system is an example of such a domain – the dynamicity and flexibility are the essential requirements for scientific experiment environment, however the amount of hardware and I/O channels involved demands applications of computer control to achieve the consistency of experimental setup. Thus, the accelerator control system environment stipulates interdisciplinary research including the methodology of end-user programming in order to handle the problems of large-scale control software development.

The one way to deal with this problem is to apply the techniques of domain modeling and form-based programming [1, 2], however this approach would probably require certain skills of the end-users, which is not always acceptable. On the other hand our experience with software maintenance for experimental physics shows that the most frequent modification requests are targeting relatively small GUI of application logic updates. These modifications could be done by end-users themselves if the appropriate tools are provided.

In this paper we propose to introduce the notion of tailoring interface in the process of developing applications for accelerator control system environment. A feasibility evaluation was performed with Web-based monitoring applications for 12 GeV Proton Synchrotron at KEK. Using this example we consider the differences between tailoring and authoring interfaces and underline design and implementation issues for developing of tailorable applications in the accelerator control domain. The application of technology of end-user tailoring can significantly reduce the time required to perform software maintenance at the time of control hardware upgrades and correspondingly decrease the overall system maintenance cost.

RUNTIME TAILORABILITY

Generic motivations for implementing tailorability in software systems are constituted by *dynamicity* and *diversity* of users' requirements. The dynamicity can be derived from the inconsistency between waterfall software development model and the development process taking part in the real world. In accordance with evolutionary design methodology [3] the requirements are not given and therefore can not be strictly analyzed. The diversity of requirements arises from the need to satisfy the requirements of heterogeneous groups of users within the single application. The possible solution to this problem is to let the users perform the required modifications by themselves. However, practical implementation of this solution requires a design to allow the system modifications at runtime and a deployment solution capable of differentiating the modifications made by each user.

Runtime tailoring is distinguished from both use and development (authoring), however it retains certain commonalties with each. An authoring interface is typically used by developers of a computer system. The interface can utilize techniques such as visual programming or form-based programming, in order to speed up the process of development. The authoring system is required to provide full control of the application and available APIs, display the composition of the system in a consistent way, and provide integration with runtime and deployment modules. In the case of authoring the software process comprises iterative steps such as deploying the application's template, identifying the required modifications through the communication with users, modifying the application, deploying the new version, and so on. Typically the users and developers of the system represent distinct and geographically distributed groups.

The intent of tailoring interface is to provide the users with the possibility to customize an application to their particular needs and work situation. An example of tailoring interface is a text processing application with customizable toolbar. In the case of tailoring, the modifications should be done by end-users and within the execution environment. For shared applications or applications deployed on the Web the system should support the persistence and authentication of changes made by each user.

The differences between the authoring and tailoring interfaces can be summarized by the following criteria: *audience* (who), *usage pattern* (how) and *context* (when).

TAILORABLE REMOTE MONITORING APPLICATION

Online monitoring application for KEK Proton Synchrotron should provide display for the beam parameters, accessible in the Java applet environment. The required *degree of tailorability* for the monitoring application is identified as the possibility for the end-users to dynamically reassign mappings of GUI components to I/O channels and customize visual preferences (color, layout) for the components. Particular I/O channels are required to display permanently, so that the tailoring functionality should be disabled for the corresponding GUI objects. The layout of the remote monitoring system is represented in the Figure 1.

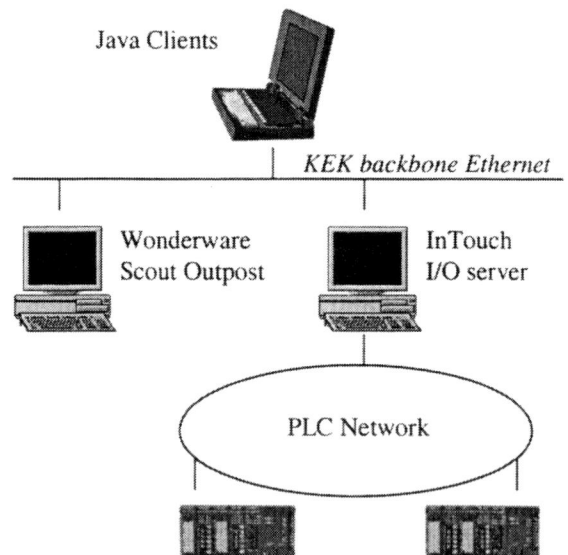

FIGURE 1. Monitoring System Outline.

PLC data are available through the commercial software – InTouch I/O server through NetDDE connections. The Scout Outpost provides CGI interface accessible from Web-based clients. Having received a CGI GET query, the Outpost retrieves the data from I/O server by NetDDE and replies the results in the form of HTML table, including I/O tag names, error codes and current values for the tags.

The monitoring application was implemented using VEDICI – a generic tailoring platform for Java Beans, which allows decoupling of tailoring interfaces and runtime components. A VEDICI application is represented as a nested hierarchy of compositional markup specifications (Bean Markup Language) with the possibility to associate an individual tailoring component with each specification. This approach allows integrating multiple tailoring interfaces within an application instance.

The implementation of dynamic monitoring components was reused from earlier implementation made with Java 1.0 for the JLC X-band High Field Experiment [4, 5]. The data update is performed by a dedicated component, which wraps Scout Outpost interface and provides a refresh manager for dynamic monitoring components. The update manager performs data polling by sending batch requests to the Scout Outpost server. The monitoring application includes an authoring tool for developing new applications and reusable *visualizers* applicable for customization of the applications at runtime (Figure 2).

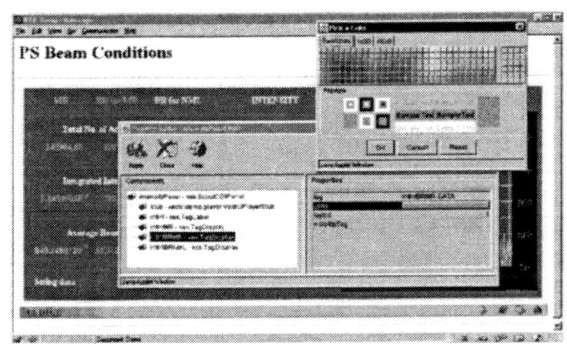

FIGURE 2. Tailoring Interface.

The visualizers provide the possibilities to browse the structure of an application (*HierarchyBrowser*), visually edit the components' properties (*PropertyEditor*) or modify the application at the source code level (*SourceEditor*). The runtime allows integrating visualizers within an application instance and, additionally, tailoring can be disabled for particular components of the application by assigning a null visualizer (*VisualizerStub*).

The users can save the customized applications in a server-side repository with authorization by username and password. After restart of browser the changes can be recovered using the same authentication scheme.

CONCLUSIONS AND FUTURE WORK

In this paper we proposed to apply the methodology of end-user tailoring in the accelerator control system environment. We considered as an example the design and implementation of Web-based monitoring applications for displaying beam conditions for 12GeV Proton Synchrotron at KEK.

In the future we would like to extend the prototype implementation by providing a rich selection of components and custom visualizers, usable in the accelerator control domain. Current implementation uses polling of CGI server that creates redundant traffic in the laboratory network. In the future version we consider replacing the polling with server push interface, which is based on the existing portable implementation of shared data channels [6].

REFERENCES

1. Mejuev, I., Abe, I., and Nakahara, K., *Nuclear Instruments & Methods in Physics Research* A **389**, 38-41 (1997).

2. Mejuev, I., Abe, I., and Nakahara K., "Object-Oriented Control System Development Using Smalltalk Language", in *International Conference on Accelerator and Large Experimental Physics Control Systems-95*, ICALEPCS'95 Conference Proceedings, 1995, Chicago, USA, pp. 713-717.

3. Floyd, C., Reisen, F.-M., and Schmidt, G., *Lecture Notes in Computer Science* **387**, 48-64 (1989).

4. Mejuev, I., Kumagai, A., Takahashi, M., Kadokura, E., Higo, T., and Takata, K., "Status of Control and Data Acquisition System for JLC X-Band High Field Experiment"; in *23rd Linear Accelerator Meeting in Japan*, Conference Proceedings, 1998, Tsukuba, Japan, pp. 367-368.

5. Higo, T., Dong, D., Fang, H., Nie, J., Gao, M., Kadokura, E., Mejuev, I., Sakai, H., and Takata, K., "High Field Experiment of 1.3m-Long X-Band Structure", in *The First Asian Particle Accelerator Conference*, APAC'98 Conference Proceedings, 1998, KEK, Tsukuba, Japan, pp. 169-171.

6. Mejuev, I., Abe, I., "Java Application for Creating a Shared Object Cash", in *International Conference on Accelerator and Large Experimental Physics Control Systems-97*, ICALEPCS'97 Conference Proceedings, 1997, Beijing, China.

Online Modeling of the Fermilab Accelerators[*]

Elliott S. McCrory, Leo Michelotti, Jean-Francois Ostiguy

Fermi National Accelerator Laboratory
PO Box 500, Batavia, IL 60510, USA

Abstract. We have implemented access to beam physics models of the Fermilab accelerators and beamlines through the Fermilab control system. The models run on Unix workstations, communicating with legacy controls software through a front end redirection mechanism (the open access server), a relational database and a simple text-based protocol over TCP/IP. The clients and the server are implemented in object-oriented C++. We discuss limitations of our approach and the difficulties that arise from it. Some of the obstacles may be overcome by introducing a new layer of abstraction. To maintain compatibility with the next generation of accelerator control software currently under development at the laboratory, this layer would be implemented in Java. We discuss the implications of that choice.

INTRODUCTION

The Fermilab accelerator control system was designed long before it was thought that behavior of these machines could be accurately modeled. In recent years, we have been able to wrap our physics modeling effort in such a way that our legacy control system can present these models to the users. These models are accessed through the Fermilab control system in four ways: The Online Model (OLM), the Open-Access Model (OAM), standalone applications and a database of physics parameters.

The OLM and the OAM rely on a computation server to do the relevant beam physics calculations. This server is connected to the client via TCP/IP. In addition, the OLM uses the database for transferring efficiently the calculation results from the server.

But our legacy wrapper has some limitations. Therefore, a standalone application framework for fitting and tuning machine parameters has been developed using our models [1]. This framework uses open software packages for the display. It was thought that much of the functionality of this effort could be incorporated into our basic modeling structure. This has not been achieved.

These models have not gained acceptance from our target audience, the physicists and operators in the Fermilab control room. First and foremost is that these models, while completely accurate, are not as robust as they need to be; for example, there is no real error recovery. Furthermore, the legacy framework in which application programs must be built limits the flexibility of these applications.

There is an effort underway to replace the controls application interface with a Java-based framework. With the appropriate specifications, it is possible to influence the outcome of this new system in order to make online models more practical and acceptable.

COMPUTATION ENVIRONMENT

The online models rely on the *MXYZPTLK/Beamline* C++ class suite [2], developed at Fermilab. Briefly, *MXYZPTLK* provides a differential algebraic framework for performing particle propagation through a beamline to any order. The programming interface to use high-order calculations is the same as the first-order calculations, which are relevant to operations. *Beamline* provides classes for representing beam line elements in a machine, and the various ways to combine these elements into hierarchical structures. The object-oriented nature of this package has allowed significant flexibility in generating practical applications. The applications presented here are an example of this flexibility.

All accesses to our online models use the same machine description and the same calculation structure. Thus, a question asked of the model through one of the access points yields the same answer as from another access point. Furthermore, if more details are needed, a special-purpose offline application can be quickly constructed from the same classes.

The computation models run on dedicated Unix workstations. Communication between the legacy (VMS) system and the Unix servers is done via an ASCII-based TCP/IP network protocol.

---

[*] Work supported by the US Department of Energy, contract # DE-AC02-76CH0-3000.

PHYSICS ANALYSIS CLIENTS

Online Database of Physics Parameters

The most widely used access to beam physics computations has been the static databases of machine and beam physics parameters. Calculating the data for these tables is an important aspect of debugging the beam physics model for a new machine. C-callable functions have been created as part of the control system API to access these tables without direct knowledge of SQL. Many online applications at Fermilab rely on these tables.

The sorts of data present in these database tables are the physical names, positions and alignment of the beam line elements in the lattice and the beam physics, like the orbit, Twiss functions, R-matrix and covariance matrix. These physics parameters are available at every element in the lattice of each machine.

Online Models

As models are developed for the various machines, the classes that perform the relevant beam physics calculation on this model are transferred to the server. Simultaneously, the classes on the operator consoles are extended to handle the specifics for this new model. Once a model is deemed to be ready, it takes a day or two of work to perform this step and create a full-fledged online model.

An example of the output of the online model in action for the 8 GeV transfer line between the Fermilab Booster and the Main Injector is shown in Figure 1.

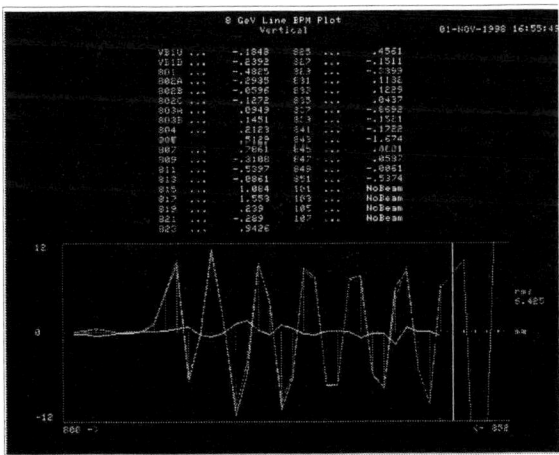

FIGURE 1. The Online Comparison between the model of the MI 8 GeV line and the BPM readings for it.

Open Access Models

A feature of the current Fermilab control system is the ability to change a user console's connection from the real hardware to an internal model. This type of model-based redirection is referred to as an Open Access Model (OAM). It was conceived as the simplest possible redirection mechanism available. At least two desirable results are accomplished through an OAM: to test algorithms in online controls applications and to provide a meaningful way to train operators. OAMs have been written in C++ for the Fermilab Main Injector and the Recycler Ring.

The emulation of the control system and of all the devices in a machine is handled in the OAM. An OAM performs essentially two actions:
1. To accept settings and interpret these settings into changes in the simulated equipment in the model, and
2. To provide model-based beam sensor readings.

Regular applications are used while the console is redirected. The user is alerted to the redirection by a prominent yellow line throughout the application display.

Fitting and Tuning With Standalone App

A different access into our modeling structure has been created for the Recycler Ring (RR). The RR is based on permanent magnet technology. The tune is changed by adjusting the strength of five electromagnetic quad families in a special insertion called the "phase trombone." A fitting scheme has been implemented to enable this calculation. Emphasis has been placed on allowing the user to guide this inherently nonlinear fitting process in a convenient way. The VMS/Unix client/server has been bypassed: the application runs directly on the Unix server and the display appears on the controls console. This work is discussed in another paper at this conference [1].

This program is a specialization of a general-purpose fitting and tuning program. It has been constructed to facilitate the calculation at hand. The general program uses the same lattice specifications and calculation engine as the online models.

LIMITATIONS

The community at Fermilab has not accepted the present system of online models. There are many reasons for this, and several things can be done to fix these problems.

The Fermilab beamline and accelerator control system is very old and generally is considered to be cumbersome for application development. While the C++ compiler under VMS is good, the hardware on which these programs are forced to run (VAX) is severely out of date. Furthermore, the application program interface (API) for the control system has essentially no abstraction, data hiding or polymorphistic capabilities; in other words, it is not object-oriented. Thus, developing online models in this environment has proven to be time consuming and challenging, out of proportion to the resulting functionality.

As for the model redirection mechanism, we have proven that it can be made to work. We have even constructed an OO framework for dealing with the programmatic complexities. But since the level of the redirection is very close to the hardware, one has to mimic the lowest-level hardware commands and bit patterns in order to realize the redirection. For example, in order to measure the FMI turn-by-turn through model redirection, the model has had to learn which bits were to be toggled in order to turn on the one-turn kicker that initiates this measurement. Moreover, the classes that mimic the BPM readbacks also have had to reproduce the 8-bit readings that the hardware is expected to generate. This is just for the tune measurement—other measurements would require similar effort to be made to work in redirection. And it is customary at Fermilab to change the way the hardware behaves in order to meet our programmatic goals, and these changes must be reflected in the redirection model. It has not been practical to keep the redirection in this context up to date.

POSSIBLE DIRECTIONS

If this effort were to continue, it would be necessary to change the architecture of the control system in order to facilitate these sorts of model couplings. It is our belief that there would need to be a hierarchy of object-oriented classes developed that represented both the physics concepts and the engineering systems embodied in the machines. An effort would need to be made to develop a level of abstraction within the controls environment that deals with these concepts. Then both the model redirection and the connection of real measurements to model predictions could be realized.

For example, one can imagine a general "BeamPosition" abstraction that would yield classes that represent the readings from BPM hardware. This abstraction should have a "model predicted" field, representing the design value of the position at this location. One could define an "Orbit" class that only has knowledge of the abstract BeamPosition class. The relevant engineering details would not be necessarily compromised since the concrete classes, which implement the BeamPosition abstraction, could have as much detail as the engineer feels is necessary. It is our opinion that a finite set of these abstractions (actually, these should be, in Java terminology, interfaces) can be invented to fully describe the equipment that is used at Fermilab. It would not be an easy job, but it would lead to enormous improvements in the flexibility and the robustness of the facility.

Having established this level of abstraction, one could then implement redirection to a model in a straightforward, efficient and practical manner. Also, these abstractions could have hooks for tying in model predictions, for example, the "design" orbit idea described above.

Our controls personnel are in a position to develop these ideas at the same time as they develop ideas on how to use the Java language properly in this context.

CONCLUSIONS

OLMs exist at Fermilab for the following machines: Main Injector 8 GeV Line, FMI, RR, Tevatron fixed target mode, Tevatron collider mode and the Antiproton Source Accumulator. New OLMs have been added quickly. Open-access models exist for the FMI and the RR.

The OAM has not been applied to more of our systems because the level at which redirection occurs is entirely too close to the hardware. In order to make the OAM work, it must be known in great detail how a front end replies to a specific type of data request.

Our long-term goal is to provide a consistent environment to construct models and to do beam physics calculations. We are prepared to track the upcoming changes in the control system to keep these models viable.

REFERENCES

1. "A Beamline Matching Application based on open source software," By J.-F. Ostiguy, these proceedings.

2. "MXYZPTLK version 3.1 User's Guide: A C++ Library for differential Algebra," By Leo Michelotti, Fermilab publications #FN-353-REV, on web as
http://fnalpubs.fnal.gov/archive/1995/fn/FN353R.html

A Beamline Matching Application based on Open Source Software

J.-F. Ostiguy*

Beam Physics Department, Fermi National Accelerator Laboratory, Batavia, IL 60510

Abstract. An interactive Beamline Matching application has been developed using beamline and automatic differentiation class libraries. Various freely available components were used; in particular, the user interface is based on FLTK, a C++ toolkit distributed under the terms of the GNU Public License (GPL). The result is an application that compiles without modifications under both X-Windows and Win32 and offers the same look and feel under both operating environments. In this paper, we discuss some of the practical issues that were confronted and the choices that were made. In particular, we discuss object-based event propagation mechanisms, multithreading, language mixing and persistence.

INTRODUCTION

Up until just a few years ago, writing software for scientific applications required the programmer to implement functionality that had little to do with his objectives. Although commercial libraries were available, high costs and licensing hassles made it more practical to reinvent user interfaces, plot widgets, etc. By dramatically improving free software localization and distribution mechanisms, the internet promotes code reuse.

In this paper, we describe an interactive lattice design application assembled with various freely available software components. The objective was not to compete with commercial products, but rather to provide an application that can be modified and adapted to meet our specialized needs. There are currently a limited number of publicly available interactive applications to perform beamline design. Popular lattice design programs like TRANSPORT and MAD have extensive capabilities, but were developed to be run in a batch-oriented environment.

The design goals were the following: (1) given a description of a beamline, allow a user to specify all aspects of a matching problem interactively (2) provide graphical feedback and allow the user to dynamically interrupt a nonlinear iteration and edit the state of variables and constraints (3) make customization as easy as possible. Although this is still a work in progress, basic features have been implemented and are fully functional; the resulting application is called BLIMP (BeamLine Interactive Matching Program). A specialized version, produced to tune the Fermilab Recycler Ring phase trombone, will be described.

---

* Work supported by DOE Contract DE-AC02-76CH0-3000

THE MATCHING PROBLEM

The matching problem is of fundamental importance for lattice designers. It can be simply stated as follows: given a beamline and a set of lattice functions specified at one extremity, determine the strength and/or longitudinal position of beamline elements necessary for the lattice functions to assume certain specified values at one or more distinct locations.

In most situations of practical importance, horizontal and vertical motion are decoupled and a beamline is to first order, completely characterized by a set of ten quantities: $\beta_{x,y}$, $\alpha_{x,y}$, $\mu_{x,y}$, $\eta_{x,y}$ and $\eta'_{x,y}$, where β and α are the familiar Courant-Snyder lattice functions, μ is the phase advance and η and η' are respectively the dispersion and its derivative with respect to the longitudinal coordinate.

CODE STRUCTURE

BLIMP is written in ANSI standard C++ and takes advantage of the facilities offered by the Standard Template Library. Variables are defined independently of basic beamline elements and can in principle involve arbitrary linear combinations of element strengths. The user can dynamically define both local and global constraints. Typically, local constraints involve equalities while global constraints involve inequalities (e.g. β function smaller than a prescribed maximum). Figure 1, is a screen shot of the user interface. BLIMP has been put together by using both internally developed and freely available software components which are now briefly described.

MXYZTPLK/Beamline *Class Libraries*

MXYZTPLK and Beamline are class libraries authored by Leo Michelotti (1, 2) at Fermilab. Although considered stable for a few years already, they continue to evolve and are maintained to keep up with the

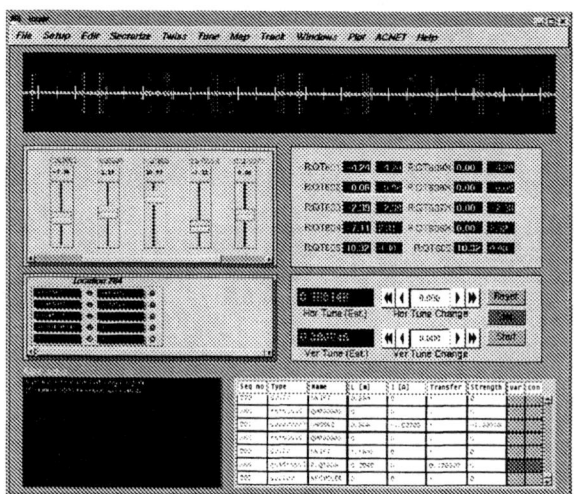

FIGURE 1. *The BLIMP user interface. The top window is a display of the beamline layout. Sliding cursors are displayed for each user-defined variable. Similarly, a custom widget is created for each local constraint.*

latest developments in the C++ language. MXYZT-PLK is a stand-alone set of C++ classes for performing automatic differentiation and differential algebra. The Beamline class library –built on top of MXYZTPLK– is a rich set of classes supporting lattice related calculations. Beamlines are represented by doubly linked lists whose nodes can either point to other beamlines or to basic elements. Beamlines can be edited, concatenated, cloned, flattened (i.e. no hierarchical structure). Most quantities of interest to accelerator physicists can be computed and both field and alignment errors can be included if necessary. Maps of arbitrary order can in principle be computed to machine precision in either four-dimensional (i.e., transverse) or full six-dimensional phase space.

Nonlinear Optimizer

Numerical nonlinear optimization problems can be classified according to (1) whether or not the objective function is expressible as a continuous, differentiable function (2) the nature of the external constraints that need to be enforced, if any. For matching problems, the objective function is usually a differentiable function of the elements' strengths and positions. In that case, Newton method has the advantage of quadratic convergence if the extremum is sufficiently close. In practice, it is expensive to compute a Hessian matrix since it involves second order derivatives. To avoid this, a standard strategy is the Davidon-Fletcher-Powell (DFP) algorithm (3) which progressively constructs the inverse of the Hessian matrix at each step of an iteration involving a sequence of one dimensional conjugate gradient searches.

BLIMP uses the DFP method as implemented in the MINUIT library from CERN (6), a good general purpose optimization library freely available for research institutions. Unfortunately MINUIT suffers from various limitations associated with its Fortran heritage. Among the most problematic issues: Fortran I/O cannot be mixed with the C/C++ I/O in a portable way; the objective function must be passed to the library as a static function. The optimization code is encapsulated into an Optimizer class; this should allow an alternative to MINUIT to be substituted with minimal side effects. A full C++ implementation would also allow the use of functor objects to completely decouple the optimizer from the rest of the code.

Graphical User Interface

The choice of a user interface toolkit has been driven by two requirements: (1) object orientation and (2) portability between various flavors of UNIX and Windows NT. The Fast Light Toolkit (FLTK) (7) satisfies both requirements and is available under the terms of the GNU Public Library License. FLTK also provides support for OpenGL (or MESA, a free compatible alternative) and BLIMP takes advantage of the facilities offered by OpenGL to efficiently display the beamline at different scales.

Events

In addition to the primitive event propagation mechanism provided by FLTK for GUI related events, BLIMP uses a generic event propagation scheme. It relies on relatively recent, but nevertheless, standard features of ANSI C++ templates. This allows the code to be structured as a collection of independent "components" in a completely generic and type-safe manner. Events can be arbitrarily complex objects and generic lists are created for each distinct published event type. Subscribers are added to each list by interested objects. To propagate an event, a publisher object need only to instantiate it and invoke its virtual propagate method. In practice, to publish and/or subscribe to a certain event type, a class need only to be derived from a Publisher or Subscriber base class (or both).

At this point, a few comments are in order. In certain commercial frameworks such as MFC, event declaration and propagation mechanisms are enforced by the framework. MFC does not rely on templates, but rather on a compex collection of macros. Some commercial compilers, like Borland C++ Builder for example, provide language level support for event propagation by introducing special additional keywords for that purpose. An intermediate approach, used by Qt (a popular C++ framework) is to extend the C++ language and use a special preprocessor to convert the original source code into standard ANSI C++.

A number of similar free generic template-based "callback" libraries can be found on the internet (4).

Persistence

A valuable feature for any interactive application is the ability to save its current state. This state is completely defined by a certain number of objects and their relation to each other, as defined by pointers and references. Unfortunately, simply saving objects in binary form is not sufficient to save the state of an application, since pointers and references are process specific. To correctly restore the state of an application, it is necessary to keep track of relations between objects and to fix all memory references a posteriori. Although it is a straightforward matter to fix explicitly declared pointers and references, correctly restoring virtual functions presents a special challenge, since virtual function pointers are not directly accessible. Fortunately ANSI C++ provides a solution. In a nutshell, the technique consists in reading a raw binary object into a block of (allocated) memory of correct size, and subsequently invoke the operator new using the ANSI C++ placement syntax. This syntax allows one to place an object in pre-allocated memory. Provided the constructor does not perform any explicit initialization, all data fields are left untouched; however, *all virtual pointers are correctly initialized*. Persistent versions of the MXYZTPLK and `Beamline` libraries have been developed based on the above described scheme.

Multithreading

Although the FLTK code is not reentrant and therefore does not directly support threads, an application can still take advantage of multithreading provided all GUI activity remains confined to a single thread. Since the `Optimizer` consumes a fair amount of CPU it is useful to dedicate a separate thread to it. An added benefit of using separate threads is that the Optimizer no longer needs to contain any specific GUI code. A threaded version of BLIMP is currently under consideration. One difficulty arises from the fact that the UNIX and WIN32 thread models are somewhat different.

APPLICATIONS

We now describe two applications that motivated the development of BLIMP.

Phase Trombone

The Fermilab Recycler ring is a new machine for antiproton accumulation and recycling which has the distinction of being the first machine to make large scale utilization of permanent magnet technology. The machine operates at a fixed energy of 8 GeV with a lattice based on fixed-field combined function magnets. The tune of the machine is adjusted (± 0.5) by varying nine electromagnetic quadrupoles grouped in five symmetric families within a region where $\eta_{x,y} = \eta'_{x,y} = 0$. Four hard constraints must be met, i.e. at the symmetry point $\alpha_{x,y} = 0$ and the two phase advances set to the desired values; an additional softer requirement is to prevent the beta functions from exceeding a maximum value.

Low Beta Insertion

In a low-beta insertion, the objective is to use a pair of quadrupole doublets or triplets to focus counter-circulating beams into a very small size interaction region. In general, the insertion has to match the lattice functions of the ring at both extremities; the phase advance is unconstrained. At the interaction point, $\beta_{x,y}$ must assume specified values and the beam envelope must go through a minimum i.e. $\alpha'_{x,y} = 0$. It is also often required for the dispersion to be as small as possible and one usually demands $\eta_{x,y} = 0$. Constraining η' may also be desirable.

Low beta insertions are notoriously nonlinear. Without experience, it is difficult for a novice to find a satisfactory solution and interactivity is certainly no substitute for experience. However, the ability to quickly experiment with different strategies and stop the iterations dynamically can be a significant advantage.

CONCLUSION

BLIMP is still a work in progress, although it is certainly already useful as it stands. The current priority is to make the application multithreaded. Also under consideration is a port from FLTK to Qt, a commercial framework much more mature than FLTK, which has only recently (Sept 2000) been released under the terms of the GPL license.

REFERENCES

1. L. Michelotti, "MXYZPLTK Version 3.1 User's Guide: A C++ Library for Differential Algebra", Fermilab Publication FN-535-REV, October 1995

2. L. Michelotti, "MXYZPLTK and `Beamline`: C++ Objects for Beam Physics", Advanced Beam Dynamics Workshop on Effects of Errors in Accelerators, their Diagnosis and Correction, AIP Conf. Proceedings No 225, 1992

3. R. Fletcher and M.J.D Powell, *"A rapidly convergent Descent Method for Minimization"*, The Computer Journal, **6**, 163-168, 1963

4. For example, see http://libsigc.sourceforge.net.

5. J. Hesse, *"EZSave for C++"*, C++ Report 12(2): 21-26, 40-41, Feb. 2000

6. F. James, "MINUIT Minimization Package Reference Manual Version 94.1", CERN Program Library D506, Computing and Networks Division CERN Geneva, Switzerland

7. Information about FLTK is available at the following URL: www.fltk.org

8. Fermilab Recycler Ring Technical Design Report, Fermilab Publication TM-1936, July 1995

Sleuth: A Quasi-Model-Independent Search Strategy for New High p_T Physics

B. Knuteson

Department of Physics, University of California at Berkeley, Berkeley, California 94720

Abstract. We present a quasi-model-independent search for the physics responsible for electroweak symmetry breaking. We define final states to be studied, and construct a rule that identifies a set of relevant variables for any particular final state. A new algorithm ("Sleuth") searches for regions of excess in those variables and quantifies the significance of any detected excess. After demonstrating the sensitivity of the method, we apply it to the semi-inclusive channel $e\mu X$ collected in 108 pb^{-1} of $p\bar{p}$ collisions at $\sqrt{s} = 1.8$ TeV at the DØ experiment during 1992–1996 at the Fermilab Tevatron. We find no evidence of new high p_T physics in this sample.

INTRODUCTION

It is generally recognized that the standard model, an extremely successful description of the fundamental particles and their interactions, must be incomplete; yet the possibilities beyond the current paradigm are sufficiently broad that the first hint could appear in any of many different guises. This suggests the importance of performing searches that are as model-independent as possible. In these proceedings I describe a novel search strategy ("Sleuth") for physics beyond the standard model that assumes nothing about the expected characteristics of new processes other than that they will produce an excess of events at high transverse momentum (p_T). The search strategy has been applied to the exclusive final states within $e\mu X$ in ≈ 108 pb^{-1} of $p\bar{p}$ collisions collected by DØ at the Fermilab Tevatron. Space limitations allow only a brief overview of the algorithm and a terse statement of the results on the final states within $e\mu X$; a more complete description can be found in the original Sleuth article. (1)

SLEUTH

We assume that the production and subsequent decay of massive new particles will yield events containing final state objects with large transverse momenta. The data are partitioned into exclusive final states using standard criteria that identify isolated high p_T electrons (e), muons (μ), and photons (γ), as well as jets (j), missing transverse energy ($\not{E}_T$), and the presence of W and Z bosons. For each exclusive final state, we consider a small set of variables that are easily related to the transverse momenta of the final state objects.

Although the details of the Sleuth algorithm are complicated, the concept is straightforward. What is needed is a data sample, a set of events modeling each background process i, and the number of background events $\hat{b}_i \pm \delta\hat{b}_i$ expected from each background process. From these we determine the region of greatest excess and quantify the degree to which that excess is interesting.

The Sleuth algorithm, applied to each final state individually, comprises the following steps:

(i) We begin by constructing a mapping from the original d-dimensional variable space into the d-dimensional unit box (i.e., $[0,1]^d$) that flattens the background distribution. We use this to map the data into the unit box.

(ii) We rigorously define the notion of a "region" about a set of N data points.

(iii) Each region R contains an expected number of background events $\hat{b}_R \pm \delta\hat{b}_R$. The probability that the background in the region fluctuates up to or above the observed number of events in the region is our first measure of the degree of interest of a particular region.

(iv) Our assumption that new physics is most likely to appear at high p_T translates to a preference for regions in a particular corner of the unit box. Such regions are searched to determine the region $\mathcal{R}$ of greatest excess.

(v) In any reasonably-sized data set, there will always be regions in which the probability for $\hat{b}_R \pm \delta\hat{b}_R$ to fluctuate up to or above the observed number of events is small. The relevant issue is what fraction $\mathcal{P}$ of *hypothetical similar experiments* contains a region as interesting as $\mathcal{R}$. In computing $\mathcal{P}$ we rigorously "take

Table 1. The number of background events expected for the populated final states within $e\mu X$. $\not{e}\mu$ denotes any process with a real muon and a jet that is mistakenly identified as an electron.

Data set	$\not{e}\mu$	$Z/\gamma^* \to \tau\tau$	WW	$t\bar{t}$	Total	Data
$e\mu\not{E}_T$	18.4 ± 1.4	26.1 ± 6.5	3.9 ± 1.0	0.011 ± 0.003	48.5 ± 7.6	39
$e\mu\not{E}_T j$	8.7 ± 1.0	3.2 ± 0.8	1.1 ± 0.3	0.4 ± 0.1	13.2 ± 1.5	13
$e\mu\not{E}_T jj$	2.7 ± 0.6	0.5 ± 0.2	0.18 ± 0.05	1.8 ± 0.5	5.2 ± 0.8	5
$e\mu\not{E}_T jjj$	0.4 ± 0.2	0.07 ± 0.05	0.032 ± 0.009	0.7 ± 0.2	1.3 ± 0.3	1

Table 2. Summary of results on the $e\mu\not{E}_T$, $e\mu\not{E}_T j$, $e\mu\not{E}_T jj$, and $e\mu\not{E}_T jjj$ channels as WW and then $t\bar{t}$ are added to the background estimate.

Data set	$\mathcal{P}$ when the background includes		
	$\not{e}\mu + Z/\gamma^*$	$\not{e}\mu + Z/\gamma^* + WW$	$\not{e}\mu + Z/\gamma^* + WW + t\bar{t}$
$e\mu\not{E}_T$	0.008 (+2.4σ)	0.16 (+1.0σ)	0.14 (+1.1σ)
$e\mu\not{E}_T j$	0.34 (+0.4σ)	0.45 (+0.1σ)	0.45 (+0.1σ)
$e\mu\not{E}_T jj$	0.01 (+2.3σ)	0.03 (+1.9σ)	0.31 (+0.5σ)
$e\mu\not{E}_T jjj$	0.38 (+0.3σ)	0.41 (+0.2σ)	0.71 (−0.5σ)
$\tilde{\mathcal{P}}$	0.03 (+1.9σ)	0.11 (+1.2σ)	0.72 (−0.6σ)

into account" the many regions that have been considered within this final state.

(vi) After obtaining $\mathcal{P}$ for each final state, the values are combined into a single quantity $\tilde{\mathcal{P}}$, which takes into account the many final states that have been considered. $\tilde{\mathcal{P}}$ takes on values in the half-open interval $(0, 1]$; the potential presence of new high p_T physics would be indicated by finding $\tilde{\mathcal{P}}$ to be small.

APPLICATION TO $e\mu X$

The $e\mu X$ data set and basic selection criteria are identical to those used in the published $t\bar{t}$ cross section analysis for the dilepton channels (2), requiring one or more isolated electrons and one or more isolated muons, each with $p_T > 15$ GeV. The number of events expected for the dominant backgrounds in the populated final states are given in Table 1. Other final states that are analyzed, but which happen to contain zero events, include $e\mu\not{E}_T jjjj$, $ee\mu\not{E}_T$, $e\mu\mu$, and $e\mu\not{E}_T \gamma$.

Top quark and W boson pair production are excellent examples of the types of processes that we would expect Sleuth to find, and we use these to test the sensitivity of the method. We find $\tilde{\mathcal{P}} > 2.0$ standard deviations (σ) in over 50% of an ensemble of mock samples when the background includes $\not{e}\mu + Z/\gamma^*$ only (i.e., all but WW and $t\bar{t}$), and we find $\tilde{\mathcal{P}} > 2.0\sigma$ in nearly 30% of an ensemble of mock samples when the background includes $\not{e}\mu + Z/\gamma^* + WW$ (i.e., all but $t\bar{t}$). The results for DØ data are shown in Table 2. It is instructive to compare Table 2 to the results obtained from a dedicated top quark search in the same $e\mu X$ data (2), in which an excess corresponding to 2.75σ is observed, recalling that Sleuth does not "know" anything about the signals (WW and $t\bar{t}$) that we have asked it to find. (1)

CONCLUSIONS

We have developed a quasi-model-independent technique for searching for new high p_T physics. After defining final states and constructing a rule that identifies a set of relevant variables for any particular final state, we systematically search for regions of excess within the space of those variables, and rigorously quantify the significance of any observed excess. Sleuth has been applied to DØ $e\mu X$ data. Removing WW and $t\bar{t}$ from the estimated background, we find indications of these signals in an ensemble of mock experiments and in the data. Including all background processes, we find no evidence of new physics at high p_T in these data.

REFERENCES

1. DØ Collaboration, B. Abbott *et al.*, *Phys. Rev. D* **62**, 92004 (2000).
2. DØ Collaboration, S. Abachi *et al.*, *Phys. Rev. Lett.* **79**, 1203 (1997).

Fast tracking in hadron collider experiments

Nikos Konstantinidis[a]* and Hans Drevermann[b]

[a]Santa Cruz Institute for Particle Physics, University of California, Santa Cruz CA 95064, USA
[b]CERN – EP Division, CH-1211 Geneva 23, Switzerland

Abstract. We present two algorithms that, when used sequentially, can reduce the combinatorics before any track reconstruction in the high occupancy tracking environments of future hadron collider experiments. The first algorithm finds the z-position of the primary physics interaction; the second selects groups of hits consistent with tracks coming from this z-position, rejecting most pile-up/noise/ghost hits. We demonstrate with examples of simulated events from ATLAS at the LHC that the algorithms are flexible, robust and efficient and at the same time fast enough to be used at the second level trigger for filtering the data before applying any tracking algorithms.

INTRODUCTION

Triggering is one of the greatest challenges at hadron collider experiments and will be even more so in the future. For example, at the LHC beams will be colliding every 25 ns and the time available for second level trigger algorithms will be about 10 ms. To make things worse, the pp (or p$\bar{\text{p}}$) interaction leading to the interesting physics process (referred to as *the physics interaction or physics event* hereafter) will be accompanied by several minimum bias interactions (referred to as *the pile-up interactions or pile-up events* hereafter) occurring simultaneously (~5 at the Tevatron Run-IIb, ~20 at the LHC design luminosity). This adds significantly to the complexity of the events.

The consequences are particularly severe for the tracking algorithms. A typical ATLAS event at the LHC design luminosity contains about 30000 silicon spacepoints[1]. This high hit occupancy (especially in the inner detector layers) leads to long execution times (due to combinatorics) and has a cost in performance (due to hit misassociation).

An approach to reduce combinatorics and hence execution time has been to apply the tracking algorithms in so-called Regions of Interest (RoI), which are rectangular slices in (η, ϕ) but opening up in z towards the beam line to account for the uncertainty on the z-position of the physics interaction (an example of an RoI in ATLAS is shown in Fig. 1b,c). The position of the RoI is determined by the first level trigger information from the outer

---

* Permanent address: CERN – EP Division, CH-1211 Geneva 23, Switzerland
[1] Hits and spacepoints have identical meaning throughout this report

detectors (calorimeters and/or muon chambers). The size depends on the type of RoI; for example, isolated electron or muon RoI are narrow, while jet RoI are wider. Restricting to RoI is natural at the trigger level and leads to some reduction of combinatorics. However, the problem of combinatorics is still present due to the need to extend the RoI in z towards the beam line, where the hit occupancy is the highest.

In this report, we are proposing a new approach, which makes use of the differences between the physics and the pile-up events in order to clean up the spacepoints before applying any tracking algorithms. There are two important differences to exploit: (a) there is a significant spread in z of the various interactions (at the LHC $\sigma_z = 6$ cm) and (b) the physics event has on average higher transverse momentum than the pile-up.

The new approach proceeds in two steps. First, the z position of the physics interaction is determined. Then, hits are rejected if they are not consistent with a track coming from the above z position. The remaining ones are grouped into track candidates.

The algorithms implementing the above ideas are described below. They are applied to example cases from ATLAS at the LHC design luminosity. These include RoI from isolated $p_T = 40\,\text{GeV}/c$ electrons, from thin QCD jets (which are the major background to electron first level triggers) and from WH events with $m_H = 100\,\text{GeV}/c^2$ and H $\to$ b$\bar{\text{b}}$ (denoted WH(100) hereafter). The size of the first two types of RoI is ($\Delta\eta = 0.2$, $\Delta\phi = 0.2$, $\Delta z = 22.4$ cm) while the WH(100) RoI are wider ($\Delta\eta = 1.0$, $\Delta\phi = 1.0$, $\Delta z = 30$ cm). The execution time of the algorithms was measured on a 600 MHz Pentium-III processor.

THE Z-FINDER

The general principle of the algorithm is summarised in the following steps:

- The RoI is divided into many small bins in ϕ.

- In a given ϕ bin, each pair of hits from different layers is used to calculate a z by linear extrapolation to the beam line (this assumes a solenoidal magnetic field, where the helix trajectories of charged tracks are straight lines in the ρ-z projection).

- A one-dimensional histogram is filled with the z calculated for each hit pair.

- The z position of the physics event is taken to be the one corresponding to the z-histogram bin with the maximum number of entries.

An example of a z-histogram from an electron RoI in ATLAS is shown in Fig. 1a.

The key point of the algorithm is the division of the RoI in small ϕ bins. This has a double significance: (a) it gives naturally more weight to high p_T tracks and therefore to the physics event as opposed to the pile-up events and (b) it reduces drastically the combinatorics, hence minimising the quadratic time behaviour of the algorithm. This is easier to understand with an example: depending on the size of the ϕ bins, the hits of high p_T tracks will be contained in one ϕ bin. Assuming that there are seven hits per track, one such high p_T track will give $7 \times 6/2 = 21$ hit pairs and therefore entries in the z-histogram. In other words, by design, there are more entries from high p_T tracks in the z-histogram, than from low p_T tracks, the hits of which fall into several ϕ bins. In order to avoid loss of efficiency due to binning effects (tracks on the boundary of bins would give fewer hit pairs) hits from a given ϕ bin are also paired with those from the two neighbouring bins.

The ϕ bin size is a parameter that can be adjusted. Its optimal value depends on the type of the RoI. In RoI from isolated electrons or muons high p_T, the ϕ binning should be as fine as possible ($0.2°$ in the ATLAS examples below), since this minimises the number of random hit pairs. In jet RoI, it has to be wider since there are more tracks with moderate p_T ($0.4°$ in the ATLAS WH(100) example).

The bin size of the z-histogram should also be kept small in order to minimise potential random fluctuations. The optimal bin size is similar to the expected resolution on the z position of the physics event.

The efficiency of the algorithm on electron RoI in ATLAS is
$$\varepsilon = 97.5 \pm 0.4\%,$$

independent of η. Most of the loss is due to detector inefficiencies and hard bremsstrahlung radiation, which reduces significantly the p_T of the track. The z position is determined with a resolution of $\sim 180 \pm 5\,\mu m$, with no tails. Similar results are achieved with WH(100) jet RoI.

The execution time as a function of the number of hits (N_h) for electron and thin QCD jet RoI is $t = 35 + 1.24 \times N_h + 0.0004 \times N_h^2$ (in μs). It can be seen that the coefficient of the quadratic term is very small; as mentioned before, this is due to the fine binning in ϕ. The average execution time for QCD jet RoI ($\langle N_h \rangle \sim 250$), which constitute the vast majority of first level trigger electron RoI, is
$$\langle t \rangle = 370\,\mu s.$$

Apart from the ϕ and z bin sizes, adjustable parameters are also the first and last detector layers to be used. This makes the algorithm flexible to use in physics cases as different as single isolated high p_T track RoI and jet RoI like in the ATLAS examples. Since the electron RoI contain only one (high p_T) track giving the z position information it is necessary to use hits from all layers in order to benefit from the combinatorics, whereas WH(100) RoI contain many tracks of moderate p_T and therefore using only the first three detector layers suffices.

The flexibility of the algorithm ensures its robustness to changes in the detector or background conditions. For instance, a study of the ATLAS electron RoI without using the hits at the first pixel layer showed that the performance of the algorithm degrades very slightly (efficiency $\sim 95\%$, z-resolution $\sim 300\,\mu m$).

THE HIT FILTER

The hit filtering algorithm is based on the fact that all hits of a track of sufficiently high p_T are contained in a small solid angle in (η, ϕ) that starts from the track's initial z position, in contrast to hits from tracks originating from different z positions. The principle of the algorithm can be described in the following steps:

- Given the z-position of the physics event, a 2D-histogram in (η, ϕ) is constructed.

- In each (η, ϕ) bin, the number N_L of different detector layers containing hits is counted. If N_L is above a given threshold all the hits in this bin are accepted, otherwise they are rejected.

- Hits from neighbouring bins are clustered into groups (this is done to eliminate binning effects). Very often, a group contains the hits of just one track.

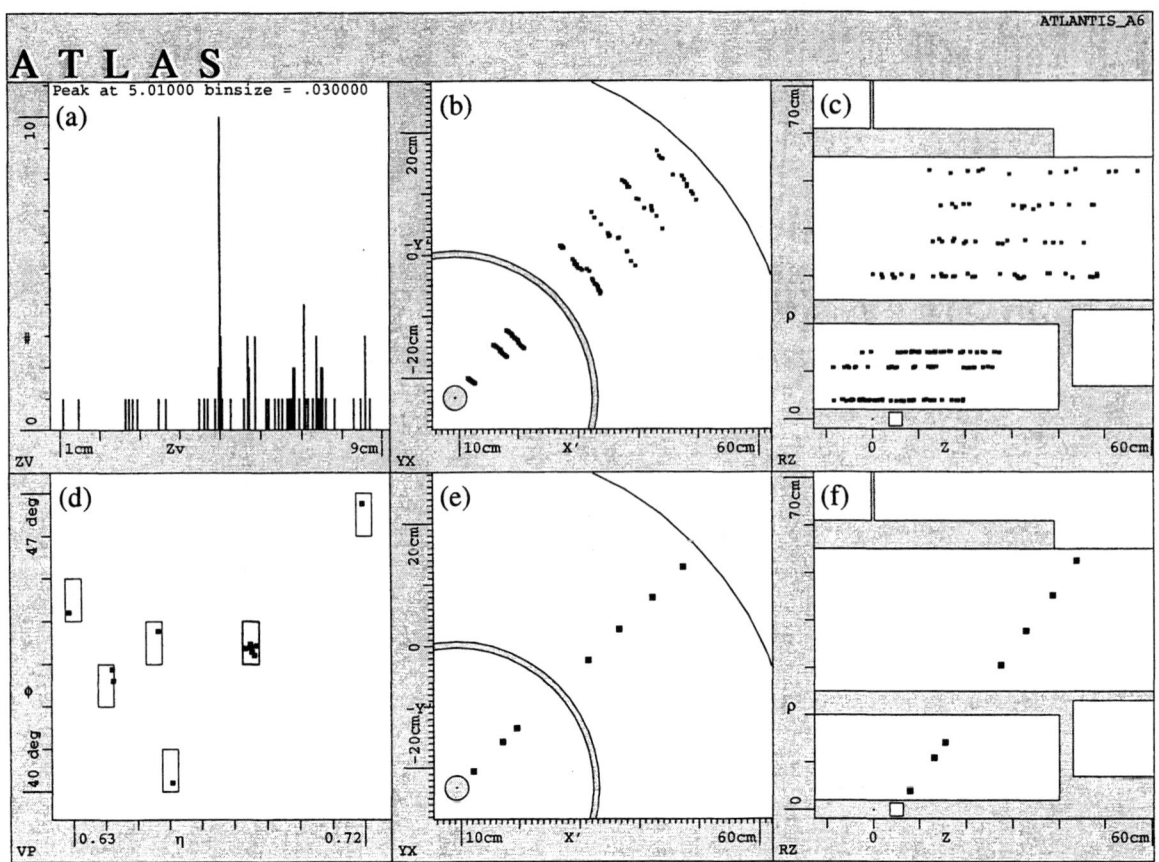

FIGURE 1. From an electron RoI: (a) the z-histogram (shown only around the initial z-position of the electron; the rest is flat), (b,c) x-y and ρ-z views of the RoI before hit filtering, (d) the part of the 2D-histogram in (η, φ) containing electron (only bins containing hits are drawn; the hits from the electron are all concentrated in one bin) and (e,f) x-y and ρ-z views of the RoI after filtering.

The size of the bins in η and φ can be adjusted according to the physics case. The size in η depends on the detector resolution in the z coordinate and the resolution on the reconstructed z position of the physics event. The size in φ determines a p_T cut-off, below which a track spans into many bins in φ and thus the algorithm starts to become inefficient. In ATLAS, for an η bin size of 0.004 and a φ bin size of 2°, the algorithm is essentially 100 % efficient for tracks with $p_T > 2\,\text{GeV}/c$, which is a cut-off commonly used for triggering.

The time behaviour of the algorithm is linear. In ATLAS, $t = 2.5 \times N_h$ (in μs). This leads to an average time of
$$\langle t \rangle = 630\,\mu s$$
for QCD jet RoI.

About 95 % of the hits are rejected by the filter. Thus, the subsequent track reconstruction can be very fast, especially since it can be restricted to the individual groups of hits. An example of applying these algorithms on an electron RoI in ATLAS is shown in Fig. 1.

CONCLUSIONS

Improving the speed of tracking algorithms is extremely important for future hadron collider experiments, where the presence of many pile-up interactions simultaneously recorded with the interesting physics event leads to a very high hit occupancy in the tracking detectors.

We propose to concentrate on the physics interaction early at the trigger level, by first finding its z position and then selecting only groups of hits that are compatible with having been produced by tracks originating from that z position. We have described two algorithms following this scheme. In examples from ATLAS, the algorithms have been shown to be both very fast and efficient. They are also general enough to be applicable to other hadron collider experiments, in a wide range of physics cases (such as single isolated tracks or jets). Finally, the algorithms are flexible and hence can be adapted easily to different detector or background conditions, a characteristic that makes them suitable to be used at triggering.

More Performance Results and Implementation of an Object Oriented Track Reconstruction Model in Different OO Frameworks

Irwin GAINES
Fermi National Accelerator Lab, Batavia, IL 60510, USA

Sijin QIAN
Brookhaven National Lab, Upton, NY 11973, USA

Abstract. This is an update of the report about an Object Oriented (OO) track reconstruction model, which was presented in the previous AIHENP'99 at Crete, Greece. The OO model for the Kalman filtering method has been designed for high energy physics experiments at high luminosity hadron colliders. It has been coded in the C++ programming language and successfully implemented into a few different OO computing environments of the CMS and ATLAS experiments at the future Large Hadron Collider at CERN. We shall report: (1) more performance result; (2) implementing the OO model into the new SW OO framework "Athena" of ATLAS experiment and some upgrades of the OO model itself.

INTRODUCTION

In AIHENP'99 at Crete (Greece), we had reported an Object Oriented (OO) model [1] for track reconstruction (with simultaneous pattern recognition and track fitting by the Kalman filtering method [2]) which was initially designed and coded in the C++ programming language in its first version in 1995 for the CMS experiment at the Large Hadron Collider (LHC) at CERN. Since 1995, along with the development of C++ language itself and our understanding about the OO concept, the model has been re-designed (or partially) a couple of times. Since 1998, the OO model and its C++ code have been successfully re-used in ATLAS (i.e., another major LHC experiment), where it has been implemented into 3 different ATLAS software environments one after another: i.e. (1) the level-2 trigger OO reference software framework, (2) the C-trig (a C-language based software framework delicate to the trigger performance study), and (3) the new ATLAS offline OO architecture "Athena". The latter two are new since AIHENP'99.

The history of the OO model (up to the international conference "Computing in Nuclear and High Energy Physics (CHEP'2000)" at Padova, Italy) has been documented in [1,3], where the main features of this model have been summarized as: (a) Class design according to the OO paradigm. The class design is based on the proven data concepts in HEP track reconstruction, so that hopefully it can be rather easily adopted by non-expert class users. (b) The OO model had successfully re-used some well encapsulated FORTRAN modules; it later had been converted into a pure C++ package in a straightforward manner. (c) The OO model is flexible enough to be re-used in different HEP experiments, only the implementation of layer class is different.

In this short report, Section 2 describes the new features of the OO package and the new performance result since AIHENP'99; Section 3 outlines the implementation of this OO model into the ATLAS new software environment "Athena". More details about the OO model can be found in [3].

NEW FEATURES OF THE OO MODEL AND NEW PERFORMANCE

The latest class diagram of the OO model for track reconstruction is shown in Fig.1. The original OO

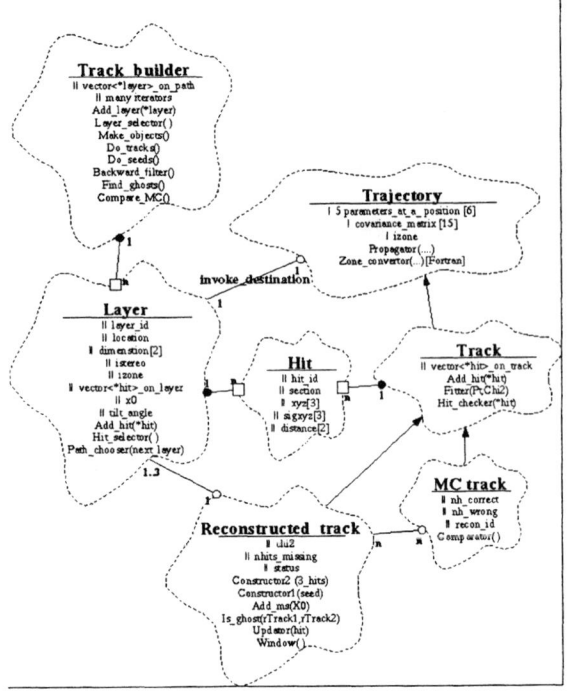

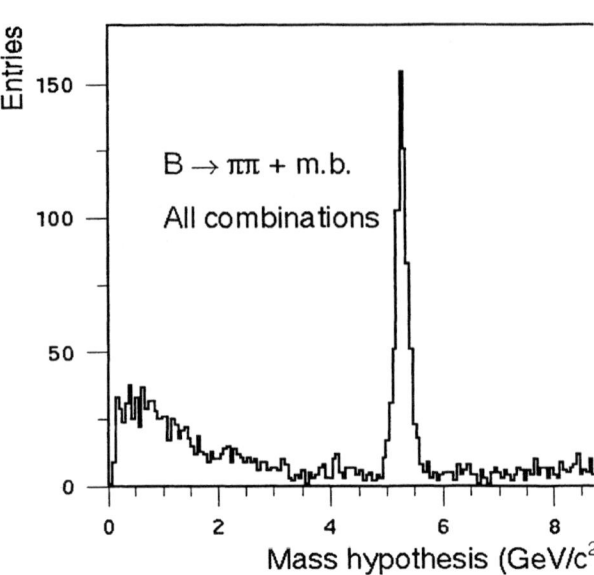

FIGURE 2. Reconstruction of the B $\rightarrow \pi\pi$ mass hypothesis in the ATLAS Level-2 trigger using the Kalman filtering method

FIGURE 1. Class diagram of the OO tracker model

TABLE 1. Statistics of the Benchmarking Measurement (for B-physics events at low luminosity)

$P_{T\,min}$ (Gev/c)	TRT Seeded		Pixel Seeded	
	Number of seeds	Execution time (millisecond)	Number of seeds	Execution time (millisecond)
0.5	<114>	<112>/event	<36>	<19>/event
1.0	<55>	<54>/event	<16>	<9>/event

model was stand alone, as it predated the development of OO software frameworks in both experiments, CMS and ATLAS. Since 1999, CMS developed a powerful reconstruction framework known as ORCA. After all necessary functionalities for input objects in ORCA became standardized in autumn of 1999, we preliminarily integrated the OO tracker model into ORCA by inputting the ORCA reconstructed hits to the model and reconstructing tracks. ORCA was continuing to develop rapidly during the past year, so the further integration has been paused to wait for the final ORCA detector and track classes. In ATLAS, in the summer of 1999, another seeding method has been implemented which uses the seeds produced by a pixel detector package and starts the Kalman filtering process from the inner-most position outwards. This results in a better performance of the OO model due to the preciseness of the pixel hits. Then, the energy loss correction was implemented at the end of 1999. Also, in the second half of 1999, the OO model had been integrated into another ATLAS level-2 trigger framework "Ctrig" which is C-based but can produce many ntuples to study the trigger performance. In Ctrig, this OO model has been extensively tested. The new performance result has contributed to the ATLAS Trigger/DAQ Technical Proposal [4] published in 6/2000. Despite the difference among two LHC experiments and different software environments, we could put the I/O code in two encapsulated functions, then the implementation structures for different environments are very similar.

Among many new performance results documented in [5], due to the limitation on the article length, here we only show some typical results about the execution speed (Table 1) and an example of B-physics study in the channel B $\rightarrow \pi\pi$ (Fig.2). The standard data sets (for testing various algorithms) are single-track (μ's, π's and electrons at different energies) events and B-physics events at low luminosity. The computer used is the 600 MHz AMD Athlon under Linux. The memory usage of the OO model is in the order of 10 Mbyte.

IMPLEMENTATION OF THE OO MODEL INTO ATLAS NEW OO ENVIRONMENT

Since the middle of 2000, ATLAS launched a new OO framework "Athena" based on the Gaudi architecture of LHC-b experiment. The design principles of Gaudi (Athena) include (a) separation

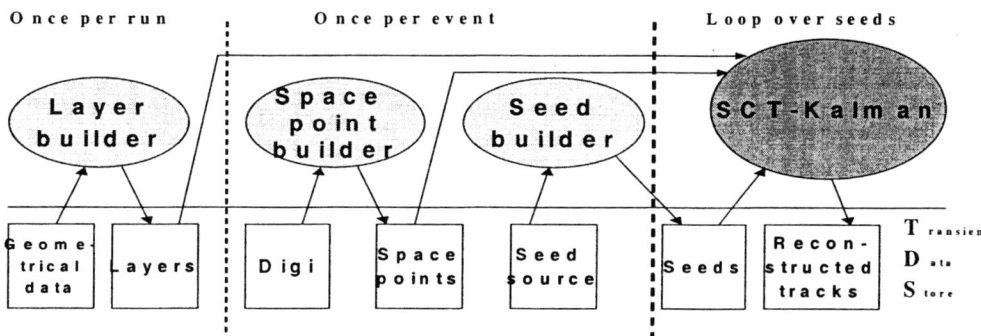

FIGURE 3. Structural diagram of Athena implementation of the OO model

between "data" and "algorithm", (b) three basic categories of data (i.e. event, detector and statistical data) and (c) separation between "transient" and "persistent" data, etc. The "Transient Data Store (TDS)" provides a means to promote the proper modularization of a package. By utilizing the TDS, we have decomposed the OO tracking package (with the name "SCT-Kalman") into a scheme (Fig.3), where the geometrical, the seeding and the 3-D space oints parts have been factored out. The intermediate objects (e.g. layers, 3-D hits and seeds, etc.) are passing through the TDS before being input into other modules. By this decomposition, a builder module can include several different modes that can be selected in the run time by the jobOptions file (a kind of run-time data control card). For example, the space point builder can have different space points building methods or an ASCII file reader, etc.; the seed builder can have TRT, pixel, muon or calorimeter seed modes, etc. Also, this scheme leaves room to accommodate the calibration correction (in the space points builder) and the alignment correction (in the layer builder) in future.

It is worthwhile to mention that along with the accumulation of our implementation experience and the improvement on the features of software framework, we are able to implement the OO model into a new OO environment in a shorter and shorter period. For instance, among 3 software frameworks in ATLAS experiment, it took a couple of years (i.e. 1998-1999) to implement the OO model into the ATLAS level-2 trigger OO reference software framework; several months (i.e. 8/1999-12/1999) into the Ctrig framework; and just a few days (one day in 8/2000 and a few weekends in 9/2000) into the ATLAS' new Athena framework preliminarily.

SUMMARY AND PROSPECTS

We are moving towards a realistic OO track reconstruction model for HEP experiments. Its memory usage is moderate, its track finding efficiency is satisfactory, and its execution speed is approaching the level-2 trigger requirement. It can be used in different HEP experiments and software frameworks with only minor modifications.

The integration of the OO model into ORCA and Athena are just the beginning, more complete implementation continues. Then more performance studies in different environments can be carried out.

ACKNOWLEDGMENTS

We are grateful to Drs. S. Wynhoff and T. Todorov for their help on the integration of the OO model into ORCA, and to the ATLAS level-2 B-physics trigger team (especially Drs. S. Gonzalez and J. Baines) for carrying out the performance study. We also appreciate the great help from ATLAS architecture team and BNL SW group (including D. Quarrie, P. Calafiura and T. Wenaus, etc.) for implementing the OO model into Athena.

REFERENCES

1. I. Gaines, P. LeBrun and S. Qian, CMS TN/96-122 (1996); I. Gaines, T. Huehn and S. Qian, in the proceedings of CHEP'97 at Berlin, Germany (4/1997); also as CMS CR/1997-018; I. Gaines and S. Qian, in the proceedings of CHEP'98 at Chicago, U.S.A (9/1998), also filed as CMS CR/1998-023; I. Gaines and S. Qian, in the proceedings of AINENP'99 at Crete, Greece (4/1999); I. Gaines, S. Gonzalez and S. Qian, in the proceedings of CHEP'2000 at Padova, Italy (2/2000).

2. P. Billoir and S. Qian, Nucl. Inst. and Meth. **A294** (1990) pp. 219 and **A295** (1990) pp. 492.
3. http://www.usatlas.bnl.gov/~sijin/oo.html
   and   http://cmsdoc.cern.ch/~sijin/oo.html
4. ATLAS HTL/DAQ/DCS group, "ATLAS High-level Triggers, DAQ and DCS Technical Proposal", CERN/LHCC/2000-17(4/2000).
   http://atlasinfo.cern.ch/Atlas/GROUPS/DAQTRIG/SG/TP/draft_tp.html
5. J. Baines et al., "B-Physics Event Selection for the ATLAS High Level Trigger", ATLAS-DAQ-2000-031 (6/2000).

   http://documents.cern.ch/cgibin/setlink?base=atlnot&categ=Note&id=daq-2000-031

Simultaneous Tracking and Vertexing with Elastic Templates

Andrew Haas

University of Washington
Seattle, WA USA

Abstract. The Elastic Templates algorithm uses simulated mean-field annealing to find near-optimal solutions of problems that are both combinatorial and continuous. In high-energy physics experiments it has been applied to the tracking problem, which is to minimize the distances between track templates and the hits belonging on each track. I will explain recent attempts to extend and improve this approach for reconstructing data from the DØ experiment. Vertex templates have been added, and the algorithm simultaneously minimizes the distances between the track templates and the vertex assigned to each track. A final step aimed at finding secondary vertices in jets is also explored.

INTRODUCTION

A challenging part of conducting a high-energy particle physics experiment is reconstructing the physical processes that took place in a collision from the signals left in the detector. In particular, the "tracking problem" is to reconstruct the paths of high-energy particles from the set of positional measurements left by ionization in the tracking chambers. The primary interaction points, or "primary vertices", must also be found, as well as any "secondary vertices" from the decay of long-lived particles inside the tracking chambers. Traditionally, these two tasks, tracking and vertexing, have been done separately, with tracking occurring before vertex finding. This paper presents a method for performing both tasks at once, thus taking advantage of all available vertex information during the track fitting stage. The algorithm has been tested on Monte Carlo data from the D0 experiment at Fermilab, and preliminary results will be shown.

ELASTIC TEMPLATES

The algorithm used is based on the elastic templates method of pattern recognition[1]. An "energy" is defined, which is the χ^2 fit of all objects under consideration. Each term in the energy is weighted by the probability, P, that the assignment is correct. This is a generalization of standard fitting, where assignments are made with either probability zero or one. Also, objects can be assigned to nothing with up to unit probability to correctly account for noise in the detector or hits from tracks with momenta too low to be reconstructed. But, there is a penalty (an addition to the energy) for unassigned objects, to bias objects to be assigned whenever the option exists.

Here, the elastic method has been extended to include more than just tracks and hits. The algorithm objects include hits, tracks, vertices, and a beam constraint for the primary vertices. The complete energy function used is:

$$E(\{P\},\{\vec{\pi}\}) = \sum_i \sum_a P_{ia} D_{ia}(\vec{x}_i, \vec{\pi}_a) + \lambda \sum_i (\sum_a P_{ia} - 1)^2$$

where the vector x is the parameters of object i, the vector π is the parameters of object a, D is the squared distance (in standard deviations) between two objects, and λ is the penalty for an object to be unassigned.

The best fit of the data to the model used is at the global minimum of the energy. This is found approximately through simulated mean-field annealing, in order to avoid the copious local minima[2]. Thermal energy is added with a Boltzman distribution at a temperature, T, and the assignment probabilities are calculated for each object under the assumption that all other objects remain in thermal equilibrium. This assumption is a good one, since changing the assignment of object a to object b often will not change the equilibrium parameters of object b by much, thus all other objects remain near thermal equilibrium with object b. The assumption becomes more valid when the temperature is lowered slowly, and in the adiabatic limit, the assumption is exact. The formula for calculating new assignment probabilities is:

$$P_{ia} = \frac{e^{(-\beta D_{ia})}}{e^{(-\beta \lambda)} + \sum_b e^{(-\beta D_{ib})}}$$

where β is the inverse temperature.

Thus, beginning at a high temperature, the system is slowly cooled, with equilibrium being restored each time the temperature is decreased. Equilibrium is found by minimizing the energy of each object with respect to the objects assigned to it. The process ends when almost all the assignments are zero or one.

TRACK SEEDING

The elastic templates algorithm can change the parameters of objects and assignments of objects, but it cannot create or destroy objects. Also, the farther an object begins from its correct (as judged by MC) position, the more likely it is to fall into a local minimum of the energy and become falsely reconstructed. Thus, a sophisticated method of initializing the objects is required, particularly for the tracks.

Primary vertex seeds are obtained from another algorithm, either from the online system or a fast offline method. These seeds are then used to divide the detector into "slices" of parameter space in phi and psudeorapidity relative to them. For each unique combination of three hits in each slice, the five parameters of the track passing through them are calculated. This set of hits is entered into the corresponding bin of a five dimensional histogram, represented as a sparse matrix. Once all combinations have been processed, the bins of the histogram are ordered from most to least occupied. Similar to a technique tried at ATLAS[3], the list of bins is then iterated through, and combinations of hits that contain a cluster already used by a hit in a bin with more

entries are removed from the bin. This process insures that each cluster is only used in one bin. Any bin with a combination of hits remaining forms a track seed, with the parameters of the track seed set to those of its histogram bin. Finally, the seed is fit to the hits in the bin, with each hit weighted by the number of times it occurs in the bin.

SECONDARY VERTEXING

For many important analyses, it is very important to determine whether a jet of particles emanates from a primary vertex on the beamline, or from a secondary vertex caused by the decay of a long-lived particle.

The total energy will always be lowered by adding a secondary vertex seed to the elastic algorithm, simply because there are more degrees of freedom. However, the expectation is that the lowering of the energy will be greater when the jet truly contains a secondary vertex. This is similar to the approach used in many analyses from the ALEPH experiment at CERN[4]. There, the χ^2 of the fit is compared with and without a secondary vertex at many trial positions. They have found that, as expected, the difference in the fit is larger for jets that really do originate from secondary decays.

After elastic fitting with only primary vertices, the algorithm looks for secondary vertices. For each jet, a secondary vertex seed is placed along the jet axis, and another elastic fit is performed. Several different seed positions are attempted for each jet, at different distances from the primary vertex. The final position and error of the secondary vertex that resulted the lowest global energy is stored, along with the difference in energy as compared to the case with no secondary vertex seed.

There are several advantages to this method. All vertices (primary and secondary) and tracks can move continuously while the difference in energy is found, tracks can change which hits are assigned to them during the fitting to the secondary vertex, and the assignment of tracks to the primary or secondary vertex is optimally calculated.

RESULTS

The seeding algorithm limits the tracking efficiency. Thus, it is tuned for high efficiency at the cost of purity, and the elastic algorithm can then discard tracks that have been assigned few hits and keep the good tracks. The efficiency is greater than for the experiment's standard tracking algorithm, and purity is still reasonable (84%). Without insisting that clusters only be used once, by the "strongest" track, far more fake seeds are created.

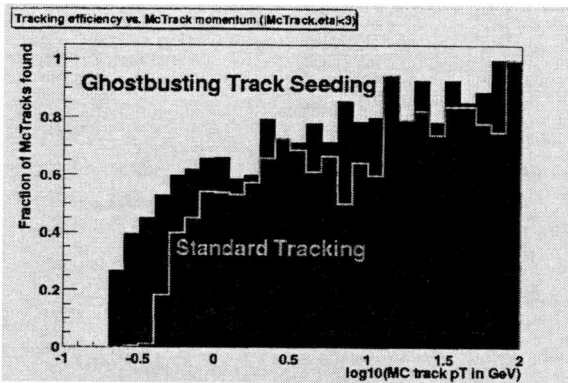

FIGURE 1. Comparison of seeding efficiency and the standard tracking algorithm, as a function of transverse track momentum, for Monte Carlo Zμμ events in the DØ detector

The resolution of impact parameter, the distance of a track from its true vertex, is a good test of the algorithm's fitting performance. As seen in the figure below, fitting to the primary vertex increases the impact parameter resolution of the tracks.

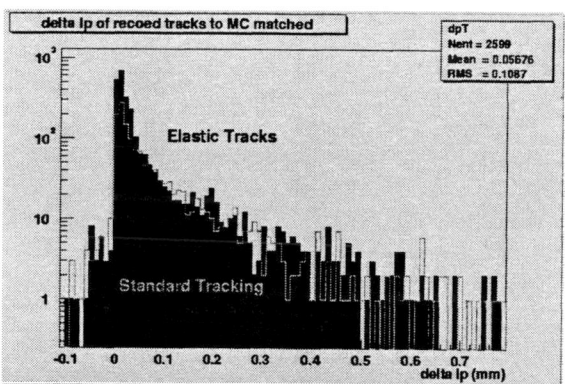

FIGURE 2. Comparison of impact parameter resolution for elastic fitting and the standard tracking algorithm for Monte Carlo Zμμ events in the DØ detector

In events with secondary vertices, such as the Zbb events plotted below, the impact parameter resolution of tracks from the secondary vertices is still better with this vertex fitting. This shows that tracks are being properly assigned to either the primary or secondary vertex seed in the event.

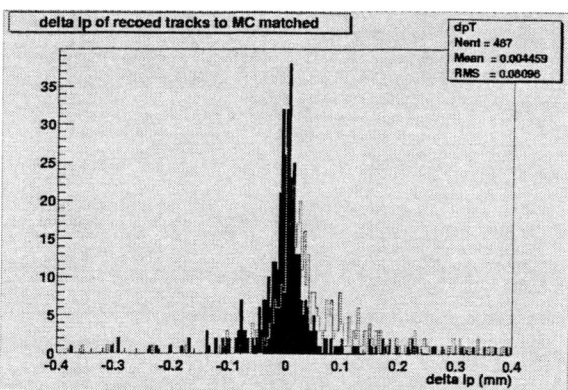

FIGURE 3. Comparison of impact parameter resolution for elastic fitting and the standard tracking algorithm for tracks from secondary b-vertices in Monte Carlo Zbb events

Lastly, the claim that secondary vertices can be detected by how much the fit improves (how much the energy decreases) was tested. As seen below, events with no real secondary vertices (Zμμ) had, on average, much smaller changes in energy when adding a secondary seed than did events that do contain real secondary vertices (Zbb).

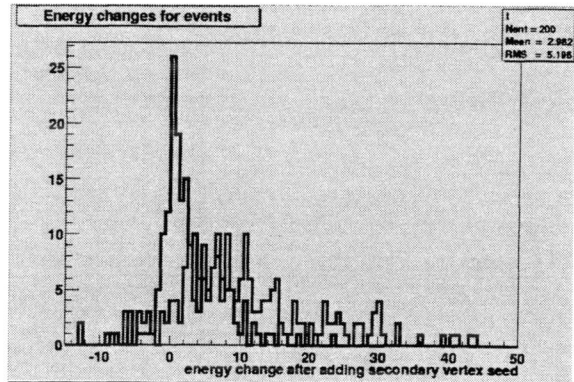

FIGURE 4. Comparison of the energy change caused by adding a secondary vertex seed for Monte Carlo Zμμ events (black) and Zbb events (blue) in the DØ detector

REFERENCES

1. M. Ohlsson, C. Peterson and A. L. Yuille, *Computer Physics Communications* **71**, 77-98 (1992).

2. M. Ohlsson, *Computer Physics Communications* **77**, 19-32 (1993).

3. Magnus Lindstrom, *Nuclear Instruments and Methods in Physics Research A* **357**, 129-149 (1995).

4. Stephen A. Armstrong, *Ph.D. Thesis*, University of Wisconsin-Madison (1998).

Vertex finding prior to tracking in magnetic field

Yatsunenko Yu. A.

Joint Institute for Nuclear Research, yay@nusun2.jinr.ru

Abstract. The "Integral mathematics model of the track pattern" allows to locate the positions of the primary intensive vertices, to subdivide the hit array into several "regions of interest" that simplify the tracking. The recognition and analysis of the jets is also possible prior to the tracking. The ideas have been successfully tested using the D0 Monte-Carlo hits data.

MATHEMATICAL MODEL OF THE TRACK PATTERN

A most suitable word in the track recognition problem is, probably, *DOUBT*. i.e. it is not clear enough : "this hit OR that hit OR..." belongs to a trajectory. In a mathematical formalism one can express these "OR" as the sum of probabilities, that was done in the "Integral mathematical model of the track pattern"[1]

$$R(c) = \sum_{n=1}^{N} \sum_{m=1}^{M_n} \exp(-(F(z_n;c)-a_{mn})^2/2\sigma^2) \quad (1)$$

where $F(z_n;c)$ is the function of the ordinate z and set $\{c\}$ of parameters which describe the trajectory; the $\{a_{mn};\ m=1,\ldots,M_n\}$ is the set of the M_n measured hits by a detector placed at $z_n (n=1,\ldots,N)$ where N is the total number of the detectors. In the space of the trajectory parameters $\{c\}$ this function has several big maxima (the real trajectories) and a lot of small ones, which reflect a noise or/and combinatorial background. The amplitude of a "good" peak makes the junction of "K" (it is number of hits along a real trajectory $\{c^*\}$) and the value of the Chi Square(χ^2):

$$R(c^*) \approx K - \chi^2(c^*)/2 \quad (2)$$

Thus, this analytical function of the trajectory's parameters includes in itself all the detected hits. However, it is *difficult* (but *possible!*) to analyze this multi-extremum function.

THE VERTEX FUNCTIONS

The general rule of the definition of the vertex function is given in thepaper[1]. The case of the magnetic field was considered in paper [2] where the definition of the helix-like trajectories was chosen as follows:

$$x(z)=x_o+r_o\sin\alpha +r_o\sin[2A\cdot(z-z_o)-\alpha]$$
$$y(z)=y_o-r_o\cos\alpha +r_o\cos[2A\cdot(z-z_o)-\alpha]$$
$$\alpha\equiv\varphi_o;\quad A\equiv eH/(2cP\cdot\cos\theta_o) \equiv eH/(2cP_L);$$
$$r_o\equiv Pc\cdot\sin\theta_o/eH\equiv P_T\cdot c/eH \quad (3)$$

The function of the vertex position (x_o,y_o,z_o) includes also all the measured hits and has the form [2]:

$$V(x_o,y_o,z_o)=\sum_{k=1}^{M_1}\sum_{n=1}^{N}\sum_{m=1}^{M_n} \exp(-[(F_{1k}-F_{mn})(z_o-z_k)/(z_k-z_n)+P_{1kmn}-F_{mn}]^2/2\sigma^2) \quad (4)$$

where,
$$P_{1kmn}\equiv \text{Arctg}[(y_{1k}-y_{mn})/(x_{1k}-x_{mn})] \quad (5)$$
$$F_{mn}(x_o,y_o)\equiv \text{Arctg}[(y_{mn}-y_o)/(x_{mn}-x_o)].$$

There is no way to the track reconstruction in eq.4, thus the $V(x_o,y_o,z_o)$ is an independent description of the track pattern. The view of the $V(x_o,y_o,z_o=32.52$ cm) is shown on Fig. 1. This was for a D0 Monte-Carlo event, the z_o-vertex was found, so this 2 dimensional plot (x_o,y_o) could be drawn. The position of the global maximum of this vertex function corresponds to the real vertex $x_o=0, y_o=0$.

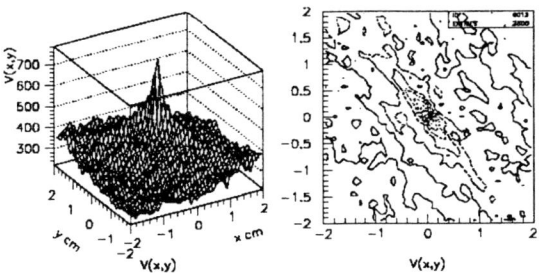

FIGURE 1. The (x,y) view of $V(x_o,y_o,z_o=32.52\text{ cm})$ (eq.3) for a D0 Monte-Carlo event. Main maximum coincides with the (x=0 y=0) vertex.

It is the principal idea of these vertex functions: most higher maxima reflect the positions of the vertices. If the variables of the vertex function are not separated (like $V(x_o,y_o,z_o)$) then only mathe-matical methods of the analysis of multi-extremum functions are to be used. If the variables can be separated or there is one variable (the beam position is known: $x_o=0, y_o=0$, thus there is one variable function - $V(z_o)$) then the well known histogram method can be used.

Thus the task of finding the intensive primary vertices consists in the finding of the z-position of the main maximum of the $V(z_o)$ histogram, then removing of all the hits associated with that maximum ("a cleaning" of the vertex,- to avoid any fake vertices) for searching for a subsequent vertex.

Such a procedure of the "cleaning" of the vertices is demonstrated in Fig. 2: there were three Monte-Carlo vertices generated for the D0 event and all of them were successively found.

ACCURACY OF THE Z-VERTEX FINDING

The parameter $\sigma(z_o)$ is determined by the detector accuracy (σ_φ, for instance), angular spreading of the particles (or longitudinal momentum P_L), number of detectors (N) and multiplicity (M):

$$\sigma(z_o) \approx (2c/eH)\, \overline{P_L} \cdot \overline{\sigma_\varphi} / \sqrt{MN} \qquad (6)$$

We analyzed 10 Monte-Carlo events generated for the D0 experiment. The histogram method has the accuracy: –0.03+/-0.11 cm, and results are better if the fitting by Gauss-like function is applied: 0.007+/-0.024 cm. It is a good starting point for the following tracking.

THE DETERMINATION OF JETS

The histogram method can be used for the analysis of jets because the kinematical parameters (eq.3) which describe a jet are separated. When the position of the primary vertex has been determined the kinematical parameters are re-calculated and booked in the one-dimensional or two-dimensional histogram. The D0 Monte-Carlo event (Fig. 3) shows that a jet can demonstrate itself as a prominent peak in the histograms of (α,θ_o).

This is a preliminary result, however it gives a way to perform a macro analysis of the jets: if the direction (α,θ_o) of a jet is found, then one can use a calorimeter (an estimation of the energy deposition); the (p_L, p_T) image of the jet gives also the estimation of the vector of the momentum (P). All these values can be used in the following precise tracking or be applied to a kind of jet trigger.

THE KINEMATICAL SEPARATION

When the parameters (α,θ_o) have been found then the associated hits can be booked in (X,Y;R,Z) histograms. It is done for each of the "sub-region" $(\alpha\pm\delta,\theta_o\pm\Delta)$. In Fig.4 are shown the initial plot(x,y) for all the hits and three such sub-regions. One can see that there is a more clear hit pattern in these sub-regions. This increases the reconstruction

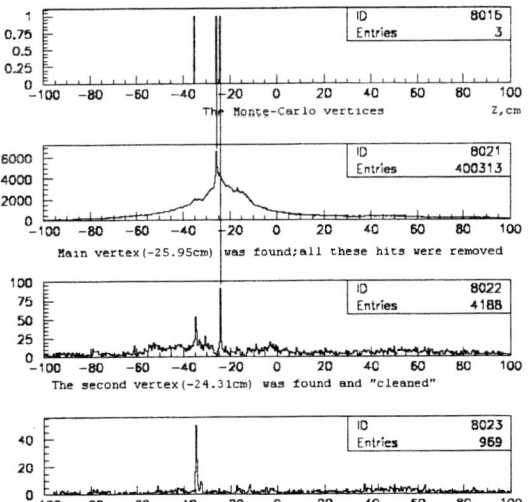

FIGURE 2. The determination and cleaning of the vertices of a D0 Monte-Carlo event.

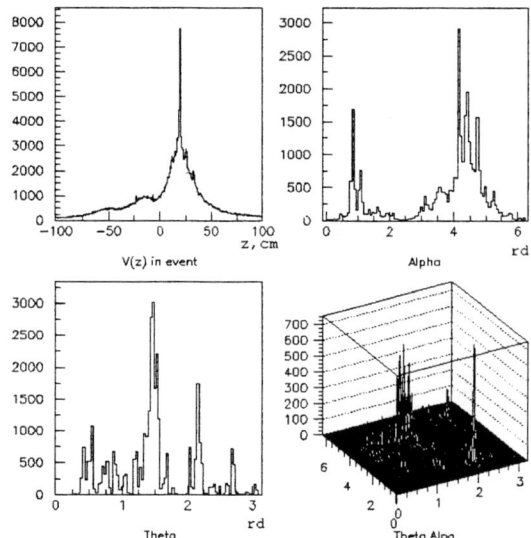

FIGURE 3. The upper plot shows the z-vertex histogram. The prominent peaks in kinematical parameters (Alpha; Theta) provide evidence for the existing of jets.

reliability and reduces the CPU-time. If there are "K" sub-regions for total number of hits "M", then each sub-region has approximately M/K hits and a typical number of operations for tracking $T_s \approx (M/K)^2 = M^2/K^2$. Then the total number of operations would be less "K^2" times in a parallel computation and K times for sequential tracking: $T_o \approx K*(M/K)^2 = M^2/K$.

CONCLUSIONS

Tracking, itself, requires sometimes several re-analysis of whole track pattern to avoid possible fake trajectories. To reduce this problem the next plan can be applied:

1) The determination of the position of the primary intensive vertices.

2) To find the rest of the hits which are not associated to the primary vertex (the procedure of "cleaning") for the following searching for a rare decay, for instance.

3) To divide of the whole hits' array into several "sub-regions" for tracking under appropriate conditions.

For the *trigger* purposes this "algebra" of a track pattern allows:

1) The global analysis of the jets.

2) The analysis of a region of interest (in P_T, for example) where the integral over this region can be examined in maximum value.

3) The analytical description of a chain of decays, which is searching for (in the "rest" of the hits) prior to the exact tracking.

ACKNOWLEDGMENTS

Thanks to the D0 experiment for the support of this report.

REFERENCES

1. Yu.Yatsunenko,"Nuclear Instruments and Methods" A287 (1990) 422-430.

2. Yu.Yatsunenko, S. Selunin. JINR,E1-92-333, Dubna, Russia),1992.

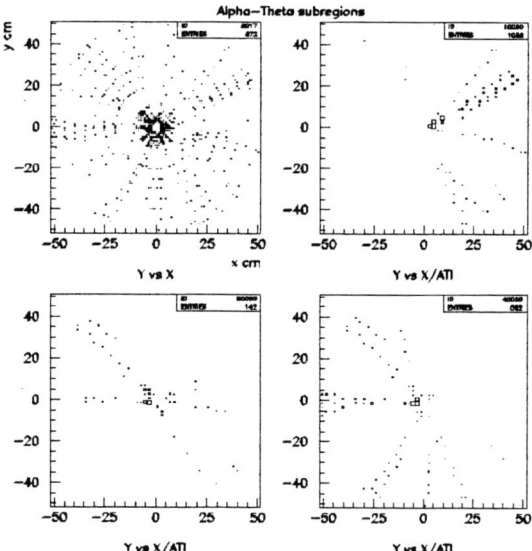

FIGURE 4. The upper plot shows all the hits on (X,Y) plane; the other are related to several (Alpha, Theta) sub-regions. The hit pattern is more clear in these sub-regions.

Singular Value Decomposition To Simplify Features recognition Analysis In Very Large Collection Of Images

F.Guillon[1], D.J.C Murray[1] and P.DesAutels[2]

[1]*SourceWorks Consulting Inc, Hull, Quebec, J8X 2J1, Canada*
[2]*Ereo Inc, Westminster, Colorado 80234, USA*

Abstract. We present and discuss the techniques used in applying the singular value decomposition (SVD) in order to resolve such difficult problems as objects and features extraction of images in the context where the number of images and related data to characterize them becomes very large. We will present preliminary results indicating the potential of such methods in analyzing data pertinent to large image collections where image complexity is high and where detailed, exact and realistic physical models of the complexities are yet to be written.

INTRODUCTION

The Internet growth has led to a tremendous amount of multimedia information becoming widely available[1] . However Internet access methodology and intelligent multimedia search engines have not kept pace with this growth. As a result of this lack of good tools for accessing and searching broadband medium, users have experienced frustration over the quality of information retrieval in general. Ereo Inc., is committed to work at improving this situation by building what could be called second generation multimedia search engine. This paper will discuss one aspect of this problem, namely the search through a database of images when the number of images becomes so large (over 1 Million images) that specific method of analysis has to be invented to deal with the similarity search problem in a large information domain (image domain in this paper) .

SIMILARITY SEARCH IN LARGE DATABASE

Recent survey[2] of the performance of the most popular Image Search Engine on Internet revealed that relevancy (satisfactory query results) was reasonable as long as the size of the database was relatively low. As the size reached the 1 Million image barrier it is obvious that relevancy decreased due to poor performances of the search engine technologies. Our effort is aimed at improving this situation namely increasing drastically the relevancy and the size of the database well above current limits. This means inventing more powerful and intelligent search tools.

Similarity through Color Histogram

Simple similarity search in a database of 1 Million images can be performed by looking at the similarity of images using their color properties. While this may not give good relevancy, it should be emphasized that this simple search is already creating a complex problem that needs to be solved. A color histogram based on the full range of color values for the RGB model[3] represents an equivalent large color vector since there are 16,777,216 possible color values (we will call this histogram: the 16M color histogram) hence similarity search using color vector requires searching for similar vectors in a high dimensional space. At run-time (for a web page search environment) current internet technology of similarity search cannot afford this high dimensional space cost so some dimensionality reduction is needed. In this paper we investigate the use of singular value decomposition (SVD) as an approach to effectively provide dimensionality reduction to the problem.

Singular Value Decomposition (SVD) Technique : Basic Idea.

SVD methods deal with solving difficult linear-least squares problems such as the *terms* in *documents* case and here *colors* in *images*. They are based on the following theorem of Linear Algebra[4]:

"Any M x N matrix **A** whose numbers of rows M is greater than or equal to its number of columns N can be written as the product of an M x N column orthogonal matrix **U**, an N x N diagonal matrix **W** of singular values and the transpose of an N x N orthogonal matrix **V**:

$$\left(A \right) = \left(U \right) \left(\begin{matrix} w_1 & & \\ & w_2 & \\ & & w_N \end{matrix} \right) \left(V \right)$$

Qualitatively the U matrix represents a vector basis for the most relevant information in the system while the eigenvalues w_i represents the variability in the information.

SVD Solution applied to Similarity Search by Colors [5]

The Singular Value Decomposition (SVD) Solution to the problem of searching a large database for similar images by colors can be applied in the following way : we can envision building a A matrix (using the notation above) corresponding to the 16M color histogram of a stack of images upon which we would like to do image color similarity comparison. In order that the SVD process can be a useful and practical empirical type of model, we need to demonstrate that we can build a A matrix that is sparse[4,6] and that the decomposition can be performed by suitable numerical optimized method.

SVD Solution : Implementation and Experiments

We used the SVD package (SVDPACKC [6]) from the NetLib repository. In particular we used an iterative method such as the Lanczos method for determining several of the largest singular triplets (singular values and corresponding left- and right-singular vectors) for large sparse matrices.

For our experiments, we used a stack of 1999 jpeg images (133 x 100 pixels). We adapted the las2 (Lanczos method) program of the package to modify the data requirements for our conditions (much larger sparse matrix). We constructed 16M (RGB color model) color histograms to build the A matrix : columns identify image through a suitable ID while rows describe a color index C (combination of R,G,B values giving one of 16M colors values). Testing showed that in order to construct a sparse matrix and being able to run the SVD package, we had to reduce the effective number of colors in the histogram from 16M to 65,536. We ran the SVD package on a Digital Alpha Linux machine (two Alpha 21264 667 MHZ CPUs with 4 MB cache, 1Gb SDRAM). Constructing the SVD sparse matrix produced the following conditions:

- 234 Mb of temporary storage to handle 1999 color histograms of 65536 color bins

- final sparse matrix formatted using only non-zero values used 78 Mb.

Requesting 200 eigenvalues produced a run time of about 20 minutes. Fig 1. shows a typical result namely the N largest singular value (eigenvalues) that retains the most relevant information contained in the color histogram of a stack of images.

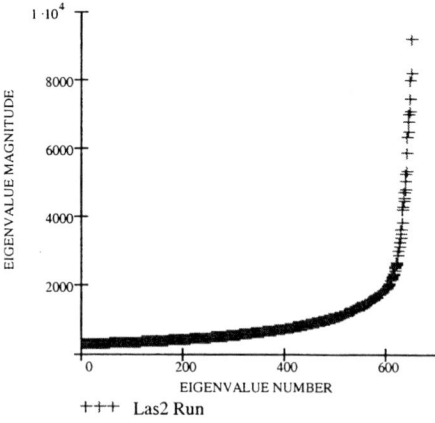

FIGURE 1. Results of a SVD run. The values that show the most variability are the retained relevant eigenvalues for this problem.

Dimensionality Reduction Results

The singular value decomposition of the (65536 x 1999) matrix A yields a new 65536 x N matrix T containing the left singular vectors or U matrix (the left singular terminology is explained in the SVD package) corresponding to the N largest singular values of A as illustrated in Fig. 1 (Fig. 1 shows the case for 700 requested eigenvalues). The N largest values retained as much information about the original histograms of the stack of images. The matrix T is used to project 65,536 dimensional colors histogram into an N-dimensional space color histograms using :

$$H_{65536}\, T = H_N$$

We can compare the similarity of images base on simple 512 color bins Histograms versus the SVD reduced N-histograms to assess the usefulness of SVD dimensionality reduction. It is found that for about N = 50 eigenvalues (the most largest values) we get the same similarity measure of image comparison using the well known P(10) metric used in text search engine development. Therefore SVD Process appears to be an efficient empirical method to reduce the size of any function describing image properties for similarity purpose.

CONCLUSION

Singular Value Decomposition Method can be an extremely valuable and simple method of analyzing data without the necessity of explicit mathematical model and it provides data size reduction as well.

This method of dimensionality reduction has been successfully proven on text and now can be considered a powerful method of similarity comparison based on image properties such as color and possibly morphology of objects (shapes) and textures contained within an image.

ACKNOWLEDGMENTS

One of us (F.G) would like to thank Ereo for financial and technical support while attending the ACAT 2000 workshop and presenting this paper.

REFERENCES

1. Gevers, T., Aldershoff, F., and Smeulders, A.W.M., "Classification of Images on Internet by Visual and Textual Information " in *Internet Imaging*, edited by G.B.Beretta and R. Schettini., SPIE Conference Proceedings 3964, Bellingham: Society of the Photo-Optical Instrumentation Engineers, San Jose, 2000, pp. 16-28.

2. Unpublished results from Ereo Inc: relevancy and database size data were taken from reliable sources in the Search Engine Internet Industry.

3. Weeks, A.R Jr., "Image Enhancement by Point Operations" in *Fundamentals of Electronic Image Processing,*, edited by E. R. Dougherty, Bellingham: Society of the Photo-Optical Instrumentation Engineers , 1998, pp. 109-120.

4. Press, W.H., Flannery, B.P, Teukolsky, S.A., and Vetterling, W.T., "Singular Value Decomposition " in *Numerical recipes in C*, Cambridge: Cambridge University Press, 1988, pp. 60-72.

5. Patent Pending under The United States Patent and Trademark Office, 2000.

6. The SVD Package labeled "SVDPACKC " for the C language can be found at the following address: http://netlib.uow.edu.au/svdpack/index.html.

The Hermite polynomials in problem of the searching for position of the global maximum.

Y. A. Yatsunenko

Joint Institute for Nuclear Research,
yay@nusun2.jinr.ru

Abstract. The symmetrization of the multi-extreme function around the objective point "u" of position of the global maximum simplifies the view of their expansion in the Hermite polynomial-series. The analysis of the fourth coefficient (*kurtosis*, *excess*) of the Hermite-series allows one to get the approach to this point "u".

The methods developed for searching the position (u) of the global maximum (the abbreviation "GM" will be used in the following) are mainly based on the Taylor-series expansion [1],[2] of the analyzed function $f(x)$

$$f(x) = f(u) + f'(u)(x-u) + \frac{1}{2}f''(u)(x-u)^2 + ... \quad (1)$$

However the Taylor-series (1) is defined at small x-vicinity of $f(u)$ ("locally") whereas the searching for the GM position foresees a large x-region where the multi-extreme character of the $f(x)$ is evinced. Thus the expansion of a $f(x)$ into the functional series should have a "non-local" character or, in other words, the coefficients of that should be determined by an integration over a x-region.

The functional series based on the Hermite polynomials [3] can be considered as the well suitable candidates for such expansion. Let us suppose for the following that the $f(x)$ is of the positive-definite character at $A < x < B$ and $f(x) \equiv 0$ for $-\infty < x < A$ and $B < x < \infty$. The direct expansion of the $f(x)$ in this Hermite-series does not give an idea for the searching of the GM position, however for the symmetrized construction (around unknown point u which is about the GM position)

$$F(x, u) = \frac{1}{2}[f(x) + f(2u - x)] \quad (2)$$

this expansion is simplified significantly

$$F(x, u) = \frac{e^{-t^2}}{\sqrt{2\pi\mu_2(F)}}[1+ \quad (3)$$

$+ 0 \cdot H_1(t) + 0 \cdot H_2(t) + 0 \cdot H_3(t) + k_4 \cdot H_4(t) + ...]$
where

$$t \equiv \frac{x-u}{\sqrt{2\mu_2(F)}}. \quad (4)$$

and $\mu_n(F)$ are the integral momenta of the $F(x,u)$ which determine the coefficients (k_n) at the $H_n(t)$ of (3)

$$\mu_n(F) = \mu_o^{-1}(F) \int F(x,u)(x-u)^n dx \quad (5)$$

Thus the $k_1 = 0, k_3 = 0$ due to the symmetrical character (around u) of the $F(x,u)$, k_2 equals zero due to the chosen type of variable t (4) and the fourth coefficient k_4 becomes a function of u

$$k_4 = \frac{1}{24}\left[\frac{\mu_4(F)}{\mu_2^2(F)} - 3\right] \equiv -\frac{E_F(u)}{12} \quad (6)$$

where the E_F is the *kurtosis* or *excess* which characterizes a sharpness of function F.

If one can analytically calculate the momenta $\mu_n(f)$ of the original function $f(x)$

$$\mu_n(f) = \mu_o^{-1}(f) \int f(x)(x-c)^n dx, \quad (7)$$

$$c \equiv \mu_1(f) = \mu_o^{-1}(f) \int f(x)x \, dx$$

then all the momenta of the symmetrized $F(x,u)$ are easily determined by these $\mu_n(f)$ and therefore $E_F(u)$ has the exact analytical form

$$E_F(u) = \frac{(c-u)^4 - 2(c-u)\gamma + \sigma^2 \cdot R_o^2}{[\sigma^2 + (c-u)^2]^2} \quad (8)$$

where $\sigma^2 \equiv \mu_2(f), \gamma \equiv \mu_3(f)$ and the parameter

$$2R_o^2 = 3\sigma^2 - \frac{\mu_4(f)}{\sigma^2} \quad (9)$$

characterizes the distance between two peaks of $f(x)$ of equal amplitude.

Thus the symmetrization gives a very simple view of the Hermite-series (3): there is one maximum (on the variable t^2) the position and the amplitude of that is determined by the value of $E_F(u)$. Therefore the extreme values of the $E_F(u)$ can determine the process of the symmetrization: if u is chosen near the GM position then the function (2) has a narrow and high peak at x in the region of this maximum otherwise $F(x,u)$ is blurred. The requirement for the extreme value of the $E_F(u)$ ($dE_F(u)/du = 0$) gives the cubic equation for u

$$2(c-u)^3\sigma^2 + 3(c-u)^2\gamma - 2(c-u)R_o^2\sigma^2 - \gamma\sigma^2 = 0 \quad (10)$$

This is the solution of the problem.

One of the three possible real roots of this equation corresponding to the minimal value of the excess (6) places u most near to the GM position of $f(x)$. Then an iterative procedure of the successive reduction of the integration region of $f(x)$ can be organized: each new x-region can be determined by the chosen u and, for instance, by the current value of σ^2. A more cautious attitude to the approximate expansion (3) requires a necessity of the analysis of all the roots of equation (10) which correspond to the minimal value of the excess (6) (there are either 2 or 1). Undoubtedly, this is a very complicated analysis, however there is the well defined rule of the termination of the iterations: if $R_o^2(9) < 0$ or the cubic equation has one root only, then one can suppose that $f(x)$ is of unimodal character in the last integration region and the *local* methods can be used for the exact determination of the GM position.

An expansion similar to (2) which should be suitable for a practical application has not been obtained yet. Nevertheless the analogous ideas about the symmetrization of the objective multidimensional function $f(\vec{r})$ (defined in a V-region)

$$F(\vec{r}, \vec{u}) = \frac{1}{2}[f(\vec{r}) + f(2\vec{u} - \vec{r})] \quad (11)$$

and the minimization of the excess can be simply used in this case. One can give the final form of the excess of the $F(\vec{r}, \vec{u})$ as the function of variable $\vec{R}(\vec{u}) = \vec{c} - \vec{u}$:

$$E_F(\vec{R}) = \frac{R^4 - 2\vec{R}\vec{\gamma} + 2\vec{R}^T\hat{K}\vec{R} + \sigma^2 R_o^2}{(\sigma^2 + R^2)^2} \quad (12)$$

where $\vec{c}$ is the center of the $f(\vec{r})$ (suppose in following that $\int_V f(\vec{r})dv = 1$) and

$$\sigma^2 \equiv \int_V f(\vec{r})(\vec{r} - \vec{c})^2 dv \quad (13)$$

$$\vec{\gamma} \equiv \int_V f(\vec{r})(\vec{r} - \vec{c})^3 dv \quad (14)$$

$$2R_o^2 \equiv 3\sigma^2 - \sigma^{-2}\int_V f(\vec{r})(\vec{r} - \vec{c})^4 dv \quad (15)$$

$$K_{mn} \equiv \sigma^2 \cdot \delta_{mn} - \int_V f(\vec{r})(x_n - c_n)(x_m - c_m) dv \quad (16)$$

($\delta_{mn} = 1$ if $m = n$ otherwise $\delta_{mn} = 0$ and $m,n = 1,...,N$; N is the dimension of vector $\vec{r}$).

Minimization of the excess (12) on the vector $\vec{R}$ leads to the equation

$$\frac{2\sigma^2 R^2 + 4\vec{R}\vec{\gamma} - 4\vec{R}^T\hat{K}\vec{R} - 2\sigma^2 R_o^2}{\sigma^2 + R^2}\vec{R} = \vec{\gamma} - 2\hat{K}\vec{R} \quad (17)$$

An attempt to find the solution based on the eigen vectors of matrix $\hat{K}$ (16) does not look as the practical one (it is under consideration, nevertheless). There can be chosen three vectors ($\vec{\gamma}, \hat{K}\vec{\gamma}, \hat{K}^{-1}\vec{\gamma}$), along of those an approximate solution might be found. The simplest one is based on the assumption that the direction of the GM position is determined by the asymmetry ($\vec{\gamma}$)

$$\vec{R}(z) = z \cdot \frac{\vec{\gamma}}{\Gamma}, \quad (\Gamma \equiv |\vec{\gamma}|) \quad (18)$$

leads to the cubic equation for parameter z

$$2(\sigma^2 - \lambda_n)z^3 + 3\Gamma z^2 - 2\sigma^2(R_o^2 - \lambda_n)z - \sigma^2\Gamma = 0 \quad (19)$$

where λ_n are the eigenvalues of matrix $\hat{K}$ (16).

However the excess (12) has the finite number of extremal values (there are 3 ones in a common case) therefore the traditional local methods of the finding of the minimal value of the excess can be effectively applied and the above given solutions can be used as the starting points.

A remark about the searching for global minimum. The naive substitution $\phi(x) = -f(x)$ is not a suitable one whereas the construction $\phi(x) = C - f(x)$ leads out this problem to the above described method of the searching for the position of the global maximum.

Despite of very approximate assumptions (12,18) the problem of searching for the GM position had been successfully solved for the two-dimensional function of the complicated view [5] as well as the effective realization of the unidimensional analysis (8) were made for the multi-extreme functions "vertex functions" (defined in previous talk and in [6]).

This "vertex function" for the straight tracks on the $XZ-plane$ has the view:

$$D(z,x) = \sum_{k=1}^{M_N}\sum_{n=1}^{N}\sum_{m=1}^{M_n} G[\frac{x - a_{kN}}{z_N - z}(z_N - z_n) + a_{kN} - a_{mn}; s_{mn}] \quad (20)$$

where N is the number of detectors and each of them has $M_n (n = 1, 2, ..., N)$ hits: $a_{mn}; (m = 1, 2, ..., M_n)$. $G(y; s) \equiv exp(-y^2/2s^2)$ is the function of the accuracy (s) of the detection. The big maxima of the vertex function correspond to the possible vertices (x, z) The example of successful determination of the position of the global maximum of such vertex functions given in Figs.1,2.

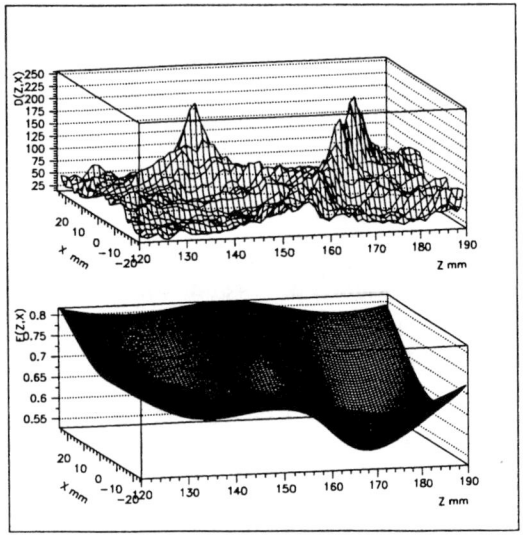

FIGURE 1. The view of the vertex function (20) is shown on upper plot. The excess (12) of this function is shown on the lower plot.

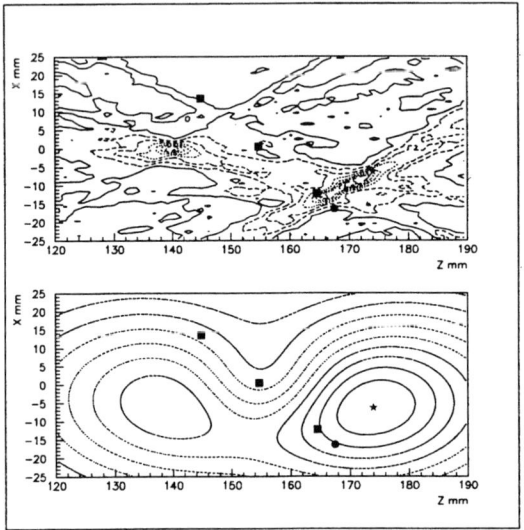

FIGURE 2. The contour view of the vertex function (20) (upper plot) and of the excess (12) (lower plot) presented on Fig.1. The "dark squares", "dark circle" are the roots of the cubic equation (13) for two eigenvalues. The root "dark circle" was used for the numerical determination of the point ("dark star") of the minimum of the excess.

REFERENCES

1. Ya.D.Sergeev, Journal of Comput. Math. and Math. Phys. (Russia) **35**, No 5, 707 (1995).

2. A.A.Goldstein and J.F.Price, Math.Comput. **25**, 569 (1971).

3. A.Korn and M.Korn. "Mathematical Handbook" (Russian edition) "Nauka", Moskow, 1977, p.775.

4. Yu.A.Yatsunenko, JINR Communication, P5-88-74, 1988, Dubna (Russia).

5. Yu.A.Yatsunenko, JINR Rapid Communication, No 5, 1989, p26, Dubna (Russia).

6. Yu.A.Yatsunenko, Nucl. Instr. and Methods (NIM), **A287**, 422 (1990).

Event Bookkeeping for CLEO-3

Jon J Thaler

Physics Department, University of Illinois at Urbana-Champaign, 1110 W. Green St., Urbana, IL 61801

Abstract. Most aspects of data analysis software share a common bookkeeping task, maintaining relationships between members of two sets of objects. To simplify the development and maintenance of this bookkeeping, CLEO has implemented a sharable package that implements all of the commonly used tools. This allows the physicists to concentrate on data analysis.

MOTIVATION

Most aspects of data analysis software share a common bookkeeping task, maintaining relationships between members of two sets of objects. For example, the tracks created by pattern recognition algorithms must be correlated with the detector hits that contribute to them. Conversely, it is important to be able to determine which tracks contain a given hit. Similar bi-directional correlations must be established and maintained in vertex reconstruction and for event kinematics. Data analysis is the identification of increasingly complex relationships within the data.

It would simplify the software environment if these relationships were managed by a single sharable package. The package would only have to be written and debugged once, and code writers would be freed to concentrate on the algorithms. To this end, the CLEO collaboration has written software, called Lattice, which is designed to perform the bookkeeping tasks described here. Lattice is written in C++, so this paper uses C++ terminology.

SPECIFICATIONS

Lattice is designed to be flexible and to have minimal impact on existing code. This makes it more likely that Lattice will be used. Lattice's internal data storage scheme is transparent to the user.

Functionality

The user can add and remove links between objects (*e.g.*, hits and tracks). Data (the Link data in figure 1) can be associated with each link (*e.g.*, hit contributions to track χ^2). The user can retrieve a list of all objects to which a given one is linked (*e.g.*, all hits on a given track), or all the objects which share links with a given one (*e.g.*, all tracks which share hits with a given track). The interface is simple, and all data is returned in STL vectors. Constraints can be specified that prevent unwanted multiple links between objects.

Flexibility

Any two sets of data can be linked. Link data is user definable. Lattices don't interact, so objects can belong to multiple Lattices. A Lattice can persist over a write/read data cycle, because it does not use memory pointers. The Lattice can be used even after the data it links has disappeared (*e.g.*, if hits are not saved after track reconstruction).

Minimal Impact

The use of Lattice incurs only a small performance penalty. Lattice is optimized for fast data retrieval, while making and deleting links is more expensive. Our experience is that this matches users' needs.

Lattice requires little rewriting of existing code. The only behavioral requirement is that data objects have identifiers, used by Lattice to distinguish objects. CLEO data objects already have this property.

Figure 1 shows the general structure. Nodes are invisible to the user. They maintain the correspondence between user data and the Lattice.

EXAMPLES

It is simplest to explain the Lattice, especially the value of connectivity constraints, with some examples.

Track Finding. Hit sharing not allowed.
The left data is the tracks; the right is the hits. Eight hits lie on three tracks. Link data is the hit contribution to χ^2. Disallowing multiple links on the right (see figure 2) imposes no hit sharing.

Track Finding. Hit sharing allowed.
Without the connectivity constraint, tracks can share hits, as shown in figure 3. Resolution of shared hit ambiguities can be deferred until later in the analysis.

Calorimeter Clustering. No overlaps.
Here, the left data is the clusters. Figure 4 shows the situation, supposing that clusters have cores and halos.

Calorimeter Clustering. Overlaps allowed.
By relaxing one connectivity constraint, clusters can share halos and cores, as shown in figure 5. This simple model does not distinguish halos and cores. To disallow core sharing while allowing halo sharing would require two Lattices, because a Lattice connects two kinds of objects, not three.

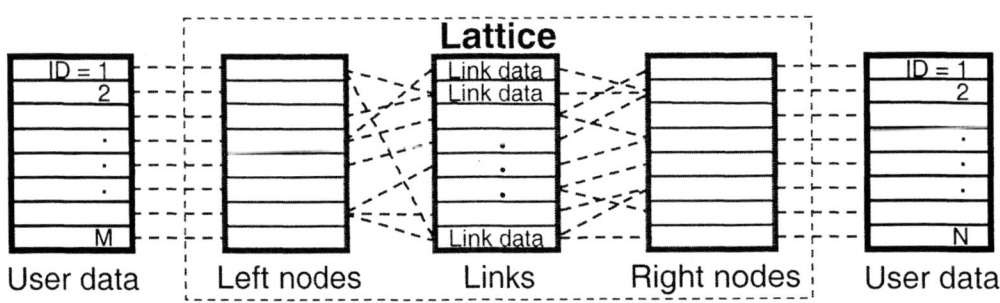

FIGURE 1. Structure of Lattice. User Data is the data to be linked. Link Data is the data associated with a link. It is a user-defined class. Identifiers do not need to be int. They do need to be sortable and unique (within a set).

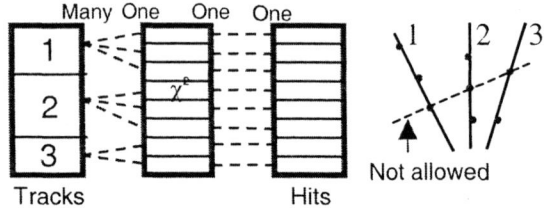

FIGURE 2. Track finding with no hit sharing. "Many" means multiple connections are allowed. "One" means single connections only.

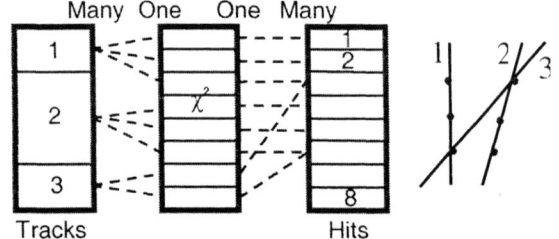

FIGURE 3. Track finding with hit sharing. One track steals hits from the others

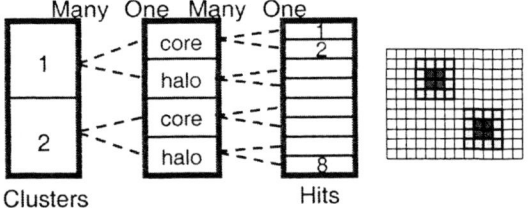

FIGURE 4. Cluster finding with no overlaps. The link data describing cores and halos must both fit into the same user defined class.

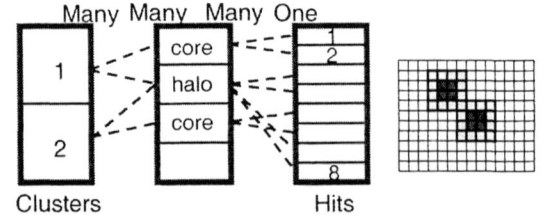

FIGURE 5. Cluster finding with overlaps. Two clusters share a halo.

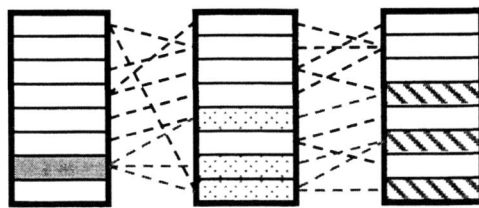

FIGURE 6. Find objects on the right that are linked to the given one on the left.

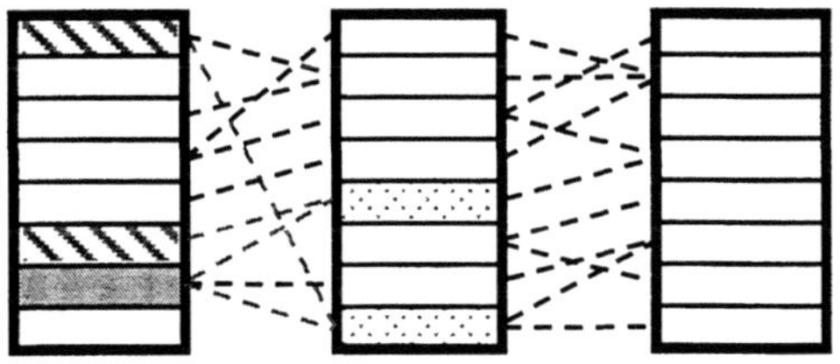

FIGURE 7. Find objects on the left that share links with the given one on the left.

INTERFACE

Due to space constraints, I only show some representative interface operations. A complete set is implemented.

Make a new Lattice:
```
pLattice = new
    Lattice<LeftData,RightData,LinkData>
    (Connectivity)
```
Lattice is a class template. You specify the left, right, and link data types, and the allowed connectivity. There are 16 possible connectivity combinations, "one" or "many" at each of the four positions (see, *e.g.*, figure 2).

Make a new link:
```
pLink = pLattice->connect
        (LeftID, RightID, LinkData&);
```
Returns a pointer to the new link. Zero is returned on failure (*e.g.*, violation of the connectivity constraint). The link data is copied into the link. Lattice does not take ownership of the user's data.

Remove an object from a link:
```
pLink = remove(LeftID, Link*);
```
A null pointer is returned on failure (*e.g.*, the object was not a member of the link).

Find objects that are linked to a given one:
```
vRight = pLattice->
            vRightGivenLeft(LeftID);
```
vRight is an STL vector of RightIDs.. This will tell you, for example, which hits lie on a given track. See figure 6. The verbose nomenclature is a consequence of the fact that LeftIDs and RightIDs might be the same type.

Find objects that share links with a given one:
```
vLeft = pLattice->
            vLeftGivenLeft(LeftID);
```
This will tell you, for example, which tracks share hits with a given one. See figure 7.

Return a reference to a link's data:
```
linkData = pLink->linkData();
```
You have write access to the link data. This allows you to update it when objects are added or removed.

STATUS

Lattice has been an integral part of the CLEO-3 software environment for two years. It is stable code. Lattice is being considered for use by two other experiments. It is not quite plug-and-play, but close.

Complete documentation is available on the web at http://web.hep.uiuc.edu/home/jjt/Lattice . The source code is available (email: jjt@uiuc.edu). I will be happy to help other experiments implement Lattice.

ACKNOWLEDGMENTS

This work was partially supported by the US Department of Energy under grant DE-FG02-91ER40677.

KID - KLOE Integrated Dataflow

The KLOE collaboration*
presented by
I. Sfiligoi

INFN LNF, via E. Fermi 40, 00044 Frascati, Italy

Abstract. KLOE is acquiring and analyzing hundreds of terabytes of data, stored as tens of millions of files. In order to simplify the access to these files, a URI-based mechanism has been put in place. The KID package is an implementation of that mechanism and is presented in this paper.

INTRODUCTION

The KLOE experiment

KLOE[1] is the experiment for which the INFN DAΦNE ϕ-factory in Frascati, Italy, was built. Its main goal is the measurement of CP violation at sensitivities of $O(10^{-4})$, but is also capable of investigating a whole range of other physics. The most interesting studies include kaon form factors, kaon rare decay and radiative ϕ decays measurements.

To achieve its goal, KLOE needs to acquire $O(10^{11})$ events, which corresponds to about 1 petabyte of data. All of them must be reconstructed and analyzed to minimize the systematic errors.

Data storage

In KLOE, events are stored as YBOS[2] files of reasonable size (up to 1 Gb at the time of writing). These files are stored first on disk and then moved to tape for long term storage and back to a (possibly different) disk area for analysis. To keep track of which files have been produced and where they currently reside, a central RDBMS[1] is used[3].

Three types of event files are present in the system:

- raw files, containing only electronic readout of the events

- datarec files, containing raw data plus some additional information gathered during data reconstruction and classification stage, such as drift chamber tracks and calorimeter energy deposits

- Monte Carlo files, containing simulated events

All the events acquired during the data acquisition will be stored as raw files and maintained for the life of the experiment. The events that will be classified as of physics interest during event classification, will be stored a second time as datarec files. Since the data reconstruction and classification process can change during the life of the experiment, several versions of the same event can reside in different datarec files, although in the long term, only the most recent version is planned to be kept.

---

* M. Adinolfi, A. Aloisio, F. Ambrosino, A. Andryakov, A. Antonelli, M. Antonelli, F. Anulli, C. Bacci, G. Barbiellini, F. Bellini, G. Bencivenni, S. Bertolucci, C. Bini, C. Bloise, V. Bocci, F. Bossi, P. Branchini, S. A. Bulychjov, G. Cabibbo, A. Calcaterra, R. Caloi, P. Campana, G. Capon, G. Carboni, A. Cardini, M. Casarsa, G. Cataldi, F. Ceradini, F. Cervelli, F. Cevenini, G. Chiefari, P. Ciambrone, S. Conetti, S. Conticelli, E. De Lucia, G. De Robertis, R. De Sangro, P. De Simone, G. De Zorzi, S. Dell'Agnello, A. Denig, A. Di Domenico, S. Di Falco, A. Doria, E. Drago, O. Erriquez, A. Farilla, G. Felici, A. Ferrari, M. L. Ferrer, G. Finocchiaro, C. Forti, A. Franceschi, P. Franzini, M. L. Gao, C. Gatti, P. Gauzzi, S. Giovannella, V. Golovatyuk, E. Gorini, F. Grancagnolo, W. Grandegger, E. Graziani, P. Guarnaccia, H. G. Han, S. W. Han, X. Huang, M. Incagli, L. Ingrosso, Y. Y. Jiang, W. Kim, W. Kluge, V. Kulikov, F. Lacava, G. Lanfranchi, J. Lee-Franzini, T. Lomtadze, C. Luisi, C. S. Mao, M. Martemianov, A. Martini, M. Matsyuk, W. Mei, L. Merola, R. Messi, S. Miscetti, A. Moalem, S. Moccia, S. Moulson, S. Mueller, F. Murtas, M. Napolitano, A. Nedosekin, M. Panareo, L. Pacciani, P. Pagès, M. Palutan, L. Paoluzi, E. Pasqualucci, L. Passalacqua, M. Passaseo, A. Passeri, V. Patera, E. Petrolo, G. Petrucci, D. Picca, G. Pirozzi, C. Pistillo, M. Pollack, L. Pontecorvo, M. Primavera, F. Ruggieri, P. Santangelo, E. Santovetti, G. Saracino, R. D. Schamberger, C. Schwick, B. Sciascia, A. Sciubba, F. Scuri, I. Sfiligoi, J. Shan, T. Spadaro, S. Spagnolo, E. Spiriti, C. Stanescu, G. L. Tong, L. Tortora, E. Valente, P. Valente, B. Valeriani, G. Venanzoni, S. Veneziano, Y. Wu, Y. G. Xie, P. P. Zhao, Y. Zhou

[1] Relational DataBase Management System

CP583, *Advanced Computing and Analysis Techniques in Physics Research: VII International Workshop*
edited by P. C. Bhat and M. Kasemann
© 2001 American Institute of Physics 0-7354-0023-7/01/$18.00

System overview

The KLOE computing environment is made of a set of UNIX servers, connected via a switching network. Logically, the servers can be divided into:

- DAQ[2] nodes
- CPU suppliers
- disk servers
- tape servers

Typically, an application runs on a CPU supplier and reads the necessary data from a remote node via a KID daemon(see next section).

THE KID PACKAGE

The KID package is made of a programming library and a not-privileged daemon, called the KID daemon. In this section, both of them are presented.

User interface

As seen in the introduction, the number of available events is very large. To select only the most interesting ones an easy to use user interface is very important; the KID approach is based on Uniform Resource Identifiers (URIs)[4].

<scheme>:<scheme-specific-part>[?<options>]

Several schemes (or protocols) are available in the main package, but user-specific ones can be added as needed.

Library

The KID programming library is written using ANSI C for compatibility reasons. Nevertheless, the object-oriented approach is followed as much as possible; the access to resources is object (or identifier) based and protocol selection is achieved through function pointers.

The external interface is very simple, having essentially only four classes of functions:

- open - get a new identifier, the only parameter is an URI

- get - read a new event from the data source (identifier-based)

- skip - discard an event from the data source (identifier-based)

- close - destroy the identifier and release any associated resource

Internally, the library is modular; the functions of the core module parses out the schema and then calls the appropriate functions of the schema-specific module. This allows a user-supplied module with its specific schema support to be put in place even at run-time.

The library is also POSIX thread safe, having mutual exclusive locks on identifiers implemented in the core module.

KID daemon

The KID daemon is a key element of the system; it allows remote applications to access local resources, such as memory buffers and local files.

It is implemented as a multi-threaded not-privileged daemon. The main thread listens for TCP/IP requests and creates a new thread for any accepted connection.

Connection specific threads read a client-supplied URI and use it to get an identifier from the KID library that will be used to read the events to be sent to the client (see Fig. 1).

To speed-up the communication, several events (typically 100 at the time of writing) are packed together and the packets are sent before the client actually asks for a new packet. Since this can be a problem for a small fraction of applications (for example event displays), the packet size can be set and the push mode reverted to pop mode at connection time.

Interface to the data handling system

Being able to access a remote resource is surely an essential facility, but it is not enough. The user still needs to know where the resource is located and how it is called. For this reason an interface to the data handling system (DH) has been created.

The interface is implemented as a KID library module. It processes the schema specific part of the URI as an SQL query, passes the resulting list of files to the cache manager (see [3] for more details) and then uses the information returned by the cache manager to open those files (that are now residing on one or more disk areas) (see Fig. 2).

---

[2] Data AcQuisition

KID standard modules

Several ready-to-use modules are present in the KID package. In this subsection, the most interesting ones are presented.

Local access modules

These modules allow the access to local resources; two of them are supplied:

- spy - allowing access to a memory buffer
- ybos - allowing access to a local file

<schema>:<resource-name>[?options]

Remote access module

This module creates a connection with a KID daemon and passes back the events to the application.

remote:<URI>@<IP-node>[?options]

DH enabled modules

The main module expects an SQL query as explained in the previous subsection.

db:<SQL-query>[?options]

However, writing general SQL queries can be quite annoying, so modules specialized in different types of queries have been created:

- dbraw - select only raw files
- dbdatarec - select only datarec files
- dbmc - select only Monte Carlo files
- dbfile - select a file based on its filename

<schema>:<where-part-of-a-SQL-query>[?options]

Support modules

Some support modules are also provided; the most used are:

- merge - get events from multiple sources
- try - find an event source from a list

<schema>:(<URI>,...,<URI>)[?options]

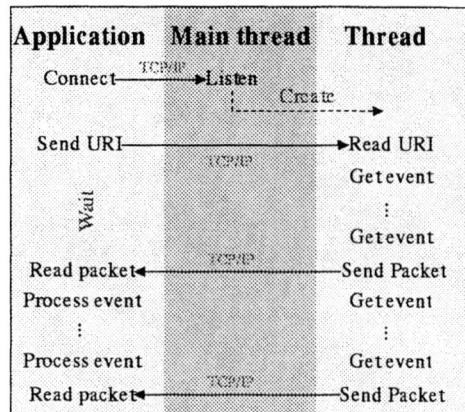

FIGURE 1. KID daemon internals

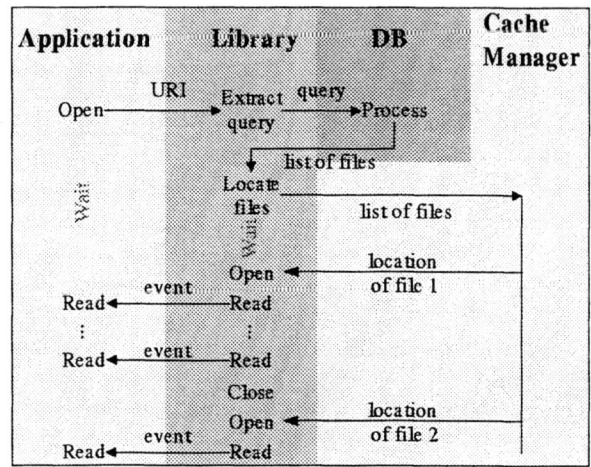

FIGURE 2. KID-DH interface

CONCLUSIONS

KID is now in use in KLOE for about 6 months. Most of the user applications and a considerable fraction of the production software is now using it.

The use of KID has simplified most of the analysis tasks and made even new types of analysis possible.

REFERENCES

1. The KLOE Collaboration, *KLOE, A General Purpose Detector for DAFNE, LNF-92/019 (IR), Frascati, 1992*
2. CDF Computing Group, *YBOS, Programmers Reference Manual, CDF Note No. 156, Batavia, 1992*
3. I.Sfiligoi, *"Data Handling in KLOE", CHEP2000 conference proceedings*
4. T. Berners-Lee, R. Fielding, L. Masinter, *"Uniform Resource Locators (URI): Generic Syntax", W3 RFC 2396, 1998*

A Component Based Approach to Scientific Workflow Management

N. Baker[1], P. Brooks[1], Z. Kovacs[2], J-M. LeGoff[2], R.McClatchey[1]

[1]*Centre for Complex Cooperative Systems, UWE, Bristol, UK BS16 1QY*
FAX No: (+44)1179 763860 Email: Richard.McClatchey@uwe.ac.uk
[2]*EP Division, CERN, Geneva, 1211 Switzerland*
FAX No: (+41)22767 8930 Email: Jean-Marie.Le.Goff@cern.ch

Abstract. CRISTAL is a distributed scientific workflow system used in the manufacturing and production phases of HEP experiment construction at CERN. The CRISTAL project has studied the use of a description driven approach, using meta-modelling techniques, to manage the evolving needs of a large physics community. Interest from such diverse communities as bio-informatics and manufacturing has motivated the CRISTAL team to re-engineer the system to customize functionality according to end user requirements but maximize software reuse in the process. The next generation CRISTAL vision is to build a generic component architecture from which a complete software product line can be generated according to the particular needs of the target enterprise. This paper discusses the issues of adopting a component product line based approach and our experiences of software reuse.

INTRODUCTION

As component technology gradually evolves and matures, system developers will gradually migrate from systems composed of interoperable objects to those composed of interoperable components. One of the main motivations for this migration is the potential of software reuse and its associated benefits of cost reduction and time to market software products. Component-based software development is concerned with constructing software artifacts by assembling prefabricated configurable building blocks. However software reuse is concerned with more than binary components. For many organizations it is the generation and application of generic software assets that are reusable across a family of target products. Binary components are just one view of the software development process. The creation and evolution of graphical models to visualize specific aspects of software artifacts is another view. What is required is some software development process that couples these high-level development approaches with implementation approaches. This paper opens with a brief discussion of the context and motivations for this research followed by an outline of software product lines. The issues and the team's experience of software reuse are discussed. The final part of the paper concentrates on the divide between object based modeling and component based development, which is preventing software reuse from reaching its full potential.

MOTIVATIONS

CRISTAL is a scientific workflow system[1] that is being used to control the production and assembly process of the CMS Electromagnetic Calorimeter (ECAL) detector at CERN Geneva. Detector production is a collaborative effort with production centres distributed across many institutes worldwide. The production process is unusual in that only one final product is manufactured; however the types of parts from which it is assembled could consist of many versions. The evolution of the detector will take many years and during this process the history of versioned parts must be captured. The ultimate detector will be part of a high energy physics experiment therefore collection and storage of manufacturing & production data is just as important as control of the process. This stored data not only gives the "as built" view of the final system but has been designed as a warehouse to provide "calibration views", "maintenance views" and other views not yet conceived by the designers. It is these specialized aspects which characterize CRISTAL as a scientific workflow system. A general workflow system is used to coordinate and manage execution of the thousands of tasks and activities that occur in any complex enterprise. Workflow management can be applied to diverse applications from banking to manufacturing. In each case the system must be capable of describing and storing the tasks and activities of the domain to be automated, executing and co-ordinating the tasks and storing the outcomes. CRISTAL has taken an object-oriented approach describing all parts, manufacturing & production tasks, manufacturing & production data using meta-modeling techniques. As a consequence of the uncertainty and specialized nature of the application the core meta-model of the CRISTAL workflow system has the potential to be applied to almost any workflow or enterprise resource management application. The motivation to achieve this potential has stemmed from requests to apply the core CRISTAL technology to bio-informatics and general manufacturing domains. However a number of problems remain to be solved in order to develop our workflow software to cope with the demands of such a diverse product family range.

SOFTWARE PRODUCT LINES

The product family problem is well known. A software product line [2,3] is a set of software systems that share a common set of features that satisfy a specific market demand. The key idea is to build shared assets that can be instantiated and combined to develop instances of the product line. Similar to a manufactured product line software products will:-

- Pertain to an application domain and market
- Share an architecture and
- Be built from reusable components

The application domain is reasonably clear in our particular example but the issues that surround a common architecture and components are less so. The following explore these issues in further depth based on our software engineering experiences.

Software Architecture

A product line software architecture[4] is the central artifact in product line engineering because it provides the framework for developing and integrating shared assets and must be common to all the products. Naturally the common user requirements map to the standard architecture but product specific requirements must map to variations provided for by the architecture. It is these specific requirement variations that define the particular product line. The problem is how to best manage and include mechanisms for this variation. [5] Discusses methods to model and capture this variation. Standard computing mechanisms to cope with variation are:-

- Alternate selection using "if then else" flow control
- Alternate selection using parameters
- And in object orientation the use of inheritance, delegation and meta-models.

In our experience in building workflow systems one of the benefits of meta-modeling is that with careful analysis and use of descriptive classes a core generic software architecture can be developed to support almost any type of workflow system. A discussion of the concepts and benefits of meta-modeling is not in the scope of this paper but more details can be found in [6].

Applying the meta-modeling approach to a product line of workflow managers (that is workflow managers for production of aircraft, cars, kitchens etc.) would necessitate describing the activities and items to configure the architecture to the particular work flow manager in the product line. Compared with a more software component based approach where the actual product line goes through a software build process where variational components are linked in or omitted according to the features of the target product. The former approach makes for a configurable adaptive architecture but the second is required to support a product line.

Reusable Components

Product line engineering practice advocates the generation and application of generic software assets that are reusable across a family of target products. It suggests analyzing common and variable product characteristics to define scope of reuse, identify reusable components with a suitable level of generality. The expected benefits are production of quality cost-effective software, rapid application development and improved maintenance. It emphasizes strategic planned reuse rather than opportunistic reuse. That is it is not just about libraries, class hierarchies or configurable architectures. This is reuse at a very high level disconnected from implementation issues. The following section discusses the implementation issues of software reuse and its application to components and component based development and the final part of the paper attempts to link the two together.

SOFTWARE REUSE

Is not a new concept as Figure 1 illustrates. Early efforts focused on small-grained reuse of software code. Our experience over the past 10 years of

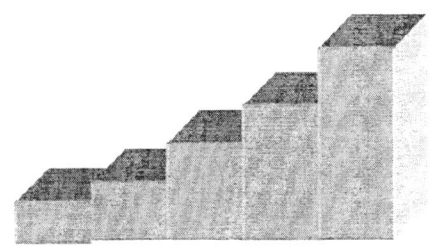

FIGURE 1. A History of Software Reuse.

building object-oriented systems has convinced us that most reuse has come from higher-level design artifacts. Very little code has been reused, except class library reuse mainly confined to client-side user interfaces. So why so little reuse at the code level? One explanation appears to be that the cost of creation and use of these small-grained assets often outweighed the modest gains. But another important factor is that the underlying software technology is moving so fast, especially true in software projects with long time scales. For example object technology has witnessed, in a short space of time Smalltalk, ADA, C++, Java, EJB, COM+, Active X and OMG CORBA.

Where we have experienced more success in reuse of software artifacts is with visual modeling languages such as Object Modeling Technique (OMT) and the Unified Modeling Language (UML)[7]. The creation and evolution of graphical models using UML has allowed us to specify, visualize, construct and document the artifacts of the software systems we have built. Building UML models has provided a structure for problem solving and allowed us to contemplate large-scale system problems. Derived from OMT, UML version 1.1 was adopted as an Object Management Group (OMG) standard in November 1997 with a recent minor version, UML 1.3, adopted in November 1999. Usually the great

thing about standards is that there are lots to choose from. However in contrast to the rapidly changing implementation software technology, UML is the universal OAD modeling standard used by OMG member organizations and Microsoft. Perhaps because of this stability we have over the years been able to reuse large-grained architectural frameworks and patterns which have been captured in UML. The term's pattern, framework, component are somewhat overloaded and the following subsections provide working definitions and discuss reuse issue experiences.

Patterns

A Pattern[8] is a solution schema expressed in terms of objects & classes for recurring design problems within a particular context. Patterns focus on reuse of abstract designs and software architecture, which is usually, described using graphical modeling notation. So in UML this specification is done using interaction, class and object diagrams. The patterns that we have reused in the construction of our workflow management system[9] have evolved out of years of proven design experience. Although made up of graphical diagrams the documentation provides a vocabulary and concept understanding amongst the team. Documentation describes heuristics for use and applicability although this is not modeled in UML. In the object oriented community well known patterns are named, described and cataloged for reuse by the community as a whole. We have not only used many well-known patterns but in the domain of workflow management discovered new patterns. It has enabled us to make use of design patterns that were proven on previous projects and is a good example of reuse at the larger grain level. UML diagrams are able to describe pattern structure but provides no support for describing pattern behavior or any notation for the pattern template. UML 1.4, which is in draft stage, will enhance the notation for patterns.

Frameworks

A framework is the term given to a more powerful and large grained object oriented reuse technique. It is a reusable semi-complete application that can be specialized to produce custom applications [10]. It specifies a reusable architecture for all or part of a system and may include reusable classes, patterns or templates. Frameworks focus on reuse of concrete design algorithms and implementations in a particular programming language. Frameworks can be viewed as the reification of families of design patterns. When specialized for a particular application then it is called an application framework and Fayad[11] identifies three categories:

- System Infrastructure where frameworks are applied to operating systems, network communications and GUI's.

- Middleware applied to ORBs and transactions

- Enterprise Frameworks which address domains such as telecommunications, business, manufacturing.

Framework requirements are defined by software vendors or standards organizations for example IBM's San Francisco Project, FASTech' FACTORYworks, and Motorola's CIM Baseline. Fingar[12] maintains that most frameworks should capture workflows since they provide the necessary modeling capabilities for constructing any business process. He states that workflow management is one of the elements common to all e-commerce applications and is essential. Many proponents of frameworks go so far as to suggest that workflow mechanisms should eliminate the need for most application programming in the workplace.

Frameworks can also be classified according to the techniques used to extend them. Whitebox frameworks rely on OO language features such as inheritance and dynamic binding. Blackbox frameworks are structured and extended using object composition and delegation.

Component frameworks are specialized frameworks that are designed to support components. D'Souza[13] describes a component based framework as a collaboration in which all the components are specified with type models; some of them may come with their own implementations. To use the framework you plug in components that fulfill the specifications. Three main industrial examples of component frameworks are OMG's Corba Component Model (CCM) Enterprise Java Beans (EJB) and Microsoft's COM+.

Components

A component is defined as a package of software that can be independently replaced. It both provides and requires services based on specified interfaces [13]. It conforms to architectural standards so that it can plug in and interoperate with other components. The granularity of components can vary from an instance of single to many classes and can be a significant part of a system, consistent with the goal of reuse. Unlike classes, components contain implementation elements such as source, binary executable or scripts. Components are binary-replaceable things and this distinction more than anything else sets them apart from classes. They package implementation and because of interface based design can be replaced. This means that when a new variant of a component is created it can replace a previous one without recompiling other components, provided it conforms to the same interface. Software developers can build applications by assembling components rather than designing and coding.

In order to gain the payoff in software reuse as advocated by product line engineering there are some difficulties to be resolved. Compared with classes and patterns which are modeled at analysis and design phase, components are modeled at implementation phase. This in our experience is a particular problem where we have so much invested in graphical models. The essence of the problem is how does a collection of classes modeled at the UML OAD level, become implementation components. This raises the follow-up question; is UML capable of component modeling? Although components have become the de facto standard for desktop development this is not the case for server development. In summary, leading component architectures have matured and evolved to support enterprise application. What is not clear is whether graphical modeling languages and tools can support the leading component architectures to deliver the goals of product-line engineering. The following section discusses these issues.

MODELING COMPONENTS

A component in UML is a software artifact that exists at runtime. The notation for modeling components in UML is shown in Figure 3. In the top part of the figure the long hand notation for a component is shown complete with attributes and operations. This particular component is realized by two interfaces, interface One and interface Two. Underneath is shown the shortened notation where almost all of the detail is hidden. The two interfaces of the component are shown as so-called "lollipops". UML components are typically found in implementation related component diagrams and deployment diagrams.

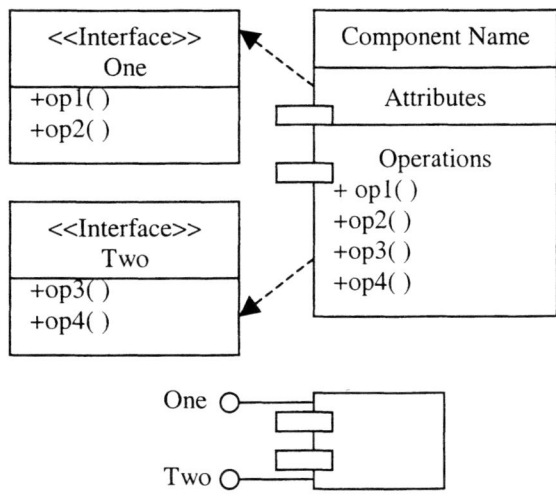

FIGURE 2. UML Component Notation

The ability to model component frameworks is just as essential as being able to model components. Although component frameworks vary they do conform to a common architectural pattern. Figure 3 adapted from Kobryn[14] illustrates this common pattern using UML notation. The pattern is represented by the UML 1.4 ellipse with dashed perimeter and contains a number of classifiers. The client represents an entity that requests a service from the component. The request is never delivered directly to the component, instead it is intercepted by two proxies, FactoryProxy and RemoteProxy. The role of these entities is very important. It allows transparent insertion of common services by the component's runtime environment or container. Proxies themselves conform to well known patterns further details of which can be found in [8]. The FactoryProxy role is concerned with creation and location whilst the RemoteProxy with operations specific to the component. Both proxies and the component itself are held in a component container. The container represents the component's runtime environment and supports common distributed services such as security, transactions and persistence. Associated with each component within the container is a Context entity which stores information such as transaction and security status. The component framework pattern shows, using an XOR, that either the component or the container is responsible for managing persistence. Kobryn[14] shows that this component framework pattern applies to both EJB as well as COM+.

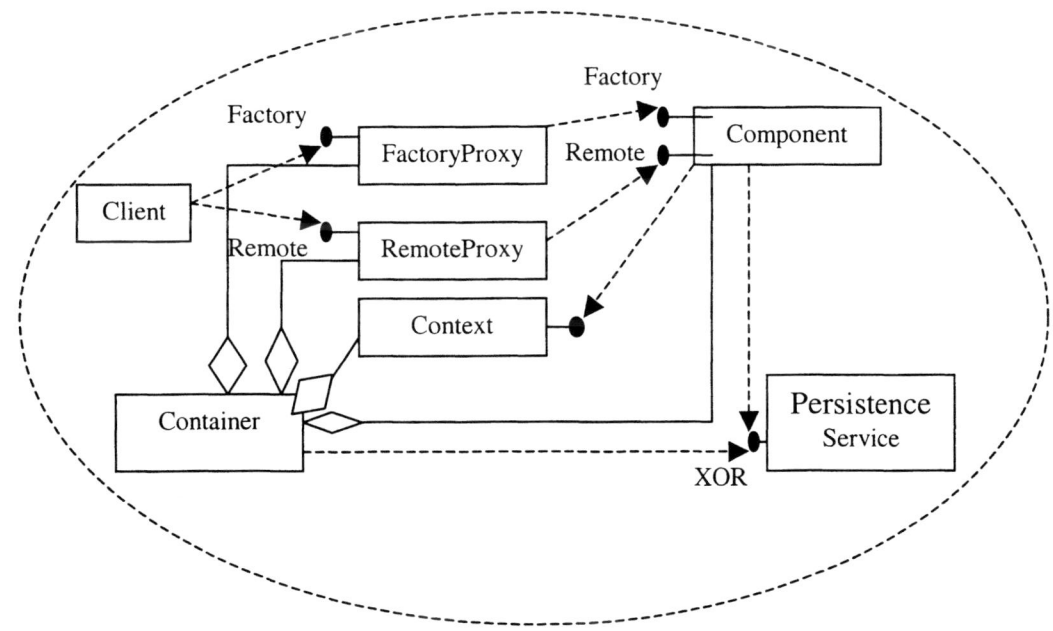

FIGURE 3. A Component Framework Pattern

Some important issues are identified when performing this comparison. Several of the classifiers in the EJB and COM+ versions of the pattern are UML stereotypes, that is user-customized extensions to the language and not part of standard UML. For example EJB interfaces can declare constants whereas in UML interfaces this is not allowed. Another issue is that when combining components with various entities that realize their interfaces it is not always clear how it should take place especially in cases of complex nesting. There are no UML constructs to support large component systems and frameworks. A very important issue is the need for support to allow developers to model components earlier in the software life cycle. Several methods have been published: Catalysis (supports component modeling), Unified Process (Limited support for components), KoBra and Pulse[15] (under development but designed to support both components and product line engineering).

CONCLUSIONS

A key enabler to product line engineering is software reuse. Reusable large grain software artifacts identified at the analysis and design phase must be evolved in a consistent manner to their realizable implementation counterparts (components) at runtime. Currently there is a semantic gap between object and component-based modeling. Although the universal modeling language UML 1.3 does provide notation for components a number of major issues have been identified which restrict support for the major component technologies. The UML community is working to overcome these limitations with minor (UML 1.4) and major revisions (UML 2.0) planned. OO Frameworks promise a new vehicle for reuse but the concepts are still being evolved. So in conclusion complete life cycle component-based product engineering is not a reality, but there are signs that progress is being made.

ACKNOWLEDGEMENTS

The authors take this opportunity to acknowledge the support of their institutes and in particular thanks to Paul Lecoq, Jean-Lious Faure, Martti Pimia and Jean-Pierre Vialle, Alain Bazan, Florida Estrella, Thierry Le Flour, Cristoph Koch, Sopie Lieunard, Steve Murray, Giovanni Organtini, Laslo Varga, Marton Zsenei and Guy Chevenier.

REFERENCES

1. Baker, N., McClatchey R., and LeGoff, J-M., " Scientific Workflow Management in a Distributed Production Environment" *EDOC'97* Workshop Proceedings, IEEE Computer Society, 1997, pp. 291-298.

2. Weiss, D., and Lai, C., Software Product-Line Engineering. Addison-Wesley, Reading Mass., 1999.

3. DeBaud, J-M., and Schmid, K., "A Systematic Approach to Derive the Scope of Software Product Lines," Proc. 21st Int'l Conf. Software Eng., ACM Press, N. Y. 1999.

4. Bayer, J., Flege, O., and Gacek, C., "Creating Product Line Architectures," Proc. 3rd Int'l Workshop Software Architectures for Product Families (IWSAPF-3), 2000

5. Keepence, B., and Mannion, M., "Using Patterns to Model Variability in Product Families," IEEE Software, IEEE Press July/August, 1999 pp. 102 -108

6. Barry, A., Baker, N., et al, Meta-Data Based Design of a Workflow System, OOPSLA'98: Workshop on Applications of Object Oriented Workflow Management Systems, Vancouver, Canada. October 1999.

7. UML Revision Task Force, OMG UML, v. 1.3, document ad/99-06-08. OMG June 1999

8. Buschmann, F., et al., Pattern-Oriented Software Architecture: A System of Patterns. Wiley, N.Y. 1996.

9. Kovacs, Z., et al., "Patterns in a Manufacturing and Production Environment", *EDOC'99* Conference Proceedings, IEEE Computer Society, 1999.

10. Fayad, M., et al. Building Application Frameworks. Wiley, N.Y., 1999

11. Fayad, M., "Introduction to the Computing Surveys' Electronic Symposium on Object-Oriented Application Frameworks." ACM Computing Surveys, Vol32, No 1, March 2000, pp. 1-11

12. Fingar, P., "Component-Based Frameworks for E-Commerce," Communications of the ACM, Vol. 43, No. 10, October 2000, pp. 61-66

13. D'Souza, D., and Wills, A. C., "Objects, Components and Frameworks with UML: The Catalysis Approach." Addison-Wesley, Reading, MA, 1999

14. Kobryn, C., "Modeling Components and Frameworks with UML," Communications of the ACM, Vol. 43, No. 10, October 2000, pp. 31-38

15. Bayer, J., et al., "Pulse: A Methodology to Develop Software Product Lines," Symp. Software Reusability' 99 (SSR'99), ACM Press, NY, 1999, pp. 122-131.

Protocols and Services for Distributed Data-Intensive Science

William Allcock, Ian Foster, and Steven Tuecke
Mathematics and Computer Science Division, Argonne National Laboratory

Ann Chervenak and Carl Kesselman
Information Sciences Institute, University of Southern California

Abstract. We describe work being performed in the Globus project to develop enabling protocols and services for distributed data-intensive science. These services include:
* High-performance, secure data transfer protocols based on FTP, plus a range of libraries and tools that use these protocols
* Replica catalog services supporting the creation and location of file replicas in distributed systems

These components leverage the substantial body of "Grid" services and protocols developed within the Globus project and by its collaborators, and are being used in a number of data-intensive application projects.

INTRODUCTION

We describe work being performed in the Globus project[1] to develop enabling protocols and services for distributed data-intensive science. We will begin with a discussion of the differences between protocols, API's, and services and why each is important. The features of our data transfer technology, GridFTP, and our replica catalog services will then be discussed. A full discussion of the protocols and API's is beyond the scope of this paper.

PROTOCOLS, API'S, AND SERVICES

There is a great deal of confusion regarding these terms. They are frequently used interchangeably, though incorrectly. Each plays a different role and provides distinct and important advantages.

A protocol defines the format of data that is sent between two systems, including the syntax of messages, character sets, and sequencing of messages. It does not specify whether this was accomplished by invoking a Java method, calling a C language function, or by someone sitting at a terminal typing the replies. The value of a protocol is that it provides *interoperability*. It allows a Java Application running on a Sun Workstation to work with a C application running on a Cray T3-E supercomputer, just as TCP/IP has enabled a heterogeneous collection of operating systems and computers to communicate over the Internet.

An Application Programmers Interface (API) specifies the interface to which an application programmer can write code. It specifies functionality, names, data types, parameter sequences, and return types. It is not an implementation, since several implementations of the same API can exist. The advantage to having a common API is *efficiency*. If the same API is supported on a variety of heterogeneous platforms, for example, as interfaces to different storage systems, then the programmer's task is greatly simplified.

Services are processes that are usually started automatically, always running, and provide basic functionality for other processes. They are also known as user agents and daemons. The E-Mail service is probably the most well known. It runs in the background. An E-Mail program (client) simply communicates with the server using a specified protocol and provides it with the data (address list, text, images, attachments, etc). The email service takes care of actually transferring the mail over the network and the service on the other end informs the appropriate client that mail has arrived. The advantages of a service are *faster application development*, *smaller programs* since each client does not need to have the service code present, and *more stable code* since the services can be more complicated to write.

DATA TRANSFER PROTOCOL: GRIDFTP

In Grid environments, access to distributed data is typically as important as access to distributed computational resources[2]. Distributed scientific and engineering applications require transfers of large amounts of data (terabytes or petabytes) between storage systems, and access to large amounts of data (gigabytes or terabytes) by many geographically distributed applications and users for analysis, visualization, etc. Unfortunately, the lack of standard protocols for transfer and access of data in the Grid has led to a fragmented Grid storage community. Users who wish to access different storage systems are forced to use multiple protocols and/or APIs, and it is difficult to efficiently transfer data between these different storage systems.

We propose a common data transfer and access protocol called GridFTP[3,4] that provides secure, efficient data movement in Grid environments. This protocol, which extends the standard FTP protocol, provides a superset of the features offered by the various Grid storage systems currently in use.

In order to make a common data transfer protocol attractive to users and developers of existing storage systems, we must provide a transfer protocol that offers a superset of the features offered by systems currently in regular use. In addition, the protocol must be extensible, in order to support future innovations by storage system users and developers.

We have observed that the FTP protocol is the protocol most commonly used for data transfer on the Internet, and the most likely candidate for meeting the Grid's needs. It is attractive in particular for the following reasons.

- It is a widely implemented and well-understood IETF standard protocol.
- There is a large code base and expertise from which to build.
- It provides a well-defined architecture for protocol extensions, and supports dynamic discovery of the extensions supported by a particular implementation.
- Numerous groups have added various extensions through the IETF. Some of these extensions are particularly useful in the Grid.
- In addition to client/server transfers (i.e. "put/get"), it also supports transfers directly between two servers, mediated by a third party client (i.e. "third party transfer").
- The separation of data and control channels onto different sockets allows for easier extensibility for parallel and striped transfers, efficiently transiting firewalls, etc.

Most current FTP implementations support only a subset of the features defined in the FTP protocol and its accepted extensions. Some of the seldom-implemented features are useful to Grid applications, but the standards also lack several features Grid applications require. We have selected a subset of the existing FTP standards and further extended them, adding the features described below. We believe that the resulting protocol is a suitable candidate for the common data transfer protocol for the grid.

Grid Security Infrastructure (GSI) and Kerberos support: Robust and flexible authentication, integrity, and confidentiality features are critical when transferring or accessing files. GridFTP must support GSI and Kerberos authentication, with user controlled setting of various levels of data integrity and/or confidentiality.

Third-party control of data transfer: In order to manage large data sets for large distributed communities, it is necessary to provide third-party control of transfers between storage servers. GridFTP provides this capability by adding GSSAPI security to the existing third-party transfer capability defined in the FTP standard.

Parallel data transfer: On wide-area links, using multiple TCP streams can improve aggregate bandwidth over using a single TCP stream. This is required both between a single client and a single server, and between two servers. GridFTP supports parallel data transfer through FTP command extensions and data channel extensions.

Striped data transfer: Partitioning data across multiple servers can further improve aggregate bandwidth. GridFTP supports striped data transfers through extensions defined in the Grid Forum draft.

Partial file transfer: Many applications require the transfer of partial files. However, standard FTP requires the application to transfer the entire file, or the remainder of a file starting at a particular offset. GridFTP introduces new FTP commands, to support transfers of regions of a file.

Support for reliable data transfer: Reliable transfer is important for many applications that manage data. Fault recovery methods for handling transient network failures, server outages, etc. are needed. The FTP standard includes basic features for restarting failed transfer that are not widely implemented. The GridFTP protocol exploits these features, and substantially extends them.

REPLICA MANAGEMENT

In this section, we present the Globus Replica Management architecture[4]. Replica management is an important issue for a number of scientific applications. For example, consider the petabytes of experimental data that will be generated by the LHC[5]. While the complete data set may exist in one or possibly several physical locations, it is likely that many universities, research laboratories or individual researchers will have insufficient storage to hold a complete copy. Instead, they will store copies of the most relevant portions of the data set on local storage for faster access.

Replica management system services include:

- Creating new copies of a complete or partial data set
- Registering these new copies in a Replica Catalog
- Allowing users and applications to query the catalog to find all existing copies of a particular file or collection of files
- Selecting the "best" replica for access based on storage and network performance predictions provided by a Grid information service

The Globus replica management architecture is a layered architecture. At the lowest level is a Replica Catalog that allows users to register files as logical collections and provides mappings between logical names for files and collections and the storage system locations of one or more replicas of these objects. We have implemented a Replica Catalog API in C as well as a command-line tool. Finally, we have defined a higher-level Replica Management API that creates and deletes replicas on storage systems and invokes low-level commands to update the corresponding entries in the replica catalog.

The basic replica management services that we provide can be used by higher-level tools to select among replicas based on network or storage system performance or automatically to create new replicas at desirable locations. We will implement some of these higher-level services in the next generation of our replica management infrastructure.

The purpose of the replica catalog is to provide mappings between logical names for files or collections and one or more copies of the objects on physical storage systems. The catalog registers three types of entries: logical collections, locations and logical files.

A logical collection is a user-defined group of files. We expect that users will find it convenient and intuitive to register and manipulate groups of files as a collection, rather than requiring that every file be registered and manipulated individually.

Location entries in the replica catalog contain all the information required for mapping a logical collection to a particular physical instance of that collection. The location entry may register information about the physical storage system, such as the hostname, port and protocol. In addition, it contains all information needed to construct a URL that can be used to access particular files in the collection on the corresponding storage system.

Each logical collection may have an arbitrary number of associated location entries, each of which contains a (possibly overlapping) subset of the files in the collection. Using multiple location entries, users can easily register logical collections that span multiple physical storage systems.

Despite the benefits of registering and manipulating collections of files using logical collection and location objects, users and applications may also want to characterize individual files. For this purpose, the replica catalog includes optional entries that describe individual logical files. Logical files are entities with globally unique names that may have one or more physical instances. The catalog may optionally contain one logical file entry in the replica catalog for each logical file in a collection.

REFERENCES

1. The Globus Project, www.globus.org
2. Chervenak, I. Foster, C. Kesselman, C. Salisbury, S. Tuecke, "The Data Grid: Towards an Architecture for the Distributed Management and Analysis of Large Scientific Datasets," to be published in the *Journal of Network and Computer Applications*.
3. Grid Forum GridFTP Introduction: http://www.sdsc.edu/GridForum/RemoteData/Papers/gridftp_intro_gf5.pdf
4. Grid Forum GridFTP Specification DRAFT: http://www.sdsc.edu/GridForum/RemoteData/Papers/gridftp_spec_gf5.pdf
5. W. Hoschek, J. Jaen-Martinez, A. Samar, H. Stockinger, K. Stockinger, "Data Management in an International Grid Project", 2000 International Workshop on Grid Computing (GRID 2000), Bangalore, India, December 2000.

Simulating Distributed Systems

Harvey B. Newman, Iosif C. Legrand

Charles C. Lauristsen Laboratory of High Energy Physics
California Institute of Technology, Pasadena, CA 91125, USA

Abstract. The simulation framework developed within the "Models of Networked Analysis at Regional Centers" (MONARC) project as a design and optimization tool for large scale distributed systems is presented. The goals are to provide a realistic simulation of distributed computing systems, customized for specific physics data processing tasks and to offer a flexible and dynamic environment to evaluate the performance of a range of possible distributed computing architectures. A detailed simulation of a large system, the CMS High Level Trigger (HLT) production farm, is also presented.

INTRODUCTION

The design and optimisation of large scale distributed systems, requires a realistic description and modelling of the data access patterns, the data flow across the local and wide area networks, and the scheduling and workload presented by hundreds of jobs running concurrently on large scale distributed systems exchanging very large amounts of data.

A process-oriented approach for discrete event simulation has been adopted because it is well suited to describe various activities running concurrently, as well as the stochastic arrival patterns typical of this class of simulations [1]. This simulation program is based on Java[TM] technology because of the support for the necessary methods and techniques needed to develop an efficient and flexible simulation frame [2].

DESIGN CONSIDERATIONS

The process-oriented approach for discrete event simulation is based on Threaded objects, or "Active Objects" (having an execution thread, program counter, stack, mutual exclusion mechanism...). They offer great flexibility in simulating the complex behaviour of distributed data processing programs. This approach offers a natural way of describing complex running programs that are data dependent and which concurrently compete for shared resources.

The MONARC simulation program [1] is built with Java[TM] technology. Java has built-in multi-thread support for concurrent processing, which can be used for simulation purposes by providing a dedicated scheduling mechanism. Java also offers good support for distributed objects architectures and for graphics. The flexible graphics tools, and facilities to analyse data interactively, are essential in any simulation project.

The tool's "simulation engine" provides a dedicated scheduling mechanism that is based on semaphores for the "Active Objects". It also provides a mechanism to dynamically add or remove objects from the system. Handling dynamically loadable modules is essential to describe complex configurations, which may change or evolve in time. The "Active Object" is the basic class that must be inherited by all the entities in the simulation, which require a time dependent behaviour. It provides the methods for synchronous and asynchronous communications with other objects, and the mechanism to communicate with the simulation engine so that it can be interrupted, suspended and resumed during execution. Objects, which extend this basic class, may implement any specific time dependent behaviour, which can be a function of messages or data received, its previous state(s), and its access to certain shared resources. In this way it is possible to implement highly non-linear processes such as caching and swapping. It also offers a means of describing the stochastic input pattern for jobs and activities in the system.

Shared resources, like CPU or I/O links, are represented in the simulation as normal objects, but access to their different update methods needs to be made, synchronised with the external "running" entities. There is a mutual exclusion mechanism when accessing unique atomic parts that avoids interruption: this guarantees the correct representation of the execution of concurrent processes.

As the number of jobs necessary to be simulated in such applications may be huge, a dedicated structure that allows "Active Objects" recycling was implemented to improve the simulation efficiency. The interrupt mechanism, implemented as an atomic (synchronised) self addressed event, for the "Active Objects" offers an effective way to simulate discrete event processes assuming a "continuous" flow in time between events which modify parts of the system. The interrupt functionality offers the possibility to simulate efficiently and accurately concurrent processes that need to share resources. As an example, the way multitasking is simulated is presented in Figure 1.

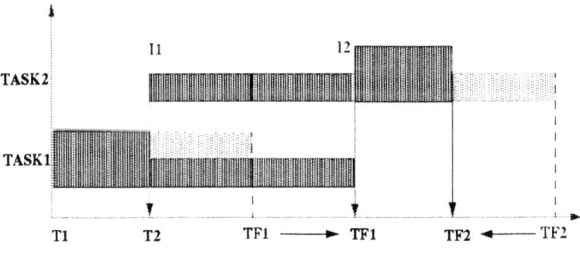

Figure 1. Modelling multitasking processing based on an "interrupt" scheme.

Referring to this figure, when a first job (Task1) starts, the time it takes is evaluated (original TF1), and this "Active" object enters into a wait state for this amount of time unless it is interrupted. If a new job (Task2) starts on the same hardware, it will cause an interrupt to the first task. Both tasks will share the same CPU power and the time to complete for each of them is re-computed assuming that they share the CPU equally or based on a running priority scheme (new TF1 and original TF2). Then both jobs will enter into a wait state and listen for other interrupts. When the first job (Task1) is finished, it creates another interrupt to re-distribute the resources for the remaining jobs. This model assumes that resource sharing is maintained between any discrete events (e.g. new job submission, job completion) that occur during the simulated time interval.

The simulation program includes a convenient set of interactive graphical tools allowing dynamic configurations as well as presentation and analysis of results. It provides a powerful development tool for evaluating and designing large scale distributed systems.

Simulation of the CMS ORCA-HLT Production Farm, Spring 2000

The simulation program was tested and validated with specific Queuing Theory problems as well as with a set of dedicated test bed measurements [3]. The simulation of a large production farm, based on Object Oriented database is presented here.

The immediate goal of the Spring ORCA/HLT production was to prepare a fully digitised sample of 2 million events with full pile-up at a simulated luminosity of 10^{34} cm^{-2}s^{-1}. This is a multi-stage process taking 2 million CPU minutes of computing and involving the transport of some 70 Tera-bytes of pile-up hits. The production starts from 2TB GEANT3 fz files, which are converted into 2TB Objectivity/DB format and stored in the HPSS system. The final results are again stored in an Objectivity/DB where they are used in reconstruction and analysis studies. This was achieved making use of a farm of 200 high performance commodity PC's running the Linux operating system. The Data Server (AMS) of Objectivity/DB was found to be particularly powerful in allowing new farm architectures to be studied. The Central Data Recording (CDR) system of IT/PDP was used to migrate data safely into HPSS, from which it is is now being accessed by Physicists studying the HLT.

The real set-up and realistic values for the hardware configuration parameters were used for the configuration of the simulation. For each type of data processing jobs the amount of I/O per event and the effective CPU per event were considered as input parameters for the simulation. The way programs read the events and all the pile-up structures was included in the simulation.

The simulation results are in good agreement with the measurements done during this CMS production. As an example, the distribution of the total time per job for the Muon Production task is compared with the measured values in Figure 2.

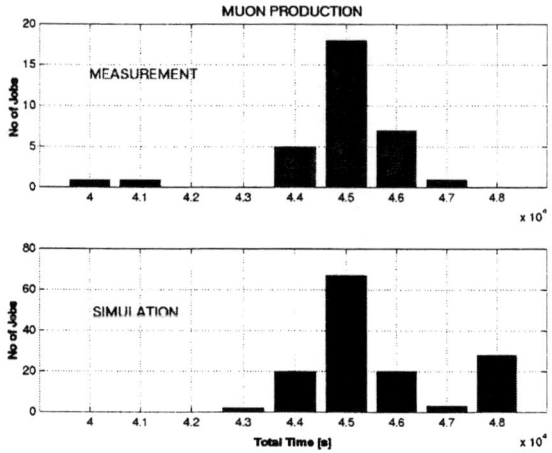

Figure 2. Comparison between the measurements and the simulation of the total time distribution for the "Muon Production" jobs.

The distribution of the job's efficiency for both types of production jobs is presented in Figure 3.

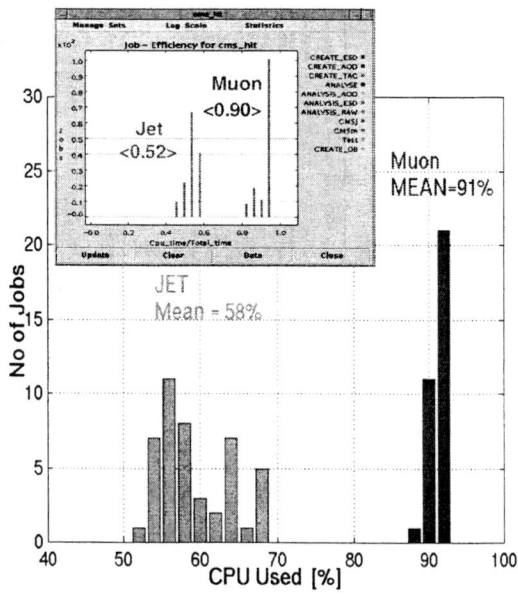

Figure 3. The distribution of the job's efficiency for the "Jet" and "Muon" production tasks.

Modelling correctly and understanding current production data processing farms is essential for the future design of the large-scale distributed systems.

SUMMARY

A CPU and code-efficient simulation approach to the problem of simulation of distributed computing systems has been developed and tested within the MONARC Collaboration. It provides a transparent way to map the distributed data processing, data transport and analysis tasks onto the simulation frame, and can describe dynamically even very complex computing models.

ACKNOWLEDGEMENTS

This work has been performed in collaboration with the MONARC project at CERN. We would like to thank David Stickland and Tony Wildish for the help in understanding the CMS production farm configuration and set-up.

REFERENCES

1. I.C. Legrand, H.B. Newman, "The Monarc Toolset for Simulating Large Network-Distributed processing Systems", Proc. of the 2000 Winter Simulation Conference. And MONARC Simulation Tool http://www.cern.ch/ MONARC/sim_tool/

2. PLOTEMY II "Heterogeneous concurrent Modeling in Java" http://www.eecs.berkeley.edu

3. Y. Morita et al. "Validation of Monarc Simulation Tools", CHEP2000, Padua, Italy (http://chep2000.pd.infn.it/, paper number 113

Symbolic Problem Solving

HELAC-PHEGAS: automatic computation of helicity amplitudes and cross sections

Aggeliki Kanaki and Costas G. Papadopoulos

Institute of Nuclear Physics, NCSR Demokritos, 15310 Athens, Greece

Abstract. HELAC-PHEGAS is a FORTRAN based package that is able to compute automatically and efficiently tree-level helicity amplitudes and cross sections for arbitrary scattering processes within the standard electroweak theory and QCD. The algorithm for the amplitude computation, HELAC, exploits the virtues of the Dyson-Schwinger equations. The phase-space generation algorithm, PHEGAS, constructs all possible kinematical mappings dictated by the amplitude under consideration. Combined with multichannel self-optimized Monte Carlo integration it results in efficient cross section evaluation.

INTRODUCTION

The need for efficient algorithms to calculate helicity amplitudes and cross sections for any process, in an automatic way, has been well recognized long time ago. Up to now, algorithms that efficiently combine helicity amplitude computation and phase-space integration have been proven successful for specific processes, like for instance four-fermion (1) production in e^+e^- collisions. On the other hand general-purpose computational packages like CompHEP (2) and GRACE (3) do not provide automatic efficient phase-space integration algorithms. Moreover the vast majority of the automatized helicity amplitude computationial algorithms, like for instance MadGraph (4), have been based on the Feynman graph representation of the amplitude which severely restricts their ability to deal with multiparticle scattering processes.

In this article we report on some developments that have lead to the construction of two programs, HELAC (5) and PHEGAS (6), that allow for an efficient and automatic evaluation of cross sections for arbitrary scattering processes.

HELAC

The traditional representation of the scattering amplitude in terms of Feynman graphs results to a computational cost that grows like the number of those graphs, therefore as $n!$ where n is the number of particles involved in the scattering process.

An alternative (7, 8) to the Feynman graph representation is provided by the Dyson-Schwinger approach. Dyson-Schwinger equations express recursively the n-point Green's functions in terms of the $1-,2-,\ldots,(n-1)$-point functions. For instance in QED these equations can be written as follows:

$$b^\mu(P) = \sum_{i=1}^n \delta_{P=p_i} b^\mu(p_i) + \sum_{P=P_1+P_2} (ig)\Pi_\nu^\mu \bar{\psi}(P_2)\gamma^\nu \psi(P_1)\varepsilon(P_1,P_2)$$

where

describes a generic n-point Green's function with respectively one outgoing photon, fermion or antifermion leg carrying momentum P. $\Pi_{\mu\nu}$ stands for the boson propagator and ε takes into account the sign due to fermion antisymmetrization.

In order to actually solve these recursive equations it is convenient to use a binary representation of the momenta involved (8). For a process involving n external particles, all momenta appearing in the computation, P^μ,

$$P^\mu = \sum_{i \in I} p_i^\mu$$

where $I \subset \{1,\ldots,n\}$, can be assigned a binary vector $\vec{m} = (m_1,\ldots,m_n)$, where its components take the values 0 or 1, in such a way that

$$P^\mu = \sum_{i=1}^n m_i p_i^\mu .$$

Moreover this binary vector can be uniquely represented by the integer

$$m = \sum_{i=1}^{n} 2^{i-1} m_i$$

and therefore all sub-amplitudes can be labeled accordingly, i.e.

$$b_\mu(P) \to b_\mu(m), \ 1 \le m \le 2^{n-1}.$$

A very convenient ordering of integers in binary representation relies on the notion of level l, defined simply as

$$l = \sum_{i=1}^{n} m_i.$$

As it is easily seen all external momenta are of level 1, whereas the total amplitude corresponds to the unique level n integer 2^{n-1}. This ordering dictates the natural path of the computation; starting with level-1 subamplitudes, we compute the level-2 ones using the Dyson-Schwinger equations and so on up to the level n one which is the full amplitude. For the spinor wave functions as well as for the Dirac matrices, we have chosen the 4-dimensional chiral representation which results to particurarly simple expressions. All electroweak vertices in both the Feynman and the Unitary gauge have been included.

The computational cost of HELAC grows like $\sim 3^n$, which essentially counts the steps used to solve the recursive equations. Obviously for large n there is a tremendous saving of computational time, compared to the $n!$ growth of the Feynman graph approach.

For QCD amplitudes colour representation and summation plays an important role. Let $1\ldots n$ denote the colour labels of quarks and $\sigma_i(1)\ldots\sigma_i(n)$ denote the colour labels of antiquarks, with $\sigma(i), i = 1\ldots n!$ being a permutation of $\{1\ldots n\}$. The colour factor is given obviously by

$$C_i = \delta_{1\sigma_i(1)} \delta_{2\sigma_i(2)} \cdots \delta_{n\sigma_i(n)}$$

Moreover the colour matrix, defined as

$$\mathcal{M}_{ij} = \sum_{\text{colours}} C_i C_j^\dagger$$

with the summation running over all colours, $1\ldots N_c$, has a very simple representation

$$\mathcal{M}_{ij} = N_c^{m(\sigma_i, \sigma_j)}$$

where $m(\sigma_i, \sigma_j) - 1$ counts how many elements of the permutations σ_i and σ_j are common. In order to extend this colour representation to QCD amplitudes we have to just consider gluons as being quark-antiquark pairs and assign to them two colour labels (i, σ_i). The colour factor and the colour matrix still has exactly the same form. The only thing one has to consider is to rewrite the known Feynman rules of QCD in a slight different way. It is worthwhile to note that exact colour summation is efficient as far as the number of equivalent gluons is smaller than $O(5-6)$. For multicolour processes other approaches have to be considered (9, 10).

The programme is also incorporating the possibility to use an extended precision by exploiting the virtues of FORTRAN90. The user can easily switch to a quadruple precision or to an even higher, user-defined precision by using the multi-precision library (11) included in HELAC. In this way, a straightforward computation of cross sections for processes like $e^-e^+ \to e^-e^+e^-e^+$ without any cut is reliably performed (12).

PHEGAS

The study of multi-particle processes, like for instance four-fermion production in e^+e^-, requires efficient phase-space Monte Carlo generators. The reason is that the squared amplitude, being a complicated function of the kinematical variables, exhibits strong variations in specific regions and/or directions of the phase space, lowering in a substantial way the speed and the efficiency of the Monte Carlo integration. A well known way out of this problem relies on algorithms characterized by two main ingredients:

1. The construction of appropriate mappings of the phase space parametrization in such a way that the main variation of the integrand can be described by a set of almost uncorrelated variables, and

2. A self-adaptation procedure that reshapes the generated phase-space density in order to be as much as possible close to the integrand.

In order to construct appropriate mappings we note that the integrand, i.e. the squared amplitude, has a well-defined representation in terms of Feynman diagrams. It is therefore natural to associate to each Feynman diagram a phase-space mapping that parametrizes the leading variation coming from it. To be more specific the contribution of tree-level Feynman diagrams to the full amplitude can be factorized in terms of propagators, vertex factors and external wave functions. In general, the main source of variation comes from the propagator factors and therefore our aim is to construct a mapping that expresses the phase-space density in terms of the kinematical invariants that appear in these propagator factors. Since in principle we need as many mappings as Feynman diagrams for the process under consideration, we have to appropriately combine them in order to produce the global phase-

Table 1. Results for several processes using HELAC-PHEGAS. In the second column the number of Feynman graphs and in parenthesis the number of steps required to solve the recursiveDyson-Schwinger equations are given.

Final states	Number of FG(DS)	$\sqrt{s}$ (GeV)	Cross section (fb)
$e^-e^+ \to u\bar{d}s\bar{c}\gamma$	90(74)	200	199.75 (16)
$e^-e^+ \to e^-\bar{\nu}_e\mu^+\nu_\mu\gamma$	108(100)	200	29.309 (25)
$e^-e^+ \to \mu^-\bar{\nu}_\mu u\bar{d}\gamma\gamma$	587(210)	500	1.730 (58)
$e^-e^+ \to \mu^-\bar{\nu}_\mu u\bar{d}c\bar{c}$	209(102)	500	0.1783 (20)
$e^-e^+ \to \mu^-\bar{\nu}_\mu u\bar{d}c\bar{c}\gamma$	2142(339)	500	0.02451 (65)
$gg \to b\bar{b}b\bar{b}W^-W^+$	960(380)	500	4.716(24)

space density. A simple and well studied solution to this problem was suggested some time ago in reference (13). It should be mentioned however that other self-adapting approaches can be used as well (14). It is important to note that although by using Feynman graphs to construct phase-space mappings we face the original $n!$ computational cost growth problem, the self-optimization cures this to a certain extent by selecting only the few mappings that dominate the phase-space density. For alternative approaches we refer to (15).

In order to describe the construction of the phase-space mappings, let us consider a typical process in which two incoming particles produce n outgoing ones. The phase space, $d\Phi_n(P = q_1 + q_2; p_1 \ldots, p_n)$, can be decomposed as follows

$$d\Phi_n = \left(\prod_{i=1}^m \frac{dQ_i^2}{2\pi}\right) d\Phi_m(P; Q_1, \ldots, Q_m)$$
$$d\Phi_{n_1}(Q_1; r_1, r_2, \ldots, r_{n_1}) \ldots d\Phi_{n_m}(Q_m; s_1, s_2, \ldots, s_{n_m})$$

where the subsets $\{r_1, r_2, \ldots, r_{n_1}\}$ up to $\{s_1, s_2, \ldots, s_{n_m}\}$ represent an arbitrary partition of $\{p_1, p_2, \ldots, p_n\}$. The above equation can be generalized recursively resulting in an arbitrary decomposition of $d\Phi_n$. Feynman graphs can be seen as a realization of such a decomposition, this latter being identified with a sequence of vertices of the graph. There are two possible cases for $2 \to n$ scattering. First, all outgoing momenta involved in the vertex are time-like,

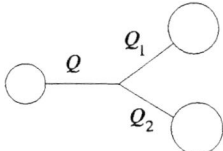

$$d\Phi_n = \ldots \frac{dQ_1^2}{2\pi} \frac{dQ_2^2}{2\pi} d\Phi_2(Q \to Q_1, Q_2) \ldots$$
$$= \ldots \frac{dQ_1^2}{2\pi} \frac{dQ_2^2}{2\pi} d\cos\theta \, d\phi \frac{\lambda^{1/2}(Q^2, Q_1^2, Q_2^2)}{32\pi^2 Q^2} \ldots$$

with $\lambda(x, y, z) = x^2 + y^2 + z^2 - 2xy - 2xz - 2yz$, and second when one of them is space-like,

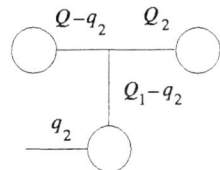

$$d\Phi_n = \ldots \frac{dQ_1^2}{2\pi} \frac{dQ_2^2}{2\pi} d\Phi_2(Q \to Q_1, Q_2) \ldots$$
$$= \ldots \frac{dQ_1^2}{2\pi} \frac{dQ_2^2}{2\pi} dt \, d\phi \frac{1}{32\pi^2 Q |\vec{q}_2|} \ldots$$

with

$$t = (Q_1 - q_2)^2$$
$$= m_2^2 + Q_1^2 - \frac{E_2}{Q}(Q^2 + Q_1^2 - Q_2^2) + \frac{\lambda^{1/2}}{Q}|\vec{q}_2|\cos\theta$$

and $(E_2, \vec{q}_2)$ being the incoming momentum q_2 in the rest frame of Q. The appropriate sequence of vertices, $\{V_1, V_2, \ldots, V_k\}$ can be chosen in such a way that a recursive construction of the phase space is realized. For instance V_1 should contain at least one incoming particle whose momentum is known. The rest of the sequence is chosen recursively: vertex V_j is characterized by an incoming momentum Q which has already been generated in one of the $\{V_1, \ldots, V_{j-1}\}$.

Following the above described algorithm we end up with an expression for the phase-space density,

$$d\Phi_n \to \prod ds_i \mathcal{P}_i(s_i) \prod dt_j \mathcal{P}_j(t_j) \prod d\phi_k \prod d\cos\theta_l$$

where s_i and t_j refer to the kinematical invariants entering the propagator factors of the graph and ϕ_k and $\cos\theta_l$ represent center-of-mass angles needed to complete the phase space parametrization. It is now straightforward to generate s_i and t_j with probability densities $\mathcal{P}_i(s_i)$ and $\mathcal{P}_j(t_j)$ that are automatically chosen accordingly to the nature of the propagating particle.

Results, demonstrating the ability of HELAC-PHEGAS to deal with multiparticle processes, are presented in table 1 (6, 16).

SUMMARY AND OUTLOOK

HELAC-PHEGAS offers a framework for high-energy phenomenology. It provides all necessary and sufficient tools for efficient, reliable and automatic computation of helicity amplitudes and cross sections. The Standard Model, including QCD, has been fully incorporated. Higher-order corrections are in principle tractable within the framework of Dyson-Schwinger equations and work is in progress in order to include electroweak corrections as described in reference (17). New physics interactions and models, icluding the Minimal Supersymmetric Standard Model and the trilinear gauge couplings will be considered in the near future.

Acknowledgments

C.G.P. would like to acknowledge Fermilab and Argonne National Laboratory for their kind hospitality.

REFERENCES

1. D. Bardin et al., "Event generators for W W physics," hep-ph/9709270 and references therein.

2. E. E. Boos, M. N. Dubinin, V. A. Ilin, A. E. Pukhov and V. I. Savrin, "CompHEP: Specialized package for automatic calculations of elementary particle decays and collisions," hep-ph/9503280.
V. A. Ilin, D. N. Kovalenko and A. E. Pukhov, Int. J. Mod. Phys. **C7** (1996) 761 [hep-ph/9612479].

3. T. Ishikawa, T. Kaneko, K. Kato, S. Kawabata, Y. Shimizu and H. Tanaka [MINAMI-TATEYA group Collaboration], KEK-92-19.
F. Yuasa et al., "Automatic computation of cross sections in HEP: Status of GRACE system," hep-ph/0007053.

4. T. Stelzer and W. F. Long, Comput. Phys. Commun. **81** (1994) 357 [hep-ph/9401258].

5. A. Kanaki and C. G. Papadopoulos, "HELAC: A package to compute electroweak helicity amplitudes," hep-ph/0002082.

6. C. G. Papadopoulos, "PHEGAS: A phase space generator for automatic cross-section computation," hep-ph/0007335.

7. F. A. Berends and W. T. Giele, Nucl. Phys. **B306** (1988) 759.

8. F. Caravaglios and M. Moretti, Phys. Lett. **B358** (1995) 332 [hep-ph/9507237].

9. P. Draggiotis, R. H. Kleiss and C. G. Papadopoulos, Phys. Lett. **B439** (1998) 157 [hep-ph/9807207].

10. F. Caravaglios, M. L. Mangano, M. Moretti and R. Pittau, Nucl. Phys. **B539** (1999) 215 [hep-ph/9807570].

11. David M. Smith, Transactions on Mathematical Software **17** (1991) 273- 283. http://www.lmu.edu/acad/personal/faculty/dmsmith2/FMLIB.html

12. F. A. Berends, C. G. Papadopoulos and R. Pittau, "NEXTCALIBUR: A four-fermion generator for electron positron collisions," hep-ph/0011031; "Four-fermion production in electron positron collisions with NEXTCALIBUR," hep-ph/0002249.

13. R. Kleiss and R. Pittau, Comput. Phys. Commun. **83** (1994) 141 [hep-ph/9405257].

14. T. Ohl, Comput. Phys. Commun. **120** (1999) 13 [hep-ph/9806432].

15. P. D. Draggiotis, A. van Hameren and R. Kleiss, Phys. Lett. **B483** (2000) 124 [hep-ph/0004047].

16. M. W. Grunewald et al., "Four fermion production in electron positron collisions," hep-ph/0005309.

17. W. Beenakker, F. A. Berends and A. P. Chapovsky, Nucl. Phys. **B573** (2000) 503 [hep-ph/9909472].

O'Mega: An Optimizing Matrix Element Generator

Thorsten Ohl

Darmstadt University of Technology, Schloßgartenstraße 9, 64289 Darmstadt, Germany
ohl@hep.tu-darmstadt.de

Abstract. I sketch the architecture of *O'Mega*, a new optimizing compiler for tree amplitudes in quantum field theory. O'Mega generates the most efficient code currently available for scattering amplitudes for many polarized particles in the standard model. A complete infrastructure for physics beyond the standard model is provided.

INTRODUCTION

Current and planned experiments in high energy physics can probe processes with many tagged—potentially polarized—particles in the final state. The combinatorial explosion of the number of Feynman diagrams contributing to scattering amplitudes for many external particles calls for the development of more compact representations that translate well to efficient and reliable numerical code. In gauge theories, strong numerical cancellations in a redundant representation built from necessarily gauge dependent Feynman diagrams lead to a loss of numerical precision, stressing further the need for eliminating redundancies.

Due to the large number of processes that have to be studied in order to unleash the potential of modern experiments, the construction of these representations must be possible algorithmically on a computer and should not require human ingenuity for each new application.

O'Mega (1) is a compiler for tree-level scattering amplitudes that satisfies these requirements. O'Mega is independent of the target language and can support code in any programming language for which a simple output module has been written. To support a physics model, O'Mega requires as input only the Feynman rules and the relations among coupling constants.

Similar to earlier numerical approaches (2, 3), O'Mega reduces the growth in calculational effort from a factorial of the number of particles to an exponential. The symbolic nature of O'Mega, however, increases its flexibility. Indeed, O'Mega can emulate both (2, 3) and produces code that is empirically at least twice as fast.

1POWS AND KEYSTONES

One Particle Off-shell Wave functions (1POWs) are obtained from Greensfunctions by applying the LSZ reduction formula to all but one line:

$$W^{q_1,\ldots,q_m}_{p_1,\ldots,p_n}(x) = \langle \phi(q_1),\ldots,\phi(q_m); out|\Phi(x)|\phi(p_1),\ldots,\phi(p_n); in\rangle . \quad (1)$$

The 1POW $W^{q,q'}_p(x) = \langle \phi(q),\phi(q'); out|\Phi(x)|\phi(p); in\rangle$ in lowest order of ϕ^3-theory, is given—for illustration—by

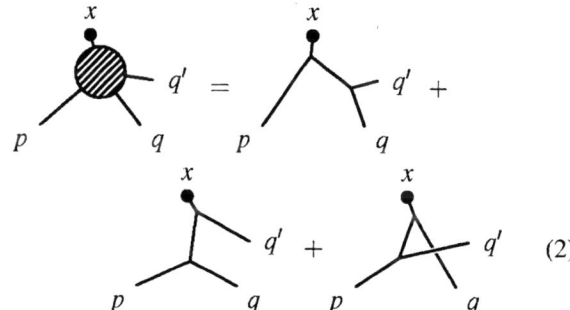

At tree-level, the set of all 1POWs for a given set of external momenta can be constructed recursively (4)

$$\text{(3)}$$

where the sum extends over all partitions of the set of n momenta. For all quantum field theories, there are—well defined, but not unique—sets of *Keystones* K (1) such that the sum of tree Feynman diagrams can be expressed as a sparse sum of products of 1POWs without double counting. In a theory with only cubic couplings this is expressed as

$$T = \sum_{i=1}^{F(n)} D_i = \sum_{k,l,m=1}^{P(n)} K^3_{f_k f_l f_m}(p_k, p_l, p_m) W_{f_k}(p_k) W_{f_l}(p_l) W_{f_m}(p_m) . \quad (4)$$

The non-trivial problem of avoiding the double counting of diagrams like (the circle denotes the keystone)

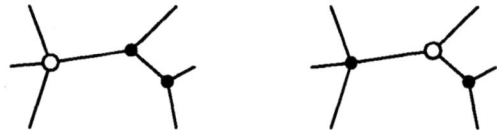

has been solved for general theories with vertices of arbitrary degrees (1).

The number of distinct momenta that can be formed from n external momenta is $P(n) = 2^{n-1} - 1$. Therefore, the number of tree 1POWs grows exponentially with the number of external particles and not with a factorial, as the number of Feynman diagrams $F(n) = (2n-5)!! = (2n-5)\cdot\ldots\cdot 5\cdot 3\cdot 1$. The equations sketched in Eqs. (3) and (4) for cubic couplings can be generalized to vertices of any order (1).

Even for vector particles and to all orders in renormalized perturbation theory, the 1POWs are 'almost' physical objects and satisfy simple Ward identities in unbroken gauge theories

$$\frac{\partial}{\partial x_\mu} \langle \text{out}|A_\mu(x)|\text{in}\rangle_{\text{amp.}} = 0 \qquad (5)$$

and well as in spontaneously gauge theories

$$\frac{\partial}{\partial x_\mu} \langle \text{out}|W_\mu(x)|\text{in}\rangle_{\text{amp.}} = \xi_W m_W \langle \text{out}|\phi_W(x)|\text{in}\rangle_{\text{amp.}} \qquad (6)$$

in R_ξ-gauge. The code for matrix elements can optionally be instrumented by O'Mega with numerical checks of these Ward identities for intermediate lines.

DIRECTED ACYCLICAL GRAPHS

The algebraic expression for the tree-level scattering amplitude in terms of Feynman diagrams is itself a tree. The much slower growth of the set of 1POWs compared to the set of Feynman diagrams shows that this representation is extremely redundant. In this case, *Directed Acyclical Graphs* (DAGs) provide a more efficient representation, as illustrated by a trivial example

$$ab(ab+c) = \qquad = \qquad , \qquad (7)$$

where one multiplication is saved. The replacement of expression trees by equivalent DAGs is part of the repertoire of optimizing compilers, known as *common subexpression elimination*. Unfortunately, this approach fails for typical expressions appearing in quantum field theory, because of the combinatorial growth of space and time required to find an almost optimal factorization.

However, the recursive definition in Eq. (3) allows to construct the DAG of the 1POWs in Eq. (4) *directly* (1), without having to construct and factorize the Feynman diagrams explicitly.

As mentioned above, there is more than one consistent prescription for constructing the set of keystones (1). The symbolic expressions constructed by O'Mega contain the symbolic equivalents of the numerical expressions computed by (2) (maximally symmetric keystones) and (3) (maximally asymmetric keystones) as special cases.

ALGORITHM

By virtue of their recursive construction in Eqs. (3), tree-level 1POWs form a DAG and the problem is to find the smallest DAG that corresponds to a given tree, (i.e. an given sum of Feynman diagrams). O'Mega's algorithm proceeds in four steps

Grow: starting from the external particles, build the tower of *all* 1POWs up to a given height (the height is less than the number of external lines for asymmetrical keystones and less than half of that for symmetrical keystones) and translate it to the equivalent DAG D.

Select: from D, determine *all* possible *flavored keystones* for the process under consideration and the 1POWs appearing in them.

Harvest: construct a sub-DAG $D^* \subseteq D$ consisting *only* of nodes that contribute to the 1POWs appearing in the flavored keystones.

Calculate: multiply the 1POWs as specified by the keystones and sum the keystones.

By construction, the resulting expression contains *no* more redundancies and can be translated to a numerical expression. In general, asymmetrical keystones create an expression that is smaller by a few percent than the result from symmetrical keystones, but it is not yet clear which approach produces the numerically more robust results.

IMPLEMENTATION

The O'Mega compiler is implemented in O'Caml (5), a functional programming language of the ML family with a very efficient, portable and freely available implementation, that can be bootstrapped on all modern computers in a few minutes.

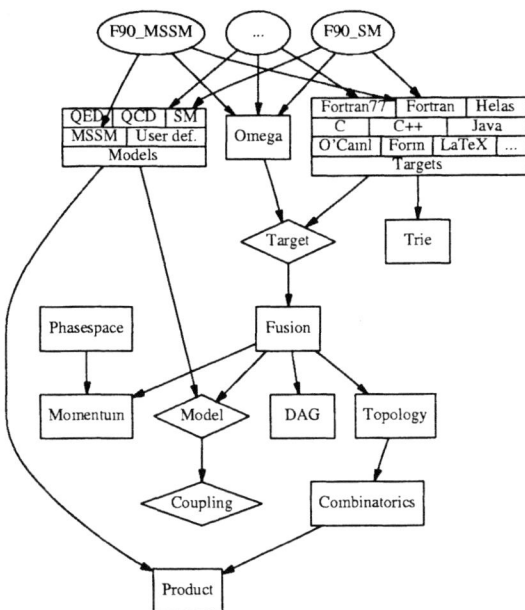

FIGURE 1. Module dependencies in O'Mega. The diamond shaped nodes denote abstract signatures defining functor domains and co-domains. The rectangular boxes denote modules and functors, while oval boxes stand for example applications.

The powerful module system of O'Caml allows an efficient and concise implementation of the DAGs for a specific physics model as a functor application (1). This functor maps from the category of trees to the category of DAGs and is applied to the set of trees defined by the Feynman rules of any model under consideration.

The implementation is concise and efficient simultaneously by exploiting the virtues of persistent data structures (6). Typically, the resources consumed by O'Mega are only a small fraction of the resources required by the compiler for the target language.

The module system of O'Caml has been used to make the combinatorial core of O'Mega demonstrably independent from the specifics of both the physics model and the target language (1), as shown in Figure 1. A Fortran90/95 backend has been realized first, backends for C++ and Java will follow. The complete electroweak standard model has been implemented (the treatment of interfering color amplitudes is still incomplete). Majorana fermions, required by supersymmetric field theories, are available (using (8)) and the MSSM is in preparation.

As mentioned above, the compilers for the target programming language are the slowest step in the generation of executable code. On the other hand, the execution speed of the code is limited by non-trivial vertex evaluations for vectors and spinors, which need $O(10)$ complex multiplications. Therefore, an *O'Mega Virtual Machine* can challenge native code and avoid compilations.

APPLICATIONS

The code generated by the Fortran90/95 backend is the most efficient code available for polarized scattering amplitudes for many particles. The results have been compared with MADGRAPH (7) for many standard model processes and numerical agreement at the level of 10^{-11} has been found with double precision floating point arithmetic. O'Mega generated amplitudes are used in the omnipurpose event generator generator WHIZARD (9). The first complete experimental study of vector boson scattering in six fermion production for linear collider physics (10) has been facilitated by O'Mega and WHIZARD.

Acknowledgments

I thank my collaborators Mauro Moretti and Jürgen Reuter. I thank Wolfgang Kilian for valuable suggestions and for "early adoption" of O'Mega. This research is supported by the German Bundesministerium für Bildung und Forschung (05 HT9RDA) and Deutsche Forschungsgemeinschaft (MA 676/6-1).

REFERENCES

1. M. Moretti, T. Ohl, and J. Reuter, (to be published), http://heplix.ikp.physik.tu-darmstadt.de/~ohl/omega/.
2. F. Caravaglios and M. Moretti, Z. Phys. **C74** (1997) 291.
3. A. Kanaki and C. Papadopoulos, DEMO-HEP-2000/01, hep-ph/0002082, February 2000.
4. H. Murayama, I. Watanabe, and K. Hagiwara, KEK Report 91-11, January 1992.
5. Xavier Leroy, *The Objective Caml system, release 3.0, documentation and user's guide*, Technical Report, INRIA, 2000 (http://caml.inria.fr/ocaml/).
6. Chris Okasaki, *Purely Functional Data Structures*, Cambridge University Press, 1998.
7. T. Stelzer and W.F. Long, Comput. Phys. Commun. **81** (1994) 357.
8. A. Denner, H. Eck, O. Hahn, and J. Küblbeck, Phys. Lett. **B291** (1992) 278; Nucl. Phys. **B387** (1992) 467.
9. W. Kilian, (to be published), http://www-ttp.physik.uni-karlsruhe.de/~kilian/whizard/.
10. R. Chierici and S. Rosati, (to be published).

A Feynman graph selection tool in GRACE system

Fukuko YUASA[a]*, Toshiaki KANEKO[b], Tadashi ISHIKAWA[a]

a)KEK, 1-1 OHO Tsukuba Ibaraki, Japan 305-0801
b)Meiji-Gakuin Univ., Kamikurata 1518, Totsuka, Yokohama, Kanagawa Japan 244-0816

Abstract. We present a Feynman graph selection tool `grcsel`, which is an interpreter written in C language. In the framework of GRACE, it enables us to get a subset of Feynman graphs according to given conditions.

INTRODUCTION

Using an automatic Feynman graph calculation package, we can generate the information of all Feynman graphs for given processes. Sometimes it is necessary to select graphs from the set of all graphs by some conditions. However, it is not so easy to select them correctly by hand when a huge number of graphs is involved, such as higher order corrections or SUSY processes.

A program `grcsel` selects out a subset of Feynman graphs from the set of graphs, generated by GRACE(1), according to given selection conditions. The output information of selected graphs is written in the same format as that of the original set. This enables us to generate Feynman amplitudes within GRACE in the same procedure as for all graphs. So we can perform cross section calculation, gauge invariance check, event generation and so on for the selected ones. `grcsel` helps us:

1. to find decay-graphs and evaluate signal/background ratio,
2. to check the accuracy of approximated calculation,
3. to confirm precision of the calculation,
4. to reduce the calculation time,
5. to develop kinematics routine.

OVERVIEW OF `GRCSEL`

`grcsel` consists of three parts: a steering-part defines basic functions of graph selections and reads input files, an interpreter-part parses and evaluates commands, and a utility-part handles subsets of particles, vertices and graphs.

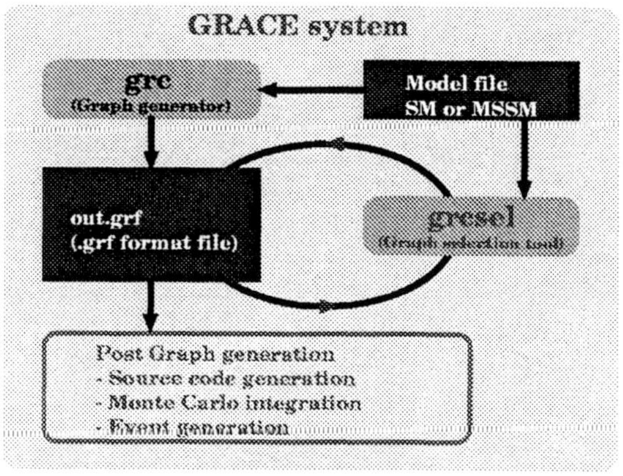

FIGURE 1.

Once a physics process and the order of calculation are fixed, Feynman graphs are generated by `grc` program with specified Feynman rules described in physics model file(2). The information on graphs generated are stored in a file named `out.grf` (we call the format of this file `.grf` format).

`grcsel` reads the physics model file and `out.grf` and selects graphs according to a kind of propagator, characteristics of graph topology, a type of vertex or a graph number. `grcsel` outputs those selected graphs in the same format as `out.grf`. Successively this output file can be used as the input to source code generation for Monte Carlo integration or event generation. We can also use `grcsel` again reading output of previous execution of `grcsel`. The schematic view of how `grcsel` works in GRACE system is shown in Fig. 1.

* E-mail:fukuko.yuasa@kek.jp

CP583, *Advanced Computing and Analysis Techniques in Physics Research: VII International Workshop*
edited by P. C. Bhat and M. Kasemann

Table 1. Basic functions to select and output graphs.

Function	Description
cutprop	Select graphs with a specified propagator.
selvlegs	Select graphs with a vertex consisting specified particles.
selvertex	Select graphs with a specified vertex.
outgset	Output a set of graphs.
renumgset	Renumber and output a set of graphs.

Running grcsel

Graph selection starts by the program grcsel:

```
grcsel
```

This program requires out.grf file by default. The graph selection commands are read through standard input, which may be given interactively or by a script file. With a script file where grcsel commands are prepared, we can redirect that file:

```
grcsel < command.in
```

To use another input .grf format file, e.g. out1.grf, instead of out.grf, we can add the filename after grcsel command as:

```
grcsel out1.grf < command.in
```

grcsel command

In a script file there are a series of grcsel commands such as declaration of variables and basic functions to specify the selection conditions or operators. grcsel has 14 basic functions in total. Three of them return a subset of graphs in accordance with specified selection condition and two functions output set of graphs. They are summarized in Table 1.

In grcsel, graphs, particles or vertices are treated as elements of a *set* of type gset, pset or vset, which are defined as a set of graphs, particles or vertices, respectively. Set variables have to be declared at first with their

Table 2. *set* operators

Operator	Function
&	Set intersection
|	Set union
~	Complement of set

types. Operations on sets are available and are shown in Table 2.

EXAMPLE

In the following example, graphs with ν_e propagator connected to the initial electron and final W^- are selected among graphs of $e^+e^- \to W^+W^-\gamma$ process. Selected graphs are output into a file named out1.grf.

```
% e+ e- --> W+ W- Photon
%
% out1.grf : with neutrino propagator
% at the vertex of initial elec-
tron and
% final W-.
%
gset gs0, gs1;

gs0 = ~[];    % all graphs

gs1 = cutprop(gs0, [``nu-e''], [0,3]);

outgset(``out1.grf'', gs1);

quit;
```

In Fig. 2, selected graphs are shown.

REMARKS

grcsel has been developed in the framework of GRACE 2.1.7.4. It can handle tree and 1-loop graphs and it supports standard and MSSM physics model. grcsel is included in a distribution kit of GRACE 2.1.7.4.

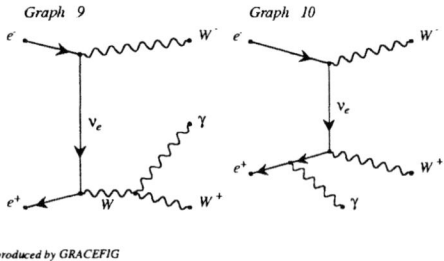

FIGURE 2.

ACKNOWLEDGMENTS

We wish to thank the members of MINAMI-TATEYA collaboration for continuous discussions and many kinds of support. We are also grateful to express our sincere gratitude to Prof. Y.Shimizu for the valuable suggestions and continuous encouragements. Authors appreciate Prof. Y.Watase for the encouragements. This work was supported in part by the Grant-in Aid (No. 12680363, 10640285, 10680366 and 11440083) of Monbu-sho, Japan.

REFERENCES

1. MINAMI-TATEYA group: GRACE manual. KEK Report 92-19.
   F.Yuasa *et al.*: Prog. Theor. Phys. Suppl. **138** (2000) 18-23.
2. T.Kaneko: Comput. Phys. Commun. **92** (1995) 127-152.

A Library of Function Classes

J. Boudreau[†], M. Fischler[‡], P. Maksimović[§]

[†] *Department of Physics and Astronomy, University of Pittsburgh, Pittsburgh PA 15260*
[‡] *Fermi National Accelerator Laboratory, Batavia Il 60510*
[§] *Department of Physics, Harvard University, Cambridge MA 02138*

Abstract. We have written a library of function classes in C++, which is distributed as part of the CLHEP foundation class libraries for High Energy Physics. The main goals of the library are to provide objects having all the mathematical behavior of functions, and to provide a mechanism for parameterizing these functions. Our functions support all the normal function operations such as addition, subtraction, multiplication, division, composition, et. cetera. The library allows programmers to compose complicated one- or multi-dimensional functions with great economy and natural semantics.

Why Function Objects?

A common computing task is to compute a function, the actual choice of which is delayed until runtime. A built-in mechanism, pointers-to-functions, exists in C, C++, and FORTRAN[1]. A function pointer can refer to a function of one or more variables, and procedures can be written with a function-pointer interface, allowing the same procedure to work with multiple functions.

However this approach is severely limited, because function pointers lack most of the mathematical behavior expected of functions of real arguments. They cannot be added, subtracted, multiplied and divided. There is no law of composition. And they cannot have state, specifically, parameter values that govern the shape of the function. In FORTRAN or C there is little that one can do to remedy this. However in C++ it is possible to create function objects which behave properly under arithmetic operations, where the arithmetic is expressed in overloaded operators. This is useful as a computing tool in that it allows clients to build a complicated function out of simpler functions without writing new subroutines or classes.

Imagine, for example, fitting data Gaussian function plus an exponentially falling background. With function objects one could construct a fitting function by writing

```
f = exp + gauss;
```

passing the result to the fitting procedure. While this may look like floating point arithmetic, the actual type of each object is "AbsFunction", the name of our data type representing function-objects. The fitting procedure itself evaluates the function using the function call operator

```
double y = f(x)
```

as many times as required in order to minimize χ^2 with respect to the *state* of the Gaussian and exponential functions). In this simple example the desired function can be produced simply without introducing new subroutines and without inheriting any base classes. Working with a class library of function objects is like embedding a symbolic manipulation program in C++. The technique is similar to one described in reference (3).

Our Implementation

Our implementation of this idea is a package called GenericFunctions(1). The package consists of:

- A small number of abstract base classes representing functions, their arguments (in case these are multidimensional), and their parameters.

- Classes which represent unary and binary operations on functions, between functions and parameters, and between parameters.

- A small set of concrete function-objects.

Both the second and the third category of classes inherit from the first. The second category provides all the machinery which makes the functions behave as described above. Functions in the third category do not have to each be taught about algebraic rules: they inherit them

---

[1] For people literate in the body of software used in particle physics, the best known example is probably the function pointer FCN used in MINUIT. For others perhaps the best example is the qsort library routine.

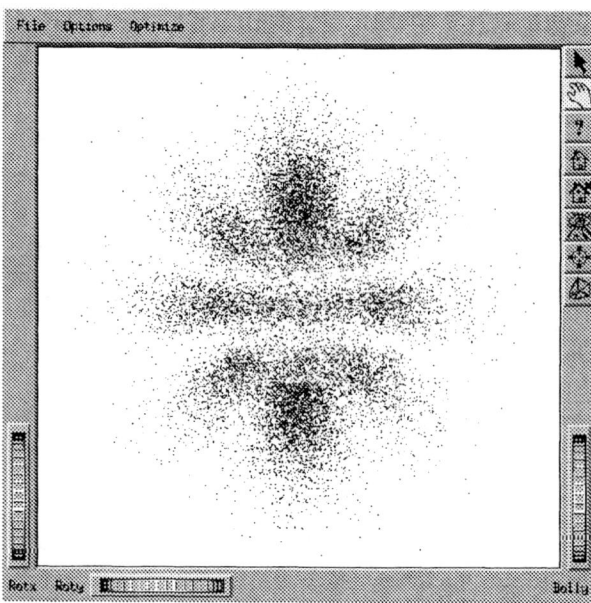

FIGURE 1. Multidimensional functions can be expressed as a direct product of lower dimensional functions. This hydrogen wavefunction was described as a direct product of lower dimensional functions.

automatically from the base classes. This category is designed to incorporate concrete function-object contributions from third parties. The recipe for extending the abstract base classes is documented in the form of a short checklist.

The operations we have defined are +, -, *, /, as well as () (function composition, allowing one to write constructs such as: ln(tang(-x/2)), with ln, tang, and x being function-objects).

Functions of more than one variable are supported, and derive from the same abstract base class as one-dimensional functions. We use the operator % to represent direct product of two functions. In some cases, n-dimensional functions may be expressed as direct products of lower dimensional functions: in two dimensions this happens whenever we may write $h(x,y) = f(x) * g(y)$. The direct product operator allows one to produce such functions easily when the lower-dimensional functions are available. Fig. 1 shows an example, a plot of a hydrogen $|\Psi|^2$ in an 5P state, a direct product of an angular wavefunction involving Legendre Polynomials and a radial wavefunction involving exponentials, power functions, and LaGuerre polynomials.

The mechanism by which functions are given state is the Parameter class. Parameters are like double precision constants, except that 1) they have upper and lower limits, and 2) they can be "connected" to other parameters; in other words, told to take their value from somewhere else. Parameters can be used in arithmetic expressions together with functions and together with built-in floating-point types. And, they can be private member data of function objects. If a function parameters are connected and the function is subsequently used in an expression, the expression is now also controlled by the parameter. More details on how to use the library are given in the package documentation(1).

We have implemented a small set of functions for the first release of the library. This now includes the functions in Table 1. Improvements and extensions are in progress.

How it works

Any expression involving functions and/or parameters is stored as an expression tree which is used later in evaluating the function. The tree for the simple expression f=sine(x)/x is shown in Fig. 2. The tree is built in parts, and each time a new operation is applied sub-trees are copied. Thus it is slow to build an expression tree. In evaluating the function, the expression tree is navigated and the return value for the entire function tree is computed recursively. This step incurs overhead from virtual function calls, but one has to compare this to the time spent in evaluating the functions. In case one is evaluating trigonometric or other higher functions the overhead is negligible. Good numerical algorithms are the key to efficiency.

Availability and Distribution

The GenericFunctions is now part of the CLHEP class library(2), organized and edited by an international organizing committee and editorial board, distributed through CERN. The GenericFunctions package lives in a namespace.

Future directions

We envision extending this library by requiring function objects to specify derivatives which allows automatic differentiation of *all* functions, simple or composite. This also permits function approximation, and allows high-speed local approximation of computationally intensive functions, using a few terms in a Taylor series. We can foresee extending the library by including operators, endowed with their own algebra. It is possible to add symbolic manipulation rules to the package, so that expressions such as $sin^2(x) + cos^2(x)$ are simplified and the ex-

Table 1. List of functions presently implemented.

Class	Implementing
Variable	the variable itself
Const	fixed constants
Power	power functions
Sqrt	square root
AnalyticConvolution	Moser-Roussarie functions
Exponential	Positive Exponentials (zero at negative lifetime)
ReverseExponential	Negative Exponentials (zero at positive lifetime)
Exp	standard exponential
Sin	sine function
Cos	cosine function
SphericalBessel	spherical Bessel functions
SphericalNeumann	spherical Neumann functions
Rectangular	rectangular functions
Periodic Rectangular	periodic rectangular function
LogGamma	natural log of Gamma function
IncompleteGamma	incomplete Gamma function
Gaussian	Gaussian or normal function
Erf	Error function
CumulativeChiSquare	Cumulative χ^2 distribution function
AssociatedLaguerre	Associated Laguerre Polynomials
AssociatedLegendre	Associated Legendre Polynomials

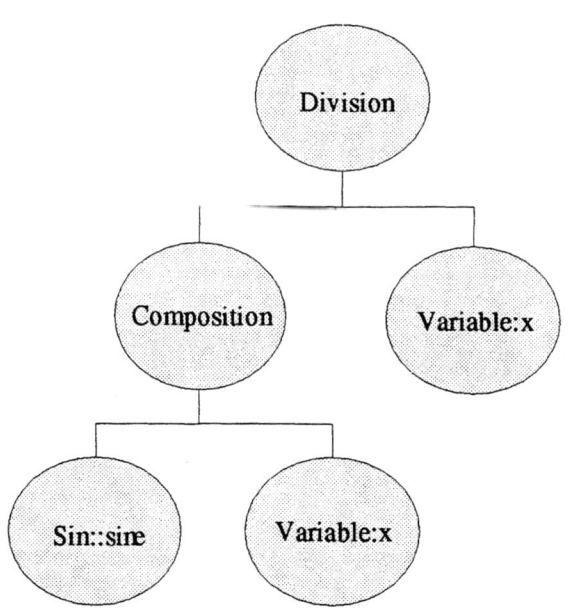

FIGURE 2. The expression tree representing the function sine(x)/x.

pression tree pruned accordingly. These goals will be addressed according to our own needs and public demand.

Acknowledgments

We wish to thank Andreas Pfeiffer and Evgueni Tcherniaev for their collaboration, and Walter Brown and Bob Jacobsen for illuminating discussions.

REFERENCES

1. *www.fnal.gov/tf/zoom/Pkg/CLHEP/docs/ GenericFunctions/0Generic.html*.

2. *wwwinfo.cern.ch/asd/lhc++/clhep/*, or from FNAL at *www.fnal.gov/tf/zoom/Pkg/CLHEP/doc/html/0CLHEP.html*.

3. Barton, J., and Nackman, L., *Scientific and Engineering C++*, Reading, Massachusetts: Addison-Wesley, 1994.

functionalObjects.h: Using Symbolic Syntax in C++ Programs

R. Nolty

California Institute of Technology, Pasadena, California 91125

Abstract. functionalObjects.h allows the C++ programmer performing common calculations to use a more symbolic syntax rather than an algorithmic syntax. This is not as ambitious as a symbolic manipulation program such as Mathematica; it is more like having the ability to drop a very simple Mathematica statement into a C++ program.

INTRODUCTION

A physicist is often faced with the task of writing a program to perform a relatively straightforward mathematical manipulation. For example, she may need to multiply a couple of multivariate functions together and integrate over one variable to obtain a new multivariate function. Using FORTRAN or C or a procedural approach with C++, the resulting code may be several hundred lines long and include calls to CERNlib routines with non-obvious names and calling sequences.

Similarly, a physicist may be reading a piece of procedural code written by someone else, and only after several hours of study be able to verify that the code is indeed performing a simple mathematical function.

Computer Algebra Systems (CAS) such as Mathematica or Maple avoid these problems; they allow the physicist to express the mathematical function to be evaluated rather than the algorithm for evaluating it. However, most of us have most of our analysis paraphernalia and infrastructure in programming languages. Results that are easily obtained in a CAS are often not useful in solving an analysis problem.

This paper documents an initial attempt to use the facilities of C++ to allow a programmer to express mathematical operations more directly in the programming language itself.

WHAT DOES IT LOOK LIKE?

A program to calculate a double definite integral,

$$\int_0^1 dy \int_0^y dx\, x*y$$

could be written like this:

```
#include <stl.h>
#include "functionalObjects.h"

functionalObjectGlobals theGlobals;

main()
{
  char* x = "x";
  char* y = "y";
  theGlobals.registerArgument(x);
  theGlobals.registerArgument(y);

  cout << "Integral: " <<
    Evaluate(Integrate(y, 0.0, 1.0,
      Multiply(y,
        Integrate(x, 0.0, y, x)))) <<
          endl;
}
```

All the action is in the last statement. Reading it from the inside out, Integrate(x,0,y,x) is a C++ function returning a functionalObject which represents the definite integral over dx, with lower limit 0 and upper limit y, of the function x. Multiply(y, Integrate(...)) is a C++ function returning a functionalObject which represents the product of the function y, and the function defined by Integrate(...). Integrate(y,0,1,Multiply(...)) is a C++ function returning a functionalObject which represents the definite integral over dy, with lower limit 0 and upper limit 1, of the function represented by Multiply(...). Finally, Evaluate(...) is a C++ function which asks its argument, a functionalObject, to evaluate its numeric value and returns that value. Thus, if the program is run, it produces the output

```
Integral: 0.124994
```

which differs by roundoff error from the exact result 1/8.

DISTINCTIVES OF THIS EFFORT

The effort to date is very preliminary; it has involved only a few days of thinking, a couple of days of coding, and a few hundred lines of C++.

Although functionalObjects.h allows symbolic functions to be expressed, it is not a symbolic manipulation program. For example, the product of one functionalObject representing the function x, and another $1/x$, would not be simplified to 1. Instead, x would be evaluated, $1/x$ would be evaluated, and their product would be evaluated. The system is designed to produce numeric results, not symbolic results.

The chief aim is to produce programs that are easy to write, easy to read, and easy to maintain.

IMPLEMENTATION

The heart of the package is the abstract base class functionalObject, which simply defines a virtual function evaluate(), returning a double. While the evaluate() function takes no arguments, the numerical value of a function may depend on a functionalArgument. For example, a functionalObject may represent sin(x). If so, the evaluate() function will check the current value of x, and return its sine. functionalArguments are managed by a global structure, functionalObjectGlobals. It has functions to declare an argument, to set the value of an argument, and to inquire the current value of an argument.

Often, more than one function may depend on the same argument. For example, a neutrino cross section and a neutrino flux may depend on the same argument, Enu. If they are evaluated at the same time, they will both query the functionalObjectGlobals for the current value of Enu (and any other arguments they depend on).

Application programmers may define classes inheriting from functionalObject to compute arbitrarily complex functions. The package also provides a few commonly-needed functions; for example, a multiplyObject is a class implementing the product of two functions (each represented by a functionalObject). It simply stores pointers to the two functionalObjects. When the multiplyObject is evaluated, it evaluates its two functions (actually it asks the two functionalObjects to evaluate themselves) and returns the product.

An integrateObject represents a definite integral. When it is asked to evaluate itself, it varies its integration variable (by communication with the functionalObjectGlobals) over its integration range and at each point evaluates its argument, which must be a functionalObject, until it has computed the definite integral.

Programs are made a bit more readable by the existence of certain functions that construct and return functionalObjects. For example, Multiply(functionalObject fcn1,functionalObject fcn2) constructs and returns a multiplyObject. Some overloads also implicitly create very simple functions. Mutliply(3.0,fcn1) creates a doubleObject which always evaluates to the double 3.0, and then returns a multiplyObject that represents the product of the new doubleObject and fcn1. Multiply("x",fcn1) creates an argumentObject that always evaluates to the current value of the functionalArgument x, and then returns a multiplyObject that represents the product of the new argumentObject and fcn1. Similarly, Integrate(char* integrationVariable, functionalObject* lowerLimit, functionalObject* upperLimit, functionalObject* integrand) constructs and returns the appropriate integrateObject representing a definite integral.

SHORTCOMINGS

As stated above, the effort is not very mature at this point. It exhibits several shortcomings, some of which could be overcome with further effort.

For some problems, a procedural approach could take advantage of peculiarities of the problem to make a much more efficient algorithm. The emphasis here is on ease of programming and ease of reading and maintaining programs, not on efficiency.

In the current implementation, functions that implicitly depend on arguments access their arguments by name. It is a bit like the early days of programming before formal parameters were invented. A function, squareX, that computes the square of x, is of no value if you want to compute the square of y. This greatly limits the ability of a programmer to develop generic functions or use functions developed by other programmers. This limitation could be overcome by making the evaluate() function accept arguments, which would be the names of functionalArguments on which the function is to depend.

No thought has been given to memory management. Some functions implicitly declare new variables on the heap, but they are never recovered when the functions go out of scope.

The operation of the package could be made more transparent. For example, an argument class could be declared whose constructor would declare the argument to the functionalObjectGlobals. Rather than using the syntax Multiply(fcn1,fcn2), an overload of the operator* could be used. However, I am somewhat reluctant to make these changes. They would make the programs appear simpler but it would not really be any simpler. I find that physicists are more comfortable when they can see how the package works, rather than having the mechanism hidden by programming gimmicks.

CONCLUSION

This effort, while very preliminary, shows that C++ has powerful facilities making it possible to express some mathematical operations more directly than has been possible in procedural languages.

The library and some example programs are available at

http://www.hep.caltech.edu/ñolty/functionalObjects/functionalObjects.tgz

in a gzipped tar archive. No license has been developed, but if there is interest I will place it under the Gnu Public License (GPL).

Large Scale Symbolic Programming with GiNaC

Alexander Frink, Christian Bauer, Richard Kreckel*

Institut für Physik, Johannes Gutenberg-Universität Mainz, Germany

Abstract. GiNaC is a free framework that embeds symbolic manipulation consistently into the C++ programming language. It deliberately neglects the split-up into a low level language and a high level language, traditional in the design of computer algebra systems. The user usually interacts with GiNaC directly in C++. GiNaC was designed to provide efficient handling of multivariate polynomials, algebras and some special functions that are needed for loop calculations in HEP. But it also bears some potential to become a more general purpose symbolic system.

INTRODUCTION

When we start a software project that relies to some extent on manipulating symbolic expressions (as opposed to quickly checking some result with our favoured Computer Algebra System (CAS)), we are usually faced with a multi-lingual situation. We start by implementing some formulae using a symbolic package and the language it provides. Then we want to get numerical results out of it which is usually done by the CAS' code-generator which produces C or FORTRAN code which we compile and let run. Sometimes, we also wish to share our work and provide an intuitive user interface for our program so others may trigger combined symbolical/numerical/graphical computations with a few keystrokes.

Even if we have mastered all those individual steps, it is not uncommon to see how the interaction of the different software packages we have been using so far makes the whole endeavor fail. It may be that our CAS' language is too restricted to formulate larger programs. It may also be that our programs or the scripts we wrote to glue everything together break at each software upgrade. It may even turn out that we cannot convince our colleagues to help us with coding in three different languages on one single project. In any case, large scale projects tend to become unmaintainable. This is a situation not uncommon in physics projects. GiNaC[1] was designed to overcome such problems by providing fundamental symbolic facilities in the C++ programming language.

---

* Authors' email addresses: Alexander.Frink@Uni-Mainz.DE, Christian.Bauer@Uni-Mainz.DE, Richard.Kreckel@Uni-Mainz.DE. This work was supported by 'Graduiertenkolleg Eichtheorien – Experimentelle Tests und theoretische Grundlagen' at University of Mainz.

[1] GiNaC stands for 'GiNaC is Not a CAS'.

HOW PROGRAMS ARE WRITTEN DOWN WITH GINAC

GiNaC deliberately denies the need for a distinction of implementation language at different steps of a project. It is entirely written in C++ and adheres to the international ISO standard (1). The user can interact with it directly in that language, freely build upon it and extend it. (Compare this with closed systems where expanding the kernel is impossible.) Here is a complete program that uses a Rodrigues repesentation $H_n(x) == (-1)^n e^{x^2} (d/dx)^n e^{-x^2}$ to compute Hermite polynomials:

```
#include <ginac/ginac.h>
using namespace std;
using namespace GiNaC;

ex HermitePoly(const symbol & x, int n)
{
    const ex HGen = exp(-pow(x,2));
    return normal(pow(-1,n)*HGen.diff(x, n)/HGen);
}

int main(int argc, char **argv)
{
    int degree = atoi(argv[1]);
    numeric value = numeric(argv[2]);
    symbol z("z");
    ex H = HermitePoly(z,degree);
    cout << "H_" << degree << "(z) == "
         << H << endl;
    cout << "H_" << degree << "(" << value << ") == "
         << H.subs(z==value) << endl;
    return 0;
}
```

When this program is compiled and called with 6 and 0.8-0.5*I as command line arguments it will readily print out the sixth Hermite polynomial together with that polynomial evaluated numerically at $z = 0.8 - 0.5i$:

```
$ c++ hermite.cc -o hermite -lginac
$ ./hermite 6 0.8-0.5*I
H_6(z) == -120-480*z^4+720*z^2+64*z^6
H_6(0.8-0.5*I) == 350.865216-267.07455999999999996*I
```

Alternatively, arbitrary length exact rational arguments are also honored, thus avoiding rounding errors:

```
$ ./hermite 6 4/5-1/2*I
```

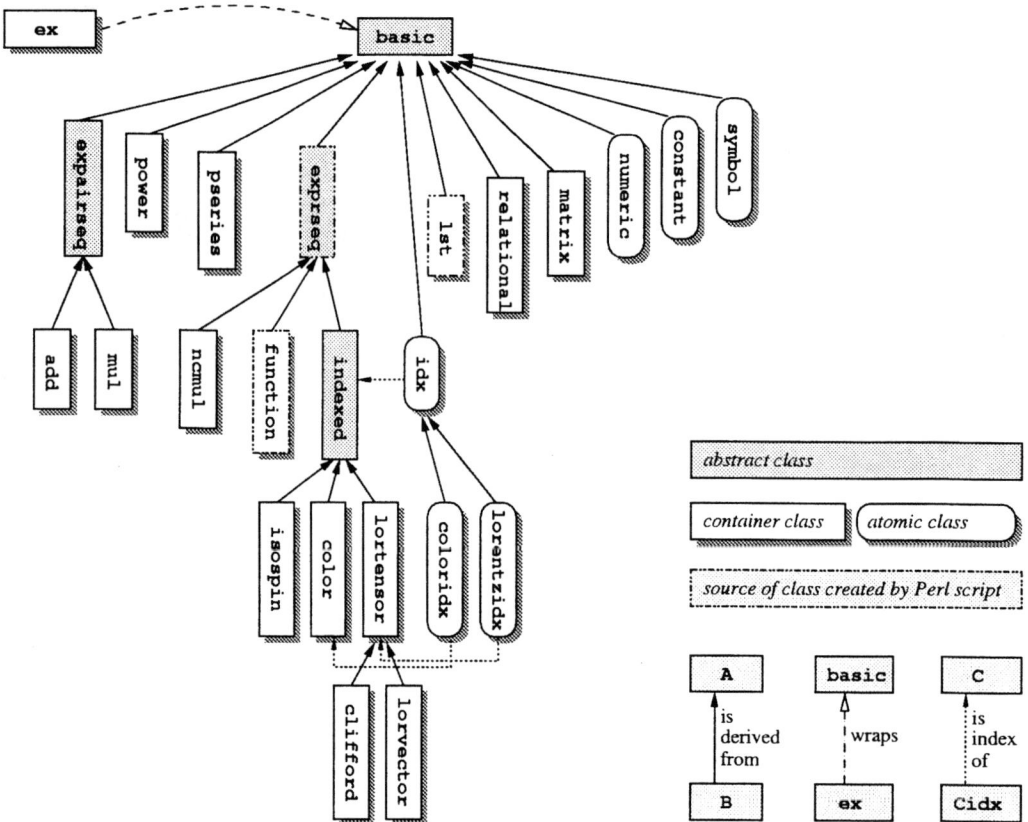

FIGURE 1. GiNaC's class hierarchy and some of the relations between the classes.

```
H_6(z) == -120-480*z^4+720*z^2+64*z^6
H_6(4/5-1/2*I) == 5482269/15625-834608/3125*I
```

Syntactically, the program shows how symbolic expressions are written down in GiNaC pretty much like common numeric terms thanks to operator overloading between some of the classes in Fig. 1.

GiNaC can be configured to work with Masaharu Goto's Cint C/C++ interpreter[2]. It is then possible to work with GiNaC interactively, defining symbolic expressions, control structures like loops and conditionals and even functions. A little example shows how one may symbolically compute the non-relativistic approximation of $\gamma = (1 - (v/c)^2)^{-1/2}$ to ten orders in v and then inverse it and get the original result back up to an $O(v^{10})$ term:

```
$ ginaccint
Welcome to GiNaC-cint (GiNaC V0.7.0, Cint V5014062)
   __,  _____   GiNaC: (C) 1999-2000 Johannes Gutenber
  (__) *         | Germany. Cint C/C++ interpreter: (C)
  ._) i N a C |  Goto and Agilent Technologies, Japan.
<------------'   with ABSOLUTELY NO WARRANTY.  For deta
Type '.help' for help.

GiNaC> symbol v("v"), c("c");}
```

---

[2] This amazing tool is also the basis of the well-known ROOT object-oriented data analysis framework (3)

```
GiNaC> ex gamma = 1/sqrt(1 - pow(v/c,2));
GiNaC> ex gamma_nr = gamma.series(v==0,10);
GiNaC> cout <<pow(gamma_nr,-2) <<endl;
1+(1/2*c^(-2))*v^2+(3/8*c^(-4))*v^4+(5/16*c^(-6))*v^6+
(35/128*c^(-8))*v^8+Order(v^10)^(-2)
GiNaC> cout <<pow(gamma_nr,-2).series(v==0,10) <<endl;
1+(-c^(-2))*v^2+Order(v^10)
```

Loops are of course written down in `ginaccint` just as they would be written for a compiled program:

```
GiNaC> for (int i=0; i<10000; i+=2) {
>         cout << bernoulli(numeric(i)) << ", ";
>       }
1, 1/6, -1/30, 1/42, -1/30, 5/66, -691/2730, 7/6, -361
7/510, 43867/798, -174611/330, 854513/138, -236364091/
2730, 8553103/6, -23749461029/870, 8615841276005/14322
, -7709321041217/510, 2577687858367/6, -26315271553053
477373/1919190, 2929993913841559/6, -26108271849644912
2051/13530, 1520097643918070802691/1806, -278332695793
01024235023/690, 596451111593912163277961/282, -560940
3368997817686249127547/46410, 495057205241079648212477
```

... and so on. Due to a limitation of Cint, however, function definitions need some special help:

```
GiNaC> //GiNaC-cint.function
next expression can be a function definition
GiNaC> const ex EulerNumber(const unsigned n)
>    {
>        const symbol xi;
>        const ex generator = pow(cosh(xi),-1);
>        return generator.diff(xi,n).subs(xi==0);
>    }
creating file /tmp/ginac26197caa
```

```
GiNaC> cout << EulerNumber(42) << endl;
Out3 = -10364622733519612119397957304745185976310201
GiNaC> quit;
```

It is thus possible to write large scripts and later compile them and link them in to the user's application.

HOW GINAC WORKS

GiNaC implements a number of symbolic classes as shown in Fig. 1. All classes are referenced by the class of all expressions `ex`. There is reference counting at work here, providing GiNaC with a transparent non-interruptive memory management. It is implemented in the interplay between class `ex` and the abstract base class `basic`, so the user who wishes to extend the system does not have to worry about memory management. For all kinds of numbers (integer, rational, float, complex) GiNaC uses Bruno Haible's super-efficient C++-library CLN (2) as a foundation class. Our class `numeric` is basically a wrapper class around CLN's class `cl_N`.

COMPARISON AND BENCHMARKS

Is GiNaC competitive? Certainly it does not feature such a load of features as some big commercial systems. But even with the tasks it can do, is it efficient? We try to answer this question with some tests. The first two try to measure the scaling behaviour when problems become very big.

The first one is a rather common three step substitute-expand consistency benchmark:

- let e be the expanded sum of n symbols squared: $e \leftarrow (\sum_{i=0}^{n-1} a_i)^2$,
- in e substitute $a_0 \leftarrow -\sum_{i=2}^{n-1} a_i$,
- expand e again, it collapses to a_1^2.

Using `ginaccint` this test may be formulated interactively in an elegant way as follows:

```
$ ginaccint
Welcome to GiNaC-cint (GiNaC V0.7.0, Cint V5014062)
    __   _____    GiNaC: (C) 1999-2000 Johannes Gutenber
  (__) *          | Germany. Cint C/C++ interpreter: (C)
  .__) i N a C   | Goto and Agilent Technologies, Japan.
  <-------------' with ABSOLUTELY NO WARRANTY. For deta
Type '.help' for help.

GiNaC> #include <sstream>
GiNaC> vector<symbol> a;
GiNaC> ex bigsum = 0;
GiNaC> for (int i=0; i<50; ++i) {
>          ostringstream buf;
>          buf << "a" << i << ends;
>          a.push_back(symbol(buf.str()));
>          bigsum += a[i];
>      }
GiNaC> ex sbtrct = -bigsum + a[0] + a[1];
GiNaC> cout << pow(bigsum,2).expand()
>                    .subs(a[0]==sbtrct)
>                    .expand() << endl;
a1^2
GiNaC> quit;
```

The results are shown in Fig. 2. The system used was an Alphaserver 8400 running at 300MHz with roughly 1GB of main memory. We have chosen this system in order to give Maple a chance since MapleV has a built in limitation to $2^{16} - 1$ terms in the representation of sums on all 32bit platforms which cause this test to break down quite early at $n = 181$. The timings show the expected n^2 scaling behavior for the GiNaC, Mathematica and MuPAD with GiNaC being the fastest of the three while the curves for Reduce and Maple are odd, since they take off very fast for small problems and become very slow at the upper end. The fundamental difference here lies in the memory management: GiNaC, Mathematica and MuPAD use reference counts, while Maple and Reduce (being Lisp-based) use a garbage collector.

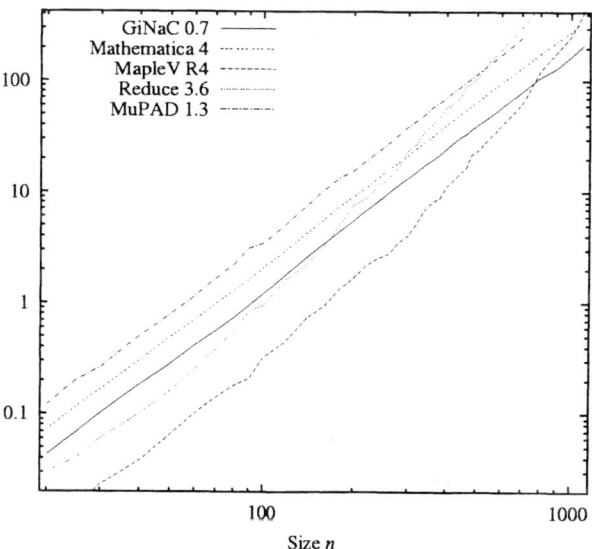

FIGURE 2. Absolute runtimes for the substitute-expand test in seconds.

The second test was done on an Intel P-III with 384MB RAM running at 450MHz. The systems were asked to expand the Gamma function around its first pole at $x = 0$ to high orders symbolically. The series expansion is: $\Gamma(x) = \frac{1}{x} - \gamma + \left(\frac{\pi^2}{12} + \frac{\gamma^2}{2}\right)x - \left(\frac{\pi^2\gamma}{12} + \frac{\gamma^3}{6} + \frac{\zeta(3)}{3}\right)x^2 + \ldots$ It can then easily be checked for consistency numerically: $\Gamma(x) = \frac{1}{x} - 0.5772157 + 0.9890560x - 0.9074791x^2 + \ldots$ Fig. 3 shows the expected breakdown of MapleV at order $n = 35$ when intermediate results exceed $2^{16} - 1$ terms. There are two curves shown for Mathematica here. This is because by default Mathematica expands the result only partially to a form comparable with the other systems. Only after applying `FunctionExpand` it gets the same result, but extremely slowly so. But since it turns

Table 1. Runtimes in seconds for the tests proposed by R. Lewis and M. Wester (only as far as applicable to GiNaC) on an Intel P-III 450MHz, 384MB RAM running under Linux. Abbreviations used: GU (gave up, like Maple's error `object too large`), CR (crashed, out of memory), NA (not available), UN (unable, a prerequisite test failed).

	Benchmark	GiNaC 0.7	MapleV R4	MuPAD 1.4.1	Pari-GP 2.0.19β	Singular 1-3-7
A:	divide factorials $\left.\frac{(1000+i)!}{(900+i)!}\right|_{i=1}^{100}$	0.20	4.11	1.13	0.37	19.0
B:	$\sum_{i=1}^{1000} 1/i$	0.019	0.08	0.10	0.041	0.54
C:	gcd(big integers)	0.25	9.92	3.01	1.65	0.11
D:	$\sum_{i=1}^{10} iyt^i/(y+it)^i$	0.78	0.13	1.21	0.20	NA
E:	$\sum_{i=1}^{10} iyt^i/(y+|5-i|t)^i$	0.63	0.05	2.33	0.11	NA
F:	gcd(2-var polys)	0.080	0.10	0.21	0.057	0.13
G:	gcd(3-var polys)	2.50	1.85	3.31	99.5	0.38
H:	det(rank 80 Hilbert)	10.0	42.9	42.5	3.97	CR
I:	invert rank 40 Hilbert	3.38	7.48	12.0	0.62	CR
J:	check rank 40 Hilbert	1.61	2.61	2.95	0.22	UN
K:	invert rank 70 Hilbert	22.1	113.8	74.0	5.90	CR
L:	check rank 70 Hilbert	9.19	24.1	14.2	1.57	UN
M_1:	rank 26 symbolic sparse, det	0.36	0.40	0.75	0.016	0.003
M_2:	rank 101 symbolic sparse, det	1903.3	GU	CR	CR	251.2
N:	eval poly at rational functions	CR	GU	CR	CR	NA
O_1:	three rank 15 dets (average)	43.2	GU	CR	CR	CR
O_2:	two GCDs	CR	UN	UN	UN	UN
P:	det(rank 101 numeric)	1.10	12.6	44.3	0.09	0.85
P':	det(less sparse rank 101)	6.07	13.6	46.2	0.38	1.25
Q:	charpoly(P)	103.9	1453.2	741.7	0.15	4.4
Q':	charpoly(P')	212.8	1435.6	243.1	CR	5.0

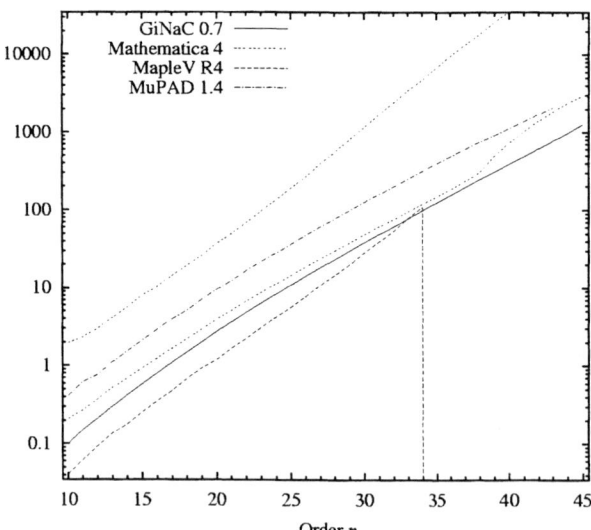

FIGURE 3. Absolute runtimes to expand the Γ-function around $x = 0$ up to order n in seconds.

out that the intermediate result is already useful to some extent (as can be seen by evaluating it numerically) we also give it credit by showing the timings for computing the series without applying `FunctionExpand`.

As yet another test, we apply GiNaC to a number of tests invented by Robert Lewis and Michael Wester (described in (4)) as far as they are applicable to GiNaC. The results are shown in Table 1.

STATUS AND AVAILABILITY

Being a special-purpose system, GiNaC aims at being a fast and reliable foundation for combined symbolical/numerical/graphical projects in C++. It is currently used as a symbolic engine for loop calculations in the HEP project (5). It may be downloaded and distributed under the terms of the GNU general public license from http://www.ginac.de/. The supporting CLN library can also be obtained from there and is also available packaged for some Linux distributions. A tutorial introduction and complete cross references of the source code can also be found there.

REFERENCES

1. American National Standards Institute, *ISO/IEC 14882-1998(E) Programming languages — C++* (1998)
2. Bruno Haible, Richard Kreckel, *CLN Class Library for Numbers*, `ftp://ftp.ilog.fr/pub/Users/haible/gnu/cln-1.1.tar.gz`, (2000)
3. Rene Brun, Fons Rademakers, *ROOT - An Object Oriented Data Analysis Framework Proceedings of AIHENP 97*, Laussanne, (1996)
4. Robert H. Lewis, Michael Wester, *Comparison of Polynomial-Oriented Computer Algebra Systems*. (Presented at the 1999 ISSAC Conference, Vancouver), available from `http://www.fordham.edu/lewis/cacomp.html`, (1999)
5. Lars Brücher, Johannes Franzkowski, Dirk Kreimer, *Comput. Phys. Commun.* **115**, (1998) 140–160.

Optimization of symbolic evaluation of helicity amplitudes

P.S. Cherzor, V.A. Ilyin and A.E. Pukhov

Skobeltsyn Institute of Nuclear Physics, Moscow State University, Moscow 119899, Russia

Abstract. We present a method for symbolic evaluation of Feynman amplitudes. We construct special polarization basis for spinor particles which produces compact expressions for tensor products of basis spinors.

Quantum amplitudes corresponded to Feynman diagrams depend on momenta of particles and their polarizations. Polarizations of bosons are described by some Lorentz vectors. Thus, in case, when only boson particles are involved, the amplitudes may be presented via dot-products and Levi-Civita tensors. However fermion polarizations are described by vectors in the spinor space, so the amplitude is expressed as a product of Dirac γ-matrices ended by two spinors (hereafter *fermion string*). Such an expression looks like *not complete evaluated* — on the step of numerical evaluation one has to perform summations over spinor and Lorentz indices. During the last few years effective algorithms for these summations were developed, e.g. [1, 2, 3, 4, 5, 6], and corresponding programs were created for automatic computation of Feynman amplitudes, e.g. [7, 8, 9, 10].

Further progress one can get, we believe, if indices summation is performed before numerical calculations. We refer on our experience with the CompHEP package [11], where the squared diagram technique is used. For processes with four particles in final state the number of squared diagrams becomes very large comparing with the amplitude, and corresponding symbolic answers are cumbersome. Nevertheless, CompHEP works effectively because it prepares analytical expression for each squared diagram in terms of dot-products, and, then, factorizes them. Also, analytical answers for diagrams can be used to resolve explicitly the cancellations between diagrams, important for gauge theories.

The idea of the fermion string symbolic evaluation is known long time ago [12]. Although a spinor can not be expressed in terms of Lorentz vectors, such a possibility exists for pairs of spinors:

$$w \otimes \bar{w}' \equiv w^i(p_1, \lambda_1) \bar{w}'_j(p_2, \lambda_2), \quad (1)$$

where p_1 and p_2 are particle momenta, λ_1 and λ_2 are their polarizations. This matrix can be expressed through the Dirac γ-matrices convoluted with vectors p_i, λ_i and some auxiliary vectors. Thus, the contribution of a fermion string, Γ, to the Feynman amplitude can be rewritten through the trace:

$$\bar{w}(p_2, \lambda_2) \Gamma w'(p_1, \lambda_1) = tr(\Gamma \, w \otimes \bar{w}'), \quad (2)$$

where this string, Γ, is some product of γ-matrices.

There are different approaches to the construction of tensor products like (1), see e.g. [13, 14, 15]. Here we propose a new method with the following advantages: 1) it produces most compact symbolic expressions for (1); 2) it gives formulae uniformly applicable in massive and massless cases, for Dirac and Majorana fermions; 3) it may be applied to interactions with the fermion number violation; 4) all phases of fermion states are fixed completely, so one can choose fermion pairs by different ways.

Spinors, corresponding to fermion external legs of the diagram, can be one of the following variants, $w \in \{u, v, u^c, v^c\}$ and $\bar{w}' \in \{\bar{u}, \bar{v}, \overline{u^c}, \overline{v^c}\}$. Here $(^c)$ denotes the operation of C-conjugation, the corresponding fermion strings appear, for example, in the case of interaction with fermion number violation [16]. Spinors u and v are defined as

$$< 0|\psi(x)|\Omega_+(p,\lambda) > \;=\; e^{-ipx} u(p,\lambda), \quad (3)$$
$$< 0|\bar{\psi}(x)|\Omega_-(p,\lambda) > \;=\; e^{-ipx} \bar{v}(p,\lambda), \quad (4)$$

where $|0>$ is vacuum state, Ω_+ (Ω_-) is one-particle fermion (anti-fermion) state. The fermion states depend on the choice of the polarization axes and their complex phases are not fixed.

The characteristic property of spinors u and v is the *completeness*:

$$\Sigma_\lambda u(p,\lambda) \otimes \bar{u}(p,\lambda) \;=\; \hat{p} + m \quad (5)$$
$$\Sigma_\lambda v(p,\lambda) \otimes \bar{v}(p,\lambda) \;=\; \hat{p} - m \quad (6)$$

where $m = \sqrt{p^2}$. One can say that any spinors, $u(p,\lambda)$ and $v(p,\lambda)$, satisfying these relations correspond to some polarization states Ω_+ and Ω_-.

By definition, for Majorana fermion one has $\psi(x) = \psi(x)^c$ and $\Omega_+(p,\lambda) = \Omega_-(p,\lambda)$. If one substitutes these identities into (3,4) one gets

$$u(p,\lambda) = C\bar{v}(p,\lambda)^T = v(p,\lambda)^c. \quad (7)$$

Thus, we conclude that spinor basis should satisfy relation (7) in order to construct an algorithm for evaluation of fermion strings with Majorana fermions. Hereafter we assume that spinors u and v satisfy this relation.

Following to [1, 2] we construct u and v spinors in arbitrary point of the momentum space as a projection of spinors defined at some auxiliary point p_0:

$$u(p,\lambda) = N(m+\hat{p})u(p_0,\lambda), \quad (8)$$
$$v(p,\lambda) = N(m-\hat{p})v(p_0,\lambda), \quad (9)$$

where $N = 1/\sqrt{2[(p_0 p) + m_0 m]}$. It is easy to check that this procedure agrees with the completeness (5, 6) at any point p if it is satisfied at the point p_0. Also, it is enough to satisfy (7) at p_0 to get this relation at any point.

Let p_0 be a massless vector, in this case expressions appearing are simplified. In particular, equations (5) and (6) become identical, and, thus, (massless) u and v spinors are basis vectors in the same spinor subspace. It is well known, that left-handed and right-handed spinors, $\xi_- = -\gamma_5 \xi_-$ and $\xi_+ = \gamma_5 \xi_+$, may be chosen as elements of this basis, and completeness reads as $\xi_- \otimes \bar{\xi}_- + \xi_+ \otimes \bar{\xi}_+ = \hat{p}_0$. Note that chiral spinors, $\xi_\pm$, are defined by these relations up to phases. Let us fix these phases.

At first, let us connect chiral spinors with massless v spinor of different polarizations taken at the point p_0:

$$\xi_+ = v(p_0,+1), \quad \xi_- = v(p_0,-1).$$

Then, from (7) one gets

$$u(p_0,+1) = \xi_+^c, \quad u(p_0,-1) = \xi_-^c.$$

Note, then, that C-conjugation transforms the subspace of u-spinors to the subspace of v-spinors. For massless p_0 these subspaces are identical and, thus, invariant under the C-conjugation. From other side C-conjugation transforms left-handed spinor to right-handed spinor. So, sum of the phases of $\xi_\pm$ can be fixed by the relation

$$\xi_- = \xi_+^c. \quad (10)$$

As a result of our construction the following expressions for u and v spinors can be derived:

$$u(p,\pm 1) = v(p,\pm 1)^c = N(m+\hat{p})\xi_\mp, \quad (11)$$
$$v(p,\pm 1) = u(p,\pm 1)^c = N(m-\hat{p})\xi_\pm, \quad (12)$$

where $N = 1/\sqrt{2(p_0 p)}$.

By means of Dirac conjugation one immediately gets

$$\bar{u}(p,\pm 1) = \overline{v(p,\pm 1)^c} = N\bar{\xi}_\mp(m+\hat{p}), \quad (13)$$
$$\bar{v}(p,\pm 1) = \overline{u(p,\pm 1)^c} = N\bar{\xi}_\pm(m-\hat{p}). \quad (14)$$

Expressions (11 - 14) reduce the problem of evaluation of matrices $w \otimes \bar{w}'$ to the evaluation of $\xi_+ \otimes \bar{\xi}_+$, $\xi_- \otimes \bar{\xi}_-$, $\xi_+ \otimes \bar{\xi}_-$ and $\xi_- \otimes \bar{\xi}_+$. The first two products can be derived directly from the definition of chiral spinors:

$$\xi_+ \otimes \bar{\xi}_+ = \frac{1+\gamma_5}{2}\hat{p}_0, \quad (15)$$
$$\xi_- \otimes \bar{\xi}_- = \frac{1-\gamma_5}{2}\hat{p}_0. \quad (16)$$

However, expressions for last two products, $\xi_+ \otimes \bar{\xi}_-$ and $\xi_- \otimes \bar{\xi}_+$, can not be derived yet, because they are sensitive to the multiplication of $\xi_\pm$ by opposite phase factors. Note that rotation of 3-dimension coordinate system around the space component of p_0, will produce the multiplication of $\xi_\pm$ by opposite phase factors. Thus, one needs to introduce additional auxiliary vectors to fix the reference frame, and (as a result) to fix phases, and (finally) to get a representation for $\xi_+ \otimes \bar{\xi}_-$ and $\xi_- \otimes \bar{\xi}_+$. Let η_1 and η_2 are space-like vectors orthogonal to p_0:

$$\eta_1.\eta_2 = 0, \quad \eta_i.p_0 = 0, \quad \eta_i.\eta_i = -1. \quad (17)$$

Three vectors (p_0, η_1, η_2) may be completed to the basis in Minkowsky space [1]. Now we need that this basis is right-handed oriented. It can be done by fixing the sign of the convolution of Levi-Civita tensor with basis vectors. We fix the basis orientation by the relation

$$\hat{p}_0 = i\gamma_5 \hat{\eta}_1 \hat{\eta}_2 \hat{p}_0. \quad (18)$$

If one constructs complex vectors $\eta \equiv (\eta_1 + i\eta_2)/2$ and $\eta^* \equiv (\eta_1 - i\eta_2)/2$, then (18) can be rewritten as

$$\hat{\eta}^* \hat{\eta} \hat{p}_0 = -(1/2)(1-\gamma_5)\hat{p}_0. \quad (19)$$

Using this equation one can check that

$$\xi_- = \hat{\eta}^* \xi_+. \quad (20)$$

Note that (16) defines ξ_- uniquely up to a phase. Thus (20) fixes the difference of phases between ξ_- and ξ_+ (whereas (10) fixes their sum). Now one can derive from (20) the second pair of products:

$$\xi_- \otimes \bar{\xi}_+ = \frac{1-\gamma_5}{2}\hat{\eta}^* \hat{p}_0, \quad \xi_+ \otimes \bar{\xi}_- = -\frac{1+\gamma_5}{2}\hat{\eta}\hat{p}_0.$$

Note that these relations are some variant of the Bouchiat-Michel equations [17, 18].

---

[1] say, by vector q_0, such that $q_0.p_0 = 1$ and $q_0.\eta_i = 0$.

Now we can write down the complete set of expressions for tensor products $w \otimes \bar{w}'$:

$$u(p,\lambda) \otimes \bar{u}(p',\lambda') = \quad (21)$$

$$= N(m+\hat{p}) \begin{pmatrix} 1-\gamma_5 & (1-\gamma_5)\hat{\eta}^* \\ -(1+\gamma_5)\hat{\eta} & 1+\gamma_5 \end{pmatrix} \hat{p}_0(m'+\hat{p}');$$

$$u(p,\lambda) \otimes \bar{v}(p',\lambda') = \quad (22)$$

$$= N(m+\hat{p}) \begin{pmatrix} (1-\gamma_5)\hat{\eta}^* & 1-\gamma_5 \\ 1+\gamma_5 & -(1+\gamma_5)\hat{\eta} \end{pmatrix} \hat{p}_0(m'-\hat{p}');$$

$$v(p,\lambda) \otimes \bar{u}(p',\lambda') = \quad (23)$$

$$= N(m-\hat{p}) \begin{pmatrix} -(1+\gamma_5)\hat{\eta} & 1+\gamma_5 \\ 1-\gamma_5 & (1-\gamma_5)\hat{\eta}^* \end{pmatrix} \hat{p}_0(m'+\hat{p}');$$

$$v(p,\lambda) \otimes \bar{v}(p',\lambda') = \quad (24)$$

$$= N(m-\hat{p}) \begin{pmatrix} 1+\gamma_5 & -(1+\gamma_5)\hat{\eta} \\ (1-\gamma_5)\hat{\eta}^* & 1-\gamma_5 \end{pmatrix} \hat{p}_0(m'-\hat{p}').$$

Here $N = 1/(4\sqrt{(pp_0)(p'p_0)})$, $m = \sqrt{p^2}$, $m' = \sqrt{p'^2}$, and dependence on the polarizations follows the rule: $\begin{pmatrix} ++ & +- \\ -+ & -- \end{pmatrix}$. The products $w \otimes \bar{w}'$ with C-conjugation may be derived now by means of relation (7).

Note, that for $m \neq 0$ the spinors u and v have spin directed along the axis $n = p/m - p_0(m/(p.p_0))$. In massless case spinors $u(p,+1)$ and $v(p,-1)$ are right-handed, whereas $u(p,-1)$ and $v(p,+1)$ are left-handed.

At this step one can conclude that in the proposed method the same set of auxiliary vectors (p_0, η, η^*) is introduced for arbitrary number of fermion strings in the amplitudes. Moreover, only one set can be used for all diagrams to fix relative phases of spinors for interference diagrams. One can compare with other methods where number of sets of auxiliary vectors depend on the number of fermion strings (see, e.g., corresponding discussion in [13]).

One can find further optimizations in the proposed technique. Indeed, vectors η and η^* in (21-24) are accompanied always by the p_0 vector. Let us define antisymmetric tensor $G_{\mu\nu} \equiv -i\varepsilon_{\mu\nu\alpha\beta}\eta^\alpha p_0^\beta$. Then one can check that $\hat{\eta}\hat{p}_0 = \frac{1}{2}G_{\mu\nu}\gamma^\mu\gamma^\nu$ and $\hat{\eta}^*\hat{p}_0 = \frac{1}{2}G^*_{\mu\nu}\gamma^\mu\gamma^\nu$. By substituting these relations in (2) one can hide the dependence on η, η^* inside the G tensor.

The G tensor has some properties allowing one to perform the indices convolutions:

$$G_{\mu\nu}p_0^\mu = 0, \quad G_{\mu\nu}G_{\mu'}{}^\nu = 0, \quad G_\mu{}^\nu G^*_{\nu\mu'} = 2p_{0\mu}p_{0\mu'},$$

$$g^{\mu\mu'}G_{\mu\nu}\varepsilon_{\mu'\alpha\beta\gamma} = 4i(g_{\nu\alpha}G_{\beta\gamma} - g_{\nu\beta}G_{\alpha\gamma} + g_{\nu\gamma}G_{\alpha\beta}).$$

As a result, one can get symbolic answer for the amplitude expressed via scalar products of particle momenta and auxiliary vector p_0, their convolutions with the Levi-Civita tensor, and particle momenta convolved with the G tensor. The use of G tensor optimizes the final expressions — if we expand G tensor via scalar products two terms appear.

Then, one can expect that the amplitude contains both G and G^* tensors. However, the product of these tensors, $G(q_1,p_1)G^*(q_2,p_2)$, can be expressed via the dot-products and Levi-Civita tensors. So, one can connect fermion legs of the diagram in such a way that amplitudes will be presented via either G or G^* tensors, but never both together.

This work was done in the general framework of the CPP Collaboration (http://wwwlapp.in2p3.fr/cpp/cpp.html), and supported by grants INTAS-CERN 99-0377 and INTAS 96-842. A.E.P. was supported also by the Programme "University of Russia" (grant 990588).

REFERENCES

1. R. Kleiss. Nucl. Phys. **B241** (1984) 61.
2. F.A. Berends, P.H. Daverveltd, R. Kleiss. Nucl. Phys. **B253** (1985) 441.
3. R. Kleiss, W.J. Stirling. Nucl. Phys. **B262** (1985) 235.
4. H. Tanaka. Comp. Phys. Comm. **58** (1990) 153.
5. A. Ballestrero, E. Maina. Phys. Lett. **B350** (1995) 225.
6. H. Tanaka et al. Nucl. Instrum. and Meth. **A389** (1997) 295.
7. T. Ishikawa et al. *GRACE manual*, KEK Report 92-19, February 1993.
8. T. Stelzer, W.F. Long. Comp. Phys. Comm. **81** (1994) 357.
9. T. Ohl. hep-ph/0011243.
10. A. Kanaki, C.G. Papadopoulos. hep-ph/0012004; Comput. Phys. Comm. **132** (2000) 306.
11. A. Pukhov et al. Preprint INP MSU 98-41/542, hep-ph/9908288
12. E. Bellomo. Nuovo Cimento, Ser.X, **21**(1961) 730.
    H. W. Fearing, R. R. Sillbar. Phys. Rev. **D6** (1972) 471.
13. A. Bondarev, hep-ph/9710398
14. E. Yehudai. hep-ph/9209293.
15. R. Vega, J. Wudka. Phys. Rev. **D53** (1996) 5286; *erratum* Phys. Rev. **D56** (1997) 6037.
16. A. Denner et al. Phys. Lett. **B291** (1992) 278; Nucl. Phys. **B387** (1992) 467.
17. C. Bouchiat, L. Michel. Nucl. Phys. **5** (1958) 416.
    L. Michel. Suppl. Nuovo Cim. **14** (1959) 95.
18. H.E. Haber. In: Proc. XXI SLAC Summer Institute on Particle Physics, ed. L. DePorcel and C. Dunwoodie, 1994, p.231.

A Feynman Diagram Analyser DIANA – Graphic Facilities

J. Fleischer and M. Tentyukov[*][†]

Fakultät für Physik, Universität Bielefeld D-33615 Bielefeld

Abstract. New developments concerning the extension of the recently introduced (1) Feynman diagram analyser DIANA are presented.

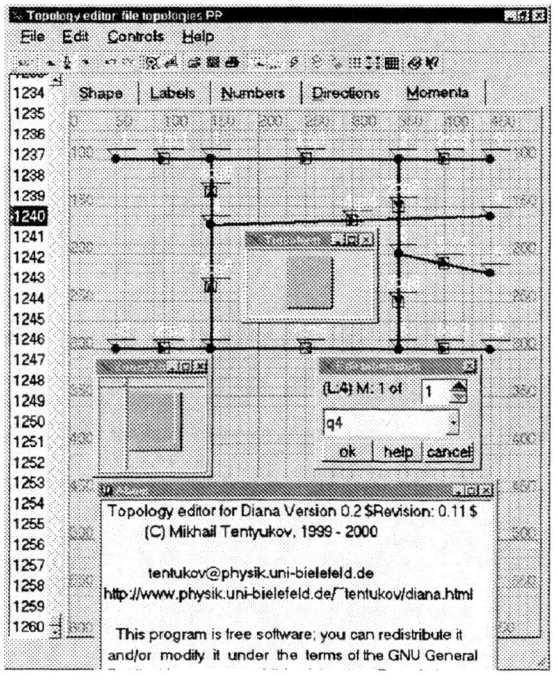

FIGURE 1. Topology editor.

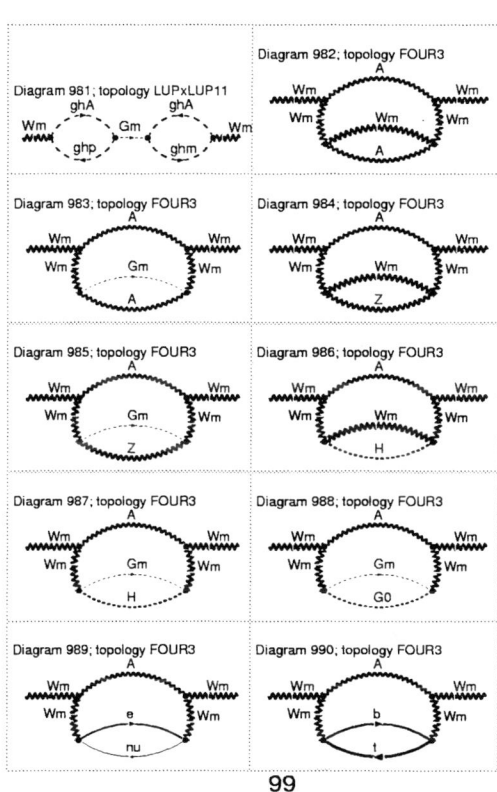

FIGURE 2. Sample of a page with diagrams arranged in columns and rows.

Recent high precision experiments require, on the side of the theory, high-precision calculations resulting in the evaluation of higher loop diagrams in the Standard Model. For specific processes thousands of multiloop Feynman diagrams do contribute. Of course, the contribution of most of these diagrams is very small. But sometimes it is not so easy to distinguish between important and unimportant diagrams. On the other hand, we often need to take into account all diagrams, to verify gauge independence, or cancellation of divergences. It turns out impossible to perform these calculations by hand. This makes the request for automation a high-priority task.

Our aim is to create a universal software tool for piloting the process of generating the source code in multiloop order for analytical or numerical evaluations and to

---

[*] Supported by DFG under FL241/4-1 and in part by RFBR #98-02-16923.
[†] On leave from BLTP JINR, Dubna, Russia

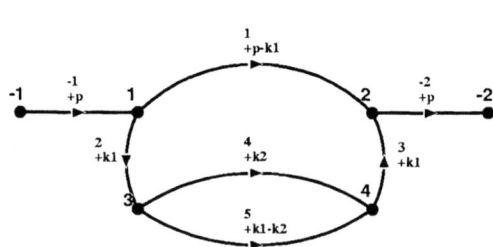

Diagram 985 topology FOUR3 (unique FOUR3) momentaset 1 (of 1)

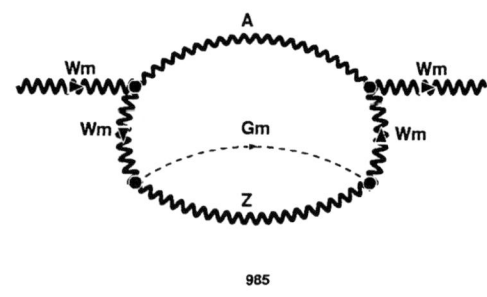

985

FIGURE 3. Details of one diagram.

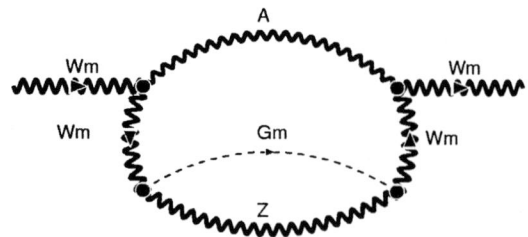

FIGURE 4. Encapsulated postscript file.

keep the control of the process in general. Based on this instrument, we can attempt to build a complete package performing the computation of any given process in the framework of any concrete model.

The project called DIANA (DIagram ANAlyser) (1) for the evaluation of Feynman diagrams was started by our group some time ago. At present, the core part is finished. The recent development of this project will be shortly described below.

DIANA has been developed for the analytic evaluation of Feynman diagrams in terms of computer algebra packages, for which we use FORM (2), but which can in principle be substituted by another language. The user has to prepare a file, which contains the model and process specifications, see details in (1). Reading this file, DIANA will generate all necessary other files and then invoke the topology editor. The purpose of the topology editor is to make the shapes for the topologies and to introduce proper integration momenta for the various topologies, Fig. 1. It is a graphical program written in C++ using the Qt widget library. For the description of the topology editor see the WEB page

http://www.physik.uni-bielefeld.de/~tentukov/topeditor.html

After all necessary files are ready, DIANA can be used to generate the FORM input and to execute the generated FORM program as well.

If the shapes of topologies are defined, DIANA is able to produce the pictorial representation of diagrams, see the WEB page

http://www.physik.uni-bielefeld.de/~tentukov/printing.html .

Three different kinds of postscript files for the diagrams can be produced.

The style "specmode.tml" (see (1) p.133) contains all the necessary function calls. Thus users of this style only need to initialize the proper postscript driver in the environment `initialization`.

The first driver permits the user to print all diagrams in one file, arranging diagrams along several rows and columns on each page, Fig. 2, according to the users request. The user must initialize the PostScript driver by means of the function

```
\initPostscript(  filename,
                  papersize,
                  orientation,
                  xmargin,
                  ymargin,
                  xleftmargin,
                  ncols,
                  nrows,
                  font,
                  fontsize   )
```

The parameters are:
1. `filename` – the output file name;
2. `papersize` – one of the possible paper sizes;
3. `orientation` – portrait or landscape;
4. `xmargin` – both left and right margins;
5. `ymargin` – both up and down margins;
6. `xleftmargin` – additional left margin;
7. `ncols` – number of columns per page;
8. `nrows` – number of rows per page;
9. `font` – the PostScript font name;
10. `fontsize` – the PostScript font size.

Example: result see Fig. 2

```
\Begin(program,routines.rtn)
\section(common,browser,regular)
\Begin(initialization)
   . . .
   \initPostscript( pictures.ps,
                    A4,
                    Portrait,
                    20,
                    20,
                    40,
                    2,
                    5,
                    Helvetica,
                    25  )
   . . .
\End(initialization)
   . . .
\End(program)
```

The second driver prints all diagrams into one postscript file, one diagram per page. The diagrams are printed together with the topology and momenta flow. Such a form is convenient not for printing, but for investigating the diagram visually by means of some postscript interpreter, e.g., by the ghostview, Fig. 3. To initialize this driver, the user has to define only the output file name by means of the function `\initInfoPS(filename)`.

Example:

```
\Begin(program,routines.rtn)
\section(common,browser,regular)
\Begin(initialization)
   . . .
   \initInfoPS(info.ps)
   . . .
\End(initialization)
   . . .
\End(program)
```

The third driver can be used to create an encapsulated postscript file containing the current diagram, Fig. 4. To use this driver the user has to invoke the function `\outEPS(filename,Height,font,fontsize)`. inside the environment `output`. If `fontsize = 0`, then the particle labels will not be printed. The width of the diagram will be defined automatically. The diagram will be scaled to fit the EPS bounding box 0 0 `Width Height`.

Example: result see Fig.4

```
   . . .
\Begin(output,\askfilename())
```

```
   \outEPS( d\currentdiagramnumber().eps,
            100,
            Helvetica,
            15   )
   . . .
\End(output)
```

No initialization is required for this driver.

By default, all propagators are depicted by solid lines. To use different kinds of lines for different particles, the user must define the type of the line. At present, DIANA supports three types of lines: "wavy", suitable for vector propagators, "spiral" usually used for representation of a gluon , and "line" is just a line (full or dashed). All of them can be directed or not, and can be of different thickness and amplitude (for "wavy" and "spiral").

The syntax of the propagator description was extended as compared to the old one (see the DIANA 1.0 manual http://www.physik.uni-bielefeld.de/~tentukov/diana_doc.tar.gz) so that it is possible to define the type of the drawing line. Let us consider, e.g., the photon propagator description:
`[A,A;a;V(num,ind:1,ind:2,vec,0);0;wavy,4,2]`
From this example we can see that before the last "]" the user can describe (optionally) how to draw the corresponding line. The syntax is:

`;linetype,parameter,linewidth`

In the above example `linetype=wavy`, `parameter=4` and `linewidth=2`

The linetype is just an abstract type of the line. It can be one of the following: `wavy, arrowWavy, spiral, arrowSpiral, line, arrowLine`. For "wavy" and "spiral" the value of the `parameter` determines the amplitude while for "line" it means the type of dashing.

Another way to define a line type is to use the function

```
   \setpropagatorline(particle,
                      linetype,
                      parameter,
                      linewidth
)
```

in the `initialization` environment, for example:

```
\Begin(initialization)
   \setpropagatorline(A,wavy,4,2)
\End(initialization)
```

REFERENCES

1. M. Tentyukov, J. Fleischer, Comput. Phys. Commun. 132 (2000) 124-141.
2. J.A.M. Vermaseren, Symbolic manipulation with FORM, Computer Algebra Netherland, 1991.

The tensor reduction and master integrals of the two-loop massless crossed box

Carlo Oleari

Department of Physics, University of Wisconsin, 1150 University Avenue, Madison WI 53706, U.S.A.

Abstract. We briefly discuss an algorithm for the tensor reduction of the two-loop massless crossed boxes, with light-like external legs, and the computation of the relative master integrals.

INTRODUCTION

The level reached nowadays by the precision measurements in high-energy scattering experiments demands the knowledge of next-to-next-to-leading theoretical amplitudes for $2 \to 2$ scattering processes.

Very recent results for two-loop scattering amplitudes for massless particle have already appeared in the literature: the maximal-helicity-violating two-loop amplitude for $gg \to gg$ [1], $e^+e^- \to \mu^+\mu^-$ and $e^+e^- \to e^-e^+$ [2], $q\bar{q} \to q'\bar{q}'$ [3] and $q\bar{q} \to q\bar{q}$ [4].

In dealing with these two-loop scattering amplitudes we have to face the problem of the tensor reduction of planar [5] and crossed double boxes [6], plus a plethora of simpler topologies [7, 8], and the computation of the relative master integrals [9, 10].

NOTATION

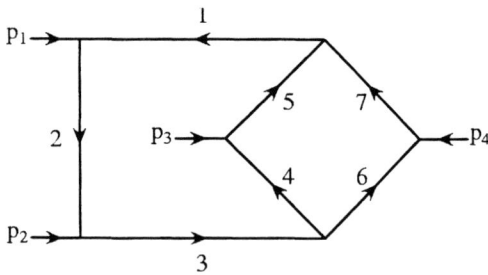

FIGURE 1. The generic two-loop crossed box.

We denote the generic two-loop tensor crossed (or non-planar) four-point function in D dimensions of Fig. 1 with seven propagators A_i raised to arbitrary powers v_i as

$$\text{Xbox}^D(v_1 \ldots v_7; s, t) [1; k^\mu; l^\mu; k^\mu k^\nu; k^\mu l^\nu; \ldots]$$

$$= \int \frac{d^D k}{i\pi^{D/2}} \int \frac{d^D l}{i\pi^{D/2}} \frac{[1; k^\mu; l^\mu; k^\mu k^\nu; k^\mu l^\nu; \ldots]}{A_1^{v_1} A_2^{v_2} A_3^{v_3} A_4^{v_4} A_5^{v_5} A_6^{v_6} A_7^{v_7}},$$

where the propagators are

$$A_1 = (k+l+p_{34})^2 + i0, \quad A_2 = (k+l+p_{134})^2 + i0,$$
$$A_3 = (k+l)^2 + i0, \quad A_4 = l^2 + i0,$$
$$A_5 = (l+p_3)^2 + i0, \quad A_6 = k^2 + i0,$$
$$A_7 = (k+p_4)^2 + i0.$$

The external momenta p_j are in-going and light-like, $p_j^2 = 0$, $j = 1\ldots 4$, so that the only momentum scales are the usual Mandelstam variables $s = (p_1+p_2)^2$ and $t = (p_2+p_3)^2$, together with $u = -s-t$. For ease of notation, we define $p_{ij} = p_i + p_j$ and $p_{ijk} = p_i + p_j + p_k$. In the square brackets we keep trace of the tensor structure that may be present in the numerator.

TENSOR REDUCTION

As it is well known, tensor integrals can be related to combinations of scalar integrals with higher powers of propagators and/or different values of D [7, 11]. This is quite straightforward to see if we rewrite the Feynman integral introducing the Schwinger parameters, diagonalizing the exponent of the out-coming integral and integrating out the loop momenta [6, 7].

The task to compute tensor integrals is then moved to the computation of scalar integrals with

- higher powers of the propagators

- in higher dimensions.

From the Schwinger representation of Feynman integrals it easy to see that integrals in D dimensions can be connected with integrals in $D+2$ dimensions (dimensional-shift) [5, 6].

In this way, tensor integrals can be directly connected to scalar integrals with higher powers of the propagators in $D = 4 - 2\varepsilon$ dimensions.

THE SCALAR CROSSED-BOX REDUCTION

The strategy to reduce the generic scalar integral to a linear combination of known ones is based on recurrence identities that relate scalar integrals with different powers of propagators. Some of these identities can be obtained using the integration-by-parts method [12] and exploiting the Lorentz invariance of the Feynman diagram [13].

Following the reduction procedure detailed in Ref. [6], any scalar crossed box with arbitrary powers of the propagators can be written as a linear combination of the following integrals, that, therefore, are called master integrals:

$$\text{Sset}^D(s) = \quad \text{—}\bigcirc\text{—} \quad (s)$$

$$\text{Tri}^D(s) = \quad \text{—}\triangleleft\!\!\square \quad (s)$$

$$\text{Xtri}^D(s) = \quad \text{—}\triangleleft\!\!\!<\quad (s)$$

$$\text{Bbox}^D(s,t) = \quad \boxed{\bigcirc\;} \quad (s,t)$$

$$\text{Dbox}^D(s,t) = \quad \boxed{\diagup} \quad (s,t)$$

$$\text{Xbox}_1^D(s,t) = \quad \boxed{\times} \quad (s,t)$$

$$\text{Xbox}_2^D(s,t) = \quad \bullet\boxed{\times} \quad (s,t),$$

where all the propagators have powers one except for the propagator with the blob, that has power two.

DIFFERENTIAL EQUATIONS FOR THE TWO MASTER INTEGRALS

The analytic expansion in $\varepsilon = (4-D)/2$ for the first master cross box $\text{Xbox}_1^D(s,t)$ was computed in Ref. [10]. We can obtain the analytic form for the second one by writing the derivative of $\text{Xbox}_1^D(s,t)$ with respect to one of the two independent physical scales (that we choose to be t), as a combination of master integrals, and solving the equation for $\text{Xbox}_2^D(s,t)$. Moreover we can verify the correctness of both the expressions of $\text{Xbox}_1^D(s,t)$ and $\text{Xbox}_2^D(s,t)$, by deriving an analogous differential equation for Xbox_2^D, and checking that the obtained identity is satisfied.

Starting from the Schwinger representation of the generic crossed box, we can differentiate with respect to t, and set the values of ν_i to reproduce the two master integrals

$$\frac{\partial}{\partial t}\text{Xbox}_1^D(s,t) = \text{Xbox}^{D+2}(1,2,1,2,1,1,2;s,t)$$
$$- \text{Xbox}^{D+2}(1,2,1,1,2,2,1;s,t),$$

$$\frac{\partial}{\partial t}\text{Xbox}_2^D(s,t) = 2\text{Xbox}^{D+2}(1,3,1,2,1,1,2;s,t)$$
$$- 2\text{Xbox}^{D+2}(1,3,1,1,2,2,1;s,t).$$

Applying the reduction formalism for the scalar integrals and the dimensional-shift, we can rewrite the right-hand sides of the system as a combination of the two master crossed boxes plus other master integrals of simpler topologies, obtained by pinching one or more of the propagators of the crossed box,

$$\frac{\partial}{\partial t}\text{Xbox}_1^D(s,t) = \frac{1}{t-u}\left[\frac{(D-4)s^2 - 4tu}{2tu}\text{Xbox}_1^D(s,t)\right.$$
$$\left. -\frac{(D-6)s}{2(D-5)}\text{Xbox}_2^D(s,t) + \text{pinchings}\right] \quad (1)$$

$$\frac{\partial}{\partial t}\text{Xbox}_2^D(s,t) = \frac{1}{t-u}\left[\frac{2(D-5)^2 s}{tu}\text{Xbox}_1^D(s,t)\right.$$
$$\left. -\frac{(D-6)(u^2+t^2)}{tu}\text{Xbox}_2^D(s,t) + \text{pinchings}\right] \quad (2)$$

Inserting the ε expansion of $\text{Xbox}_1^D(s,t)$ computed in Ref. [10] and the ε expansions of the sub-topologies listed in Refs. [5, 7, 14, 15] into Eq. (1), and solving it with respect to $\text{Xbox}_2^D(s,t)$, we obtain, in the physical region $s > 0$, $t,u < 0$,

$$\text{Xbox}_2^D(s,t) = \Gamma^2(1+\varepsilon)\left\{\frac{G_1(t,u)}{s^3 t} + \frac{G_2(t,u)}{s^2 t^2} + \right.$$
$$\left. \frac{G_1(u,t)}{s^3 u} + \frac{G_2(u,t)}{s^2 u^2}\right\},$$

where

$$G_1(t,u) = s^{-2\varepsilon}\left\{\frac{6}{\varepsilon^3} + \frac{1}{\varepsilon^2}(32 - 6T - 6U)\right.$$
$$+ \frac{1}{\varepsilon}(1 - 12\pi^2 - 24T + T^2 - 24U + 16TU + U^2)$$
$$- 43 - 18T + 13T^2 + \frac{8}{3}T^3 - 18U + 16TU$$
$$+ 11T^2 U + 13U^2 - 20TU^2 + \frac{8}{3}U^3 + \pi^2\left(17T\right.$$

$$+17U - \frac{112}{3}\Big) - 122\zeta(3) + 62T\operatorname{Li}_2\left(-\frac{t}{s}\right)$$
$$- 62\operatorname{Li}_3\left(-\frac{t}{s}\right) + 62\operatorname{S}_{1,2}\left(-\frac{t}{s}\right)$$
$$+ i\pi\Bigg[\frac{1}{\varepsilon}(16 + 6T + 6U) - 34 - 9\pi^2 - 6T - 10T^2$$
$$- 6U + 14TU - 10U^2\Bigg]\Bigg\},$$

$$G_2(t,u) = s^{-2\varepsilon}\Bigg\{-\frac{2}{\varepsilon^4} + \frac{1}{\varepsilon^3}\left(-8 + \frac{5}{2}T + \frac{7}{2}U\right)$$
$$+ \frac{1}{\varepsilon^2}\left(-\frac{29}{2} - \frac{5}{12}\pi^2 + 7T - T^2 + 20U - 4TU\right.$$
$$\left. - U^2\right) + \frac{1}{\varepsilon}\Bigg[-\frac{1}{2} + 17T + 2T^2 - \frac{T^3}{3} + \frac{\pi^2}{6}(14$$
$$+ 5T - 29U) + 13U - 28TU - 4U^2 + 3TU^2$$
$$- U^3 + \frac{19}{2}\zeta(3) - 2T\operatorname{Li}_2\left(-\frac{t}{s}\right) + 2\operatorname{Li}_3\left(-\frac{t}{s}\right)$$
$$- 2\operatorname{S}_{1,2}\left(-\frac{t}{s}\right)\Bigg] + \frac{37}{2} + \frac{37}{40}\pi^4 + 7T - 5T^2$$
$$- \frac{22}{3}T^3 + \frac{2}{3}T^4 + 5U - 20TU + \frac{8}{3}T^3U - 2U^2$$
$$+ 24TU^2 - T^2U^2 - 8U^3 - \frac{4}{3}TU^3 + \frac{4}{3}U^4$$
$$+ \frac{\pi^2}{6}\left(79 - 22T - 5T^2 - 200U + 76TU + 25U^2\right)$$
$$+ (68 - 13T - 33U)\zeta(3) + \left(10\pi^2 - 32T + 17T^2\right.$$
$$+ 12TU\Big)\operatorname{Li}_2\left(-\frac{t}{s}\right) - 36\operatorname{S}_{2,2}\left(-\frac{t}{s}\right)$$
$$+ (28T - 6U - 32)\operatorname{S}_{1,2}\left(-\frac{t}{s}\right) - 26\operatorname{S}_{1,3}\left(-\frac{t}{s}\right)$$
$$+ (32 - 60T - 12U)\operatorname{Li}_3\left(-\frac{t}{s}\right) + 86\operatorname{Li}_4\left(-\frac{t}{s}\right)$$
$$+ i\pi\Bigg[\frac{2}{\varepsilon^3} + \frac{1}{\varepsilon^2}(11 - T + U) + \frac{1}{\varepsilon}\left(1 - \frac{31}{6}\pi^2 - 10T\right.$$
$$\left. - 2T^2 + 4U - 2TU - 2U^2\right) + 11 + 4T - 2T^2$$
$$+ \frac{10}{3}T^3 + \frac{\pi^2}{3}(-65 + 28T - U) + 2U - 8TU$$
$$- 8U^2 + 2U^3 - 89\zeta(3) + (14T + 18U)\operatorname{Li}_2\left(-\frac{t}{s}\right)$$
$$- 32\operatorname{Li}_3\left(-\frac{t}{s}\right) + 44\operatorname{S}_{1,2}\left(-\frac{t}{s}\right)\Bigg]\Bigg\},$$

$T = \log(-t/s)$, $U = \log(-u/s)$, and where we used Nielsen's generalized polylogarithms $S_{n,p}$ defined by

$$S_{n,p}(x) = \frac{(-1)^{n+p-1}}{(n-1)!\,p!}\int_0^1 dt\,\frac{\log^{n-1}(t)\log^p(1-xt)}{t}$$
$$n,p \geq 1,\quad x \leq 1$$

Expressions for $\text{Xbox}_2^D(s,t)$ in the other two kinematic regions, $t > 0$, $s,u < 0$ and $u > 0$, $s,t < 0$ can be easily obtained through the analytic continuation of the polylogarithms and the logarithms.

It is a strong check of the whole formalism that inserting the analytic expansion of $\text{Xbox}_1^D(s,t)$ and the derived expression of $\text{Xbox}_2^D(s,t)$ into Eq. (2), we obtain an equality identically satisfied.

Acknowledgements. This work has been done in collaboration with C. Anastasiou, T. Gehrmann, E. Remiddi and J. B. Tausk.

REFERENCES

1. Z. Bern, L. Dixon and D.A. Kosower, JHEP **0001** (2000) 027, hep-ph/0001001.
2. Z. Bern, L. Dixon and A. Ghinculov, hep-ph/0010075.
3. C. Anastasiou, E.W.N. Glover, C. Oleari and M.E. Tejeda-Yeomans, hep-ph/0010212.
4. C. Anastasiou, E.W.N. Glover, C. Oleari and M.E. Tejeda-Yeomans, hep-ph/0011095.
5. V.A. Smirnov and O.L. Veretin, *Nucl. Phys.* **B566** (2000) 469.
6. C. Anastasiou, T. Gehrmann, C. Oleari, E. Remiddi and J.B. Tausk, *Nucl. Phys.* **B580** (2000) 577, hep-ph/0003261.
7. C. Anastasiou, E.W.N. Glover and C. Oleari, *Nucl. Phys.* **B575** (2000) 416, *Erratum-ibid* **B585** (2000) 763, hep-ph/9912251.
8. C. Anastasiou, E.W.N. Glover and C. Oleari, *Nucl. Phys.* **B565** (2000) 445.
9. V.A. Smirnov, *Phys. Lett.* **B460** (1999) 397.
10. J.B. Tausk, *Phys. Lett.* **B469** (1999) 225.
11. O.V. Tarasov, *Phys. Rev.* **D54** (1996) 6479, *Nucl. Phys.* **B502** (1997) 455.
12. F.V. Tkachov, *Phys. Lett.* **100B** (1981) 65; K.G. Chetyrkin and F.V. Tkachov, *Nucl. Phys.* **B192** (1981) 159.
13. T. Gehrmann and E. Remiddi, *Nucl. Phys.* **B580** (2000) 485, hep-ph/9912329.
14. R.J. Gonsalves, *Phys. Rev.* **D28** (1983) 1542.
15. G. Kramer and B. Lampe, *J. Math. Phys.* **28** (1987) 945.

Gauge Invariant Classes of Feynman Diagrams and Applications for Calculations

E.E. Boos

Skobeltsyn Institute of Nuclear Physics, Moscow State University, Moscow 119899, Russia

Abstract. In theories like SM or MSSM with a complex gauge group structure the complete set of Feynman diagrams contributing to a particular physics process can be split into exact gauge invariant subsets. Arguments and examples given in this paper demonstrates that in many cases computations and analysis of the gauge invariant subsets are important.

The increase of collider energies requires computations of processes with more particles in the final state and with better precision (NLO, NNLO etc). At LEP1 the basic processes were 2 fermions (γ) production; LEP2 deals basically with 4 fermion (γ) processes; Tevatron, LHC and LC in many cases need in an analysis of the processes with 5,6,8 and so on fermions in the final state, for example, top pair production with decays - 6 fermions; single top production in W-gluon fusion mode - 5 fermions; strongly interacting Higgs sector in hadronic collisions- $pp \to q\bar{q}W^+W^-$ - 6 fermions; study of Yukawa coupling in $pp(e^+e^-) \to t\bar{t}H$ - 8 fermions etc. Typically for the processes with multi-particle final states number of contributing diagrams is large. For hadronic collisions not only number of diagrams but also number of partonic subprocesses is very large.

One of the problem in process computations is a gauge cancellation among many diagrams. Any method of calculation should preserve gauge invariance which is rather complicated in theories like SM with not a simple gauge group. The well known statement from the quantum theory of gauge fields is that the whole set of Feynman diagrams contributing to any physics process is exactly gauge invariant. However, in practical calculations it remains amazing how gauge cancellations take place. But in general the complete set of Feynman diagrams contributing to a particular physics process can be split in to exact gauge invariant subsets, and the gauge cancellations occur in each of the subset. It will be demonstrated that in many cases the idea to split the complete set of diagrams on exact gauge invariant subsets (1) could be useful in practice:

- any physics approximation should be based on gauge invariant classes of diagrams

- better precision of computations in many cases

- better understanding of physics parameters like running couplings or scales (QCD scale, ISR scale etc). Often different part of the same process need different values because different kinematical regions might be important

- any MC generator needs well behaved and compact matrix elements and the gauge invariance gives that

For simple cases it is very easy to find gauge invariant subclasses of diagrams. Take, for instance, the Bhabha-scattering. In that case the two s-channel and two t-channel SM diagrams are separately gauge invariant as immediately follows if one substitutes the final e^+e^- pair by the $\mu^+\mu^-$ pair. In fact, it gives the simplest example of so called "Flavor Flip" introduced in (1). There are two kinds of flips: Gauge and Flavor, which correspond to permutations the $2 \to 2$ subdiagrams with on- and/or off-shell legs (see the exact definitions of flips in (1)). For a concrete physical process one can get one Feynman diagram from another by a sequence of gauge and flavor flips. In correspondence to all the diagrams for the process one may put some graph called Forest. The Forest is a graph with each vertex representing a diagram and the edges given by the flips (gauge and/or flavor) of four-point sub-diagrams. The Gauge Forest is such a Forest or part of the Forest in which the points connected by the only gauge flips. The connected components of the Gauge Forest are called Groves. The general theorem has been proved in (1) by the mathematical induction method:

The Forest F(E) for an external state E consisting of gauge and matter fields is connected if the fields in E carry no conserved quantum numbers other than the gauge charges. The Groves are the minimal gauge invariant classes of Feynman diagrams.

A very simple forest as an example is shown in Fig. 1 for the process $u\bar{d} \to c\bar{s}\gamma$. All diagrams are connected

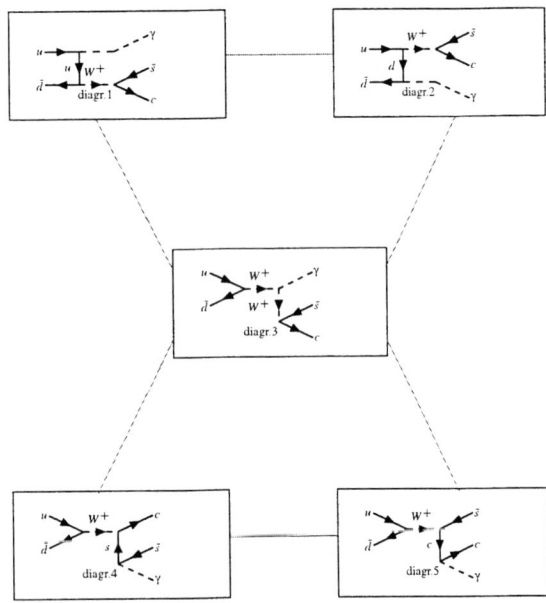

FIGURE 1. The Forest for the process $u\bar{d} \to c\bar{s}\gamma$ is the Grove.

by the gauge flips passing through the middle diagram 3. Therefore all of them form one Grove - one gauge invariant set of diagrams.

More complex examples of the Forests and Groves are shown in (1). They have been obtained by means of the program *bocages* (2) in which the algorithm of tree diagrams generation based on the flips has been realised. Table shows how the number of diagrams for various processes with 6 external fermions splits into minimal gauge invariant subsets (Groves).

E	Σ	classes
$u\bar{u}u\bar{u}u\bar{u}$	144	$18 \cdot 8$
$u\bar{u}u\bar{u}u\bar{u}\gamma$	1008	$18 \cdot 24 + 36 \cdot 16$
$u\bar{u}u\bar{u}d\bar{d}$	92	$4 \cdot 11 + 6 \cdot 8$
$u\bar{u}u\bar{u}d\bar{d}\gamma$	716	$4 \cdot 95 + 6 \cdot 24 + 12 \cdot 16$
$\ell^+\ell^- u\bar{u}d\bar{d}$	35	$1 \cdot 11 + 3 \cdot 8$
$\ell^+\ell^- u\bar{u}d\bar{d}\gamma$	262	$1 \cdot 94 + 3 \cdot 24 + 6 \cdot 16$
$\ell^- \nu d\bar{u}d\bar{d}$	20	$2 \cdot 10$
$\ell^- \nu d\bar{u}d\bar{d}\gamma$	152	$2 \cdot 76$
$\ell^+\ell^-\ell^- \nu d\bar{u}$	20	$2 \cdot 10$
$\ell^+\ell^-\ell^- \nu d\bar{u}\gamma$	150	$2 \cdot 75$
$\ell^- \nu \ell^+ \bar{\nu} d\bar{d}$	19	$1 \cdot 9 + 2 \cdot 4 + 1 \cdot 2$
$\ell^- \nu \ell^+ \bar{\nu} d\bar{d}\gamma$	107	$1 \cdot 59 + 2 \cdot 12 + 2 \cdot 8 + 2 \cdot 4$
$\ell^- \bar{\nu}\ell^+ \nu \ell^+\ell^-$	56	$4 \cdot 9 + 4 \cdot 4 + 2 \cdot 2$
$\ell^- \bar{\nu}\ell^+ \nu \ell^+\ell^-\gamma$	328	$4 \cdot 58 + 4 \cdot 12 + 4 \cdot 8 + 4 \cdot 4$
$\ell^+ \nu \ell^- \bar{\nu}\nu\bar{\nu}$	36	$4 \cdot 6 + 6 \cdot 2$
$\ell^+ \nu \ell^- \bar{\nu}\nu\bar{\nu}\gamma$	132	$4 \cdot 26 + 2 \cdot 6 + 4 \cdot 4$
$\nu\bar{\nu}\nu\bar{\nu}\nu\bar{\nu}$	36	$18 \cdot 2$

The familiar LEP2 gauge invariant classes (CC09, CC10, CC11, etc.) appear here automatically. However now we know that these classes are not only gauge invariant but they are minimal classes.

Let us take as an example the CC20 process (so called "single W") $e^+e^- \to e^-\nu d\bar{u}$ which splits to t- and s-channel CC10 gauge invariant subclasses. By means of the CompHEP (3) one can compute the contributions of the classes and their interference as shown in the Table below (4) (quark phase space cuts: $E_q \geq 3$ GeV, $M_{ud} \geq 5$ GeV and the lepton phase space cut: $\cos\theta_e \geq 0.997$).

$\sqrt{s}$	$\sigma(CC10-t)$	$\sigma(CC10-s)$	$\sigma(t-s interf.)$
quark phase space cuts, no ISR			
190	147(0)	680(1)	5(0)
350	635(1)	420(1)	21(0)
500	1127(2)	270(0)	19(0)
800	1981(4)	143(0)	16(0)
lepton and quark phase space cuts, no ISR			
190	116(0)	2(0)	0.0(0)
350	513(1)	7(0)	0.2(0)
500	928(2)	10(0)	0.3(0)
800	1671(4)	15(0)	0.4(0)

The CC10 t-channel part contains the single W boson production. It grows with the collision energy and it starts to dominate the CC10 s-channel part (W boson pair production) at about 320 GeV. One should stress a few points here

- a good precision of computations is obtained only if one splits the complete CC20 set of diagrams to CC10 subsets because in that case one can use different kinematical variables of integration for different subsets with different mapping of singularities (the interference contribution is small)

- the "overall" scheme of the W-boson width treatment could be used only for separate classes, otherwise there will be an artificial suppression of the CC10 t-channel part by the factor related to the second W pole

- for the CC10 s-channel part obviously a scale of the order of energy should be used for the electromagnetic α and ISR while the CC10 t-channel part has a very small characteristic virtuality of the soft virtual photon, and therefore a typical scale for the corresponding α and ISR should be taken much smaller, of the order electron momentum transfer (4, 5)

The same general statements are also true for the process of so called "single Z" production $e^+e^- \to e^+e^-\nu\bar{\nu}$. In this case there are 56 Feynman diagrams which split to 10 gauge invariant classes or 10 Groves $56D = 4*9D + 4*4D + 2*2D$ (6). Two classes of

4 diagrams each contribute to the single Z as given in the second column of the Table below. In the Table the contributions of the gauge invariant subsets in fb are given at the energy $\sqrt{s}$ =200 GeV. First row - with angular cuts, second row - no angular cuts for e^-, e^+. The following angular and lepton energy cuts are used: $\cos\theta_{e^-} \geq 0.997$ and $\cos\theta_{e^+} \leq 0.997$ and $E_l \geq 15$ GeV.

	18W	8Z	9W$^+$W$^-$	4ZZ
θ_e, E_l	36.1	16.4	0.91	0.02
only E_l	106.6	153.6	240.5	44.9

If the energy and angular cuts are applied still there are significant contributions from both single W (first column) and Z (second column) productions. One should be careful in interpretation of experimental measurements in this channel. (Contributions of other gauge invariant subsets are very small and we do not show them here).

One more example of applications of the gauge classes is related to the method of simplification of flavour combinatorics in the evaluation of hadronic processes (7). Here a serious computational problem is the large number of partonic subprocesses due to a presence of many quark partons with different flavors in the colliding hadrons and contributions of many additional diagrams for each subprocess because of the CKM quark mixing. However in the approximation when CKM matrix is reduced to the CK matrix without a mixing with the 3d quark generation

$$V_{CKM} \Longrightarrow \begin{pmatrix} V & 0 \\ 0 & 1 \end{pmatrix}, \quad V = \begin{pmatrix} \cos\vartheta_c & \sin\vartheta_c \\ -\sin\vartheta_c & \cos\vartheta_c \end{pmatrix}$$

where ϑ_c is the Cabbibo angle and neglecting masses of the quarks from the first two generations $M_u = M_d = M_s = M_c = 0$ the problem can be simplified drastically. In this case diagrams contributing to the process can be split into the gauge invariant classes with different topologies of the incoming and outgoing quark lines. Then one can make a rotation of down quarks in all vertices of Feynman diagrams thus, transporting the mixing matrix elements from the diagrams to the parton distribution functions. As a result a number of rules for a convolution with quark distribution functions appear depending on the topology of the gauge invariant class (see details in (7); the method has been realized in the CompHEP version for hadron collisions V41.10).

In this talk we have discussed the method which allows splitting the complete set of Feynman diagrams contributing to a physics process into gauge invariant subclasses. Well known gauge invariant classes of diagrams like CC10, CC11, CC09 etc naturally appear in such an approach. It was demonstrated the above classes are the minimal invariant classes ("Groves"). For a concrete physical process one creates the graph - "Forest" in which vertices represent diagrams and edges show the connection between diagrams by possible flips, flavor and gauge. The vertices of the graph (diagrams) connected by the only gauge flips form connected subgraphs, "Groves", and the corresponding diagrams form minimal gauge invariant classes. The flavor flips connect diagrams from different gauge invariant classes.

Separation into gauge invariant classes in some cases allows to better understand properties of processes, to get better precision of calculations, to make in tree level computations a natural choice of characteristic scales for ISR, structure functions, running couplings etc. it also makes possible reasonable approximations, and leads to a simpfication of flavor combinatorics etc.

In some cases for processes with multi-particle final states the number of gauge invariant classes is much smaller than the number of physical reactions (8). So one can compute, in principle, amplitudes for gauge invariant subclasses of diagrams and then compute processes by taking different combinations of that amplitudes for classes.

The analysis was done for the tree level Feynman diagrams. A consideration at loop level is in progress (9).

The author thanks the Organising Committee of the ACAT2000 Conference for kind hospitality. I would like to thank my collaborators and coauthors T. Ohl, M. Dubinin, and V. Ilyin. The work was partly supported by the RFBR-DFG 99-02-04011, RFBR 00-01-00704, CERN-INTAS 99-377, and Universities of Russia 990588 grants.

REFERENCES

1. E. Boos and T. Ohl, Phys.Rev.Lett. **83** (1999) 480; E. Boos and T. Ohl, hep-ph/9909487.
2. T. Ohl, Objective Caml program *bocages* (1999, unpublished).
3. E.E. Boos et al., preprint INP MSU 94-36/358 and SNUTP-94-116, hep-ph/9503280; P. Baikov et al., Proc. of the Xth Int. Workshop on High Energy Physics and Quantum Field Theory, ed. by B. Levtchenko and V. Savrin, Moscow, 1996, p. 101.; A. Pukhov et al, Preprint INP MSU 98-41/542, hep-ph/9908288.
4. E. Boos and M. Dubinin, hep-ph/9909214, Talks given at the LCWS 99 and QFTHEP 99 Conferences.
5. Y. Kurihara et al., hep-ph/0011276.
6. M. Gruenwald, G. Passarino et al., in: Reports of working groups on precision calculations for LEP2 Physics, ed.by R. Pittau, CERN Yellow Report 2000-0009, hep-ph/0005309.
7. E. Boos, V. Ilyin et al., JHEP **0005** (2000) 052.
8. E. Boos and T. Ohl, Phys. Lett. **B407** (1997) 161.
9. E. Boos and T. Ohl, in preparation

A Parallel Version of FORM 3

D.Fliegner*, A.Rétey**, J.A.M. Vermaseren[†]

*Abt. Nichtlineare Dynamik, MPI für Strömungsforschung, D-37073 Göttingen, Germany
**Institut für Theoretische Teilchenphysik, Universität Karlsruhe, D-76128 Karlsruhe, Germany
[†]NIKHEF, P.O. Box 41882, 1009 DB, Amsterdam, The Netherlands

Abstract. The parallel version of the symbolic manipulation program FORM for clusters of workstations and massive parallel systems is presented. We discuss various cluster architectures and the implementation of the parallel program using message passing (MPI). Performance results for real physics applications are shown.

1. The Sequential Version of FORM

FORM (1) is a program for symbolic manipulation of algebraic expressions specialized to handle very large expressions in an efficient and reliable way. It is used non-interactively by executing a program that contains several parts called *modules*. The execution of each of these module is divided into several steps:

- **Compiling:** the input is translated into an internal representation.

- **Generating:** for each term of the input expressions the statements of the module are executed. In general this generates a lot of terms for each input term.

- **Sorting:** all the output terms that have been generated are sorted and equivalent terms are summed up.

The special properties of FORM result from the fact that with a few exceptions it allows *local* operations on *single* terms only (e.g. replacing parts of a term by another term or a whole expression, multiplying a term by a certain expression). As a consequence expressions can be handled as "streams" of independent terms that can be read in from a file and processed sequentially. The generation of terms is done in a way that the output terms drop out term by term. The output terms are partially sorted and stored in intermediate buffers and temporary sort files on disk. At the end of a module all the existing sorted patches are merged into a single sorted output stream of terms which is written to a file and used as input source of terms for the next module. The procedure is not limited by the memory of the system, but the maximum size of the sort files, i.e. the total disk space. This is what makes FORM particularly useful for large symbolic problems. Obviously, the restriction to local operations also allows a straight forward parallelization.

2. Evaluation of Parallel Platforms

For the parallelization of FORM, message passing and a corresponding library MPI(CH) (2) was chosen, because of its availability on a wide range of different architectures. It was clear from the beginning that the performance of regular networking hardware would not be sufficient. To give an impression of the performance of special communication hard– and software we show results for various combinations that we have used instead:

- Digital Unix Alpha cluster with 32bit-Myrinet (3)
  MPICH/ParaStation II (4)
  (TTP Universität Karlsruhe (5))

- Compaq Linux Alpha Cluster 64bit-Myrinet
  MPICH/ParaStation II
  (ALICE Universität Wuppertal (6))

- IBM SP2, MPI (RZ Universität Karlsruhe)

- Compaq Tru64 Unix Alpha SMP server
  MPICH using shared memory segments
  (TTP Universität Karlsruhe)

The MPI(CH) libraries have been examined using the Pallas MPI benchmarks (PMB) (7), which determines the throughput and latency of the message passing operations under different circumstances. Here we only present the results for the simplest possible single transfer benchmark, a ping-pong test: one process sends a message of n bytes to another process, which immediately sends that message back. In contrast to more complicated benchmarks there is no concurrency with other messages passing activity during this test, i.e. the bandwidth and latency are measured under optimal conditions. A Fast Ethernet gives a maximum throughput of 100 Mbit/s and a minimum latency of about 200 μs.

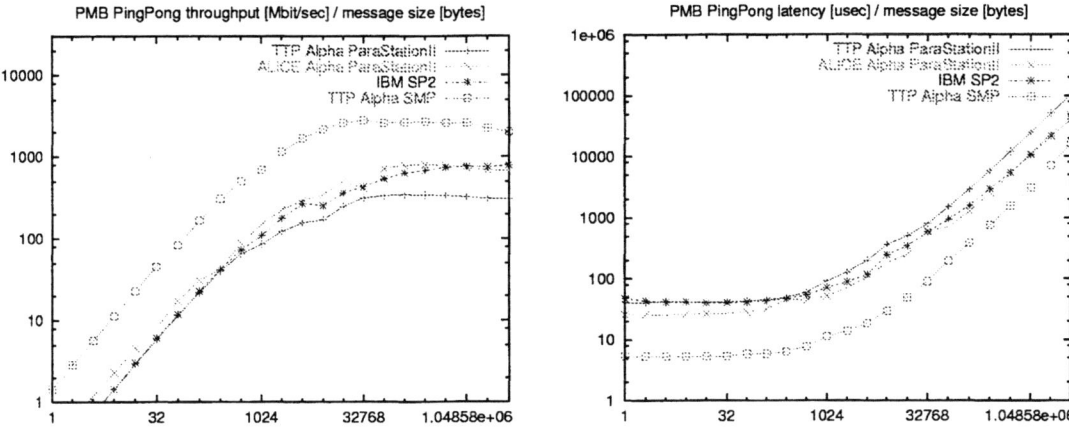

FIGURE 1. MPI(CH) bandwidth and latency results for the PingPong benchmark.

As can be seen in figure 1 the ParaStationII MPICH implementation on the 64bit Myrinet network yields about the same maximum throughput as the IBM SP2 (800 Mbit/s), but the maximum bandwidth on the Alpha SMP server is still considerably larger (more than 2.5 Gbit/s). In figure 1 (right) the measured latency is shown. The minimum latency ($25\mu s$) of the ParaStationII MPICH implementation using the 64bit Myrinet hardware is smaller than the minimum latency on the IBM SP2. The Alpha SMP server yields an even smaller latency (less than $7 \mu s$) by using shared memory segments.

3. The Parallel Version: ParFORM

In a first step of the parallelization only the sorting procedure was parallelized. In this version the generated output terms are distributed by a master process among an arbitrary number of slave processes, which do a partial sorting. Finally the sorted patches are merged together by the master process.

In a second step also the generation of terms was parallelized. The input terms are distributed among all processes before generating the output terms. The distribution of terms is organized such that the master process sends a patch of input terms to each slave process at the beginning of the module and then waits for the slaves to ask for new terms whenever they have finished with their last patch of input terms. This concept also yields a decent load leveling automatically. In the current implementation the number of terms in a patch can be chosen at the start of the program. Of course choosing a too coarse grained distribution will result in the danger of running into worst cases, where all the work sits in only one of the input patches, and only one processor is busy. On the other hand, a fine grained distribution causes more overhead. The best setting turns out to be strongly dependent on the problem that is run: figure 2 shows the speedup for a combinatorics problem that amounts to ideal input for the parallel version. Beginning with the second processor a linear speedup is achieved even if single input terms ($N = 1$) are distributed. If a too coarse grained distribution is chosen ($N = 1000$ input terms per patch) the speedup is lost. In typical real physics applications (figure 3 shows the result for a complicated 3-loop Feynman diagram) things look different: the speedup is usually not linear anymore and there is obviously an optimum input patch size somewhere between $N = 10$ and $N = 100$.

Figure 4 shows the result for a SMP system with processors that are much faster than those of the clusters shown in figure 3. In spite of the faster inter-process communication through shared memory segments the speedup is smaller. This is a generic problem that can only be solved by omitting message passing and using threads instead.

4. Conclusion & Outlook

The parallel version of FORM based on MPI can be used on massive parallel computers, clusters of workstations and SMP systems. With only a few exceptions all the features of the sequential version are already operational in the parallel program as well. For ideal input programs ParFORM indeed yields a linear speedup beginning with the second processor. There is a considerable speedup in real physics applications, if the problem size is not too small. ParFORM is currently being optimized for huge clusters of workstations (e.g. ALICE (6)). We also work on a parallel version for SMP systems (e.g. TERAS (8)) using threads instead of message passing). The final aim is a version for clusters of SMP systems.

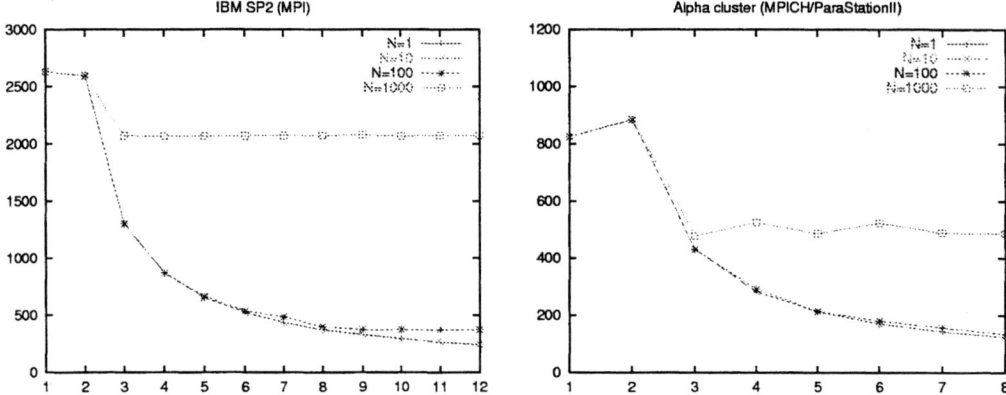

FIGURE 2. Runtimes [sec] vs. number of processors for ideal input on a IBM SP2 (left) and a Digital Alpha cluster using ParaStationII and a 32bit Myrinet network (right). Shown are the runtimes for different granularities of the input-term distribution.

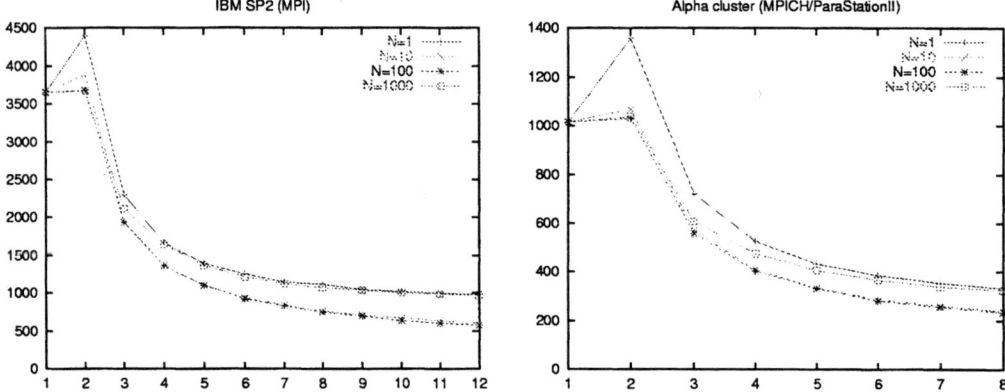

FIGURE 3. Runtimes [sec] vs. number of processors for a real physics application on a IBM SP2 (left) and a Digital Alpha cluster using ParaStationII and a 32bit Myrinet network (right). Shown are the runtimes for different granularities of the input-term distribution.

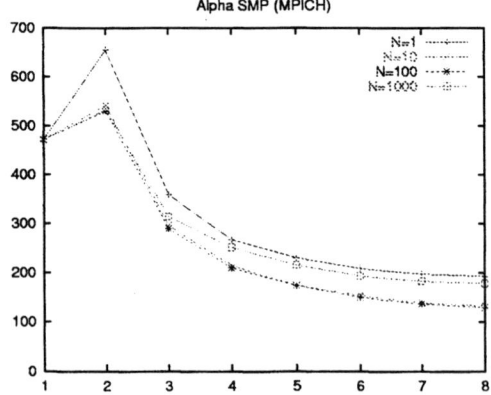

FIGURE 4. Runtimes [sec] vs. number of processors for the physics application of figure 3 on a Compaq Alpha SMP Server.

REFERENCES

1. FORM homepage: http://www.nikhef.nl/form
2. MPI(CH) homepage: http://www.mcs.anl.gov/mpi
3. Myricom homepage: http://www.myri.com
4. ParaStation homepage: http://parastation.ira.uka.de
5. TTP homepage: http://www-ttp.physik.uni-karlsruhe.de
6. ALICE homepage: http://alice.iai.uni-wuppertal.de
7. Pallas MPI Benchmarks: http://www.pallas.de/PMB2
8. TERAS homepage: http://www.sara.nl

Two-loop Calculations in the MSSM with FeynArts

S. Heinemeyer

HET, Brookhaven Natl. Lab., Upton, New York 11973, USA

Abstract. Recent electroweak two-loop corrections in the Minimal Supersymmetric Standard Model (MSSM) are reviewed. They have been obtained with the help of the programs *FeynArts* and *TwoCalc*, making use of the recently finished MSSM model file for *FeynArts*. Short examples of how to use the two codes together with the analytic result for the $O(G_F^2 m_t^4)$ corrections to the ρ-parameter in the MSSM are presented.

INTRODUCTION

Theories based on Supersymmetry (SUSY) are widely considered as the theoretically most appealing extension of the Standard Model (SM). The Minimal Supersymmetric Standard Model (MSSM) predicts the existence of scalar partners $\tilde{f}_L, \tilde{f}_R$ to each SM chiral fermion, and spin-1/2 partners to the gauge bosons and to the scalar Higgs bosons, where two Higgs doublets are present in the MSSM. So far, the direct search of SUSY particles at present colliders has not been successful. One can only set lower bounds of $O(100\,\text{GeV})$ on their masses, see Ref. (1). An alternative way to probe SUSY is to search for the virtual effects of the additional particles. The experimental precision has to be matched with high precision theoretical predictions for the various precision observables. The most prominent role in this respect is played by the ρ-parameter, see Ref. (2):

$$\rho = \frac{1}{1-\Delta\rho}, \quad \Delta\rho = \frac{\Sigma_Z}{M_Z^2} - \frac{\Sigma_W}{M_W^2}. \quad (1)$$

The radiative corrections to the vector boson self-energies, $\Sigma_{Z,W}$ constitute the leading, process independent corrections to many electroweak precision observables, such as the W boson mass, M_W, where

$$\delta M_W \approx M_W/2 \; c_W^2/(c_W^2 - s_W^2)\,\Delta\rho, \quad (2)$$

with $c_W^2 = 1 - s_W^2 = M_W^2/M_Z^2$. Within the MSSM the corrections have so far been restricted to $O(\alpha\alpha_s)$, see Ref. (3). In order to match the accuracy obtained in the SM and the (prospective) experimental uncertainties, the leading two-loop corrections to $\Delta\rho$ at $O(G_F^2 m_t^4)$ are desirable. These corrections are the lowest order contributions involving non-SM particles with a m_t^4 dependence. In the limit of a large SUSY scale, $M_{SUSY} \gg M_Z$, the SUSY particles decouple from the observables, leaving the two Higgs doublets of the MSSM active. First results for this contribution have recently been calculated in Ref. (4).

Feynman-diagrammatic two-loop calculations involve a large number of diagrams. In the MSSM the additional problem of a proliferation of scales is apparent. Therefore, in order to perform the calculation as presented in Ref. (4), the use of computer algebra programs is inevitable. In particular, we made use of the amplitude generator *FeynArts*, see Ref. (5), where the MSSM model file has recently been completed. The reduction of the amplitudes to scalar integrals have been performed with the program *TwoCalc*, see Ref. (6).

TECHNIQUES FOR TWO-LOOP CALCULATIONS

In order to calculate the $O(G_F^2 m_t^4)$ corrections to the ρ-parameter, the diagrams in Fig. 1 have to be evaluated.

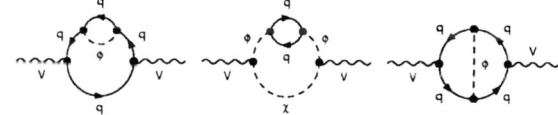

FIGURE 1. Generic diagrams for $O(G_F^2 m_t^4)$ corrections to $\Delta\rho$. ($V = Z, W, f = t, b, \phi, \chi = h, H, A, H^\pm, G, G^\pm$)

The diagrams and the corresponding amplitudes have been generated with the help of the *Mathematica* program *FeynArts* and the recently accomplished MSSM model file. This model file contains all relevant information about the MSSM particles and vertices. (Only counter term vertices are missing at present.) *FeynArts* has been checked in several ways to insure its reliability. Apart from many self-energy calculations, whole processes like $e^+e^- \to t\bar{t}$, $e^+e^- \to H^+H^-$, $q\bar{q} \to t\bar{t}$ and $e^+e^- \to \tilde{\chi}^{0,+}\tilde{\chi}^{0,-}$ have been calculated by hand and with *FeynArts*. Perfect agreement has been found for all processes, see Refs. (5, 7).

How easy the evaluation of the amplitudes has become with *FeynArts* is demonstrated in the following sequence from a *Mathematica* session where the Z boson self-energy amplitude, corresponding to Fig. 1, has is obtained. (A detailed guide can be found in Ref. (5).)

(start of *Mathematica* session)

```
> <<FeynArts.m;
```
($\to$ loading *FeynArts* into *Mathematica*)

```
> se2 = CreateTopologies[2, 1->1,
        ExcludeTopologies->Internal];
```
($\to$ two-loop topologies are created)

```
> V2V2 = InsertFields[se2,
         V[2] -> V[2],
         Model->"MSSM",
         ExcludeParticles -> {...} ];
```
($\to$ fields are inserted into the topologies, incoming field V[2] = Z and outgoing field V[2] are specified, the model is chosen to be the MSSM)

```
> Paint[V2V2];
```
($\to$ the diagrams are painted (optional), see Fig. 1)

```
> V2V2A = CreateFeynAmp[V2V2];
```
($\to$ Feynman diagrams are converted into amplitudes)

```
> V2V2A >> V2V2.amp;
```
($\to$ result for the two-loop amplitude of the Z self-energy is saved in the file V2V2.amp)
(end of *Mathematica* session)

The further evaluation of the amplitudes has been performed with the *Mathematica* program *TwoCalc*, see Ref. (6). It performs the reduction of the self-energy amplitude at the two-loop level to a basic set of scalar integrals. The application of *TwoCalc* is demonstrated in the following sequence from a *Mathematica* session where the Z boson self-energy amplitude is processed.

(start of *Mathematica* session)

```
> <<TwoCalc.m;
```
($\to$ loading *TwoCalc* into *Mathematica*)

```
> amp =<<V2V2.amp;
```
($\to$ the amplitude obtained with *FeynArts* is loaded)

```
> SetOptions[ TwoLoop,
              CollectFunction -> 0];
```
($\to$ options are set. CollectFunction allows to choose between DREG and DRED)

```
> res = TwoLoopSum[amp,
        SelfEnergyPart -> 1];
```
($\to$ the amplitude is reduced to scalar integrals)

```
> res = Collect[res,
        {A0[_], T[__]}, Simplify];
```
($\to$ the result is simplified (optional))

```
> res >> V2V2.res;
```
($\to$ the result for the Z boson self-energy amplitude in terms of scalar integrals is saved in V2V2.res)
(end of *Mathematica* session)

The analytically obtained result now consists of scalar integrals at the one- and at the two-loop level. In the case presented here, since the external momentum is set to zero, these are the functions A_0 and T_{134} at the one- and two-loop level, respectively, see Refs. (8, 9).

The pure two-loop diagrams have to be supplemented with the corresponding one-loop diagrams with counter term insertion. In order to obtain the $O(G_F^2 m_t^4)$ corrections an expansion of the amplitudes up to $O(m_t^4/M_W^4)$ had to be performed. All remaining $M_W = M_Z c_W$ have been set to zero. Furthermore we had to apply the MSSM sum rules for Higgs boson masses, especially implying for the case $M_Z, M_W \to 0$ that the lightest MSSM Higgs boson has the mass $m_h = 0$ at tree level. After adding up all contribution, the expansion of the result in terms of $\delta = (4-D)/2$ leads to a finite result in the limit $\delta \to 0$. All relevant details can be found in Ref. (4).

RESULTS FOR $\Delta\rho$ AND M_W

After extracting the prefactor of m_t^4/M_W^4 and setting M_W to zero, besides $s_\beta = \tan\beta/\sqrt{1+\tan^2\beta}$ only two mass scales remain: the top quark mass, m_t, and the mass of the $\mathcal{CP}$-odd Higgs boson, M_A. The result for $\Delta\rho$ can be conveniently expressed in terms of $a \equiv m_t^2/M_A^2$:

$$\Delta\rho_1^{\text{SUSY}} = 3\frac{G_F^2}{128\pi^4}m_t^4\frac{1-s_\beta^2}{s_\beta^2 a^2} \times$$

$$\left\{ \text{Li}_2\left(\left(1-\sqrt{1-4a}\right)/2\right)\frac{8}{\sqrt{1-4a}}\Lambda \right.$$

$$-2\text{Li}_2\left(1-\frac{1}{a}\right)\left[5-14a+6a^2\right]$$

$$+\log^2(a)\left[1+\frac{2}{\sqrt{1-4a}}\Lambda\right] - \log(a)\left[2-20a\right]$$

$$-\log^2\left(\frac{1-\sqrt{1-4a}}{2}\right)\frac{4}{\sqrt{1-4a}}\Lambda$$

$$\left. +\log\left(\frac{1-\sqrt{1-4a}}{1+\sqrt{1-4a}}\right)\sqrt{1-4a}(1-2a) \right.$$

$$-\log(|1/a - 1|)(a-1)^2$$
$$+\pi^2\left[\frac{2\sqrt{1-4a}}{-3+12a}\Lambda + \frac{1}{3} - 2a^2\frac{s_\beta^2}{1-s_\beta^2}\right]$$
$$\left.-17a + 19\frac{a^2}{1-s_\beta^2}\right\} \quad (3)$$

with $\Lambda = 3 - 13a + 11a^2$. As a consistency check, in the limit of $M_A \to \infty$, $a \to 0$, we obtain

$$\Delta\rho_1^{\mathrm{SUSY}} = 3\frac{G_F^2}{128\pi^4}m_t^4\left[19 - 2\pi^2\right]. \quad (4)$$

This is the SM result with $M_H \to 0$. It shows that the MSSM decouples to the SM limit (also at the two-loop level) when the new scale M_A is made large.

With the help of eq. (2) the shift in the W boson mass can be evaluated. The pure SUSY contribution turns out to be small, see Ref. (4). More important, however, is the effective shift from the SM result, which has been used also for the evaluation of the $O(G_F^2 m_t^4)$ corrections within the MSSM (where the SM Higgs boson mass has been set to m_h, the mass of the lightest MSSM Higgs boson), and the new MSSM result, eq. (3). In Fig. 2 this effective change in M_W is shown as a function of m_h.

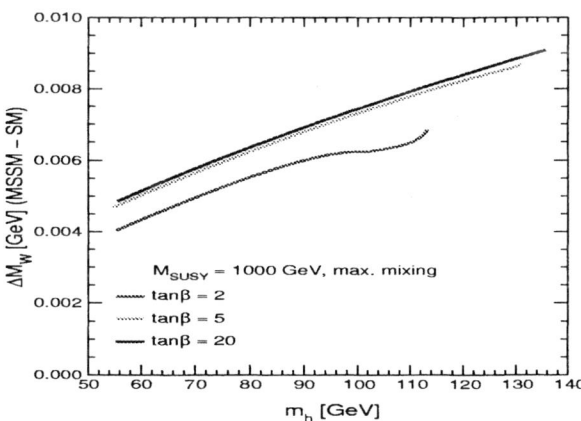

FIGURE 2. The effective change in M_W is shown as a function of m_h for $\tan\beta = 2, 5, 20$.

For the evaluation of m_h in the MSSM we made use of the currently most accurate calculation based on a Feynman-diagrammatic calculation, see Ref. (10), where the result is implemented into the Fortran code *FeynHiggs*, see Ref. (11). The common SUSY mass scale has been set to $M_{SUSY} = 1000\,\mathrm{GeV}$, the off-diagonal entry in the scalar top mass matrix has been set to $X_t = 2M_{SUSY}$ ("maximal mixing"). The Higgs mixing parameter has been fixed to $\mu = 200\,\mathrm{GeV}$. For $m_h \gtrsim 90\,\mathrm{GeV}$, the current LEP2 limit as given in Ref. (12), this results in an effective change in M_W of 5 – 9 GeV.

CONCLUSIONS

We presented the calculation of the SUSY contributions of $O(G_F^2 m_t^4)$ to the ρ-parameter. In order to obtain these two-loop corrections, the computer algebra programs *FeynArts* and *TwoCalc* have been used. Examples of the handling of these programs were given. A compact analytical result in terms of m_t^2/M_A^2 has been derived. The effective change from the previously used SM result to the new MSSM result is an upward shift in M_W of about 5 – 9 GeV.

ACKNOWLEDGMENTS

S.H. thanks T. Hahn, C. Schappacher and other members of the TP, Universität Karlsruhe, Germany, for their effort put into *FeynArts* and the new MSSM model file. S.H. gratefully acknowledges the collaboration with G. Weiglein, with whom the results presented here have been obtained.

REFERENCES

1. Part. Data Group, *Eur. Phys. Jour.* **C15** (2000) 1.
2. M. Veltman, *Nucl. Phys.* **B123** (1977) 89.
3. A. Djouadi, P. Gambino, S. Heinemeyer, W. Hollik, C. Jünger and G. Weiglein, *Phys. Rev. Lett.* **78** (1997) 3626; *Phys. Rev.* **D57** (1998) 4179.
4. S. Heinemeyer and G. Weiglein, *in preparation*; S. Heinemeyer, talk given at the RADCOR2000, radcor2000.slac.stanford.edu/talks/heinemeyer.ps
5. T. Hahn, hep-ph/0012260; (see www.feynarts.de).
6. G. Weiglein, R. Scharf and M. Böhm, *Nucl. Phys.* **B416** (1994) 606.
7. see www.hep-processes.de.
8. G. Passarino and M. Veltman, *Nucl. Phys.* **B160** (1979) 151.
9. A. Davydychev and J. Tausk, *Nucl. Phys.* **B397** (1993) 123; F. Berends and J. Tausk, *Nucl. Phys.* **B421** (1994) 606.
10. S. Heinemeyer, W. Hollik and G. Weiglein, *Phys. Rev.* **D58** (1998) 091701; *Phys. Lett.* **B440** (1998) 296; *Eur. Phys. Jour.* **C9** (1999) 343.
11. S. Heinemeyer, W. Hollik and G. Weiglein, *Comp. Phys. Comm.* **124** (2000) 76; hep-ph/0002213; (see www.feynhiggs.de)
12. S. Andringa et al., ALEPH 2000-074 CONF 2000-051, DELPHI 2000-148 CONF 447, L3 Note 2600, OPAL Technical Note TN661.

Very Large Scale Computing

CompHEP-PYTHIA interface: integrated package for the collision events generation based on exact matrix elements

A.S Belyaev, E.E. Boos, A.N. Vologdin, M.N. Dubinin, V.A. Ilyin, A.P. Kryukov,
A.E. Pukhov, A.N.Skachkova, V.I. Savrin, A.V. Sherstnev, S.A. Shichanin

Skobeltsyn Institute of Nuclear Physics, Moscow State University, Moscow 119899, Russia

Abstract. CompHEP, as a partonic event generator, and PYTHIA, as a generator of final states of detectable objects, are interfaced. Thus, integrated tool is proposed for simulation of (almost) arbitrary collision processes at the level of detectable particles. Exact (multiparticle) matrix elements, convolution with structure functions, decays, partons hadronization and (optionally) parton shower evolution are basic stages of calculations. The PEVLIB library of event generators for LHC processes is described.

In the widely used generators PYTHIA [1], ISAJET [2] and HERWIG [3] data bases of matrix elements of hard subprocesses are built in. It means that matrix elements are stored as formulas. Furthermore, the matrix element squared $|M|^2$ is represented by means of some function modelling the behaviour of the integrand to get effective Monte-Carlo integration and events generation. Thus, as one can see, mainly $2 \to 2$ subprocesses are included in these data bases.

However, the generation of events with 3, 4 and more bodies in the final states of hard subprocesses is needed for the Tevatron, LHC and future linear collider physics. One can note, in particular, that for such states there is no possibility to construct simple analytical formulae to match singular behaviour of $|M|^2$. Multidimensional phase space (4 dimensions in the 3-body case plus 2 dimensions in case of hadron collisons for convulation with parton distributions, 7+2 dimensions in the 4-body case etc) with untrivial regions corresponding to the singularities of the martrix element leads to complicated symbolic structures. Thus, one needs a new approach to the generation of events at the partonic level.

Partonic level final states with top quarks, Higgs bosons and intermediate vector bosons, like Wtj, ttH, Wbb, $ttbb$ and $tttt$, give practical examples of multidimensional phase space integration. Then, multiple production of light quarks, gluons, leptons and photons also assumes the evaluation of multiparticle final states if one is interested in effects at high p_T, large invariant masses etc. This problem stands as an important one especially if one should evaluate precisely background processes.

The singular behaviour of the phase space integrand is connected with singular behaviour of propagators in Feynman diagrams: some masses are extremely small (light quarks and leptons) or even zero (gluons and photons), while other masses are of order 100 GeV (M_W, M_Z, M_{top} ...) and collison energies are of hundreds or even thousands of GeV. As a result, huge energy scale difference for the parameters involved produces serious computational problems. One has to regularize the integration measure to smooth the singularities (see the discussion in [4]).

To get more partons in final state one can exploit, for example, the QCD *parton-shower* generation. However, this method is good only for soft regimes (small p_T, small angles etc.) and fails in case of hard production of these extra partons.

We propose to use programs created for automatic computation of matrix elements, like CompHEP [5], GRACE [6] and MADGRAPH [7][1] as a tool for generation of data base of hard subprocesses for generators like PYTHIA, ISAJET and HERWIG. In particular, at the step of the evaluation of hard subprocess the phase space grid is adapted to match a singular behaviour of $|M|^2$. So, we develop the *two stage* approach:

1. CompHEP produces cross sections and proper phase space grid. This information is stored in special data base, **PEVLIB**, on the hard disk;

2. events generated by CompHEP are used as an input to generators PYTHIA, ISAJET and HERWIG, for further decaying and hadronization of final partonic states.

---

[1] see also programs presented on this Workshop by T. Ohl and C. Papadopoulos [8, 9].

In this paper we present the interface to provide an automatic input of CompHEP events to PYTHIA. This interface is under the development still and new options are assumed to be realized, in particular, automatic addition of new events in the data base if the number of events already stored in the existing sample is not enough, and the regeneration of events in case of change of physical parameter set.

Some comments why we propose to separate the *matrix element* and *decay/showers/hadronization* steps of the computations:

- it is easy to automate the interface;

- more flexibility of the computation model is reached:

    - it allows to develop/implement new options in these two steps independently (what corresponds to the standard theoretical approach: at the "matrix element" step quantum effects are evaluated, interference etc, while the second step corresponds to the probability processes);

    - it gives a possibility to input partonic events in different programs for the second step (PYTHIA, ISAJET, HERWIG etc), as well as to create partonic events by different programs, e.g. CompHEP, GRACE, MADGRAPH etc. Of course the standardization of the partonic event files is necessary.

In PYTHIA we use the subroutine PYUPEV to input CompHEP events as an external process.

CompHEP generates events (unweighted in v.41) and writes them to the file

events_N.txt

where N is the number of working session. This is the text file with a header, where information about the subprocess is given, cross section value is written, and some information about the beams is presented (in particular what PDF set was used and what is the QCD scale). Then, events are written (one event - one line).

Then, a command

mixPEV

is used to mix several subprocesses in one event flow according to their relative weight, $\sigma_i/\Sigma_j\sigma_j$, where σ_i is the cross section of the i-th subprocess. This command, mixPEV, randomizes also the position of events from different subprocesses in the final event flow.

As a result of the command mixPEV the file Mixed.PEV with the events from mixed subprocesses is created. In this file headers of all subprocesses contributed are listed in the beginning and then events are written. In the end of each line (event) the information about color flows is given, allowing the user to switch on, for example, Lund Fragmentation Model or other models using color flows in the $N_c \to \infty$ limit. This new option is realized in CompHEP v.41. The event line includes also: the number of subprocess, to which the event belongs, and components of the particles momenta (for in-coming particles only their z-components).

When the command mixPEV is completed the protocol file, Prt.PEV, is created.

The CompHEP code is available from the following addresses:

http://theory.npi.msu.su/comphep/

/afs/cern.ch/cms/physics/COMPHEP/V_41.10.tar.gz

The CompHEP-PYTHIA interface code and command mixPEV are available from:

/afs/cern.ch/cms/physics/PEVLIB/cpyth

where one can find the Fortran code in directory interf45 (for interface with PYTHIA 5.7/JETSET 7.4) and interf46 for interface with PYTHIA v.6.x. The code for the command mixPEV is placed in the directory c_source (the mix.c file).

One should compile Fortran routine from the directory interf46 (or correspondingly from interf45) and link with the PYTHIA object code. For some platforms a user should comment the dummy routine PYUPEV in the original PYTHIA code.

For more details on the PYTHIA switches and other comments we refer the reader to the file main.f, where many comment lines are given with corresponding explanations.

Using the CompHEP-PYTHIA interface we create the PEVLIB library of partonic events for LHC processes. Each process is stored in the subdirectory in

/afs/cern.ch/cms/physics/PEVLIB/

In each process subdirectory there is the README file where some details are given concerning the partonic events computed.

Among processes already stored in PEVLIB are the following:

- SingleTop (see details about the number of events generated and subprocesses in the corresponding README file);

- $(H \to \tau^+\tau^-) + 2\,tag\,jets$. Here electroweak background events can be found (about 10^5 with Z off-shell and $4 \cdot 10^5$ with Z on-shell), and QCD background (about 65000 events);

- $(H \to b\bar{b}) + t\bar{t}$. Here signal events are generated for $M_H = 100$, 115, 120 and 130 GeV, as well as the events for background processes $t\bar{t}b\bar{b}$ (about 300000 events) and $t\bar{t} + 2\,jets$ (about 2 millions events);

This work was supported by the INTAS-CERN 99-0377 grant and by the Programme "University of Russia" (grant 990588), and done in the general framework of the CPP Collaboration (http://wwwlapp.in2p3.fr/cpp/cpp.html).

REFERENCES

1. T. Sjostrand, Comp. Phys. Comm. 82 (1994) 74

2. F.E.Paige et al., *ISAJET 7.40: a Monte Carlo event generator for pp, p̄p, and e^+e^- reactions*, BNL-HET-98-39, Oct 1998, hep-ph/9810440.

3. G.Marchesini et al., *HERWIG VERSION 5.9.*, hep-ph/9607393.

4. V.A. Ilyin, D.N. Kovalenko, A.E. Pukhov, Int. J. Mod. Phys. C7 (1996) 761.

   D. Kovalenko, A. Pukhov, Nucl. Instr. and Meth. A389 (1997) 299.

5. A.Pukhov et al, *CompHEP v.33 User's manual*, Preprint INP MSU 98-41/542, hep-ph/9908288

6. T. Ishikawa et al. *GRACE manual*, KEK Report 92-19, February 1993.

7. T. Stelzer and W.F. Long. Comp. Phys. Commun. **81** (1994).

8. T. Ohl, *O'Mega: An Optimizing Matrix Element Generator*, hep-ph/0011243.

   T. Ohl, *O'Mega&WHIZARD: Monte Carlo Event Generator Generation For Future Colliders*, hep-ph/0011287.

9. A. Kanaki and C.G. Papadopoulos, *HELAC-PHEGAS: automatic computation of helicity amplitudes and cross sections*, hep-ph/0012004; Comput. Phys. Commun. **132** (2000) 306.

Integration of GRACE and PYTHIA*

K. Sato,[a] S. Tsuno,[a] J. Fujimoto,[b] T. Ishikawa,[b] Y. Kurihara[b] and S. Odaka[b,†]

[a]*Institute of Physics, University of Tsukuba, Tsukuba, Ibaraki 305-8571, Japan*
[b]*High Energy Accelerator Research Organization (KEK), Tsukuba, Ibaraki 305-0801, Japan*

Abstract. We have successfully developed a technique to integrate an automatic event-generator generation system GRACE and a general-purpose event generator framework PYTHIA. The codes generated by GRACE are embedded in PYTHIA in the created event generator program. The embedded codes give information on parton-level hard interactions directly to PYTHIA. The choice of PDF is controlled by the ordinary parameter setting in PYTHIA. This technique enables us to create easy-to-handle event generators for any processes in hadron collisions. Especially, in virtue of large capability of GRACE, we can easily deal with those processes containing many (four or more) partons in the final state, such as multiple heavy particle productions. This project is being carried out as a collaboration between the Japanese Atlas group and the Minami-Tateya group, aiming at developing event generators for Tevatron and LHC experiments.

GRACE [1] is a software package for automatic calculation in high energy physics, developed by the Minami-Tateya group. The core part of the package is a program to generate Feynman diagrams relevant to specified initial and final states. It then generates FORTRAN codes for calculating the corresponding cross section on the basis of the amplitude of each diagram. The GRACE package also includes an integration and event-generation program called BASES/SPRING [2]. Hence, it provides us with a very powerful environment for developing event generators for studies of high energy physics.

Since the basic building block is the amplitude, GRACE has an advantage in those processes composed of many coherent diagrams. (The number of elements increases quadratically if the calculation is based on the matrix element.) This turns out to be an advantage in multi-body production processes, since in general the number of contributing diagrams increases as the number of final-state particles increases. The program `grc4f`[3] is a good example. This is known to be the most reliable event generator at the tree level for four-fermion productions in electron-positron collisions at LEP2 energies.

In future hadron-collision experiments such as LHC, interesting processes (Higgs-boson productions, SUSY-particle productions etc.) in many cases result in multiple production of heavy particles (Z/W and/or top quark) or cascade decays. They produce many particles in the final state, and are composed of many coherent diagrams. For example, we have to take into account 144 diagrams in total to evaluate the process $pp \to H^0 b\bar{b} + X \to b\bar{b}b\bar{b} + X$. Detailed studies of such processes require an exact (coherent) treatment of the whole production and decay reactions. GRACE is expected to be a powerful tool for such studies.

However, GRACE can treat perturbative hard-scattering reactions only. We need to add non-perturbative partonic structures of hadrons (PDF) and QCD evolutions in the initial and final states, in order to construct realistic event generators. The most straightforward way to accomplish it is to connect GRACE to a general purpose event generator, such as PYTHIA [4], ISAJET [5] and HERWIG [6]. We chose PYTHIA in this work.

There would be two ways for the connection. One way is to interface them using an external data file, in which hard-scattering event data provided by GRACE are described. Such a system can be flexible and portable. The coding may be easier since only the I/O routines are necessary to be coded. However, since the generation procedure is separated to two steps, a special care or an appropriate software assistance is necessary to keep the consistency in the parameters. This method is used in GRAPE [7], a previous work of the Minami-Tateya group for electron-proton collisions. A similar method is also adopted in the CompHEP for hadron collisions [8].

The second way is to embed the GRACE output codes in PYTHIA. Although the coding may be more complicated, the event generation can be a single-step job in this method. Since necessary parameters can be set through the parameter setting in PYTHIA, any inconsistency in the parameter choice can be automatically avoided.

---

* Presented by S. Odaka.
† E-mail: shigeru.odaka@kek.jp

We chose the latter method in the present work. All the program codes describing the hard scattering are embedded in the subroutine PYUPEV, prepared by PYTHIA to install user-defined processes. This subroutine is called during the event generation if it is requested using PYUPIN in the initialization stage.

For the event generation, it is important to give an appropriate definition of "kinematics", the mapping of kinematical variables to a set of uniform random numbers. We have tried two methods for it. In the first method, the mapping is fully defined by user-defined analytic functions. The CPU time may be saved if the functions are appropriately defined. The integration/event generation package BASES/SPRING is fully utilized in the second method, where a variable-grid mapping is implemented. Since the grids are optimized by BASES, users need not to care about very details of the mapping. Of course, in both methods, users need to choose an appropriate set of variables to avoid non-diagonal singularities.

The subroutine PYUPEV that we have coded can be separated to two parts, an initialization stage and an event generation stage. The initialization stage is called by PYUPIN once in each job. The maximum of SIGEV (see below) is searched to set the return argument SIGMAX in the functional-mapping method. In the grid-mapping method, BASES is called to estimate the total cross section and SIGMAX is set to be equal to it. The grids are optimized here.

PYTHIA calls the second stage in the event-generation loop with a frequency corresponding to SIGMAX. In this stage, first of all, one of the "sub-processes" is chosen. In most cases in hadron collisions, every "process" of interest is composed of several incoherent sub-processes, since hadrons are composite and parton species in the final state are hard to identify.

The following procedure is different for two different methods of "kinematics". In the functional-mapping method, a set of kinematical variables defining an event is determined from a set of uniform random numbers according to the defined "kinematics". The differential cross section of this event is calculated using the GRACE output codes. The return argument SIGEV is calculated from the differential cross section, PDF and the Jacobian of the "kinematics".

Instead, the event generator SPRING is called in the grid-mapping method. SPRING generates an event using the GRACE output codes, PDF and the grid information optimized by BASES. SIGEV is always equal to SIGMAX.

In both methods, the above procedure is followed by a Lorentz boost to the laboratory frame and a determination of the color flow. It should be noted that the color flow can be determined automatically using the information from BASES.

After returning to PYTHIA, an event sampling is done according to the weight SIGEV/SIGMAX. The sampling is dummy (i.e., all events are accepted) in the grid-mapping method. After that, the initial- and final-state parton radiations are added by PYTHIA to simulate the effect of QCD evolution, resulting in an underlying hadronic activity and a finite transverse momentum of the hard-scattering system.

The coding has been refined by developing actual event generators. So far, we have developed generators for the following processes:

$$pp(\text{or}\, p\bar{p}) \to q\gamma + X, \tag{1}$$

$$pp(\text{or}\, p\bar{p}) \to Wg + X \to \mu\nu g + X, \tag{2}$$

$$pp(\text{or}\, p\bar{p}) \to H^0 W + X \to b\bar{b}\mu\nu + X, \tag{3}$$

$$pp(\text{or}\, p\bar{p}) \to b\bar{b}(\text{QCD})W + X \to b\bar{b}\mu\nu + X, \tag{4}$$

$$pp(\text{or}\, p\bar{p}) \to H^0 b\bar{b} + X \to b\bar{b}b\bar{b} + X, \tag{5}$$

$$pp(\text{or}\, p\bar{p}) \to b\bar{b}b\bar{b}(\text{QCD}) + X. \tag{6}$$

The results were compared with those from other existing generators, PYTHIA built-in generators and CompHEP, to examine the coding. We found reasonable agreements in all cases.

As an example, the performance of a developed event generator is compared with that of a PYTHIA built-in generator in Table 1 for one of the simplest cases, Process (2). In this study, the weak boson was not made to decay and the simulations of both initial- and final-state parton radiations were turned off. The two "kinematics" methods were tried.

The cross section is in very good agreement, showing that the integration of GRACE and PYTHIA is done successfully. The functional-mapping method gives a better performance in CPU time than the grid-mapping method, as expected. Although both GRACE + PYTHIA generators consume appreciably longer CPU time than the PYTHIA built-in generator, the difference is not very serious since the simulation of parton radiations and hadronization, which are not implemented in this study and common to all three generators, takes much longer time.

The most noticeable advantage of the GRACE + PYTHIA system is in the fact that using this technique we can easily develop event generators for those processes which are not and/or hard to be implemented in PYTHIA. Process (5) is one of such processes. The total cross section estimated by using the developed generator is shown in Table 2 for three assumed Higgs-boson masses, and compared with the result of CompHEP. We have applied the grid-mapping method only in this case. The agreement with CompHEP is quite good.

Table 1. Performance of a developed event generator for the process $pp \to Wg + X$ ($\sqrt{s} = 14$ TeV, $p_T(g) \geq 5$ GeV/c). The results are compared with those of a PYTHIA built-in generator. A Linux-PC with 300 MHz-Pentium II is used. The weak boson is not made to decay. The initial- and final-state parton radiations are not simulated. The full simulation consumes another 45 minutes for 100k events.

	Functional mapping	Grid mapping	PYTHIA(ISUB=16)
Total cross section (nb)	63.36± 0.20	63.43± 0.13	63.17± 0.20
Generation efficiency (%)	19	35	19
CPU time for 100k events (min)	12.5	20.3	4.6

Table 2. Comparison between GRACE and CompHEP for the process $p\bar{p} \to H^0 b\bar{b} + X \to b\bar{b}b\bar{b} + X$ at $\sqrt{s} = 2$ TeV. The total cross section is evaluated for three cases of the Higgs-boson mass.

$M(H^0)$ (GeV)	GRACE (fb)	CompHEP (fb)
80	6.006	6.083
120	0.989	1.002
160	0.357	0.356

We plan to make some improvements in order to make the development easier. So far, routines interfacing PYTHIA and GRACE are written by hand. Since the functionality of these routines are almost common to all generators, we will be able to make them automatically generated by changing the libraries referred by GRACE.

Processes in hadron collisions are in most cases composed of several incoherent sub-processes. The present GRACE system cannot handle such processes automatically. Some modifications and additions by hand are necessary now. We would like to automate these tasks.

The difference between the sub-processes is, in most cases, the difference in quark species in the initial and/or final states. If we can treat quark masses and their couplings as variables, these sub-processes can share an identical code for the calculation. The Cabbibo-Kobayashi-Masakawa matrix can be implemented automatically if such a treatment is realized.

In summary, we have established a technique to embed GRACE output codes in PYTHIA. This technique allows us to develop new hadron-collision event generators easily. We have applied the technique to some two-, three- and four-body production processes. Obtained results are in good agreement with the results from existing generators. The generator for four b-quark productions, Processes (5) and (6), will be released in a few months. Since GRACE is expected to be advantageous in multibody production processes, we would like to go to five- or six-body production processes as the next step. We also have an automatic next-to-leading order (NLO) calculations for hadron-collision processes in our view.

This project is being carried out as a collaboration between the ATLAS-Japan group composed of Japanese members of the ATLAS collaboration at LHC, and the Minami-Tateya group, aiming at developing event generators for Tevatron and LHC. K.S., S.T. and S.O. are from the ATLAS-Japan group, and J.F., T.I. and Y.K. from the Minami-Tateya group. There are many other people from these two groups who have contributed to this work. Among them, the authors wish to thank here Y. Takaiwa, S. Kawabata and K. Kato for their educational contributions, and T. Abe for discussions.

REFERENCES

1. T. Ishikawa et al., GRACE manual, KEK Report 92-19 (1993); F. Yuasa et al., Prog. Theor. Phys. Suppl. 138, 18 (2000); hep-ph/0007053.
2. S. Kawabata, Comput. Phys. Commun. 41, 127 (1986); Comput. Phys. Commun. 88, 309 (1995).
3. J. Fujimoto et al., Comput. Phys. Commun. 100, 128 (1997).
4. T. Sjostrand, Comput. Phys. Commun. 82, 74 (1994). We used an updated version (6.138) for the present work.
5. H. Baer, F.E. Paige, S.D. Protopopescu and X. Tata, hep-ph/0001086.
6. G. Marchesini et al., Comput. Phys. Commun. 67, 465 (1992); G. Corcella et al., hep-ph/9912396.
7. T. Abe, hep-ph/0012029, to appear in Comput. Phys. Commun.; T. Abe et al., Proc. Workshop on Monte Carlo Generators for HERA Physics (Plenary Starting Meeting), DESY-PROC-1999-02 (1999) p. 566.
8. V.A. Ilyin, A.E. Pukhov and A.N. Skachkova, talk in this workshop; E.E. Boos et al., Proc. the Second Int. Workshop on Software Engineering, ed. D. Perret-Gallix (World Scientific, Singapore, 1992) p. 665.

Algorithm for Computing Excited States in Quantum Theory

X.Q. Luo[1],* H. Jirari[2], H. Kröger[2], and K. Moriarty[3]

[1] *Department of Physics, Zhongshan University, Guangzhou 510275, China*
[2] *Département de Physique, Université Laval, Québec, Québec G1K 7P4, Canada*
[3] *Department of Mathematics, Statistics and Computer Science, Dalhousie University, Halifax, Nova Scotia B3H 3J5, Canada*

Abstract. Monte Carlo techniques have been widely employed in statistical physics as well as in quantum theory in the Lagrangian formulation. However, in the conventional approach, it is extremely difficult to compute the excited states. Here we present a different algorithm: the Monte Carlo Hamiltonian method, designed to overcome the difficulties of the conventional approach. As a new example, application to the Klein-Gordon field theory is shown.

INTRODUCTION

There are two standard formulations in quantum theory: Hamiltonian and Lagrangian. A comparison of the conventional approaches is given in Table 1.

Monte Carlo (MC) method with importance sampling is an excellent non-perturbative technique to calculate path integrals in quantum theory. In the last two decades, it has successfully been applied to Lagrangian lattice gauge theory [1, 2, 3]. In the standard Lagrangian MC method, however, it is extremely difficult to compute the spectrum and wave function beyond the ground state. On the other hand, the standard Hamiltonian formulation is capable of doing it.

Recently, we proposed an algorithm to construct an effective Hamiltonian from Lagrangian MC simulations in Ref. [4]. We called it the MC Hamiltonian method. The advantage, in comparison with the standard Lagrangian MC approach, is that one can obtain the spectrum and wave functions beyond the ground state. It also allows to do thermodynamics. In this paper, we briefly review what we have done, and present some new results.

ALGORITHM

Effective Hamiltonian

Let us review briefly the basic ideas of our approach. The (imaginary time) transition amplitude between an initial state at position x_i, and time t_i, and final state at

Table 1. Comparison of the conventional methods.

Formulation	Hamiltonian	Lagrangian
Approach	Schrödinger Eq. $H\|E_n> = E_n\|E_n>$	Path Integral $<O> = \frac{\int d[x] O[x] \exp(-S[x]/\hbar)}{\int [dx] \exp(-S[x]/\hbar)}$
Algorithm	Series expansion, variational approx., Runge-Kutta ...	MC simulation with importance sampling
Advantage	Both ground state, & excited states can be computed.	It generates the important configurations
Problem	Difficult for many body systems.	Difficult for excited states.

x_f and t_f is related to the Hamiltonian H by

$$M_{fi} = <x_f,t_f|x_i,t_i> = <x_f|e^{-H(t_f-t_i)/\hbar}|x_i>$$
$$= \sum_{n=1}^{\infty} <x_f|E_n> e^{-E_n T/\hbar} <E_n|x_i>, \quad (1)$$

where $T = t_f - t_i$. According to Feynman's path integral formulation of quantum mechanics (Q.M.), the transition amplitude is also related to the path integral:

$$M_{fi} = \int [dx] \exp(-S[x]/\hbar)|_{x_i,t_i}^{x_f,t_f}. \quad (2)$$

The starting point of our method, as described in detail in Ref. [4] is to construct an effective Hamiltonian H_{eff} (finite $N \times N$ matrix) by

$$M_{fi} = <x_f|e^{-H_{eff} T/\hbar}|x_i>$$

* Email: stslxq@zsu.edu.cn

$$= \sum_{n=1}^{N} <x_f|E_n^{eff}> e^{-E_n^{eff}T/\hbar} <E_n^{eff}|x_i>. \tag{3}$$

The eigenvalues E_n^{eff} and wave function $|E_n^{eff}>$ can be obtained, by diagonalizing M using a unitary transformation

$$M = U^\dagger D U, \tag{4}$$

where $D = diag(e^{-E_1^{eff}T/\hbar},...,e^{-E_N^{eff}T/\hbar})$. Once the spectrum and wave functions are available, all physical information can also be obtained.

Since the theory described by H, whose basis in Hilbert space is infinite, is now approximated by a theory described by a finite matrix H_{eff}, whose basis is finite, the physics of H and H_{eff} might be quite different at high energy. Therefore we expect that we can only reproduce the low energy physics of the system. This is good enough for our purpose. In Refs. [4, 5, 6, 7], we investigated many 1-D, 2-D and 3-D Q.M. models (Table 2) using this MC Hamiltonian algorithm. We computed the spectrum, wave functions and some thermodynamical observables. The results are in very good agreement with those from analytical and/or Runge-Kutta methods.

Table 2. Q.M. systems, investigated by the MC Hamiltonian method using the regular basis.

System	Potential
Q.M. in 1-D	$V(x) = 0$
	$V(x) = \frac{1}{2}m\omega^2 x^2$
	$V(x) = -V_0 sech^2(x)$
	$V(x) = \frac{1}{2}x^2 + \frac{1}{4}x^4$
	$V(x) = \frac{1}{2}|x|$
	$V(x) = \begin{cases} \infty, & x<0 \\ Fx, & x\geq 0 \end{cases}$
Q.M. in 2-D	$V(x,y) = \frac{1}{2}m\omega^2 x^2 + \frac{1}{2}m\omega^2 y^2$
	$V(x,y) = \frac{1}{2}m\omega^2 x^2 + \frac{1}{2}m\omega^2 y^2 + \lambda xy$
Q.M. in 3-D	$V(x,y,z) = \frac{1}{2}m\omega^2 x^2 + \frac{1}{2}m\omega^2 y^2 + \frac{1}{2}m\omega^2 z^2$

Basis in Hilbert Space

To get the correct scale for the spectrum, the position state $|x_n\rangle$ (Bargman states or box states) at t_i or t_f should be properly normalized. We denote a normalized basis of Hilbert states as $|e_n\rangle$, $n = 1,...,N$. In position space, it can be expressed as

$$e_n(x) = \begin{cases} 1/\sqrt{\Delta x_n}, & x \in [x_n, x_{n+1}] \\ 0, & x \notin [x_n, x_{n+1}] \end{cases} \tag{5}$$

where $\Delta x_n = x_{n+1} - x_n$.

The simplest choice is a basis with $\Delta x_n = const.$, which is called the "regular basis". In Refs. [4, 5, 6, 7], the regular basis is used. For many body system or quantum field theory, the regular basis will encounter problem. For example, in a system with a 1-D chain of oscillators (see later), if the number of oscillators is 30, the minimum non-trivial regular basis is $N = 2^{30} = 1073741824$, which is prohibitively large for numerical calculations.

Guided by the idea of importance sampling, in Refs. [8, 9], we proposed to select a basis from the Boltzmann weight proportional to the transition amplitude between $x_i' = 0$ at $t_i' = 0$ and $x_f' = x_n$ at some t_f'. In one free particle case, the distribution is just a Gaussian

$$P_{basis}[x_n] = \frac{1}{\sqrt{2\pi}\sigma} \exp(-\frac{x_n^2}{2\sigma^2}), \tag{6}$$

where $\sigma = \sqrt{\hbar t_f'/m}$. We call such a basis the "stochastic basis".

Matrix elements

As explained above, the calculation of the transition matrix elements is an essential ingredient of our method. The matrix element in the normalized basis is related to $<x_{n'},t_f|x_n,t_i>$ by

$$\begin{aligned} M_{n'n} &= <e_{n'},T|e_n,0> \\ &= \int_{x_{n'}}^{x_{n'+1}} dx' \int_{x_n}^{x_{n+1}} dx'' \frac{<x',t_f|x'',t_i>}{\sqrt{\Delta x_{n'} \Delta x_n}} \\ &\approx \sqrt{\Delta x_{n'} \Delta x_n} <x_{n'},t_f|x_n,t_i>, \end{aligned} \tag{7}$$

where $<x_{n'},t_f|x_n,t_i>$ can be calculated using MC simulations as follows.
(a) Discretize the continuous time.
(b) Generate free configurations $[x]$ between $t \in (t_i, t_f)$ obeying the Boltzmann distribution

$$P_0[x] = \frac{\exp(-S_0[x]/\hbar)}{\int [dx] \exp(-S_0[x]/\hbar)} |_{x_n,t_i}^{x_{n'},t_f}, \tag{8}$$

where $S_0 = m\dot{x}^2/2$.
(c) Measure

$$<O_V> = \int [dx] \exp(-V(x)/\hbar)|_{x_n,t_i}^{x_{n'},t_f} P_0[x] \tag{9}$$

The path integral in Eq. (7) is then

$$\begin{aligned} <x_{n'},t_f|x_n,t_i> &= <O_V> \times \sqrt{\frac{m}{2\pi\hbar T}} \\ &\quad \times \exp[-\frac{m}{2\hbar T}(x_{n'} - x_n)^2]. \end{aligned} \tag{10}$$

QUANTUM FIELD THEORY

The main purpose of the algorithm is to study many body systems and quantum field theory beyond the ground state. As an example, we consider a chain of N_{osc} coupled oscillators in 1 spatial dimension. Its Hamiltonian is given in Ref. [10] as

$$H = \sum_{j=1}^{N_{osc}} \frac{1}{2}[p_j^2 + \Omega^2(q_j - q_{j+1})^2 + \Omega_0^2 q_j^2], \quad (11)$$

where p_j and q_j are the momentum and displacement of the j-th oscillator respectively. This model is equivalent to the Klein-Gordon field theory on a (1+1)-dimensional lattice.

The spectrum of the system is analytically known:

$$\begin{aligned} E_n &= \sum_{n_k} (n_k + \frac{1}{2})\hbar\omega_k, \\ \omega_k &= \sqrt{\Omega^2(2\sin k/2)^2 + \Omega_0^2}, \end{aligned} \quad (12)$$

where $n_1, ..., n_{N_{osc}} = 0, 1, ...$, and $k = 2\pi l/N_{osc}$ with l an integer between $-N_{osc}/2$ and $N_{osc}/2$.

We generate a stochastic basis according to Eq. (6) with N=1000 configurations $[q_1, ..., q_{N_{osc}}]$ for the initial and final states for $N_{osc} = 9$ oscillators. For the adjustable parameter σ, we set $t'_f = t_f$ for simplicity. (Of course, one should study systematically the dependence of the results on σ). Table 3 compares the spectrum from the MC Hamiltonian with the analytical results for the first 20 states with $\Omega = 1$, $\Omega_0 = 2$, $m = 1, \hbar = 1$, and $T = 2$. They agree very well with the exact ones.

SUMMARY

In this paper, we have tested the MC Hamiltonian method with a stochastic basis in a many body Q.M. system with a chain of coupled oscillators: the Klein-Gordon field theory on a (1+1)-dimensional lattice. The results are very encouraging. We believe that the application of the algorithm to more complicated body systems and quantum field theory will be very interesting.

ACKNOWLEDGEMENTS

X.Q.L. is supported by the National Science Fund for Distinguished Young Scholars (19825117), National Science Foundation, Guangdong Provincial Natural Science Foundation (990212) and Ministry of Education of China. H.K. and K.M. are supported by NSERC, Canada.

Table 3. Comparison of the spectrum of the chain of 9 coupled oscillators, between the MC Hamiltonian method with a *stochastic* basis and the analytic ones.

n	E_n^{eff}	E_n^{Exact}
1	10.904663192168	10.944060480668
2	12.956830557334	12.944060480668
3	12.985023578737	13.057803869484
4	13.044311582647	13.057803869484
5	13.299967341242	13.321601993380
6	13.345480638394	13.321601993380
7	13.552195133687	13.589811791733
8	13.585794986361	13.589811791733
9	13.680136748933	13.751084748745
10	13.744919087477	13.751084748745
11	14.984737011385	14.944060480668
12	15.012353803145	15.057803869484
13	15.057295761044	15.057803869484
14	15.108904652020	15.171547258300
15	15.125356713561	15.171547258300
16	15.187413290039	15.171547258300
17	15.308536490102	15.321601993380
18	15.396255686587	15.321601993380
19	15.420708031412	15.435345382196
20	15.432823810789	15.435345382196

REFERENCES

1. M. Creutz, Quarks, Gluons and Lattices, Cambridge University Press, Cambridge (1983).
2. H. Rothe, Lattice Gauge Theory: an Introduction, World Scientific, Singapore (1992).
3. I. Montvay and G. Münster, *Quantum Fields on a Lattice*, Cambridge University Press, Cambridge (1994).
4. H. Jirari, H. Kröger, X.Q. Luo, and K. Moriarty, Phys. Lett. **A258**, 6 (1999).
5. X.Q. Luo, C. Huang, J. Jiang, H. Jirari, H. Kröger, and K. Moriarty, Physica **A281**, 201 (2000).
6. X.Q. Luo, J. Jiang, C. Huang, H. Jirari, H. Kröger, and K. Moriarty, Nucl. Phys. **B(Proc. Suppl.)83-84**, 810 (2000).
7. J. Jiang, C. Huang, X.Q. Luo, H. Jirari, H. Kröger, K. Moriarty, Commun. Theor. Phys. **34**, 723 (2000).
8. H. Jirari, H. Kröger, C. Huang, J. Jiang, X.Q. Luo, and K. Moriarty, Nucl. Phys. **B(Proc. Suppl.)83-84**, 953 (2000).
9. C. Huang, J. Jiang, X.Q. Luo, H. Jirari, H. Kröger, and K. Moriarty, quant-ph/9912051.
10. E. Henley and W. Thirring, Elementary Quantum Field Theory, McGraw-Hill (1962).

Simple Scaling for Faster Tracking Simulation in Accelerator Multiparticle Dynamics

J. A. MacLachlan

Fermi National Accelerator Laboratory, Box 500, Batavia IL 60510

Abstract. Macroparticle tracking is a direct and attractive approach to following the evolution of a phase space distribution. When the particles interact through short range wake fields or when inter-particle force is included, calculations of this kind require a large number of macroparticles. It is possible to reduce both the number of macroparticles required and the number of tracking steps per unit simulated time by employing a simple scaling which can be inferred directly from the single-particle equations of motion. In many cases of practical importance the speed of calculation improves with the fourth power of the scaling constant. Scaling has been implemented in an existing longitudinal tracking code; early experience supports the concept and promises major time savings. Limitations on the scaling are discussed.

INTRODUCTION

Multiparticle tracking programs have a long history and established utility for modeling the evolution of the longitudinal phase space distributions for particles in accelerators as the particles respond to the rf in acceleration or bunch manipulation. The macroparticle distribution can be used to approximate the evolution of the beam current distribution or fourier spectrum throughout the process being modeled. Collective behavior of the beam is modeled by calculating the beam current every time step and applying the forces arising from it to the single particle motion. Such calculations raise questions about the number of macroparticles needed and the relevant bandwidth for quantities calculated in frequency domain which require careful attention. It is very easy to generate spectacular spurious instabilities by an insufficient number of macroparticles. Recent studies of high brightness injectors and the so-called factory accelerators have renewed interest in these questions.

SCALING CONCEPT

The objective of the scaling is to allow more rapid calculation of the time evolution of the energy and rf phase of particles in a synchrotron or storage ring. A more complete exposition of the principles has been published.[1] What follows is an effort to stimulate interest and possible further developments. Gerasimov[2] has developed scaling rules for Fokker Planck simulations of beam cooling which are likewise used to accelerate macroparticle modeling. Perhaps similiar tactics will work elsewhere.

A suitable single turn map for the macroparticles is[3]

$$\varphi_{i,m} = \varphi_{i,m-1} + \frac{2\pi h \eta}{\beta_s^2 E_s} \varepsilon_{i,m-1} \quad (1)$$

$$\varepsilon_{i,m} = \varepsilon_{i,m-1} + eV(\varphi_{i,m} + \phi_{s,m}) - eV(\phi_{s,m}) \; , \quad (2)$$

where φ is a phase difference between a particle and the synchronous phase ϕ_s, likewise ε is the energy difference between a particle and the synchronous energy E_s, i labels particles, and m labels turns.

By inspection of the map parameters it is plausible that the phase space motion can be accelerated by scaling the phase slip factor η and the potential by the same factor. The potential is not necessarily just that from the rf system; the collective potential produced by the beam current enters identically.

There is a reduction by a factor λ^{-1} in the computing time by speeding up the clock in the scaled calculation. However, scaling up the time means that frequencies associated with the motion, like the rf frequency for example, are also scaled up. A consequence of the frequency scaling is that a resonant potential in the problem, like a higher order cavity mode for example, must be entered into the calculation with its resonant frequency scaled up by the same factor. If this is done, the result of scaled and original mappings are practically indistinguishable from

---

* Work supported by the U.S. Department of Energy under contract No. DE-AC02-76CH03000.

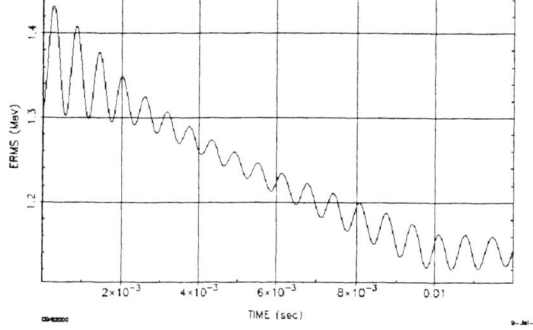 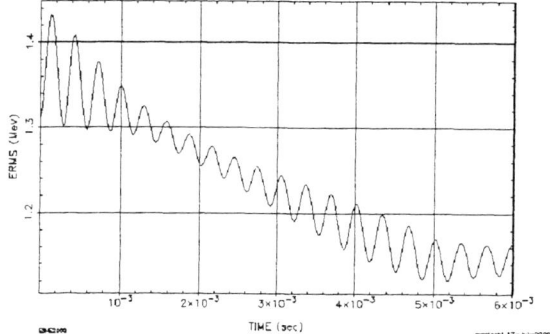

FIGURE 1. (left) The rms energy spread *vs.* time in the ring specified in Table II for a beam intensity of $2 \cdot 10^{13}$ protons including both an h=3 HOM and the perfectly conducting wall space charge force **FIGURE 2.** (right) The same conditions used for Fig. 1 except scaled with $\lambda = 2$ Note that the time scale is $t' = t/\lambda$, one half of that in Fig. 1.

one another for a reasonable choice of λ when the comparison is made at a scaled time $t' = t/\lambda$.

When broadband impedance or the direct interparticle forces are included, the consequences of the frequency scaling are different and very advantageous. When the effects of beam pipe impedance must be covered over some particular frequency bandwidth, or, more or less equivalently wake fields down to some minimum range, the charge distribution must be divided into bins sufficiently narrow to represent detail in the distribution to that scale. The bandwidth of interest must be spanned by fourier components of the beam current. If beam circulation frequency is $f_\circ$ and the required bandwidth is W, the charge distribution must be divided into $2W/f_\circ$ bins for the finite transform. However, in the scaled system $f_\circ$ is λ times higher; the number of bins can be reduced by a factor λ^{-1}. The number of harmonics is reduced, but the frequency range covered is the same. What has been sacrificed is sensitivity to features in the frequency dependence of the longitudinal impedance $Z_\parallel$ on the scale of $f_\circ$. However for space charge there is no sacrifice at all. In frequency domain the equivalent impedance is

$$\frac{Z_{sc}}{n} = -i \frac{Z_\circ g}{2\beta\gamma^2} , \qquad (3)$$

where n is the harmonic number, $Z_\circ = \sqrt{\mu_\circ/\varepsilon_\circ} = 377\Omega$, $\gamma = E/m_\circ c^2$, and g is the geometric factor for the beam tube. The evaluation of this term is not compromised by more widely spaced frequency sampling. For wideband impedances is general, there will be little loss in information for reasonable choices of λ.

What has been gained is the opportunity to reduce the the number of macroparticles required for a given level of numerical noise in the mapping. It has been shown that when a three point difference formula for the derivative of the linear charge density is used in calculating the space charge force in time domain, the number of macroparticles can be reduced by a factor λ^{-3} when the number of bins is reduced by the factor λ^{-1}.[4] An argument based on sampling in the frequency domain shows that more generally this result applies to any source of longitudinal impedance with a frequency dependence featureless on the scale of $\Delta f \sim$ few $\times f_\circ$.[5] Thus, the big payoff comes not by speeding up the clock but by reducing the number of particles. However, if the important impedance is a narrow resonance, the macroparticle count can not be reduced by scaling, leaving only the first power effect. Nonetheless, for a broad class of applications the gain from scaling will be a factor λ from the reduced number of iterations and an additional factor of λ^3 from the reduced macroparticle count.

NUMERICAL COMPARISON

The effectiveness of the scaling is illustrated by comparing an un-scaled and a scaled calculation of the evolution of the energy spread in a storage ring. The parameters, similliar to those of the Los Alamos Proton Storage Ring, are given in the Table. The collective potential is generated by a hypothetical spurious cavity mode at h=3 and the perfectly conducting wall force from $2 \cdot 10^{13}$ protons. To evaluate the space charge force, the gradient of the charge distribution is calculated; there must be enough macroparticles to produce a smooth distribution. An insufficient number results in large local fluctuation of the force and spurious breakup of the beam into small clumps. In Figs. 1 and 2 the evolution of the rms energy spread is shown as the charge is raised from zero to a final value of $2 \cdot 10^{13}$ protons over .012 seconds. The un-scaled example in Fig. 1 needed $2 \cdot 10^6$ macroparticles and took 32 hours of processor time on a Sun Ultra 2. The plot in

Fig. 2 was generated from tracking with the scaling $\lambda = 2$ and used $2.5 \cdot 10^5$ macroparticles; it required 1.9 hours on the same computer, just a little better than λ^4 times faster. The choice of two for λ was arbitrary; too large a value can give incorrect results.

UTILITY

If the scaling factor λ can be significantly larger than one without reducing the accuracy of the result, much time can be saved in macroparticle modeling. One limitation on the choice of λ has been noted, *viz.*, that the spacing of harmonics of the circulation angular frequency $\omega_\circ$ should not obscure important features of the frequency dependence of $Z_\parallel(\omega)$. Returning to the difference equations eqs. 1 and 2, one sees that scaling by integral λ is equivalent to a multi-turn map. This indicates that an additional feature of the scaling with $\lambda > 1$ is the introduction of an artificial increase in the effective time step. This artificial discretization error leads to inaccurate macroparticle trajectories that can be readily apparent in extreme cases. The parameter to control is the synchrotron oscillation phase advance between energy increments. When this is below 0.01π or so, the calculation should be at least qualitatively reasonable. Another condition, which is generally the same as the limitation on synchrotron oscillation phase advance per iteration is that rf phase slip per iteration should be small with respect to a bunch length, and energy increment per iteration should be small with respect to the beam energy spread. The scaling described is so simple to implement and so innocuous in typical applications that it seems reasonable to employ it when calculation time for the un-scaled case is more than a few minutes. The precaution of considering the granularity of the sampling of $Z_\parallel(\omega)$ requires forethought, but the test for the appropriateness of the iteration step and the adjustment of the frequency of any narrow resonances can easily be handled by the modeling code.

The discussion has used mostly a frequency domain description for simplicity. The example, however, is a pure time domain calculation using a slightly modified version of the ESME tracking code.[6] A version now available, called ESME2001, incorporates λ scaling as an option; it appears compatible with all other features in both frequency and time domain. The scaling of potentials, frequency, and time are automatic, but no tests for the appropriateness of the scaling constant have been included.

Table 1. Storage ring parameters used in tracking examples

Parameter	Symbol	Value	Units
Circumference	C	90.261	m
momentum	p	1459.7	MeV/c
transition energy/$m_\circ c^2$	γ_T	3.08	
slip factor $\gamma_T^{-2} - \gamma^{-2}$	η	.18694	
rf peak voltage	V_{rf}	5	kV
rf harmonic	h	1	
circulation frequency	$f_\circ$	2.7940	MHz
HOM cavity mode frequency ($h \approx 3$)	f_{res}	8.4658	MHz
resonant impedance	R_{sh}	300	Ohm
quality factor	Q	100	

ACKNOWLEDGEMENTS

The author is grateful to Zaira Nazario, an undergraduate intern from the University of Puerto Rico at Rio Piedras, for brief but intense work on this subject, especially on the demonstration of invariance of the Vlasov equation under scaling.

REFERENCES

1. J. A. MacLachlan and Z. Nazario, "Scaling for faster macroparticle simulation in longitudinal multiparticle dynamics", Phys. Rev. ST Accel. Beams **3** 114401 (2000)

2. A. Gerasimov, "Macroparticle Simulation for Stochastic Cooling", FERMILAB-Pub-94/376 (1994)

3. J. MacLachlan, "Difference Equations for Longitudinal Motion in a Synchrotron", Fermilab note FN-529 (1989)

4. J. Wei, "Longitudinal Dynamics of the Non-Adiabatic Regime on Alternating Gradient Synchrotrons", PhD dissertation for State University of New York at Stony Brook (1990)

5. J. A. MacLachlan, "Particle Tracking in E-φ Space for Synchrotron Design and Diagnosis", FERMILAB-Conf-92/333 (1992)

6. The current documentation, containing references to the underlying principles, is most accessible on the ESME web page `www-ap.fnal.gov/ESME`

Adaptive Mesh Simulations Of Astrophysical Detonations Using the ASCI Flash Code

B. Fryxell[a], A.C. Calder[a], L.J. Dursi[a], D.Q. Lamb[a], P. MacNeice[b], K. Olson[a,b],
P. Ricker[a], R. Rosner[a], F.X. Timmes[a], J.W. Truran[a], H.M. Tufo[a], and M. Zingale[a]

[a] *ASCI Center for Astrophysical Thermonuclear Flashes, University of Chicago, Chicago, IL*
[b] *NASA/Goddard Space Flight Center, Greenbelt, MD*

Abstract. The *Flash* code was developed at the University of Chicago as part of the Department of Energy's Accelerated Strategic Computing Initiative (ASCI). The code was designed specifically to simulate thermonuclear flashes in compact stars (white dwarfs and neutron stars). This paper will give a brief introduction to the astrophysics problems we wish to address, followed by a description of the current version of the *Flash* code. Finally, we discuss two simulations of astrophysical detonations that we have carried out with the code. The first is of a helium detonation in an X-ray burst. The other simulation models a carbon detonation in a Type Ia supernova explosion.

INTRODUCTION

The Flash Center at the University of Chicago is one of five Level I centers established by the Department of Energy under the ASCI Strategic Alliances Program. The goal of the center is to build a next-generation simulation code for studying Astrophysical Thermonuclear Flashes. These flashes, which result from explosive nuclear burning, are thought to be the cause of a number of observed astrophysical events, including X-ray bursts, novae, and Type Ia supernovae. The complex behavior exhibited by these objects makes them valuable laboratories for the numerical study of a wide range of basic physics phenomena. They are also valuable for what they can teach us about the universe, such as the formation of the elements and the cosmic distance scale.

ASTROPHYSICAL THERMONUCLEAR FLASHES

The three events we wish to model, X-ray bursts, novae, and Type Ia supernovae, all have a great deal in common, although each presents unique computational challenges. Each event begins with a compact star, either a white dwarf (for novae and supernovae), or a neutron star (for X-ray bursts) in orbit around a "normal" star. If the orbital radius is sufficiently small, matter from the normal star will be accreted onto the surface of the compact object over a relatively long period of time compared to the time scale of the flash itself. Due to the orbital angular momentum, the matter cannot fall directly onto the surface of the compact object, but instead accumulates in a disk around it. The angular momentum is slowly diffused outward by poorly understood dissipation processes, allowing the matter to fall onto the surface of the star. This process of accretion may take place at the equator or may be directed to the magnetic poles in the case of highly magnetized stars.

The accreting material that reaches the compact object forms a thin layer on the surface. Once this layer has reached a critical depth, the action begins. For X-ray bursts and novae, a thermonuclear runaway will start at the base of the accreted layer. Due to the extreme temperature sensitivity of the nuclear reactions, any small perturbation will cause the runaway to initiate at a single point rather than simultaneously over the entire surface. A sub-sonic flame front (deflagration) or a supersonic detonation will then propagate outward from this point and consume the entire layer. Little if any material will be ejected from an X-ray burst, due to the enormous gravitational field of the neutron star. However, observations of novae do show an ejected shell of matter that engulfs the companion star, forming a common envelope binary system. These events cause little change to the underlying star, allowing them to recur many times from the same source.

Type Ia supernovae are quite different from the other two events. Here, the accreted material compresses the entire star, causing a runaway to initiate at or near its center. A detonation would result in complete burning to iron group elements, and the intermediate mass elements seen in supernova spectra would not survive. Therefore, burning must propagate as a deflagration front, at least at early times. At late times, the ejected matter becomes supersonic. This has led theorists to suggest that the deflagration undergoes a transition to a detonation at some point during the explosion. In any event, the entire star will be blown apart by the explosion, so that recurrence is impossible.

THE FLASH CODE

Flash [3] is a sophisticated multi-physics simulation code. It was designed to be portable across a wide variety of computers and to run efficiently on massively parallel architectures. An object-oriented framework was added to make maintenance and modification as easy as possible. The code is modular, so that new physics routines and algorithms can be added with a minimum of effort. Parallelization of the code is accomplished using the Message Passing Interface (MPI) library. The Hierarchical Data Format library HDF-5 is used for parallel I/O. The code has been highly optimized and recently achieved a sustained performance of 0.238 TFlops. We were awarded the 2000 Gordon Bell Prize for this accomplishment [1].

At the heart of the *Flash* code is an Adaptive Mesh Refinement (AMR) toolkit, called PARAMESH [4]. AMR concentrates grid points in regions where they are most needed. This capability provides reduced time to solution and decreased memory usage, both of which enable higher resolution simulations.

Simulation of astrophysical thermonuclear flashes requires a wide variety of physics modules. One module solves Euler's equations for compressible gas dynamics using the Piecewise-Parabolic Method [2]. This method was specifically designed for accurate and efficient solution of flows containing discontinuities, such as shock waves, and is therefore ideal for the events we wish to model. The equation of state required for our simulations contains a mixture of partially degenerate and relativistic electrons, an ideal gas component for the ions, and radiation. Modules exist both for analytic computation of the thermodynamic quantities and their derivatives and for a thermodynamically consistent table look up algorithm [6]. Nuclear energy generation is computed by solving a reaction network [5]. The typical networks used by *Flash* contain 10-20 isotopes and more than 100 reactions. The network is represented by a system of linear equations, which are solved using standard linear algebra packages.

The code also contains modules for diffusive processes, such as radiative transfer and thermal conduction. Self-gravity is computed by solving Poisson's equation. A module for including magnetic field effects is currently being integrated into *Flash*. Future planned additions include a level set front tracking scheme to prevent unphysical diffusion across sharp interfaces and a sub-grid model for treating the effects of turbulent motions that cannot be resolved on the grid.

SIMULATION OF ASTROPHYSICAL DETONATIONS

This section discusses two *Flash* code calculations of astrophysical detonations. The first, a simulation of an X-ray burst on a neutron star, follows the evolution of a 2 km wide section of the stellar surface in an attempt to determine what observational consequences might result from the flash. The second simulation is a study of a very small section of a detonation that might occur in a Type Ia supernova.

X-ray Bursts

As mentioned above, an X-ray burst results from the accretion of material from a binary companion star onto a neutron star. For the parameters we considered, instability occurs after a 110 m thick layer of helium has accumulated onto the stellar surface. At this time the density at the base of the layer is 2×10^8 g cm^{-3} and the temperature is 1×10^8 K. For comparison, the density of the accreting material was only 1×10^{-5} g cm^{-3}. The size of the computational domain was 2,000 m × 1,500 m with a minimum grid resolution of 1 m.

The runaway initiates at a point at the interface between the accreted material and the underlying neutron star. A detonation front propagates outward from this point, both vertically toward the surface of the accreted layer and laterally around the circumference of the star. The detonation cannot propagate downward toward the center of the star, since there is no fuel there to burn. The detonation approaching the surface of the helium layer dies out due to the lower density and temperature and turns into a stationary deflagration front. The shock wave races ahead into the low density regions of the atmosphere. The detonation propagates laterally through the accreted layer at a constant speed of 1.3×10^9 cm s^{-1}, covering the entire star on a time scale of approximately 3 ms. A series of surface waves are excited by the explosion, which propagate around the star at approximately the same speed as the detonation. The entire atmosphere undergoes violent oscillations with a period of 50 μs. Material is pushed outward at a velocity of 1/3 the speed of light, reaching a height of approximately 10 km above the surface. For a more detailed description of the calculation see [9].

See http://flash.uchicago.edu/research/gallery.html for an assortment of color snapshots and movies of this simulation.

Cellular Detonations

The second set of calculations is relevant to those models of Type Ia Supernovae in which a deflagration to detonation transition is thought to occur. The purpose of this simulation is to determine the detailed structure of the detonation front. Detonations in laboratory experiments exhibit a very complex three-dimensional cellular structure. In order to make a complete model for Type Ia Supernovae, it is necessary to understand what observational consequences this structure might have.

In this case, the compact object is a white dwarf, and the material through which the detonation propagates is assumed to be pure carbon at a density of 1×10^7 g cm^{-1}, the density at which the transition to detonation is thought to occur. The size of the computational grid is 12.8 × 12.8 × 256 cm. The minimum grid resolution is 0.05 cm.

The front propagates for approximately 50 cm before the cellular instability becomes noticeable. At this point the front becomes non-planar, and a series of weak shock waves form perpendicular to the primary shock. Very high pressure spots, called triple points, develop where these shocks collide. These triple points act as tiny explosions, which sustain the propagation of the detonation. The composition distribution behind the front shows both under-reacted and over-reacted regions. Burning continues for a distance of approximately 20 cm behind the front, at which point the solution becomes one-dimensional.

At this density, the cell size is only a few cm, which is much too small to have any observation consequences. However, as the detonation approaches the surface of the white dwarf and the pre-shock density decreases to 1×10^6 g cm^{-1}, the cell size will become comparable to the size of the star. When this happens, it is unlikely that burning will continue to completion and the observed distribution of elements produced by the explosion will be very different from that predicted by one-dimensional theories. The other noticeable effect is in the speed of the detonation front, which is 1-2% slower than predicted by one-dimensional theory. This slight difference is unlikely to have any major impact on supernova models.

See http://flash.uchicago.edu/~fxt/cell3d.html for results of this calculation. A more complete description of the calculation can be found in [7,8].

CONCLUSION

Although it is still under development, the *Flash* code is one of the most sophisticated astrophysics codes currently available. The combination of state-of-the-art numerical algorithms, adaptive mesh refinement, and efficient parallel implementation gives us the capability to attack problems that could only be dreamed of before. The *Flash* code has allowed us to produce the highest resolution, most accurate multi-dimensional simulations of X-ray bursts on neutron stars and detonations in Type Ia supernovae ever performed. Results from other simulations can be viewed at http://flash.uchicago.edu. The code will continue to evolve, as new physics modules are added, making the study of a wider range of astrophysical problems possible.

ACKNOWLEDGMENTS

This work was supported by the Department of Energy under Grant No. B341495 to the Center for Astrophysical Thermonuclear Flashes at the University of Chicago and by NASA/Goddard Space Flight Center.

REFERENCES

1. Calder, A.C., Curtis, B.C., Dursi, L.J., Fryxell, B., Henry, G., MacNeice, P., Olson, K., Ricker, P., Rosner, R., Timmes, F.X., Tufo, H.M., Truran, J.W., and Zingale, M., "High-Performance Reactive Fluid Flow Simulations Using Adaptive Mesh Refinement on Thousands of Processors," in *Supercomputing 2000 Proceedings*, (2000).
2. Colella, P., and Woodward, P.R., *J. Comp. Phys.*, **54**, 174-201, (1984).
3. Fryxell, B., Olson, K., Ricker, P., Timmes, F.X., Zingale, M., Lamb, D.Q., MacNeice, P., Rosner, R., Truran, J.W., and Tufo, H.M., *Ap. J.*, in press, (2000).
4. MacNeice, P., Olson, K.M., Mobarry, C., deFainchtein, R., and Packer, C., *Comp. Phys. Comm.* **126**, 300-354, (2000).
5. Timmes, F.X., *Ap. J. Suppl.* **124**, 241-263, (2000).
6. Timmes, F. X. and Swesty, F. D., *Ap. J. Suppl.* **126**, 501-516, (2000).
7. Timmes, F.X., Zingale, M., Olson, K., Fryxell, B., Ricker, P., Calder, A.C., Dursi, L.J., Tufo, H.M., MacNeice, P., Truran, J.W., and Rosner, R., *Ap. J.* **543**, 938-954, (2000).
8. Timmes, F.X., Zingale, M., Olson, K., Ricker, P., Fryxell, B., Calder, A.C., Dursi, L.J., Truran, J.W., and Rosner, R., *Ap. J.*, submitted, (2001).
9. Zingale, M., Timmes, F.X., Fryxell, B., Lamb, D.Q., Olson, K., Calder, C., Dursi, L.J., Ricker, P., Rosner, R., MacNeice, P., and Tufo, H.M., *Ap. J. Suppl.*, in press, (2000).

Matrix Distributed Processing and FermiQCD

Massimo Di Pierro

Fermilab, Batavia, IL 60510, USA

Abstract. Matrix Distributed Processing (MDP) is a collection of classes and functions written in C++ for fast development of efficient parallel algorithms for the most general lattice/grid application. FermiQCD is an Object Oriented Lattice QCD application of MDP, under development at Fermilab.

INTRODUCTION

It is believed that, down to the smallest observed length scale, fundamental interactions in nature are local. This means that the equations one writes to describe the physical world are, in the majority of cases, local differential equations or systems of local differential equations. They can be non-linear, strongly coupled and stochastic but, if they describe a fundamental interaction, they are also local.

With very few exceptions, these equations do not have an exact analytical solution, therefore they must be solved numerically. This is done by discretizing the space on which the equations are defined and applying iteratively the appropriate algorithm.

The most general local differential equation contains derivatives which, after discretization, becomes quasi-local terms. For example

$$\frac{d^n}{dx^n}\phi(x) \to \sum_{k=0}^{n} \frac{(-1)^k}{(2a)^n} \binom{k}{n} \phi(x+(n-2k)a) \quad (1)$$

where a is the lattice spacing introduced in the discretization process. "Quasi-local" here means that a local term (the n-th derivative in x) becomes a linear combination of non-local terms localized within a radius na from x.

The typical iterative algorithm that solves a local differential equation has the form

$$\text{ITERATE} \quad \forall x : \phi(x) = H(x,\phi(y)) \quad (2)$$

where H is some function of the position, x, and of the value of the field, $\phi(y)$, in some neighborhood of x within $|x-y| \leq na$.

The exact form of H is not completely determined by initial differential equation, since there are different inequivalent ways to discretize it. A difference in the discretization procedure means a difference in the convergence speed and a difference in the discretization errors (that vanish with $a \to 0$).

Finding the numerical solution can be very costly but these algorithms can be very efficiently parallelized (using a supercomputer and/or a cluster of workstations). This is because one can partition the space x on which the field $\phi(x)$ is defined over different CPUs. Each CPU applies the algorithm, eq. 2, to the local sites and this can be done in parallel. Because of the quasi-locality of the function H it is necessary that each process maintains an updated copy of the field variables $\phi(y)$ for each y in the neighborhood of the local sites x. Each CPU will distinguish between local sites $\{x\}$ (the sites stored by the CPU), boundary sites $\{y\}$ (sites that are not local but a local copy exists because they must be accessed) and hidden sites (sites that do not affect the computation performed by that particular CPU).

For every parallel algorithm to work it is necessary to keep the boundary sites updated, i.e. if a field variable at a particular site is modified by one of the CPU, its copies (maintained by different CPUs) have to be modified accordingly. This requires communication among the different CPUs.

Matrix Distributed Processing (MDP) (1) provides the tools to implement this kind of algorithms on a computer in an easy and object oriented way. It also provides some basic classes for matrix manipulations, statistical analysis and a random number generator.

Communications in MDP are based on Message Passing Interface (MPI) which is *de facto* a standard for parallel applications. MPI calls are hidden inside the basic classes that constitute MDP and are invisible to the user.

EXAMPLE

As a first example of an application, let us consider here the following problem:

Problem: *Solve numerically, in U, the following equation*

$$\nabla^2 U = \cos(U+V) \quad (3)$$

where $U(x)$ and $V(x)$ are fields of 3×3 matrices defined on a four dimensional space x with the topology of a torus T^4. $V(x)$ is initialized with random $SU(3)$ matrices. (In this example $U(x)$ plays the role of the field $\phi(x)$ of the last section.)

Solution: The first step is to discretize the space on which the fields are defined by approximating it with a N^4 lattice (with $N = 8$). The second step consists in writing down a discretized form of eq. 3, using eq. 1. In dimensionless units (defined by imposing $a = 1$) one obtains

$$\begin{aligned} U(x) &= H(x,U) \\ &\equiv \frac{1}{8}\big[\cos(U(x)+V(x)) + \\ &\quad U(x+\hat{0})+U(x-\hat{0}) + \\ &\quad U(x+\hat{1})+U(x-\hat{1}) + \\ &\quad U(x+\hat{2})+U(x-\hat{2}) + \\ &\quad U(x+\hat{3})+U(x-\hat{3})\big] + O(a) \end{aligned} \quad (4)$$

where $x \pm \hat{n}$ is $y = (x_0 \pm \delta_{0,n}, x_1 \pm \delta_{1,n}, x_2 \pm \delta_{2,n}, x_3 \pm \delta_{3,n})$.

The third and usually non-trivial step is writing a computer program that implements, in a parallel way, the recursive relation of eq. 4.

Here is how this can be implemented using MDP:

```
01: #include "MDP_Lib2.h"
02: #include "MDP_MPI.h"
03: int main(int argc, char **argv) {
04:    mpi.open_wormholes(argc, argv);
05:    int box[4]={8,8,8,8};
06:    generic_lattice space(4,box);
07:    Matrix_field U(space,3,3);
08:    Matrix_field V(space,3,3);
09:    site x(space);
10:    forallsites(x) {
11:       U(x)=0;
12:       V(x)=space.random(x).SU(3);
13:    };
14:    U.update();
15:    V.update();
16:    for(int i=0; i<100; i++) {
17:       forallsites(x)
18:          U(x)=0.125*(cos(U(x)+V(x))+
19:                      U(x+0)+U(x-0)+
20:                      U(x+1)+U(x-1)+
21:                      U(x+2)+U(x-2)+
22:                      U(x+3)+U(x-3));
23:       U.update();
24:    };
25:    V.save("V_field.dat");
26:    U.save("U_field.dat");
27:    mpi.close_wormholes();
28:    return 0;
29: };
```

- lines 1,2 read the MDP libraries;
- lines 4 and 27 open and close the communication channels among the parallel processes;
- line 6 defines the object `space` belonging to the class `generic_lattice` with size specified by the `box`; (by default a generic lattice has the topology of a torus but the user can specify a different topology. The user can also specify on which processor each lattice site is stored. MDP optimizes the communications accordingly)
- lines 7,8 define the two fields of matrices `U(x)` and `V(x)`;
- line 9 defines a variable `x` of class `site` defined on the `space`;
- lines 10-13 initialize the fields in parallel;
- lines 14,15 take care of the communication to update the copies of the boundary sites;
- lines 16-24 perform 100 iterations of the algorithm, eq. 4; each iteration is automatically parallelized over the available CPUs;
- lines 25,26 save the input and output fields.

Many lattice/grid problems can be solved in a similar way. MDP provides some of built-in field classes and the user can easily define his/her own field class which inherit the standardized `update`, `load` and `save` member functions. The standard `load`/`save` functions guarantee the portability of data to different platforms (both parallel and non-parallel).

MDP also features a parallel random number generator, i.e., one random generator for each lattice site, that insures reproducibility of computations independently on the way the lattice is partitioned.

FERMIQCD @ FERMILAB

Fermilab is using MDP to develop a general purpose Object Oriented Lattice QCD application (2), called FermiQCD[1]. The typical problem in QCD (Quantum Chromo Dynamics) is that of determining the correlation functions of the theory as a function of the parameters. From the knowledge of these correlation functions one can extract hadron masses and matrix elements and compare them with experimental results. This provides both

---

[1] FermiQCD can be downloaded from:
http://thpc16.fnal.gov/fermiqcd.html

a useful check of the theory (QCD in particular) and also a unique way to extract some of the fundamental parameters of the Standard Model (for example the CKM matrix elements).

On the lattice, each correlation function is computed numerically as the average of the corresponding operator applied to elements of a Markov chain of gauge field configurations. Both the processes of building the Markov chain and of measuring operators involve quasi-local algorithms.

Some of the main features of FermiQCD are the following:

- it supports an arbitrary number of lattices in each parallel program and an arbitrary number of fields defined on each lattice;
- each lattice can have an arbitrary dimension, arbitrary topology and arbitrary partitioning;
- some of the basic built-in fields are:
  `gauge_field`,
  `fermi_field`,
  `staggered_field`,
  `scalar_field`;
- `gauge_fields` are in the adjoint representation of $SU(N_c)$ for an arbitrary N_c.

The basic parallel algorithms implemented in FermiQCD are (3):

- heatbath Monte Carlo to create the Markov chain of gauge field configurations;
- $O(a^2)$ improved heathbath Monte Carlo;
- minimum residue inversion and stabilized biconjugate gradient inversion for the fermionic matrix;
- ordinary and stochastic fermionic propagators;
- ordinary fermionic actions: Wilson, Clover ($O(a)$ improved) and D234 ($O(a^2)$ improved);
- staggered fermionic actions: Kogut-Susskind, Lepage ($O(a^2)$ improved).

Moreover FermiQCD is able to read existing Lattice QCD data in the CANOPY/ACPMAPS format, in the UKQCD format and in the MILC format.

Here are few examples of FermiQCD Object Oriented capabilities (compared with examples in the standard textbook notation for Lattice QCD)

1) **QCD:** (algebra of Euclidean gamma matrices)

$$A = \gamma^\mu \gamma^5 e^{3i\gamma^2} \qquad (5)$$

FermiQCD:

```
Matrix A;
A=Gamma[mu]*Gamma5*exp(3*I*Gamma[2]);
```

2) **QCD:** (multiplication of a fermionic field for a spin structure)

$$\forall x: \quad \chi(x) = (\gamma^3 + m)\psi(x+\hat{\mu}) \qquad (6)$$

FermiQCD:

```
/* assuming the following definitions
generic_lattice space_time(...);
fermi_field chi(space_time,Nc);
fermi_field psi(space_time,Nc);
site x(space_time);
*/
forallsites(x)
    chi(x)=(Gamma[3]+m)*psi(x+mu);
```

3) **QCD:** (translation of a fermionic field)

$$\forall x, a: \quad \chi_a(x) = U(x,\mu)\psi_a(x+\hat{\mu}) \qquad (7)$$

FermiQCD:

```
forallsites(x)
    for(a=0; a<psi.Nspin; a++)
        chi(x,a)=U(x,mu)*psi(x+mu,a);
```

ACKNOWLEDGMENTS

I wish to acknowledge the University of Southampton (UK) where the MDP project started and to thank the following members of the Fermilab Theory Group: E. Eichten, J. Juge, A. Kronfeld, P. MacKenzie and J. Simone, for many suggestions and comments about FermiQCD. I also acknowledge I borrowed many Lattice QCD algorithms from existing CANOPY, MILC and UKQCD programs; I thank here the authors for letting me study their codes.

The project FermiQCD is supported by the U.S. Department of Energy under contract No. DE-AC02-76CH3000.

REFERENCES

1. M. Di Pierro, Matrix Distributed Processing 1.0, hep-lat/0004007 (Tutorial and Licence for MDP)
2. M. Di Pierro, From Monte Carlo Integration to Lattice QCD, hep-lat/0009001
3. H. J. Rothe, Lattice Gauge Theories, Wold Scientific Lectures Notes in Physics vol.43

A Terabyte Analysis Machine For SDSS Data

James Annis[†], Gabriele Garzoglio[‡], Kurt Ruthsmandorfer[†], and Chris Stoughton[†]

[†]*Experimental Astrophysics Group and* [‡]*Computing Division, Fermilab, Batavia, IL 60510*

Abstract. The Sloan Digital Sky Survey data has driven us to design the Terabyte Analysis Machine, a high-I/O compute cluster built on a cluster of commodity computers, a fibre channel network, and a large disk tower; on Linux, on a new cluster file system called GFS, and on the sophisticated SX database. Our initial implementation is a 7 dual processor nodes with 500 Gb of local disk and 730 Gb of global disk. We expect to use the TAM to search the first 1000 sq-degrees of the SDSS for clusters of galaxies.

INTRODUCTION

The 2.5 Terabytes of data from the finished SDSS will consist of catalogs, atlas images, and binned sky. From the 200 Gb of pre-2000 high quality data we have learned the value of the ability to look at the atlas images, about exactly how slow a supposedly I/O limited task like A-star hunting can be, and about just how much compute power it takes to find clusters of galaxies or to measure on all the atlas images weak lensing optimized ellipticity parameters. Our experiences have driven us to design the Terabyte Analysis Machine[1] (TAM); a machine with an architecture designed to allow minimum I/O time for I/O-bound problems and to bring significant compute power to bear for compute bound problems when presented with terabyte scale data sets.

We need access to terabytes of data, but we are not the first. In the 1990's the database world settled [2] on clusters sharing neither memory nor disk, only the network; both the high energy physicists and movie makers adopted that model. More interestingly, Landsat presented the remote sensing community with terabyte scale spatially coherent data; the group at the University of Maryland responded with an exploration of an I/O limited analysis machines ([3], [4], [5]) leading them to the Realm,[2] their next generation compute cluster. We can use this experience to guide our design.

DESIGN NOTES

The basic insight on optical astronomical data is that all of the measurements we do are local: even cluster finding operates on scales small compared to the whole sky. We are firmly in the "embarrassingly parallel" regime, a pleasant surprise: divide and conquer strategies can turn large jobs into multiple small ones and provide naturally linear speedup[3] and scaleup[4]. It is also the ideal situation for exploiting large numbers of cheap, fast commodity processors, disk, and memory. Astronomy is perpetually cash starved; it is rare that the better solution is also the cheaper one, but this is the case.

> Design Note 1. Money -is- an object: use commodity computers and learn to cluster.

We are not inverting maps or matrices, our problems require message passing between CPUs at a period closer to a kilosecond than to a microsecond.

> Design Note 2. Take full advantage of the embarrassingly parallel regime: we can do without fast, low latency inter-node communications, and probably without message passing parallel programming.

We do, however, need access to terabytes of data. This is far larger than can fit into memory and we will need to store out-of-core and intermediate data onto disk, preferably local disk, where bandwidth is higher than global

---

[1] A machine is one or more computers tightly organized to perform a particular task: the computers of Google are definitely a machine (see [1]), whereas most Condor clusters (http://www.cs.wisc.edu/condor) are not.

[2] http://realm.umd.edu

[3] performing the same calculation n times faster as the machine gets n times bigger

[4] performing in the same time an n times larger calculation on an n times larger machine

disk and guaranteed not to interfere with I/O requests from other processors.

> Design Note 3. We want significant disk local to each machine. Local disk can transform I/O bound problems into compute bound ones.

For these I/O dominated problems, the initial partitioning of the data to local disk is important. The partitioning cannot be write once, partly because we are still taking data, partly because the simple partitioning that is optimal for small numbers of disk becomes sub-optimal for more disks, but fundamentally because the optimal partition depends on the question being asked. We will need to the ability to redistribute data amongst local disk.

> Design Note 4. We want a global store of data.

Finally, we note that observational optical astronomers analyze data in two ways: using quick interactive analyses (the main stay of the field) and long pipeline jobs applied to large data sets. Far and away the best way for astronomers to develop is to have the data available interactively in the same way it is available to the pipeline jobs. Furthermore, when trying to use the compute power gets in the way of doing the science, astronomers chafe: a machine that is fast but inconvenient to use is worse than a slow machine that is easy to use.

> Design Note 5. Thought cycles of astronomers are the scarcest resource of all: design the machine for efficient large scale analyses, but preserve reasonable interactive ease of use.

CURRENT TECHNOLOGY

These design notes, given the current technology, imply a high I/O compute cluster, able by cleverness to do fast I/O bound searches, and able by brute force to perform quick compute bound problems.

> Tech Note 1: Beowulf clusters are commodity items.

The rise of the Internet economy has created a demand for commodity clusters of commodity computers and many vendors now sell turnkey systems at a price varying by $\lesssim 20\%$. Dual CPU boxes are the current maximal performance/dollar point, and the dimensions of the SDSS camera suggest 6 is a magic number.

> Tech Note 2: The current free interconnect is fast Ethernet.

The only significant cost is a switch which gathers the 100-BaseT connections into a Gigabit Ethernet uplink, and this is at the half-node level.

> Tech Note 3: IDE disk is cheap and nearly as fast as SCSI.

One or two IDE disk per node is close to free, (1/10 of a node cost); local disk is not a problem.

> Tech Note 4: There are many ways to make a global store; we are attracted to a high speed serverless design built on a fibre channel network.

The fibre channel network allows multiple nodes to access the same disks over 100 Mbyte/s links. The open source file system GFS[5] controls multiple node access via disk firmware file locks to prevent file system corruption and uses journaling to prevent corruption in the case of a node going unresponsive during I/O.

> Tech Note 5: For most of our uses, a load sharing system is better than a batch system.

Just beyond the state of the art is using Mosix with GFS. Mosix[6] is a kernel modification that allows processes to transparently slide from one machine to another based on machine load.

THE FERMILAB IMPLEMENTATION

Our baseline machine consists of 6 compute nodes and 1 interactive node. All are dual 600 Mhz Pentium IIIs with 1 Gb of RAM and 72 Gb of local IDE disk.

Initially we planned 16×50 Gb disk packaged in two JBODs, which with simple partitioning could allow up to 200 Mbytes/s I/O from the global store. As we had the funds and disliked the idea of retrieving from tape the data on a lost disk, we instead turned to a hardware RAID controller with 10×73 Gb disk.

Our initial testing of GFS has been encouraging. We ran tests with multiple processes reading large files off of a single logical volume on the RAID disk set, both with the nodes accessing the logical volume via GFS, and with one node acting as a NFS server of the logical volume to the other nodes. Using NFS (without any fine tuning), we achieved only 10 Mbytes/sec aggregate throughput. Using GFS we achieved 90 Mbytes/sec, limited by the rate allowed by the single RAID hardware controller.

---

[5] Global File System; http://www.globalfilesystem.org
[6] http://www.mosix.org

THE TERABYTE ANALYSIS MACHINE

The TAM is meant to enable science; one analysis it is optimized for happens to be the architect's own: cluster finding need access to overlapping sky area and brute compute power to examine large areas of sky. In general, one of TAM's strengths is in bringing large amounts of CPU to bear on easily accessible datasets.

Beyond this, we expect to integrate TAM with SX, the SDSS science archive [6] in two different ways. First, we will access data over sockets connected to the SX database cluster, a cluster optimized for raw I/O speed

In this use, TAM is the analysis cluster attached to the SX database cluster. In a second mode, we will run a SX server on the TAM itself. While accessing the data from this server is a possibility, what really drives us is the exploration of a facility for optimal re-partitioning and re-indexing of the entire SX dataset. For cluster finding, there are efficient algorithms for finding the kth nearest neighbor. In general these involve re-partitioning and re-indexing of the data, and it usually is faster to just do a straight scan of the existing database. However, if one is in the code/run/debug/explore cycle of scientific programming, speed counts. Our plan is to set up a facility to perform the re-partitioning and re-indexing of the data onto the local IDE disk, and leave that version of the data there for a few weeks while the scientist works; a variation of disk caching. For the kth nearest neighbor this will be called the distance machine, but we expect that the infrastructure we build will easily allow the swapping in of modules implementing other database efficient algorithms for other specific problems.

Lastly, having access to all of the data opens up all of the possibilities of the visualization database of Kent: color GIF images of the SDSS data, with position links to the NED database, to the photo output files and spectra, and full resolution reconstructed frames. This is a highly useful tool for understanding the data.

THE TERABYTE ANALYSIS MACHINE AND SDSS DATA

In the year 2000, 2.5 Terabytes is 50 disks and $50k, so the budgets of interest top out at $100k. An astronomy department committed to serious SDSS analysis (perhaps together with an aggressive junior faculty startup funds) is capable of building a machine optimized for SDSS work. While one option is to wait and let the compute power and storage space come to you[7], but in reality the scale of interesting problems also grows with time. Good compute nodes cost $\sim$ $5k, and large fast disks $\sim$ $1k, roughly independent of time, and roughly scaling with the problems of current interest. The machine we are implementing, 7 nodes with 500 Gb of local disk and 730 Gb of global disk, runs $\sim$ $80k; about $\sim$ $110k with hardware RAID.

REFERENCES

1. Brin, S. and Page, L. 1998 *WWW7/Computer Networks*, 30(1-7): 107-117
   (http://dbpubs.stanford.edu:8090/pub/1998-8)

2. deWitt, J. and Gray, J. *Communication of the ACM* 35(6), 1992
   (http://research.microsoft.com/-gray/CacmParallelDB.doc)

3. Acharya, A., Uysal, M., Bennett, R., Mendelson, A., Beynon, M., Hollingsworth, J., Saltz, J., and Sussman, A. 1996., In *Proceedings of the Fourth ACM Workshop on I/O in Parallel and Distributed Systems*, May 1996.
   (http://www.cs.umd.edu/projects/hpsl/html-papers/iopads96.html)

4. Chang, C., Moon, B., Acharya, A., Shock, C., Sussman, A., and Saltz, J. 1997. In *Proceedings of the 13th International Conference on Data Engineering*, Apr. 1997.
   (http://citeseer.nj.nec.com/chang97titan.html)

5. Shock, C. Chang, C., Moon, B., Acharya, A., Davis, L., Saltz, J. and Sussman, A., 1998. *Parallel Computing*, 24(1):65-90, Jan. 1998.
   (http://citeseer.nj.nec.com/shock98design.html)

6. Szalay, A., Kunszt, P., Thakar, A., Gray, J., Slutz, D., and Brunner, R.J. 1999. astro-ph/9912382.
   (http://arXiv.org/abs/astro-ph/9912382)

7. Gottbrath, C., et al. Bailin, J., Meakin, C., Thompson, T., and Charfman , J.J. 1999. astro-ph/9912202.

---

[7] highly recommended in this regard is [7]

A Large Linux-PC Farm for Online Event Reconstruction and Data Management at HERA-B

A. Gellrich*

Department of Physics, Humboldt University Berlin, Germany

Abstract. HERA-B enters a new regime with respect to data rates and volumes in HEP pointing towards the LHC era. The experiment exploits a Linux-PC farm for full online event reconstruction with 200 CPUs which has been in operation in 2000. Conceptual aspects of the Farm and the data handling in HERA-B as well as installation and running experiences after 6 months of full-time operation are presented and discussed.

INTRODUCTION

HERA-B (1, 2) is a fixed target experiment at DESY with high sensitivity for rare heavy flavor decays and QCD-studies. The HERA-B detector is a forward spectrometer with roughly 600000 readout channels. Particles are produced by a thin wire target which is inserted into the beam halo of HERA's 920 GeV proton ring. The detector consists of a vertex detector, several tracking devices, and particle identification systems. It is optimized for tracking in a high occupancy environment.

Taking into account the small production rates for heavy flavors compared to inelastic events in hadroproduction, a very high initial event rate is needed to produce sufficient numbers of interesting events. The HERA-B wire target can be steered to produce events at a rate of 10 MHz with several superimposed interactions. HERA-B writes events of roughly 150 kB size at rates of typically 20 Hz to archive which leads to data volumes of a few ten TB per year. The high input rate and the small signal to background ratio require a trigger and data acquisition system which provides sufficient filtering with a background suppression of 10^6 and moreover is able to process event data as they come along online. Sufficient processing power is obtained by introducing PC-farms into the trigger and data acquisition system. HERA-B makes use of a level 2 trigger farm consisting of 240 CPUs and a level 4 online reconstruction farm, in the following simply called Farm, with 200 CPUs. Due to the still small fraction of $O(10\%)$ of interesting physics events in the finally recorded samples a clear concept for the data management is needed.

HERA-B has entered a new regime with respect to data rates and volumes in HEP approaching conditions which will be present in the LHC era.

FARM

As the last stage of HERA-B's four level trigger and data acquisition system (3, 4) a large PC farm was built. Its purpose is to carry out full event reconstruction online before data are logged on disk and archived on tape. Re-processing is severely restricted due to the resources which would be needed to read back data from archive (tape) and the CPU power for processing. Retrieving a data volume of 30 TB from tape at a typical bandwidth of 4 MB/s takes already more than 100 days.

The tasks of the Farm can be summarized as follows:

- Fully reconstructed events are provided online.
- Event classification is performed.
- A final event selection called level 4 trigger is done.
- Detailed data quality monitoring is carried out.
- Data to be used for calibration and alignment is derived and collected.
- Data logging and archiving is prepared.
- The vast computing power can be used for event data re-processing.

Detailed studies of various implementation scenarios had been carried out before a solution based on standard off-the-shelf components was taken. For information on studies in this context see (5, 6).

The Farm exploits 200 Pentium-III/500 MHz processors housed in commodity dual-processor PCs with

---

* Supported by the German ministry of education and science.

256 MB SDRAM and 13.1 GB (E)IDE-disks. The PCs are grouped into so-called mini-farms with 15 nodes connected via switched Fast-Ethernet. Up-links are realized by GigaBit-Ethernet. The hardware was completed end of 1999 and was put into operation in 2000.

Linux was chosen as the operating system since it has become the main development platform in HEP as well as in HERA-B. The farm-PCs are set up workstation-like. With an average event processing time of 4 s a rate of 50 Hz can be processed with 200 CPUs.

Within the farm nodes tasks are realized as separate Unix processes, using Inter-Process Communication tools for data and message exchange. Between nodes home-made message passing systems based on the Internet protocols UDP and TCP/IP are applied. For archiving data to tape OSM[1] software is used.

Event reconstruction software is housed in a frame program (*ARTE*). This frame program is also used for offline purposes such as Monte Carlo simulation and physics analysis (7). Although C/C++ has become the standard programming language, parts of the software are still written in FORTRAN. Data structures are based on ZEBRA-tables which are defined by a data description language. Recorded events consist of such tables. The Linux operating system on the Farm allows to use *ARTE* without modifications. I/O is done by means of shared memory rather than files. Offline developed software can be used directly on the Farm. The traditional separation between online and offline software could be overcome which eases and speeds up developments considerably.

Data quality monitoring and collection of information for calibration and alignment are major benefits of on-line processing. To achieve the designed detector performance in particular of the trigger system a fast feedback system is mandatory which updates calibration and alignment constants online. To make use of the full statistics, which is distributed over 200 nodes, software called a Remote Histogramming Package was developed. It allows to gather locally filled histograms by central processes. The package provides HBOOK-like functions and can also be used offline.

In periods without data taking, in particular in long shutdown periods, the system can be used for data reprocessing. From the Farm and logging point of view there is no principal difference between online running and re-processing. The involved software is the same: Instead of reading out the detector, raw event data are retrieved from tape and distributed to the Farm.

---

[1] Open Storage Manager

DATA MANAGEMENT

HERA-B must deal with a large data volume of 30 TB per year which needs to be inspected accumulatively over the lifetime of the experiment. Only a small fraction of events and/or information per event can be stored on random access media. Data management starts with the logging of event data to disk coming from the data acquisition system through the Farm. It incorporates consecutive writing of events to files, archiving of files, copying of files to commonly available disks, and retrieving of files from tape libraries (*staging*).

Normal user's physics analysis must be restricted to the reconstruction information of the events. Only a small fraction of $O(1\%)$ of raw data can be provided on disk. Three main types of event data files are distinguished:

- DST[2]: Reconstruction output plus raw data.
- MINI: Reconstruction output only (5% of DST).
- micro: Not yet defined subset based on ROOT (8).

It is planned to keep all MINI-files on disk which leads to a data volume of 1.5 TB per year. In addition, raw data information (DST) of roughly 1% of events can be stored on disk per year (300 GB/y).

At HERA-B calibration and alignment constants as well as slow control information are stored in a home-made database system based on Berkeley-DB (9). For event data management no commercial database is used since developments were not far enough advanced when the implementation was planned in 1995. HERA-B uses so-called index files as event directories. To keep data manageable, unique files names, convenient files sizes (400 MB per DST-file), and well-defined directory structures are exploited. Naming conventions allow to identify the source of data and reflect the logging mode. In normal data taking events are subsequently stored run-wise. Re-processing can be done run-by-run or for selected data sets.

Each event contains a classification bit mask which is assigned in the last part of the reconstruction. It distinguishes up to 32 categories (classes) of events. For each run an index file is written which contains one entity per event, consisting of the event number, date and time of data taking, and the classification. All index files are collected in a run catalogue. Index files and run catalogue allow for direct event access and are therefore crucial tools for efficient analyses which are usually based on small subsets of events. Special selections, in particular at the DST-level, are obtained by means of the event classification.

---

[2] DST: Data Summary Tape

In order to make some DST-files temporarily available on disk, a so-called *staging* tool has been developed and included into the analysis frame program. If a user request a file which is not already in a dedicated disk pool (currently 206 GB), data will automatically be retrieved from tape. To make most efficient use of resources, files are kept on disk as long as space is available for further copies. The oldest file is deleted first.

SUMMARY

Installation, commissioning, and running of a large multi-processor system, exploiting standard components and Linux, turned out to be easy and efficient with respect to costs and manpower. The system was running stably and reliably during the run period 2000 and has been used for re-processing in the following shutdown. Designed performance values have been clearly exceeded in all fields. Input and output data rates up to 10 MB/s were reached to disk (logging). Archiving to tape was done at a sustained rate 4 – 5 MB/s.

Compared to pseudo-online or offline approaches HERA-B benefits from immediately available fully reconstructed events for physics analysis, online event classification, and in particular from a comprehensive data quality monitoring system. Moreover, calibration and alignment constants are already derived online for some detector components which allowed for an online feedback. More detectors will be included in the coming months.

In 2000 a total data volume of more than 8 TB was logged and archived. Although this is only a fraction of the planned data volume of 30 TB per year, data management has become a major issue. Due to the small signal to background ratio of $O(10\%)$ even at the end of the trigger chain, HERA-B must cope with large event samples which need to be carefully inspected during physics analysis. Since typically only 1 TB can be stored on random access media, only parts of the full information of all events will be available on disk. Due to rather conservative approaches in condensing information, e.g., track selection, MINI-files still make up roughly 30% of DST-files.

The early phase of understanding the detector performance required rather frequent access to raw data (DST-files) in 2000. *Staging* has been a valuable tool to make DST-files temporarily available on disk. It turned out that in particular studies to understand detector hardware are based on small subsets of the data. Certain runs are favored by many detector groups. Physics analysis has started to move towards MINI-files which allows for immediate access to all events. First results were obtained.

HERA-B looks forward to the running period 2001 which will start after the luminosity upgrade shutdown of HERA end of summer.

REFERENCES

1. T. Lohse et al., *Proposal, DESY-PRC* **94/02** (1994).

2. E. Hartouni et al., *Design Report, DESY-PRC* **95/01** (1995).

3. M. Dam et al., *Higher Level Trigger Systems for the HERA-B Experiment*, IEEE Transactions on Nuclear Science Vol. **45**, No. 4 (1998).

4. A. Gellrich and M. Medinnis, *Higher Level Triggering Software*, Nucl. Instr. Meth. **A408** (1998) 173-180.

5. A. Gellrich et al., *The Processor Farm for Online Triggering and Full Event Reconstruction of the HERA-B Experiment at HERA*, CHEP '95, Rio de Janeiro, Brazil, 1995.
A Test System for the HERA-B Online Trigger and Reconstruction Farm, DAQ 96, Osaka, Japan, 1996.
A Prototype System for the Farm of the HERA-B Experiment at HERA, CHEP '97, Berlin, Germany, 1997.
The Fourth Level Trigger Online Reconstruction Farm of HERA-B, CHEP '98, Chicago, USA, 1998.
A Linux-PC Farm for Online Event Reconstruction at HERA-B, 11th IEEE NPSS Real Time Conference, Santa Fe, USA, 1999.
The Linux-PC Farm for Online Event Reconstruction of HERA-B, 1st LCB Event Filter Farms Workshop at 3rd LHC Computing Workshop, Marseille, France, 1999.
Full Online Event Reconstruction at HERA-B, CHEP 2000, Padova, Italy, 2000.

6. http://www-hera-b.desy.de/subgroup/farm/welcome.html

7. H. Albrecht, *The Computing Model for HERA-B* CHEP '97, Berlin, Germany, 1997.

8. http://root.cern.ch/

9. A. Amorim et al., *The HERA-B database management for detector configuration, calibration, alignment, slow control, and data classification*, CHEP 2000, Padova, Italy, 2000.

Client and Event Driven Data Hub System at CDF

Ben Kilminster[†], Kevin McFarland[†], Tony Vaiciulis[†]
Hiroyuki Matsunaga[*], Makoto Shimojima[*]

[†] *Department of Physics, University of Rochester, Rochester, NY 14627*
[*] *University of Tsukuba, Tsukuba, Ibaraki 305, Japan*

Abstract. The Consumer-Server Logger (CSL) system at the Collider Detector at Fermilab is a client and event driven data hub capable of receiving physics events from multiple connections, and logging them to multiple streams while distributing them to multiple online analysis programs (consumers). Its multiple-partitioned design allows data flowing through different paths of the detector sub-systems to be processed separately. The CSL system, using a set of internal memory buffers and message queues mapped to the location of events within its programs, and running on an SGI 2200 Server, is able to process at least the required 20 MB/s of constant event logging (75 Hz of 250 KB events) while also filtering up to 10 MB/s to consumers requesting specific types of events.

INTRODUCTION

CDF Run II Physics with the TeVatron

The Collider Detector at Fermilab (CDF) as well as Fermilab's proton/anti-proton particle acceleration facilites have undergone significant improvements in the pursuit of a better understanding of physics at the smallest scale. Due to the creation of the Main Injector and other improvements, the colliding proton/anti-proton beams in the Tevatron will deliver twenty times the luminosity of the previous data taking run. This means that there will be 7 million beam crossings per second in the CDF detector. CDF has risen to this challenge with faster electronics readout, and more extensive tracking, thus increasing the number of readout channels ten-fold. The combination of a better accelerator and detector improvements leads to a maximum 1 KHz digitization rate of 250 KB size events.

Data from the Detector

The data rate is reduced by a series of trigger systems, which use detector information to evaluate events. The third filter is a PC farm called Level 3 which uses event reconstruction software to lower the rate from 250 MB/s to a more desirable 20 MB/s to be logged for further analysis. The system responsible for this is the Consumer-Server/Logger (CSL). Its primary responsibilities are to (Figure 1):

1. Receive detector event data at a rate of 20 MB/s from multiple senders

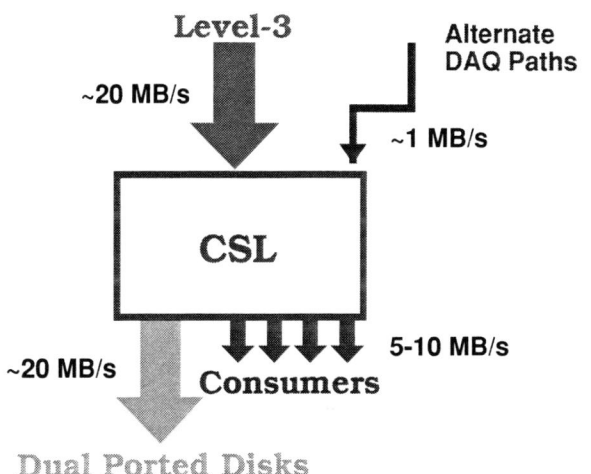

FIGURE 1. Rates in/out of CSL

2. Write events to multiple files depending on trigger information and event type.

3. Send up to fifty percent of the data to multiple external programs (consumers) which monitor the integrity of the data. Typically they monitor luminosity, trigger efficiencies, and detector channel activity. Consumers may select events based on trigger information.

4. Receive (independent from main data stream) a few MB/s of data from various detector sub-systems such as diagnostics data or calibration data.

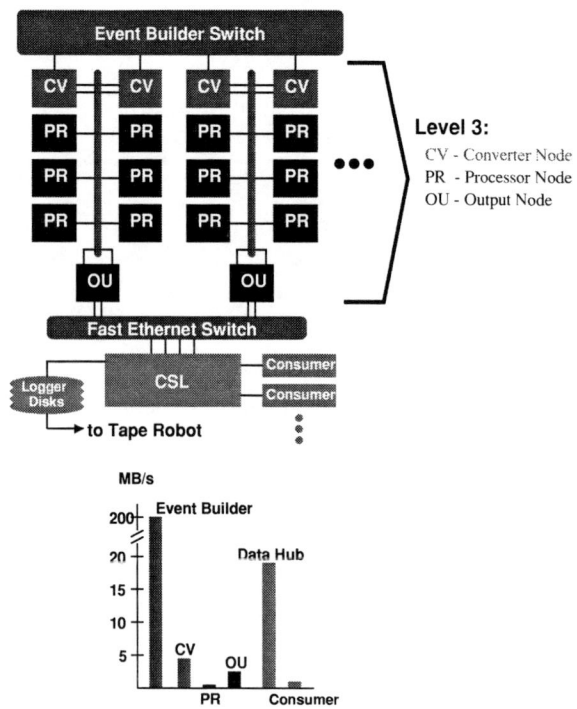

FIGURE 2. Assembled events are sent from the *event builder*, by means of an ATM switch, to *converter* nodes in Level 3, then distributed to available *processor* nodes running physics software (farm), collected and sent to the CSL by output nodes, then logged and distributed to *consumers*. The bottom chart shows typical rates through each component.

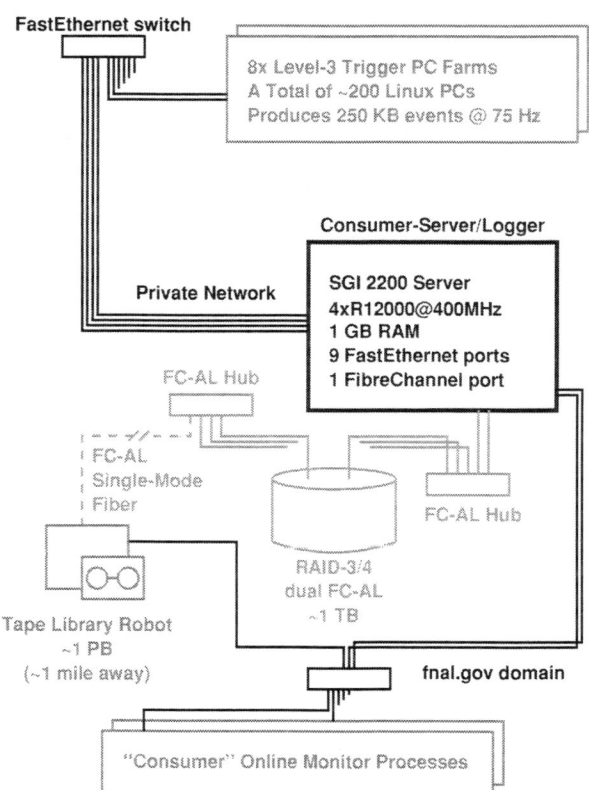

FIGURE 3. Files are written to 1 TB dual controlled/ported RAIDs

Figure 2 shows various data acquisition components, and their relative processing rates.

SYSTEM HARDWARE

Events processed by the Level 3 Farm are received by the CSL from a FastEthernet switch via four FastEthernet ports. The CSL runs on a single SGI 2200 Server with four 400 MHz R12000 processors. It logs data via its FibreChannel port to a series of dual ported RAIDs. Files are then read out from the arrays and sent by single-mode fiber to a robot which writes data to a 1 PB tape archive (Figure 3). Because the connection to the disk arrays is done with two independent fiber channel-arbitrated loops, file output from the CSL is not dependent on readout to tape. This, combined with the 1 TB RAID, mean that data may be stored successfully even if communication to the tape robot is down for as much as eight hours.

CSL SOFTWARE

The CSL consists of independently running C programs which communicate through System V IPC shared memory segments and message queues. Collectively, these control the movement of events to logger and consumer output.

Typical operation is as follows (Figure 4): A receiver parent process waits for connections from an event sender. When a connection is made, a receiver child process is spawned to receive events. Consumer sending processes are also spawned for each connected consumer. After receiving an event, the receiver child writes it to one of a hundred shared memory buffers and puts a message on the logger message queue representing the buffer. The logger, after finding a buffer message in its queue, logs the event to the appropriate data files and puts the message on the distributor queue. The distributor checks a consumer request table and either puts the buffer message on the consumer-send queue or returns the message to the receiver queue. The consumer-send process looks for a message on its queue and sends the corresponding event to the appropriate consumers. After this the event has been fully processed and the message is returned to

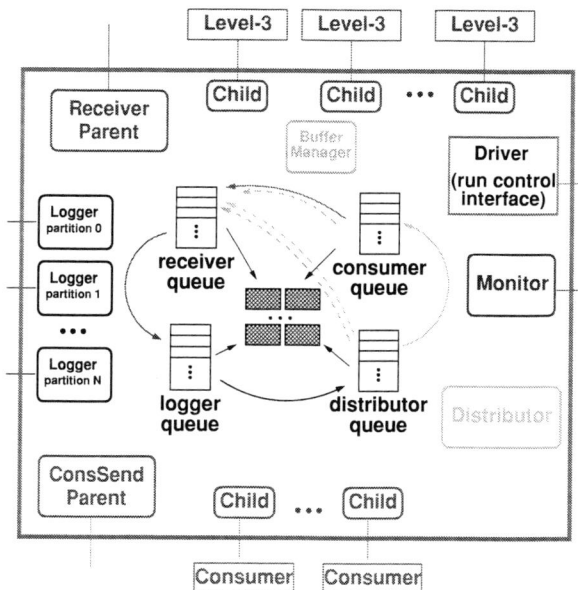

FIGURE 4. Software Design - Events are stored in shared memory, and messages are circulated through message queues from receiver to logger to distributor to consumer sender, and back to receiver. The Driver contols and communicates the CSL state, while the monitor records and sends out rate and process information.

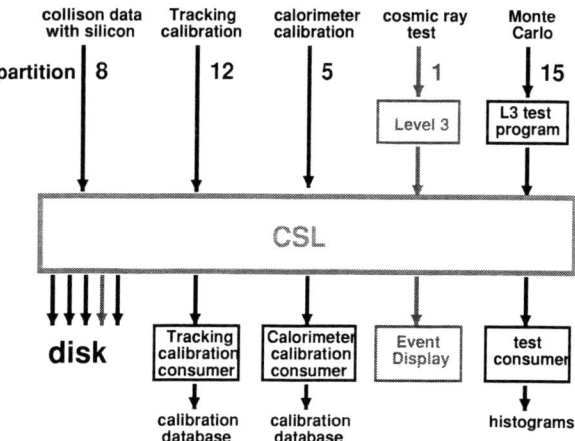

FIGURE 5. Data in the CSL may come from different paths of sub-systems called partitions. Here is an example of an early operations mode. To handle multiple input and output connections, receiver, logger, and consumer child processes are spawned off for each new partition within the CSL.

the receiver queue so that another event may be received. A buffer manager process watches over the receiver message queue to ensure that there are always buffers available for incoming events by removing messages from the consumer queue. In this way, only as many events are sent to consumers as can be done without interfering with the steady input/output rate of 20 MB/s as required.

Each process writes out rate and client information to shared memory and a central monitoring process assembles and sends the information via SmartSockets (www.talarian.com) to a java-based display running externally. A driver process controls the state of the CSL and the interface to the central run control system.

A partition is a sub-set of CDF detectors and/or data acquisition components. The CSL must handle events coming from different partitions. For robustness, a separate logger process handles each partition. Multiple receiver and consumer-send children may join a partition dynamically during the course of a run (Figure 5).

Since the CSL must log and send events to consumers based on physics information, yet not waste time probing their internal structure, events passing through Level 3 are "stamped" with a header structure containing needed event information. When the CSL receives the event, it reads and removes the header, and then adds the information to the event's representative message. Events can therefore be input as just a byte-input stream and written out in sequential ROOT format without actually decoding the event.

CSL STATUS

In a recent CDF commissioning run several months before the start of data taking, the fully operational CSL has exceeded rate specifications by logging 25 MB/s of average-sized events while also sending to consumers.

ACKNOWLEDGMENTS

We gratefully acknowledge contributions from the ROOT Team, CDF Task Force, CDF Online Group, and Rob Kennedy.

The COMPASS Computing Farm Project

M. Lamanna

CERN, 1211 Geneva 23 (Switzerland) and INFN, Padriciano n.99 34012 Trieste (Italy)

Abstract. The COMPASS experiment at CERN is building a large facility for off-line computing in close collaboration with the CERN IT division: the COMPASS Computing Farm. The motivations, the experience, and the plans for deploying a 2,000 Spec INT95 computing farm are discussed. The main technical points are the use of Intel-based Linux PCs for both the I/O and the CPU intensive tasks, and the use of an object data base to store all the data.

INTRODUCTION

COMPASS is a fixed-target experiment with a wide physics programme at the CERN SPS. The apparatus will perform a number of different measurements, in different configurations, notably using both muon and hadron beams in the 100-300 GeV range at very high intensities(1).

The high data rate (over a few months per year) to be processed and the need for a flexible software environment to cope with the experiment's different measurements have pushed the COMPASS collaboration to design the off-line analysis software from scratch and to build a dedicated facility for the off-line computing, namely the COMPASS Computing Farm (CCF).

The COMPASS experiment has started commissioning the detector and taking data in the Summer 2000. All detector systems were present on the floor (although with a limited numbers of components) and a full chain from the front-end system, through the online data acquisition, to the off-line system was tested.

The off-line system should cope with a number of constraints, namely the high data acquisition rate (35 MB/s) and the very large data sample (10G events, 30 kB each, 300 TB/y) to be reconstructed almost online.

COMPASS decided to use the Central Data Recording (CDR) to record all data: the online system does not write events on tape at the experiment site, but sends them over a few km of dedicated network to the computer centre, where the CCF, the tape servers, and the corresponding high-speed tape drives are located.

The estimated computing power to reconstruct all the events at the speed of the data acquisition is 2,000 Spec INT95, which will be provided by some 100 PCs. The choice of network technology is Gigabit and Fast Ethernet. A disk pool of a few TB is being setup: presently most of the disk space is SCSI, but EIDE servers are being deployed.

The last requirement is the use of a data base layer integrated with a storage manager to handle the very large sample of data. The solutions for the data storage are still under evaluation, together with the development of the C++ reconstruction framework. During Summer 2000, the data have been kept in both native and Objectivity/DB format. CASTOR, a hierarchical storage manager being developed at CERN has been used to manage all data(3) (originally HPSS has been used).

THE DATA AND ANALYSIS MODELS

COMPASS decided to adopt a computing model such that all the data are reconstructed in a single reconstruction facility in parallel with the data taking.

COMPASS decided to write the bulk of the reconstruction programs in C++. The reconstruction and analysis program CORAL (COmpass Reconstruction and AnaLysis; *http://coral.cern.ch*) has been designed using object-oriented techniques. It uses an object data base called Objectivity/DB for both event and calibration data.

The large quantity of data (300 TB/year) means that all the data may not be disk resident at the same time. The limitation in total disk size can be made transparent with the use of a Hierarchical Storage Manager (HSM).

The data flow model should foresee different scenarios; the most relevant ones are the DAQ mode and the quasi-online reconstruction mode.

In the DAQ mode, events are written on the online farm event builder disks in raw format ($\approx$ 10 streams) to the CCF and then populate *in parallel* the data bases ("stage1"). The original files are not kept, and the corresponding data bases are moved to the HSM.

In the quasi-online reconstruction mode, the data follow the same path as in the DAQ mode, but they are read by the CPU clients when they are still on the disk pool. The events are dispatched to clients via the Objectivity/DB object server (AMS).

The model proposed for the CCF is a farm which uses as much as possible "commodity solutions", mainly PCs and mass-market network technologies.

At the beginning of 1998 we started with the idea of deploying a number of mid-range RISC servers for I/O plus a cluster of cheap PCs. In this document, *data servers* refer to the I/O intensive core of the farm, and *CPU clients* the number-crunching PCs (Fig.1).

This model is also well suited for the collaborating institutes; COMPASS institutes from Germany and Italy have shown interest and some prototypes already exist. These analysis farms will be fed with data samples from the CCF, to perform the final analysis of specific reactions; some of the data not present in these remote farms (RAW data samples, calibrations) can be accessed via wide-area network.

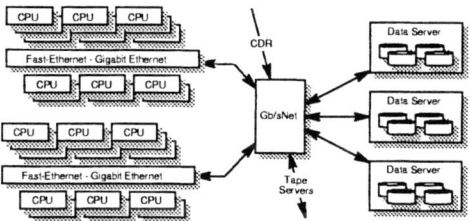

FIGURE 1. A schematic of the CCF. The CPU clients are on the left, on the right the data servers (holding the farm disk pool). The lines represent the main network connections and data flows: the CDR line represents the data flux from the experiment to the farm; the Tape Servers line represents the data exchange between the farm and the tape infrastructure.

THE COMPONENTS OF THE CCF

The decision to shape the CCF in a modular environment was taken very early, on the basis of both technical and economical considerations. The deployment of "independent" units in all CCF areas (instead of single high-end devices) best addresses the problem of the performance to price ratio for each area, gives the possibility to increase the configuration in small steps, to share hardware and resources within similar projects and, most importantly, to remove single points of failure.

The outcome is that all solutions should either be tested against others (e.g. Windows NT vs. Linux) or at least allow for possible future comparisons (i.e. use generic solutions or exploit only generic functionalities from a software component). Obviously, the tests should be performed if possible in a realistic environment to assess compatibility and scalability issues.

The CPU clients are Intel PCs. For the data server part, we tested DEC and Sun servers, and Intel PCs (2 Pentium II CPUs). Linux PCs (with Fast Ethernet interface) became eventually the solution also for data servers.

THE TESTS

An online farm prototype and the corresponding software to generate mock data with the appropriate characteristics (data rate, events size) have been developed to test the CCF. This farm has been set-up at the experiment site and uses the full network infrastructure which will be used in the experiment.

The CCF has been tested in the DAQ mode in a set-up with 11 data servers. The main result is that the behaviour scales with the number of data servers: the 11 data servers can sustain about 35 MB/s for many hours (Fig.2).

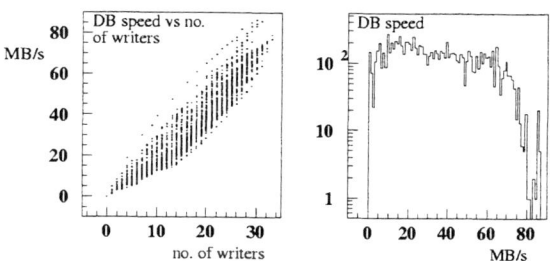

FIGURE 2. The "stage1" subsystem in the 35 MB DAQ test (Events conversion into objects in Objectivity/DB data bases). The rate of data in MB/s is shown as a function of the number of active parallel writers and as a histogram.

The sustained 35 MB/s figure means that at the same time the following activities coexist, all of them at 35 MB/s: the raw data are input into the CCF and written to the data servers disk; the same flux is input into the data bases; data bases are read from disk and copied to a remote system to mimic the HSM transfer. All quantities are measured over a test period of $\approx$ 8 hours.

The Quasi-Online mode is presently under test; configurations with 5 data servers and about 10 MB/s are run to test the behaviour of the multi-threaded version of the AMS (v.5.2.1). The system is stable but the AMS behaviour is not yet satisfactory because of time-out problems, which could be recovered at the expense of resubmitting some reconstruction jobs.

The calibration data base (CDB) is an important component of the CORAL, because it will be in charge to distribute calibration constants in a consistent way to all nodes processing data in parallel. The library is also in evolution (the CERN port of the original BaBar calibration data base is being completely rewritten)(5). To test the functionality, the performances, and the implementation, we perform dedicated tests (using CORAL clients).

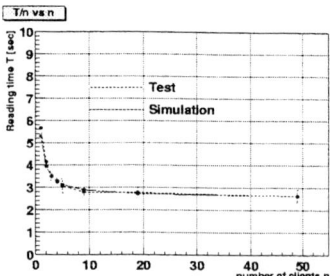

FIGURE 3. The plot shows the results of a set of tests of CDB access, compared with a simulation. The mean reading time per client vs. the number of clients (up to 50) is plotted.

In Fig. 3, the performances of the CDB are shown as measured versus the number of parallel clients. The data base access speed is ≈75% of the theoretical network speed (7-8 MB/s; for a Fast Ethernet network AMS server). The measurement is compared with some preliminary results of a simulation made using the MONARC toolkit(4).

PILOT RUN

All the data have been recorded successfully and made available to the collaboration using CASTOR (more than 1 TB of data for detector debugging).

The observed data base input rate is consistent with the previous tests. Since the actual data rate from the experiment was lower than the design value (due to the detector commissioning activities and lower number of channels), this was achieved accumulating data for hours and then processing them at the maximum speed (Fig. 4): the speed per server is consistent with the previous tests.

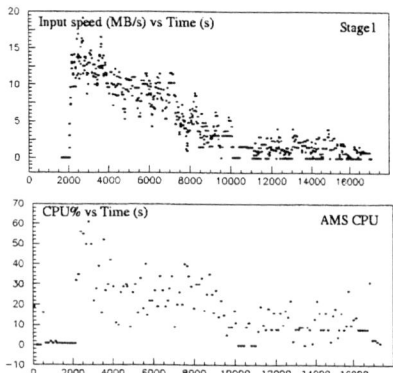

FIGURE 4. The data base input speed vs. time is shown (Stage1). The process is started after a period of accumulation of data; rapidly all the data are processed and the speed becomes equal to the input one. The corresponding AMS CPU consumption is shown.

All the chain down to a scan of the data (similar to quasi-online processing) has been demonstrated with real data. The performances are still somewhat poor as in the mock data tests (some bugs in the Objectivity/DB have been pinned down since then and our code and data organisation are also being improved).

CONCLUSIONS

The mock data tests gave us the possibility to select the appropriate hardware for every option, to shape the architecture and the control software in a modular way, and to test many prototypes in different realistic conditions.

The main message from the 2000 run is that the control software is adequate also for data acquisition. The mock data results have been cross-checked with the experience with the real data; no major new problem has been found.

The experience with the Objectivity/DB package is rather positive, but open problems remain. The usage of Objectivity/DB within the CORAL framework is well-understood and sufficiently isolated to consider it a separate interchangeable component.

The experience with CASTOR was very positive and the replacement of HPSS straightforward. The control software is still being simplified giving to CASTOR full responsibility of the various data pools present in the system. The CASTOR-Objectivity/DB interface was successfully used for the first time.

I would like to acknowledge the support and the help of many groups in the CERN IT division (especially the PDP and the DB groups), the COMPASS Collaboration, and the COMPASS Offline Group.

REFERENCES

1. COMPASS, COMPASS Proposal, CERN/SPSLC/96-14, SPSLC/P297, March 1, 1996; COMPASS, Addendum 1, CERN/SPSLC/96-30, SPSLC/P297 Add. 1, May 20, 1996.

2. M. Lamanna "The COMPASS Computing Farm project", Proceedings of the CHEP 2000 Conference, Padova, February 2000, edited by M. Mazzucato, p.576.

3. CASTOR, developed in the IT/PDP group, *http://wwwinfo.cern.ch/pdp/castor/* the HSM interface has been developed in the IT/DB group.

4. I. LeGrand "Simulating Distributed Systems", contribution to this conference.

5. Calibration DB, developed in the IT/DB group, *http://wwwinfo.cern.ch/db/objectivity/docs/conditionsdb*.

A Data Grid Prototype for Distributed Data Production in CMS

Mehnaz Hafeez[a,b], Asad Samar[a], Heinz Stockinger[b,c]

(a) California Institute of Technology, Mail Code 256-48, 1200 E. California Blvd., Pasadena, CA 91125, U.S.A.
(b) CERN, European Organization for Nuclear Research, CH-1211 Geneva 23, Switzerland
(c) Inst. for Computer Science and Business Informatics, Univ. of Vienna, Rathausstr. 19/9, A-1010 Vienna, Austria

Abstract. The CMS experiment at CERN is setting up a Grid infrastructure required to fulfil the needs imposed by Terabyte scale productions for the next few years. The goal is to automate the production and at the same time allow the users to interact with the system, if required, to make decisions which would optimise performance.

We present the architecture, design and functionality of our first working Objectivity file replication prototype. The middle-ware of choice is the Globus toolkit that provides promising functionality. Our results prove the ability of the Globus toolkit to be used as an underlying technology for a world-wide Data Grid. The required data management functionality includes high speed file transfers, secure access to remote files, selection and synchronisation of replicas and managing the meta information. The whole system is expected to be flexible enough to incorporate site specific policies. The data management granularity is the file rather than the object level.

The first prototype is currently in use for the High Level Trigger (HLT) production (autumn 2000). Owing to these efforts, CMS is one of the pioneers to use the Data Grid functionality in a running production system. The project can be viewed as an evaluator of different strategies, a test for the capabilities of middle-ware tools and a provider of basic Grid functionalities.

INTRODUCTION

In autumn 2000 distributed High Level Trigger (HLT) studies take place at both sides of the Atlantic. Regional Centres in France, Great Britain, Italy, Russia and the USA will produce data that have to be replicated to and from CERN as well to/from other Regional Centres. Based on this requirement we have developed a software tool called Grid Data Management Pilot (GDMP) that supports data replication in a Data Grid environment. The first production prototype is used by the CMS experiment to transfer Objectivity [9] database files between Regional Centres.

GDMP can be seen as a mature production software as well as an evaluator of existing Grid technologies.

Let us elaborate on the aspect of being an evaluator. The concept of a Grid [2] system is rather new in the HEP community and has its roots in the Particle Physics Data Grid [7] as well as currently started Grid projects like GriPhyN [5] and DataGrid [1, 6]. Up to now, no HEP experiment has been using Grid tools based on the Globus [3] middle-ware in order to do data production in a distributed way. We consider this as a pioneer step of CMS in the direction of Data Grids in production systems. Our aim has been to take as much as possible of the tools provided by Globus in order to have a replication system that is based on a single middle-ware system.

DISTRIBUTED DATA PRODUCTION IN CMS

CMS has planned four trigger levels with increasing complexity and sophistication in the trigger algorithm. These trigger levels are called level-1, 2, 3 and 4. In some cases triggers at level-2, 3 and 4 are grouped together and called the *higher level triggers (HLT)*.

At LHC, data will be collected by the CMS detector where level-1 trigger has only a small latency protected by a pipeline. The trigger rate at level-1 is around 1 billion events/sec. It is not possible to store all these data, so the HLT will reduce the rates to manageable numbers. Firstly, at level-1 the trigger rate is reduced from 1 billion events/sec to 100,000 events/sec. After level-2 and level-3 the trigger rate is reduced to 100 events/sec. The studies of HLT using simulated data requires large datasets. One million simulated events after level-1 is equivalent to 10 seconds of LHC running.

For the HLT studies, the CMS collaboration has organised five groups according to the physics channels namely: egamma, muon, jetmet, btau and level 1 trigger. For the current production more than 6 million events are requested for the simulation. This corresponds to roughly 6 TB of data. A number of collaborating CMS institutes is involved in the current production such as: INFN, Caltech, IN2P3, Fermilab, CERN, Helsinki Institute of Physics, Moscow, Bristol and Pakistan.

The CMS data production consists of following steps:

1. simulation of the detector response for a given physics channel,

2. digitisation of the above data with event pile-up,

3. reconstruction of physics objects such as tracks, clusters, jets, etc. using the data from step 2.

The preparations of distributed data production have been going on since last year. A data transfer production system based on Perl scripts, HTTP and secure shell copy transfers has been put in place [10]. The purpose of these scripts is to allow a user to see the contents of an Objectivity's federated database catalogue from anywhere with WWW access, without the need to run any part of the Objectivity software locally. Furthermore, the script software allows for data transfer of not only Objectivity but any kind of file, for instance Zebra-fz or Root files.

These Perl scripts have been the initial input for the architecture and design of GDMP. The idea was to extend the functionality by introducing a security environment and providing a fully automated replication mechanism based on a Grid infrastructure using Globus. The current architecture of GDMP presented in this paper is based on Objectivity, i.e. GDMP is an asynchronous data replication tool for replicating Objectivity files over the wide area network.

GDMP can be seen as a Grid-enabled successor of the initial Perl scripts. However, GDMP does not replace the initial software system completely. Since GDMP is currently based on Objectivity's native file catalogue to initiate file replication, it is only used for replicating Objectivity files. The Perl scripts are deployed for fz-file transfers and are still a fall-back solution to GDMP. In a future release, GDMP will incorporate a replica catalogue [4] provided by the Globus toolkit which allows for flexibility in file transfer. Once the replica catalogue is used, GDMP will support file replication for any kind of file and can replace the initial software based on Perl completely.

THE GDMP ARCHITECTURE

GDMP is a multi-threaded client-server system that is based on the Globus toolkit. The software consists of several modules that closely work together but are easily replaceable. In this section we describe the tools used from

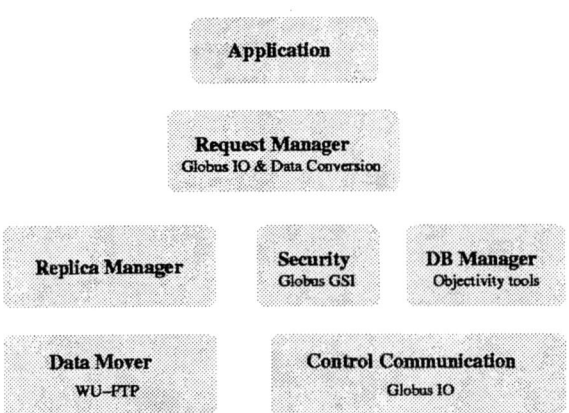

FIGURE 1. Logical datamodel of the tool

Globus as well as the modules and the software architecture of GDMP. The core modules are Control Communication, Request Manager, Security, Database Manager and the Data Mover. An application which is visible as a command-line tool uses one or several of these modules.

Since GDMP is focusing on the Data Grid functionality, only the following limited set of the entire Globus toolkit is used by GDMP.

- Globus GSI for security

- Globus IO and data conversion for communication

- Globus thread library for the multi-threaded GDMP server

Figure 1 shows the current architecture and all the modules of the software. A detailed description of the modules can be found in [8]. All the software modules are written in C++ and run on Solaris 2.6, 7 and Linux RedHat 6.1.

The GDMP server is a daemon constantly running on sites which produce data or want to export their data to other sites.

The server itself uses the communication module for receiving requests from application clients and a thread pool to handle multiple clients concurrently. For each client one thread is used.

The server uses a dedicated server certificate and thus a proxy on its own which is included in the GDMP software distribution. Thus, when a server is started, the required Grid proxy is gained automatically.

Since all the access to data has to be done in a secure environment, a client has to be authorised and authenticated before it can request a service from the server. We use the *single login* procedure which is available through Globus, i.e. once a client has successfully got the proxy on one machine, it can send requests to any server without any further password entered (provided the local client is authorised to access the server).

REPLICATION POLICIES, DATA MODEL AND APPLICATIONS

Currently, GDMP is restricted to replicate only Objectivity files due to the use of the native Objectivity federation catalogue to handle files in GDMP. We will replace the Objectivity file catalogue by the Globus Replica Catalogue [4] in order to support a flexible replication model.

The GDMP architecture is based on the subscription model where each site that wants to get notified about changes at other sites subscribes to a remote site. In principle, a site, where Objectivity files are written, has to trigger the GDMP software which notifies all the "subscriber" sites in the Grid about the new files. In detail, a data production site announces newly written data by publishing its catalogue.

The "subscriber" (destination) sites receive a list of all the new files available at the source site and can determine themselves when to start the actual data transfer. The data transfer is done with a WU-FTP server and an NC-FTP client.

In principle, a site only needs three commands to participate in the automatic replication process. A site can subscribe to another site by issuing the command `gdmp_host_subscribe`. A production site announces that new files are available by using the tool `gdmp_publish_catalogue`. The consumer can then decide when to start the file transfer from the producer to the consumer site with the tool `gdmp_replicate_file_get`.

GDMP allows a partial-replication model where not all the available files in a federation are replicated. This is achieved by applying a filter on the file catalogue. This allows for a partial replication model where the producer as well as the consumer can limit the amount of files to be replicated.

FAULT TOLERANCE AND FAILURE RECOVERY

In the current version of GDMP, each site is itself responsible for getting the latest information from any other site in case of a site failure. A site can recover from the site failure by issuing the command `gdmp_get_catalogue` and receiving the entire catalogue information from another site.

In case of a broken connection, re-sending of several files is not needed since as soon as a file arrives safely at the destination site, the file is attached and the file entry is deleted from the import catalogue immediately. Only the file which is currently been sent when the network connection breaks, has to be resent. Since the implementation of WU-FTP has a "resume transfer" feature, not even the entire file has to be transfered but only the part of the file that is still missing since the last check point in the file. This allows for an optimal utilisation of the bandwidth in case of network errors.

CONCLUSIONS

We have been developing a file replication tool that allows for secure and fast data transfers over the wide-area network in a Data Grid environment. With our production software we have proved that Globus can be used as a middle-ware toolkit in a Data Grid. This has been a pioneer step in the direction of a Data Grid and to the best of our knowledge first software approach where a wide-area replication tool based on Globus is used in a production system. Furthermore, this work can also be regarded as an evaluator of Grid tools and thus has valuable input for other Data Grid activities like DataGrid, PPDG and GriPhyN.

REFERENCES

1. The CERN DataGrid Project: http://www.cern.ch/grid/
2. Ian Foster and Carl Kesselman (editors), The Grid: Blueprint for a New Computing Infrastructure, Morgan Kaufmann Publishers, USA, 1999.
3. The Globus Project: http://www.globus.org
4. The Globus Data Grid effort: http://www.globus.org/datagrid/
5. The GriPhyN Project, http://griphyn.org
6. Wolfgang Hoschek, Javier Jaen-Martinez, Asad Samar, Heinz Stockinger, Kurt Stockinger, Data Management in an International Data Grid Project, to appear in *1st IEEE, ACM International Workshop on Grid Computing (Grid'2000)*, Bangalore, India, Dec. 2000.
7. The Particle Physics Data Grid (PPDG), http://www.cacr.caltech.edu/ppdg/
8. Asad Samar, Heinz Stockinger. Grid Data Management Pilot (GDMP): A Tool for Wide Area Replication, to appear in *IASTED International Conference on Applied Informatics (AI2001)*, Innsbruck, Austria, February 2001.
9. Objectivity Inc., http://www.objectivity.com
10. Tony Wildish. Accessing Objectivity catalogues via the web. http://wildish.home.cern.ch/wildish/Objectivity/scripts.html

Object Level Physics Data Replication in the Grid

Koen Holtman

*California Institute of Technology,
Mail Code 256-48, 1200 E. California Blvd., Pasadena, CA 91125, U.S.A.*

Abstract. To support distributed physics analysis on a scale as foreseen by the LHC experiments, 'Grid' systems are needed that manage and streamline data distribution, replication, and synchronization. We report on the development of a tool that allows large physics datasets to be managed and replicated at the granularity level of single objects. Efficient and convenient support for data extraction and replication at the level of individual objects and events will enable for types of interactive data analysis that would be too inconvenient or costly to perform with tools that work on a file level only.

Our tool development effort is intended as both a demonstrator project for various types of existing Grid technology, and as a research effort to develop Grid technology further. The basic use case supported by our tool is one in which a physicist repeatedly selects some physics objects located at a central repository, and replicates them to a local site. The selection can be done using 'tag' or 'ntuple' analysis at the local site. The tool replicates the selected objects, and merges all replicated objects into a single single coherent 'virtual' dataset. This allows all objects to be used together seamlessly, even if they were replicated at different times or from different locations.

The version of the tool that is reported on in this paper replicates ORCA based physics data created by CMS in its ongoing high level trigger design studies. The basic capabilities and limitations of the tool are discussed, together with some performance results. Some tool internals are also presented. Finally we will report on experiences so far and on future plans.

INTRODUCTION

We report on the development of a prototype tool, in the Caltech/CMS group, for object level physics data replication in the grid. The creation of this tool is part of a longer development effort [1, 2] in which we aim to develop software tools and middle-ware that allows for large scientific datasets to be managed and replicated at the granularity level of single objects, with sizes of 100 bytes to 1 MB, rather than at the level of multi-MB files. The development of the tool was largely an integration effort, in which existing software was combined together with code that incorporated the results of previous R&D. We discuss the parts that were integrated below.

- To manipulate and store physics objects, the Objectivity/DB object database is used; this corresponds to the current CMS software development strategy. However the tool architecture does not depend strongly on the use of Objectivity: any middleware that provides portable object streaming to and from files could have been used.

- The data to be replicated in the prototype are ORCA [3] based physics objects, created by CMS in its ongoing high level trigger design studies. For our largest prototype test to date, we used the *SimEventBody* objects for 140,000 ORCA events. Each of these objects is close to a MB in size, the complete 'source' dataset used was 105 GB, divided over 103 Objectivity/DB database files.

- We use Globus middleware [5] to implement wide-area authentication without passwords sent in the clear, and also Globus GSI FTP [7] as a mechanism for fast, tunable data transport. This way, the WAN network efficiency of our tool is equal to that what can be achieved with a tuned file based replication tool. We intend to follow future improvements in the Globus FTP space. We re-use parts of GDMP [6] as a middleware layer on top of Globus to implement authenticated, message based client-server communication.

- We use the results of existing research [2] as a basis for the object indexing mechanisms used by the tool. Our tool implements large parts of the design of [2] – the biggest omission is that that the top level catalog structures are not synchronised across sites yet. Also, the 'chunk' level of granularity was replaced with a more fluid granularity level of event ID ranges.

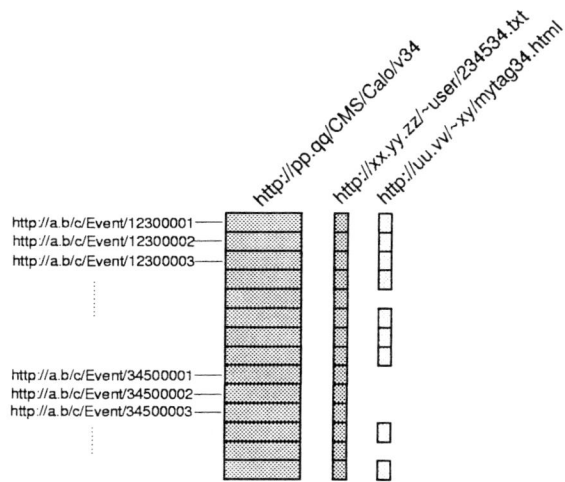

FIGURE 1. Logical datamodel of the tool

DATA MODEL

The data model defined by the tool corresponds to a sparse SQL table (see Fig. 1), where every table element is a physics object – a piece of physics data that can potentially be replicated. The rows correspond to events and the columns to the different types of physics objects (for example calorimeter raw data, tag object of user X, tag object of user Y) which may be present for the event.

Every row, every distinct event, is identified by an event ID, which is an ASCII prefix string combined with a 64-bit integer. For the topmost event (row) in Fig. 1, the prefix string is *http://a.b/c/Event/* and the integer is 12300001. The intention is then whenever a party (an experiment, a Monte Carlo production group inside an experiment) creates events, this party must choose a new unique prefix, and can then proceed to independently assign 64-bit event numbers to its events. The prefix can, for example, be an URL pointing to a page inside a website maintained by the party. Every column, every type of physics object, is identified by an ASCII string. Again, by using strings that are URLs pointing into private webspaces, multiple parties can create and fill multiple columns without naming conflicts.

All physics objects in the data model are read-only. This is an important restriction in terms of how the tool can be used. If data is to be changed, this can only be done by versioning: adding and filling a new column, and then instructing physics jobs to request objects from the new column in stead of the old one. Changing the instructions for the physics jobs will, in practice, involve updating the metadata structure used by the jobs to translate high level descriptions of a column ('the latest certified version of the production tag') to a specific column ID.

To access the local objects which have been replicated by the tool, a physics job must initialise an iterator object which is provided by the tool. The iterator is initialised with a column ID and an event list. The event list is a compact representation of a set of event (row) IDs. After initialisation the job calls the 'next' method of the iterator to retrieve the subsequent objects that were specified by the column ID and event list. If objects from multiple columns are needed for each event, multiple iterators can be initialised: their 'next' methods are guaranteed to visit the events in the same order.

TOOL INTERNALS

We explain the tool internals by following the most basic replication use case of the tool as shown in Fig. 2. In this most basic case, a central site has a one 'object with the full observations' for each event. Also, for each event, a summary object (tag object) is created. Both the user workstation and the central site have their own independent Objectivity/DB federation for storing objects. The only link between these federations is that their schema (internal class definitions) are identical with respect to the classes of any objects to be replicated. The summary objects are copied to the user workstation, and they are used to isolate an 'interesting' subset of the events, this set is saved as an event list. Then, the object replication tool is used to replicate the 'full observation' objects for these events to the user workstation. When the replication operation is finished, the full objects can be analysed by iterating over them.

The internal actions of the object replication tool, performing step 4 in Fig. 2, are as follows. The tool is started on the user workstation, with an event list and a column ID as its command line arguments. The tool connects to an object replication server (also implemented in this project) that runs on the central site. The connection uses TCP/IP, and Globus (GSI) user authentication is used to control access to the server. The tool will then check if any of the requested objects are already on the user work-

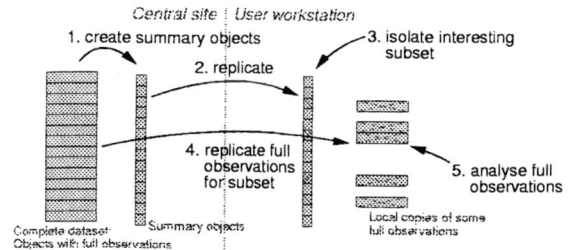

FIGURE 2. Most basic replication use case

station, if they were already replicated in a previous invocation of the tool. After having isolated the set of objects which is not yet present locally, the tool will request these objects from the server. The server locates the requested objects in its local database files, and extracts (copies) them into new Objectivity/DB database files, which are created on the server in a temporary area. The maximum size for each of these files is a server-side configuration parameter. When a file has been filled, it it sent to the user workstation using GSI FTP [7], a tunable FTP implementation which makes use of Globus authentication for security. The tool supports use of multiple FTP transfers in parallel, this is important to get good performance on a wide-area link. On arrival of a database file, the tool attaches it to the local Objectivity/DB federation catalog. The tool also adds a reference to the file to its own catalog, which is used to find replicated objects when an iterator is initialised. A specialised negotiation protocol is used between the tool and the server to make sure that the new database files have Objectivity database ID numbers which are unique at the user workstation side. When the replication operation is complete, the server deletes all database files it created in its temporary area.

When the tool is used a second time to to request the replication of a new, maybe partially overlapping, set of objects, it will transport only those objects which are not yet available locally, and will take care of seamlessly integrating all replicated objects on the user workstation. Thus the tool minimises the overhead associated with changing (refining) the cut predicate that selects the objects to be analysed on the user workstation. This allows the physicist using the tool to work in a fluid, iterative way, much more fluid than with most existing tools, where the cost of extending the object set to be analysed on the local workstation is often that the entire set has to be extracted and moved again.

PERFORMANCE RESULTS

Some performance tests were done to validate the functioning of the tool. Replication of *SimEventBody* objects from a server at Caltech to a user workstation at the Supercomputing 2000 floor in Dallas could be done at 1.2 MB/s, using 4 parallel FTPs (each with a 200 KB send buffer size), which maximised the share of the provided internet capacity we could use. Replication from a server on a 600 Mhz Dell workstation to a client on the same workstation, using the local loop-back interface, could be done at 5.8 MB/s, in that case the bottleneck was the speed of the disks in the workstation.

CONCLUSIONS

Coding of the prototype tool and server took some 2 person months after the design was completed. The tool was implemented on both Linux and Solaris, the related object replication server on Linux, Solaris, and partially (the data-intensive parts) on Windows 2000. The implementation on Windows 2000 took serious effort, compared to the time spent resolving the differences between Linux and Solaris, and our design was constrained significantly because only small parts of the Globus toolkit (the FTP clients) were available on Windows 2000. The integration effort was relatively painless, though some unexpected, nondeterministically occurring middleware bugs were encountered. A resolution to these bugs is still pending at the time of writing, which is about 1 month after these bugs were found. Temporary workarounds could be found, but this experience does suggest that, when a production system is being built that relies on large pieces of middleware, at least a few months of buffer time should be put into the planning to allow for the resolution of unexpected middleware bugs.

Future development plans in this effort are to integrate the current tool more closely with the CMS ORCA physics analysis system. Also, a system for peer-to-peer object granularity replication between many sites will be developed, which will make use of the Globus Replica Catalog to implement a global view of the contents of all sites.

REFERENCES

1. Holtman K., van der Stok P., Willers I. "Towards Mass Storage Systems with Object Granularity." In *Proc. of the Eighth NASA Goddard Conference on Mass Storage Systems and Technologies*, Maryland, USA, March 27-30, 2000, p. 135-149.

2. Holtman K., Stockinger H. "Building a Large Location Table to Find Replicas of Physics Objects." In *Proc. of CHEP 2000*, Padova, Italy, February 2000.

3. http://cmsdoc.cern.ch/orca/

4. http://www.objy.com/

5. http://www.globus.org/

6. Samar A., Stockinger H. "Grid Data Management Pilot (GDMP): A Tool for Wide Area Replication in High-Energy Physics." To appear in *Proc. of IASTED International Conference on Applied Informatics (AI 2001)*, Innsbruck, Austria, February 2001.

7. http://www.globus.org/datagrid/deliverables/gsiftp-tools.html

SAM for D0 - a Fully Distributed Data Access System

I. Terekhov, V. White, L. Lueking,
L. Carpenter, H. Schellman[†], J. Trumbo, S. Veseli, M. Vranicar

Fermi National Accelerator Laboratory, Batavia, IL 60510
[†] *Northwesterm University, Evanston, IL, 60208*

Abstract. The SAM (Sequential Access through Meta-data) system is being built as a distributed cache management and data access layer for the D0 experiment at Fermilab. The innovation of the project is the fully distributed architecture of the system which is designed to be deployable worldwide. It uses a central database for the meta-data and a hierarchy of CORBA servers for the actual data movement. SAM provides distributed disk caching, data routing and replication and therefore has components attractive to the Grid.

OVERVIEW

The SAM (earlier, alternative expansions include Sequential Access Method) system is the data access layer for the D0 experiment at Fermilab. During the next several years, the D0 experiment will store a total of about 1 PByte of data, including raw detector data and data processed at various levels. However, it is not the mere size of the data that presents the challenge; as any modern High Energy Physics collaboration, D0 is a truly distributed institution: it includes over 550 scientists from some 65 institutions scattered worldwide.

The SAM project (1) has taken an innovative approach to the data management by designing and implementing the system, from the outset, as a distributed mutlitiered client-server system deployable worldwide. In the present paper, we concentrate on the distributed nature of the architecture, and leave out the few D0-specific aspects of the projects. This presentation is in the spirit of the Particle Physics Data Grid, where SAM is the FNAL's participant (2).

Fundamentally, the goal of SAM as the data access system is to facilitate physicist access to their data, while efficiently using the scarce hardware resources such as Mass Storage Systems (MSS), networks and computers with their disks, CPU's etc. The bulk of the data is planned to be stored in the MSS's, primarily Fermilab's Enstore (5). The primary functionality of SAM is to provide a globally distributed disk cache; this distributed caching, which involves Grid features such as file routing and replication, is the focus of our paper. For a more general overview of the project, see (3).

THE APPLICATIONS SERVED BY SAM

In the sequential mode, a user application (a physics analysis or reconstruction program, or a Monte-Carlo event generator) processes data in a stream, accessing each data unit exactly once, the order of data units in the stream being irrelevant. The units of data are laid out sequentially in files. Consider file data retrieval, or consumption, in SAM (production, or storage is symmetric.) The application requests a next file from the stream, using a call to a server called project master (4). The application is notified via a callback when a file is ready, i.e., has been physically placed (or staged) on a local disk [1].

The application then uses the standard UNIX system calls to open, read and close the file, at which point it requests another file. The delivery of data to consumers is done by project master and other servers asynchronously with respect to the actual consumption. In a smoothly functioning system, the next file will have been delivered by the time the consumer finishes its current file. There are other interesting aspects of the process, for which we refer the reader to (4); for example, the consumer may in fact be spread across multiple processes perhaps running even on different machines (thus making a SAM consumer a distributed concept itself).

---

[1] In the Grid, other mechanisms exist to access files. Globus (6) provides specialized library to handle URL-based file names, Condor (7) allows transparent redirection of local file operations to remote servers.

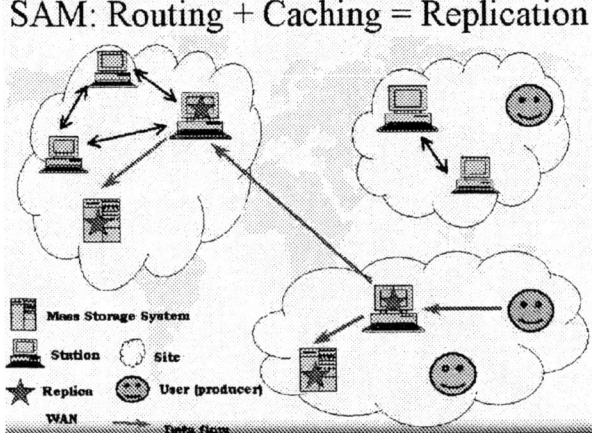

FIGURE 1. File routing while storing data with SAM.

DATA ROUTING IN SAM

For our presentation, it is important that the user reads (writes) a local file, pulls it from (pushes to) the SAM system, by name, and is not concerned with the way that the system relocates the file. The functionality of the distributed caching is to move the data through the disk cache, making multiple copies if necessary, between the user applications and the MSS's. In the most interesting case, a file may come from (or go to) a remote MSS, thus requiring multi-stage transfers of files for a given user. We say that such transfers require *routing*, which is an important aspect of SAM's distributed caching. There are at least two reasons for such routing:

- The MSS may not be directly accessible from the user's location, because of a slow or unreliable network connection or low-performance disk;
- It is in general desirable to leave a copy of a file at "nearby" location(s) for subsequent access

For the purposes of routing, we consider the distributed caching network, exemplified in Figure 1. Nodes on this network are MSS's and *stations*. A station in SAM is a collection of local hardware resources, such as disk and CPU, used for data processing. It is a distributed concept in itself (in general, it spans a relatively homogeneous local cluster), but it is beyond the scope of this paper so we consider the station a point-like node on the global network. Nodes are connected with Wide-Area Network (WAN) links.

As shown in the figure, when a user requests to store a file in an MSS, the file in general travels through one or more intermediate stations. At each station, disk cache allocation takes place. This allocation is subject to the local station's rules (we will elaborate below). While the file resides at an intermediate disk awaiting further transfer it actually is a part of the cache and is of course available for retrieval at this (and other) stations.

By the time of file reaching its final destination, there may be multiple copies of it left on caches on the path. As a result, by routing the file through the network of stations and using caching at each node, SAM performs the dynamic *data replication* service, an important aspect of the Grid. By configuring the routing information stored in each station the global system administrators have control over how the replication takes place.

DISK CACHING AND DATA REPLICATION

We stress the fact that the data replication in SAM is dynamic and naturally occurring. Each file replica is a part of the respective station's cache and as such is subject to the local caching policies which we will briefly describe now. A station continuously replaces files that are not in use by newly requested files. (Since multiple groups and types of users access the station's resources, multiple, perhaps different, algorithms are possible for different virtual parts of the cache.) A cached file is in use when it is either being routed on the distributed cache network, or is being served to the local projects (SAM's term for the activity of data consumption)(4). Thus, in general, a replica will be deleted if it has not been in use long enough and if the disk space is claimed for other purposes. Similar to the routing, by configuring the cache replacement algorithms and allocating cache for different user groups or types[2], the local system administrators control the lifetime of a replica and therefore affect the global replication process.

Occasionally, such dynamic caching and replication may be undesirable for some critical applications requiring that all their files be local for immediate access. SAM provides means for an administrator to lock a dataset on disk statically. Static replication is an effective reservation tool, but in general it sacrifices overall efficiency as it may result in unused data occupying the cache.

Implicit in our discussion so far has been a global *replica catalog*, also known as a catalog of file locations. This catalog is updated every time a file enters or leaves disk cache (or an MSS copy is created). The catalog is consulted in the course of file retrieval, for finding the best replica. SAM implements this catalog as a part of the meta-data database (the "M" in "SAM").

---

[2] SAM categorizes data access by *access modes*. These types, or modes of access play a primary role in global resource management, not covered here (1).

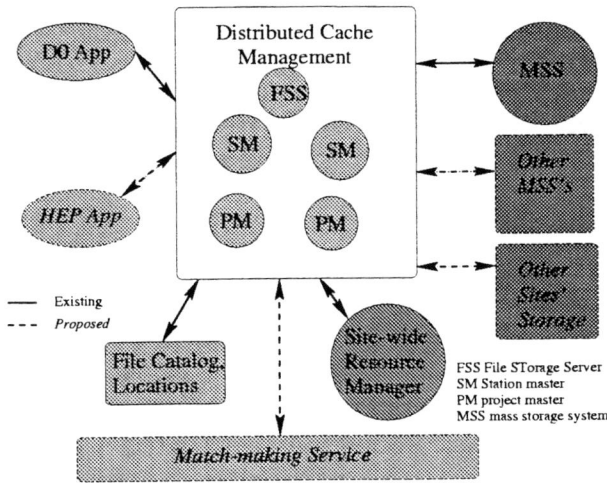

FIGURE 2. SAM's distributed caching in the grid picture.

SAM IN THE GRID

SAM is a highly modular and flexible system. In Figure 2, we show how its distributed caching framework can be used in the Grid. First, it can serve arbitrary application performing sequential access to the data (on the left). Second, it may use multiple abstract MSS's for transferring files (on the right). Third, SAM can be interfaced to an external replica catalog, such as Condor's match-making service (7) (at the bottom).

SUMMARY

In this paper, we have described distributed caching in the Sequential Access through Meta-data (SAM). We have outlined caching with the emphasis on file routing and replication. SAM is a real system, with many components having been in production since year 1999. At the time of writing this paper, SAM is however in active development; we encourage the reader to check SAM's web pages for its evolving design and project status.

REFERENCES

1. The SAM project home page http:d0db.fnal.gov/sam/.
2. The Particle Physics Data Grid, Fermilab page http://grid.fnal.gov/ppdg/.
3. L. Lueking *et al*, The Data Access Layer for D0 Run II: Design and Features of SAM, presented by V. White at *The International Conference on Computing in High Energy and Nuclear Physics* (CHEP 2000), February, 2000, Padova, Italy.
4. I. Terekhov and V. White, Distributed Data Access in the Sequential Access Model at the D0 Experiment at Fermilab, in *The Ninth IEEE International Symposium on High Performance Distributed Computing*, Pittsburgh, Pennsylvania, August, 2000 in Proceedings.
5. The Enstore project home page http://www-isd.fnal.gov/enstore/.
6. The Globus project home page http://www.globus.org/.
7. The Condor project home page http://www.cs.wisc.edu/condor/.

The Control Actions Framework

Paolo Calafiura

Lawrence Berkeley National Laboratory, 1 Cyclotron Road, Berkeley, CA 94720

Abstract. We have developed a prototype application framework that controls the execution of HEP software modules. The control actions framework is based on the well-known "hooks" architecture that many HEP developers are familiar with since the Fortran days. This design focuses on the critical association among the module instances and the controller component and features strongly typed associations, extensive run-time configurability and minimal compile-time dependency.

INTRODUCTION

Several HEP application frameworks[1,2] follow the "hooks" architecture, in which a controller exposes "hooks" at well determined stages of the application (for example when a new record has been read). The "physics" modules provide corresponding "callback" methods that the controller invokes to control the physics module execution. In this architecture, the application behaves like an event-synchronized set of Finite State Machines (FSM). The controller (an FSM itself) drives the physics modules FSM and keeps them synchronized.

This architecture is simple, familiar to the developers and the users, and works remarkably well for a large fraction of the data analysis and reconstruction tasks. Unfortunately some of the existing HEP frameworks that implement this architecture introduce a tight coupling among the controller and the physics modules: for each hook exposed by the framework the developers of the controllers must provide a callback method. This can be tedious and error prone to code. What is worse, this coupling makes the framework very rigid: introducing a new hook quickly becomes an expensive operation: every existing "physics" module must be edited and/or recompiled to implement the matching callback method.

We want to support a more dynamic approach that allows the module developers to provide callbacks only for the actions (hooks) their modules must react to, and to introduce new actions as needed without triggering massive recompilations. The framework will notify modules only about the actions they are registered for, and it will control the order in which the modules are run.

This dynamic association among the modules and the event loop is clearly related to the well-known Observer pattern [3]. But the observable/observer association is a one-to-many relationship. At the core of the hooks architecture there are actually two many-to-many associations:

- an event source (a data file, the operating system, the command prompt) triggers many actions and an action can be triggered by many sources

- each module performs zero or more actions and each action is performed by one or more module.

DESIGN AND IMPLEMENTATION

The focal point of our design is the Control Action concept. The Control Action provides the proverbial extra level of indirection we need to break up the two many to many sources-action-modules associations. The Action is an Observer of the event sources and at the same time an Observable[1] that notifies the physics module to run their matching callback method. For example (Fig. 1) a "FileClosed" Action observes one or more event sources waiting for a file to be closed. When a source broadcasts a FileClosed message, the Action takes control of the execution flow and runs those modules that registered their interest in the "FileClosed" event. After all registered modules have been notified the Action returns control to the event loop. Each action instance has a distinct type

```
class Init :
    public ControlAction<Init> {};
class newRecord:
    public ControlAction<newRecord> {};
class Finalize :
    public ControlAction<Finalize> {};
```

Note the use of static polymorphism[4] to tag each ControlAction with its own type. Since the Action knows its type it sends a Typed Message [5] to the modules it schedules. Only modules that provide action handlers (callbacks) of matching type can be hooked to an Action. For example a typical physics module will implement three action handlers to hook up to the "Init", "newRecord" and "Finalize" framework states:

```
class HitFinder :
    virtual public TActionHandler<Init>,
    virtual public TActionHandler<newRecord>,
    virtual public TActionHandler<Finalize> ….
```

TActionHandler<Action> is an interface class with a "run" method that HitFinder must implement. For example TActionHandler<Init>::run() will be the "Init" callback method.

HitFinder can not be hooked up to, say, the FileClose action: the compiler would issue an error message

```
"testSource.cc", line 32: error:
argument of type "HitFinder*" is
incompatible with parameter of type
"TEventObserver <FileCloseAction> *"
```

---
[1] such an entity is sometimes called a Repeater.

This is one of the main advantages of implementing the ControlAction as a Typed Message: this technique allows the compiler to check (to a large extent) the consistency of the actions-modules associations. At the same time a module has no compile-time dependency on the actions that it does not handle. In particular a new Action can be introduced at any time without recompiling any existing module. Finally this framework allows to configure at run time the event sources, the active actions and the module instances that hook up to them.

DISCUSSION AND CONCLUSION

The Typed Message pattern is not the most common choice for the hooks architecture implementation. For example, the dynamic proxy classes introduced in Java 1.3 [6] follow the classic "event registry" approach: a hook registry class maintains a table to match hooks and providers (actions and sources) and a table to match hooks and callbacks. Some of the more sophisticated HEP control framework [7] also use the registry approach to control module execution. These frameworks are based on traditional dynamic polymorphism and deal only with abstract hooks and callbacks. Hence a mismatch among hooks and callbacks types will only be uncovered at run time, using a potentially inefficient dynamic cast. On the other hand an event registry is more flexible than the typed messages we are using. One could for example dynamically load a library with a new set of actions into a running framework and start using them right away.

As always, the Control Action Framework is no silver bullet. There are certainly application domains (on-line monitoring and control come to mind) where a traditional event-registry system would be more appropriate. On the other hand we believe that the Control Action Framework is a better (safer and faster) choice for the simulation and reconstruction domains, which have a reasonably stable set of internal states and a large and fast changing pool of modules hooked up to them.

We have developed a functional prototype of the framework, which implements all the features we have described[2]. The prototype and more details are available from the project web site:

http://electra.lbl.gov/Atlas/framework/
controlstates/actiondesign.html

---
[2] and a few more which have to do with the user interface and with the data model interaction.

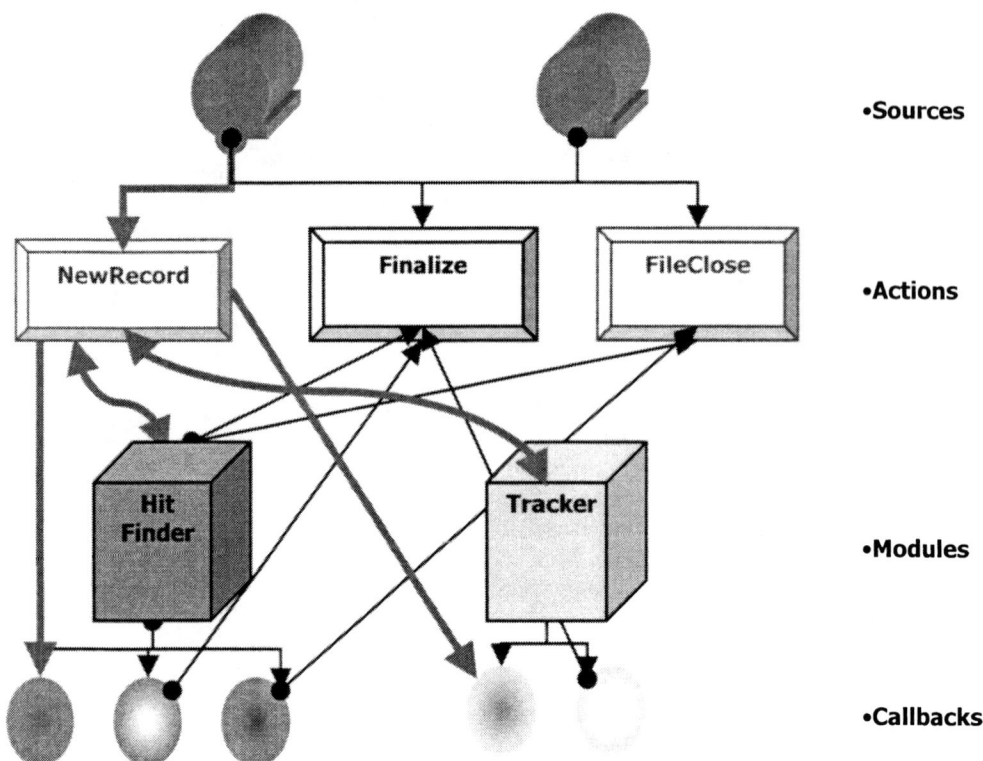

FIGURE 1. The Control Action Framework

ACKNOWLEDGMENTS

We would like to thank for their comments, suggestions and criticisms: Jim Kowalkowski, Vincenzo Innocente, Charles Leggett, Pere Mato, John Milford, Dave Quarrie, Marjorie Shapiro, Lassi Tuura, Craig Tull and Laurent Vacavant.

REFERENCES

1. Sexton-Kennedy, E., "A Users Guide to the AC++ Framework" http://www-cdf.fnal.gov/upgrades/computing/projects/framework/

2. Cattaneo, M. et al, "GAUDI- The Software Architecture and Framework for building LHCb Data processing Applications" in *CHEP2000 Proceedings*, http://chep2000.pd.infn.it/abs/abs_a152.htm

3. Gamma, E. et al. *Design Patterns, Elements of Reusable Object-Oriented Software*, Addison-Wesley, 1995

4. Barton, J. and Nackman, L. *Scientific and Engineering C++: An Introduction with Advanced Techniques and Examples*, Addison-Wesley, 1994

5. Vlissides, J., *Pattern Hatching: Design Patterns Applied*, Addison-Wesley, 1998.

6. Modi, T. "Understanding the Dynamic Proxy Classes in Java 1.3," Java Report, SIGS Publications, New York, NY, Feb. 2001.

7. Kowalkowski, J. et al, "D0 Offline Reconstruction and Analysis Control Framework" in *CHEP2000 Proceedings*, http://chep2000.pd.infn.it/abs/abs_a230.htm

POSTER SESSION

EMS: A Framework for Data Acquisition and Analysis

J.M.Nogiec, J.Sim, K.Trombly-Freytag, and D.Walbridge

Fermi National Accelerator Laboratory[*], *Batavia, Illinois 60510*

Abstract. The Extensible Measurement System (EMS) is a universal Java framework for building data analysis and test systems. The objective of the EMS project is to replace a multitude of different existing systems with a single expandable system, capable of accommodating various test and analysis scenarios and varying algorithms. The EMS framework is based on component technology, graphical assembly of systems, introspection and flexibility to accommodate various data processing and data acquisition components. Core system components, common to many application domains, have been identified and designed together with the domain-specific components for the measurement of accelerator magnets. The EMS employs several modern technologies and the result is a highly portable, configurable, and potentially distributed system, with the capability of parallel signal data processing, parameterized test scripting, and run-time reconfiguration.

INTRODUCTION

Reusability is one of the great promises of object-oriented technologies, which can measurably increase productivity in software development. The objective of the EMS project is to provide reuse in the test and data analysis application domain, *i.e.,* to replace a multitude of different existing systems with a single expandable system, capable of accommodating different test and analysis scenarios and varying algorithms.

The EMS uses a component-based framework that allows for reuse of both domain-specific and general-purpose components. Reuse of domain components provides for the highest possible level of reusability, because these components represent collections of related classes that work together to support specific domain-oriented functionality.

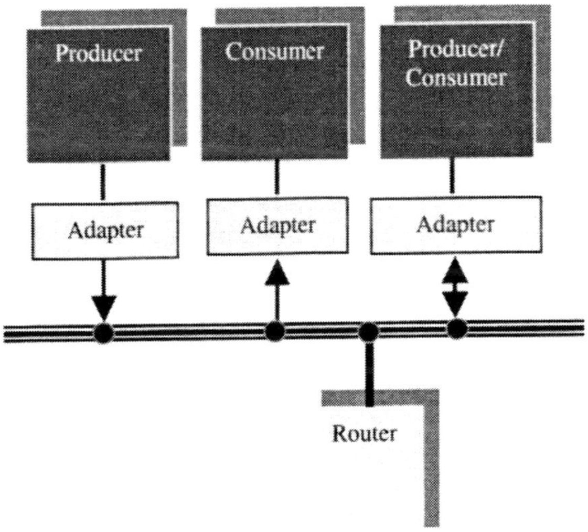

FIGURE 1. Communication over a software bus.

ARCHITECTURE

The EMS employs a component-based framework that creates a foundation from which developers can build platform-independent applications. Within the system, components communicate through messages (events) exchanged over a software bus (Figure 1).

[*] Operated by the Universities Research Association under contract with the U.S. Department of Energy

Various communication patterns are supported including unicast, multicast, and broadcast. The communication links/patterns are enforced by the router component and can be defined externally from the communicating components to form routing tables. Both contents-based (subscriber/provider model) and address-based communication is provided with both the source routing and routing tables methods. The bus conveys four independently routed categories of

CP583, *Advanced Computing and Analysis Techniques in Physics Research: VII International Workshop*
edited by P. C. Bhat and M. Kasemann
2001 American Institute of Physics 0-7354-0023-7

events: data, controls, debug information, and exceptions.

The core system consists of an architectural framework supporting communication and system assembly, and a set of core components.

All components are Java Beans connected to the bus via adapter objects. In order to send and receive messages over the bus, each component has to implement appropriate communication interfaces. These interfaces allow for exchanging of control, data, debug, and exception events. A hierarchy of abstract components implementing necessary interfaces provides a foundation for building all components.

Depending on their role in communication, components can be producers or consumers. A producer of one event can also be a consumer of some other event or events. Interactions between two communicating components are shown in the collaboration diagram (Figure 2).

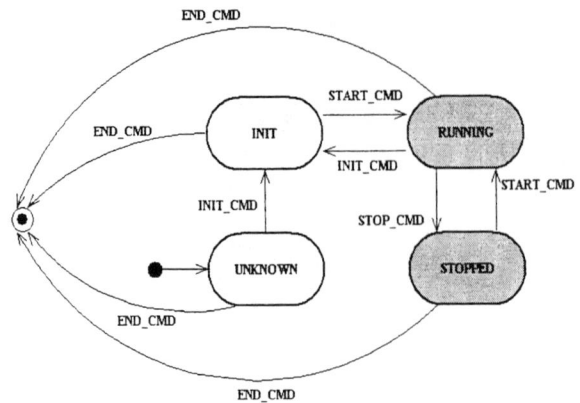

FIGURE 3. Standard component states.

Some of the components are part of the graphical user interface whereas others have no visible interface. Notable core GUI components include the property editor and control panel and the traffic monitor (Figure 4).

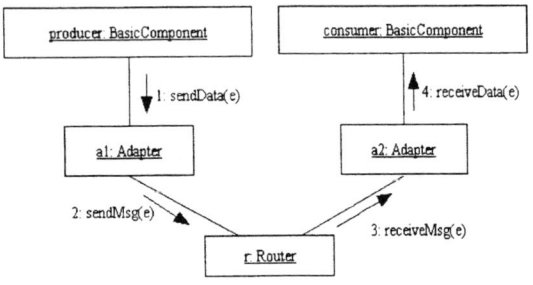

FIGURE 2. Component collaboration diagram.

COMPONENTS

Components have well defined sets of states. Transitions between states are caused by the internal behavior of the component and can be enforced externally by sending appropriate control signals to the component. A standard set of states is shown in Figure 3.

The system components can be divided into the following categories: general-purpose core components (*e.g.*, traffic monitor, property editor), application framework components (*e.g.*, measurement framework), and highly specialized application components (*e.g.*, harmonics measurement application).

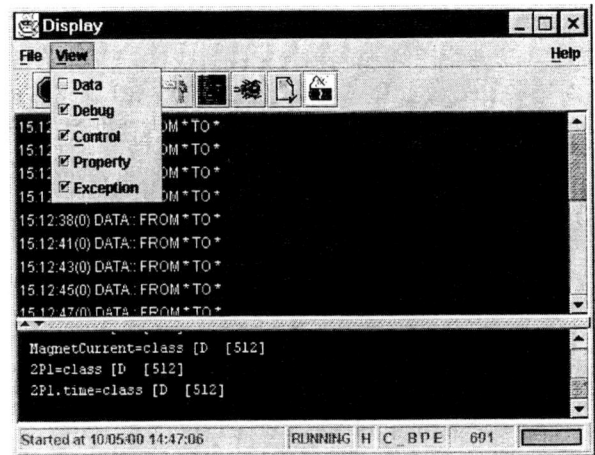

FIGURE 4. The Traffic Monitor GUI.

The property editor and control panel enables run-time manipulation of components' properties, whereas the traffic monitor provides run-time information on bus communication. Notable measurement platform components include the graphical data display (Figure 5) and the numerical data display. These components are capable of displaying all data generated by any other system component.

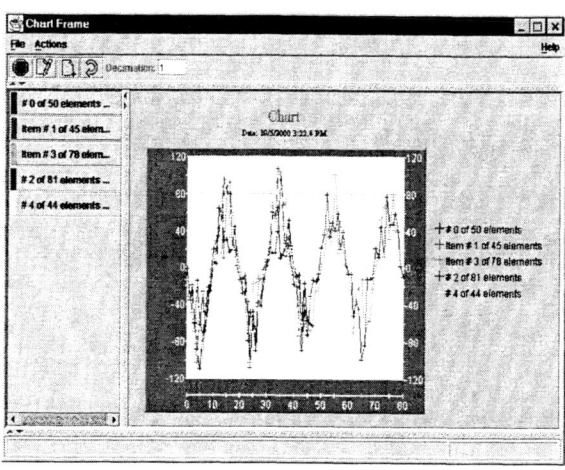

FIGURE 5. The Graphical Data Display GUI.

OBJECT REPRESENTATION

The eXtensible Markup Language (XML) is a diverse language that can be customized to enable programmers to exchange and display information. By representing Java domain objects in XML one can achieve data exchange with non-Java platforms and have a portable and human-readable representation. XML can also be translated into an object model, and *vice versa*. These XML features convinced the authors to use XML as a standard way of externally representing objects in EMS. It has been applied to describe the system configuration, calibrations, parameters, and results.

Since the lifetime of the object frequently exceeds the life of the application process, one has to provide a mechanism to support object persistence, that is, the storage of objects in a non-volatile memory. In the application of EMS to measurement systems, object persistence is implemented via Object Design's Persistent Storage Engine (PSE Pro). This product provides a seamless binding to the Java language and drastically reduces the amount of code required to manage persistent objects. Other possible solutions to the object persistency problem include serialization, object databases, and relational databases.

XML descriptions of data and object parameters can be used to generate objects. Such information contained within an object can also be written to an XML document. In addition, objects can be stored in an object database and subsequently restored from the database. Thus, during its lifetime an object can exist in one or more of these forms: an XML document, an object, or as a persistent object stored in a database.

SYSTEM USE

EMS-based systems can be assembled from pre-existing components without traditional programming. First, an XML description of the system is prepared using an XML editor, and then the system is generated according to this description. The XML description contains, in addition to definitions of components and their properties, routing information and definitions of the initial control signals.

The state of the system, including components' properties, can be manipulated at run-time and saved at any time as an XML document.

The next version of EMS, which is currently under development, will enable the user to define routing configurations graphically and allow for easy dynamic reconfiguration.

CONCLUSIONS

The EMS is an extensible, vendor-independent framework with characteristically shallow inheritance hierarchies. This framework allows for adding new services and new application platforms while maintaining a consistent set of APIs.

It replaces traditional application development involving coding of individual applications with assembly of systems from a set of available components. Therefore, it effectively separates the development process from the application building process and allows users with limited programming skills to create new or modified systems.

Integrated debugging, exception and event handling, dynamic reconfiguration, and run-time property manipulation uniquely characterize the core EMS framework.

ACKNOWLEDGEMENTS

The authors would like to thank John Tompkins for his continuous encouragement and support of the idea of component-based development and frameworks. Thanks also to Joe DiMarco, Hank Glass, Phil Schlabach, and George Velev for comments and contributions to the magnetic measurement platform built on top of EMS.

Effects of Limited Resources in 3D Real-time Simulation of an Extended ECHO Complex Adaptive System Model

Dana M. Dominiak[*], Frank Rinaldo[♀], Martha W. Evans[¶]

[*]*Department of Computer Science, Illinois Institute of Technology, Daniel F. and Ada L. Rice Campus, 206 East Loop Drive, Wheaton, IL 60187-8489 USA*
[♀]*Comnet International Co, 3030 Warrenville Road, Suite 320, Lisle, IL 60532 USA*
[¶]*Department of Computer Science, Illinois Institute of Technology, Stuart Building, Chicago, IL 60616 USA*

Abstract. An evolutionary model of adaptive agents called 'ECHO' was proposed by John Holland. ECHO is a first step toward mathematical theory in the field of complex adaptive systems. Researchers in numerous disciplines have used the existing ECHO simulation both to model and to explain complex system behaviors. This paper describes the effects of limited resources in a 3D simulation of an extended Holland ECHO model. In this simulation, adaptive agents move about the ECHO terrain and interact with other agents in real-time. Adaptive agents are bred using a genetic algorithm. The model's environment contains limited resources, represented as symbols. Elaborate relationships are developed by the agents to utilize resources through both competition and cooperation. Researchers have a better tool by which to identify and explain complex adaptive system behavior by observing the emergence of complexity first hand.

INTRODUCTION

The genetic algorithm (GA) has been more successful finding solutions to complex, nonlinear problems from the "bottom-up" than traditional "top-down" methods. However, a challenge in the search of a complex solution space has always been forcing a GA to find something other than local optima. This is caused by evolution's tendency to quickly fill a niche and therefore effectively abandon the search for alternative solutions.

A complex adaptive system (CAS) describes a model containing intricacy and fluidity found in natural systems. As relatively simple agents interact in a dynamical setting, global behaviors will emerge. These global effects emerge from more simple, local interactions. CAS have typically been used to model biological and other natural systems, which are greatly different than the specific minimum/maximum search problems that the GA has traditionally been used to solve.

Complex and living systems are difficult to understand by top-down analysis. Instead, they are better understood by synthesis [1]. Therefore, such models should be implemented as computer simulations [2]. Understanding of biological processes in particular is one of the primary purposes of studying CAS, however, the organization of nature in the broader sense is an even greater justification for such models. CAS theory seeks to identify rules and processes which describe all complex systems, such as economies, ecosystems, animal and insect population behaviors, immune systems, embryogenesis, etc. Control or prediction of complex systems might become possible if "lever points" and mechanisms can be found which allow for a better understanding of these systems [3].

It is hoped that by discovering the processes involved in creating complexity from adaptive simulation, the rules governing real living systems will be discovered [4, 3, 5, 6].

Creating a system that resembles natural systems is a main purpose for implementing such a simulation. It is likely that the causes of emergence will be identified with such a 'constructionist' [6] tool, eventually leading to theory, and ultimately, to control and prediction of natural systems.

Additionally, it is often more cost-effective to develop many simple agents than it is to develop a single complicated agent [1, 7]. Therefore, any improved CAS techniques are valuable to industry and scientists in the hopes of evolving better methods to solve difficult machine problems and facilitate natural system simulations.

THE ECHO MODEL

The ECHO model was initially proposed by John Holland [3], father of genetic algorithms, as a vehicle for carrying out "thought experiments" on CAS. Numerous ECHO computer simulations have been subsequently implemented. These simulations capture the fluidity and complexity of Holland's model better than pure cognitive exercises. A screen image of the extended ECHO model discussed in this paper can be seen in Figure 1.

Most ecological models omit evolutionary methods [8]. However, ECHO consists of a population of agents that are modified over time by a GA. The agents adapt and evolve by selective pressures imposed by their environment, including other agents.

The agents' world in ECHO consists of limited resources in the form of symbols: {a, b, c, d, e, … }. Competitive and cooperative relationships will be developed among agents in order to utilize these limited resources [8].

Non-linearities inherent in complex systems are difficult to model analytically [8]. Therefore, simulations such as ECHO facilitate the analysis of these phenomena. Such a model will help to identify and clarify the properties of CAS, hopefully leading to mathematical theory.

ECHO agents must compete for limited resources. Each site in the world might contain agents, an inflow of resources, or both.

Interactions are dictated by offense and defense tags, which are possessed by each agent. The model possesses an implicit fitness measure, an important characteristic for inaugurating complexity [3, 9].

ECHO was used by environmental researchers to model food web complexity [10]. In a paper describing this model, researchers were able to show that individual behaviors (i.e. local rules) influenced the global food web's complexity. It was proved that explicitly deriving differential equations was not necessarily the most facile or accurate method for modeling food web complexity.

EXTENSIONS TO ECHO

Mechanisms and procedures were added to Holland's basic ECHO model to emphasize spatially oriented characteristics and physical hierarchies inherent in natural systems. A primary goal of the initial ECHO design was stripping away unnecessary details while maintaining diverse complexity. A principal goal of our extended model was retaining this design criterion.

While Holland's ECHO models are an extraordinary step toward CAS theory, they are far from replicating the complexity present in actual natural systems. This lucidity was no accident, and helps to identify lever points and mechanisms of CAS by utilizing the simplest of approaches. However, a critical role in the evolution of adaptive agents has recently been shown to be the environment [4, 9]. Therefore, perhaps the minimal addition of some environmental factors to ECHO are well justified.

The extensions to ECHO include agent resource recycling, resource adhesion tags, resource hierarchies, and resource decay.

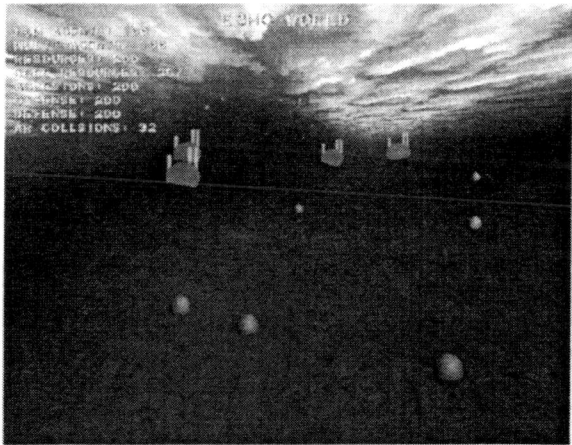

FIGURE 1. A Screen of the Extended ECHO Visualization

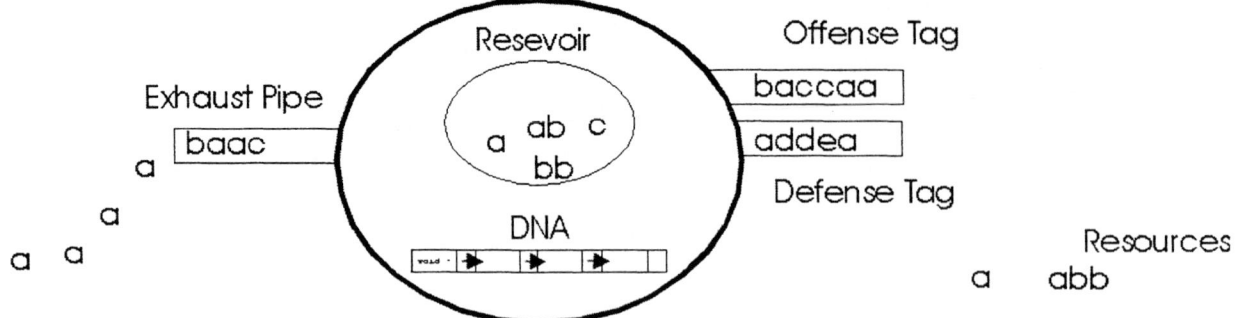

FIGURE 2. An Extended ECHO Agent Showing Tagging and New Exhaust of Resources.

The extended ECHO agent is pictured in Figure 2. The agent contains the classic ECHO offense and defense tags, as well as a new exhaust tag, allowing resources to be recycled back in the environment by a method other than direct agent interaction. The evolutionary model implemented here using a classic GA and tagged agents has led to the emergence of both competitive and cooperative relationships such as parasitism and symbiosis.

TABLE 1. Evolutionary Techniques

General Technique	Implementation
Randomness	Mutation, Simulated Annealing
Co-evolution	Parasitism, Symbiosis, Competition, Cooperation, Arms Races
Speciation	Tagging, Multiple Populations

CONCLUSION

This CAS visualization tool allows the observation, interpretation and explanation of emergent and complex phenomena found in natural systems. Such a tool allows researchers to observe complex interactions among simple agents, re-run interesting evolutionary models from a known initial condition, and observe the relationship of resources among adaptive agents. Such a tool is a good first step in the identification, prediction, and control of emergent complex phenomena.

ACKNOWLEDGMENTS

Thanks to Cristian Soulos and Pascal Pochol for their work on the 3D engine and low-level system routines. Research funding and hardware resources provided by Webfoot Technologies, Inc.

REFERENCES

1. Bonabeau, E. W., and Theraulaz, G., "Why Do We Need Artificial Life?" in *Artificial Life: An Overview*, edited by C. G. Langton, 1997.

2. Resnick, M., "New Paradigms for Computing, New Paradigms for Thinking", in *Computers and Exploratory Learning*, edited by A. diSessa, C. Hoyles, and R. Noss, 1995, pp. 31-43.

3. Holland, J., *Hidden Order: How Adaptation Builds Complexity*, Reading, Massachusetts: Addison-Wesley Publishing Company Inc., 1995.

4. Dawkins, R., *The Selfish Gene*, New York: Oxford University Press, 1989.

5. Keely, B. I., "Evaluating Artificial Life and Artificial Organisms", in *Artificial Life V*, edited by C. G. Langton and K. Shimohara, Proceedings of the Fifth International Workshop on the Synthesis and Simulation of Living Systems, 1997, pp. 264-271.

6. Resnick, M., "Learning About Life", *Artificial Life: An Overview*, edited by C. G. Langton, 1997, pp. 229–241.

7. Maes, P., "Modeling Adaptive Autonomous Agents", *Artificial Life: An Overview*, edited by C. G. Langton, 1997, pp. 135–162.

8. Hraber, P. T., Jones, T. and Forrest, S., "The Ecology of ECHO", in *Artificial Life*, 3(3), 1997, pp. 165-190.

9. Kawata, M. and Toquenaga, Y., "From Artificial Individuals to Global Patterns", in *TREE*, (9)11, 1994, pp. 417-421.

10. Schmitz, O.J. and Booth, G., "Modeling Food Web Complexity: The Consequence of Individual-based Spatially Explicit Behavioral Ecology on Trophic Interactions", Yale University, 1996.

High Performance Visual Display for HENP Detectors

Michael McGuigan, Gordon Smith, John Spiletic

{mcguigan,smith3,spiletic}@bnl.gov
Information Technology Division, Brookhaven National Lab
Upton, NY 11973

Valeri Fine, Pavel Nevski

{fine,nevski}@bnl.gov
Physics Dept., STAR/ATLAS,Brookhaven National Lab
Upton, NY 11973

Abstract. A high end visual display for High Energy Nuclear Physics (HENP) detectors is necessary because of the sheer size and complexity of the detector. For BNL this display will be of special interest because of STAR and ATLAS. To load, rotate, query, and debug simulation code with a modern detector simply takes too long even on a powerful work station. To visualize the HENP detectors with maximal performance we have developed software with the following characteristics. We develop a visual display of HENP detectors on BNL multiprocessor visualization server at multiple level of detail. We work with general and generic detector framework consistent with ROOT, GAUDI etc, to avoid conflicting with the many graphic development groups associated with specific detectors like STAR and ATLAS. We develop advanced OpenGL features such as transparency and polarized stereoscopy. We enable collaborative viewing of detector and events by directly running the analysis in BNL stereoscopic theatre. We construct enhanced interactive control, including the ability to slice, search and mark areas of the detector. We incorporate the ability to make a high quality still image of a view of the detector and the ability to generate animations and a fly through of the detector and output these to MPEG or VRML models. We develop data compression hardware and software so that remote interactive visualization will be possible among dispersed collaborators. We obtain real time visual display for events accumulated during simulations.

INTRODUCTION

In this paper we discuss the construction of general high end visualization tools for HENP detectors and data.

The BNL visualization server is a multiprocessor SGI Onyx2 graphics supercomputer with the ability to smoothly visualize large data sets with optimized performance and advanced OpenGL features. Such systems are unusual in HENP and too expensive for small institutions but are beginning to appear in central locations such as CERN and Fermilab. It is important that the facility be in close proximity to the experimental data to avoid slow down from the transfer of large data sets.

Modern detectors like STAR and ATLAS have several groups designing graphics software specifically for the detector. To achieve maximum value to these groups and to avoid porting many independent applications to the BNL visualization server we shall work in as general and generic framework of detector design as possible. The code should be able to read STAR and ATLAS or other experiment's detectors in the same framework using a translation of descriptive GEANT geometry.

Our focus will be on optimizing and enhancing the OpenGL performance for high end graphics on the BNL visualization server. In addition the ability of the BNL visualization server to drive a sophisticated stereoscopic visualization system is very helpful in understanding the depth information for the detector. This is not gratuitous use of stereo. With a flat display it can be difficult to judge spatial relationships between tracks and events, not to mention subtle but important effects such as resolution tails, tracks, etc.

Visualization is extremely important for debugging detector simulation codes. The complexity of the modern detector is enormous. Simply viewing several levels of structure together with tracks through these levels will bring even powerful workstations to a standstill. Also interactive visualization is necessary to view the effects of changes in the code.

VISUALIZATIONS

The visualizations are illustrated with the accompanying figures. In Figure 1 the stereoscopic visualization of the STAR detector at RHIC is depicted. In this case stereoscopy was achieved through two separate BARCO projectors, one for the left eye and one for the right. Polarized filters are placed over the projectors and the image is rear projected through a special screen that preserves polarization. The viewer wears inexpensive polarized glasses to observe the stereo effect. Theoretical simulations of the formation of the quark gluon plasma in gold on gold collisions can be observed in theatre as well [1].

In Figure 2 the visualization of the ATLAS detector for the LHC is shown. This detector is a thousand times as complex as STAR and will be used to explore physics beyond the Standard Model. The visualization represents an exploded view of the detector so that the internal structure becomes visible. It is also possible to merge the detector with event simulators as we are working with a OpenGL viewer from the ROOT physics data analysis package [2].

In Figure 3 we show the visualization of actual gold on gold collisions at RHIC. The color indicates level of ionization with higher values toward red and lower values toward blue. Over a thousand tracks are represented here interpolating through sixty thoudand hits in the STAR detector. The viewer is the examiner display from OpenInventor which has several nice features including stereo, transparency and clipping controls. OpenInventor is a visualization toolkit originally developed at SGI and has recently been made open source by them. Specialized classes for high energy physics have been developed from the HEPVIS project [3].

CONCLUSION

In summary, we have developed a general high end visual display for detector and simulations on the multiprocessor visualization server at BNL.

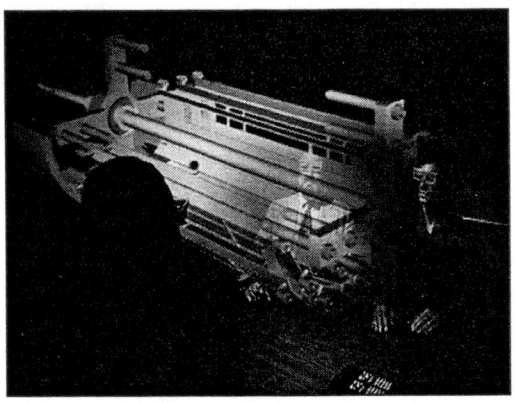

FIGURE 1. STAR detector interactive stereoscopic display.

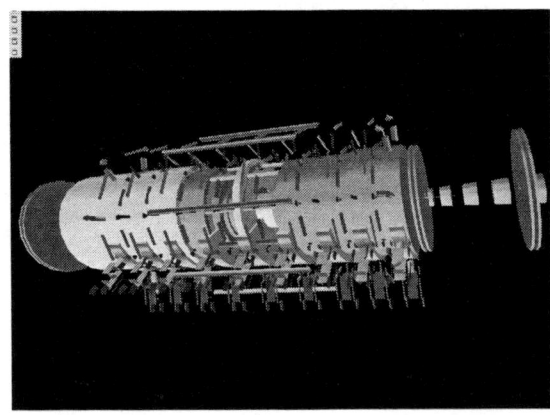

FIGURE 2. Atlas detector with 20 million distinct elements. Three levels of detail are shown.

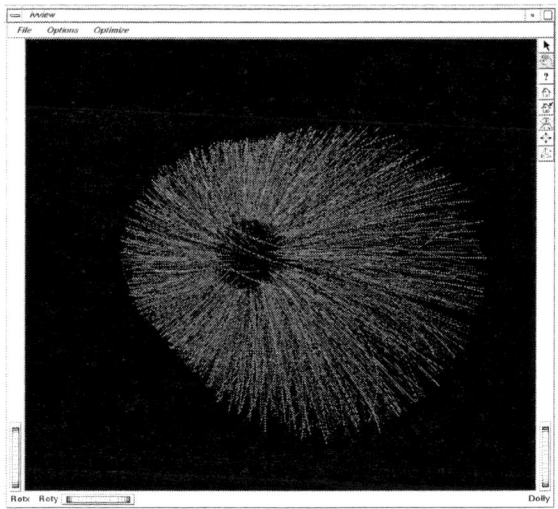

FIGURE 3. STAR event of RHIC collision. Display of event is through OpenInventor.

ACKNOWLEDGMENTS

We thank S. Murtagh, T. Hallman, M. Murtagh, J. Berger, J. S. Lange for help on various phases of the project. Work supported uder DOE contract DE-AC02-98CH10886.

REFERENCES

1. McGuigan, M., Smith, G. and Longacre, R., *VDE 2000 Proceedings*, 2000, pp. 123-127.

2. Fine, V., Brun, R. and Rademaker, F., *ACAT 2000 Proceedings*, 2000.

3. Kallenbach, J., *VDE 2000 Proceedings*, 2000, pp. 71-75.

ROOT OO model to render multi-level 3-D geometrical objects via an OpenGL

Rene Brun[1], Valeri Fine[2,3], Fons Rademakers[1]

[1]CERN - European Organization for Nuclear Research - CH-1211 Geneva 23, Switzerland
[2]Brookhaven National Laboratory, P.O. Box 5000, Upton, NY 11973
[3]Joint Institute for Nuclear Research, Dubna, Russia

Abstract. This paper presents a set of C++ low-level classes to render 3D objects within ROOT-based frameworks. This allows developing a set of viewers with different properties the user can choose from to render one and the same 3D objects.

INTRODUCTION

Modern accelerators like the Relativistic Heavy Ion Collider (RHIC) at the Brookhaven National Laboratory and the coming Large Hadron Collider (LHC) at CERN enable physicists to study the fundamental constituents of matter more closely than ever before. They accelerate beams of gold nuclei or protons to nearly the speed of light before smashing them together and creating hundreds of new particles.

Because of the sheer size and complexity of the new detectors, and the huge amounts of data modern detectors are expected to produce, 3D visualization is becoming a major component of any High Energy Physics Framework.

ROOT 3D CLASS LIBRARY

The ROOT package [1] provides two sets of classes. One of them is the "end-user" set. It is to give the user a tool to define his / her 3D object model. It is optimized for two kinds of geometry, namely for so-called "detector geometry" and "event geometry". The "detector geometry" is to describe HEP detectors. It assumes the detector definition is provided in terms of GEANT3 shapes. "Event geometry" is a collection of 3D markers and 3D polylines. ROOT end-user applications and end-user 3D OO models must not depend on the 3D rendering package and hardware / software platforms.

A 3-D OO ROOT model allows a user to create and render rather complicated 3-D objects with the various 3-D viewer classes. At present the ROOT user can use `TBrowser`, `TVirtualX`, `X3D` and OpenGL layers to draw ROOT 3D objects like `TNode`, `TShape`, `TVolumeView`, `TVolume`, `TPolyLine3D`, `TPolymarker3D` etc. These ROOT classes allow the creation of hierarchical 3D objects. Such an organization allows creating an effective OpenGL model to render the original object by a limited number of user-defined geometry levels. By default the system renders three levels of the hierarchy. This allows rendering very complicated objects (for example the ATLAS detector [2] OO model contains about 30×10^6 nodes) but still can be drawn and manipulated in a reasonable time with a simple PC.

ROOT low-level 3D classes

To achieve the "ROOT" goal – the ROOT end-user application and the end-user 3D OO model must not depend on the 3D rendering package and hardware / software platforms – ROOT 3D viewer's classes are not designed to render the generic OpenGL "pictures". It is optimized to render 3D models based on ROOT classes mentioned above. However the "Open Inventor" viewer does allow mixing the ROOT 3D class and those from Open Inventor meta-files. Figure 1 shows a UML class diagram of ROOT low-level 3D rendering classes.

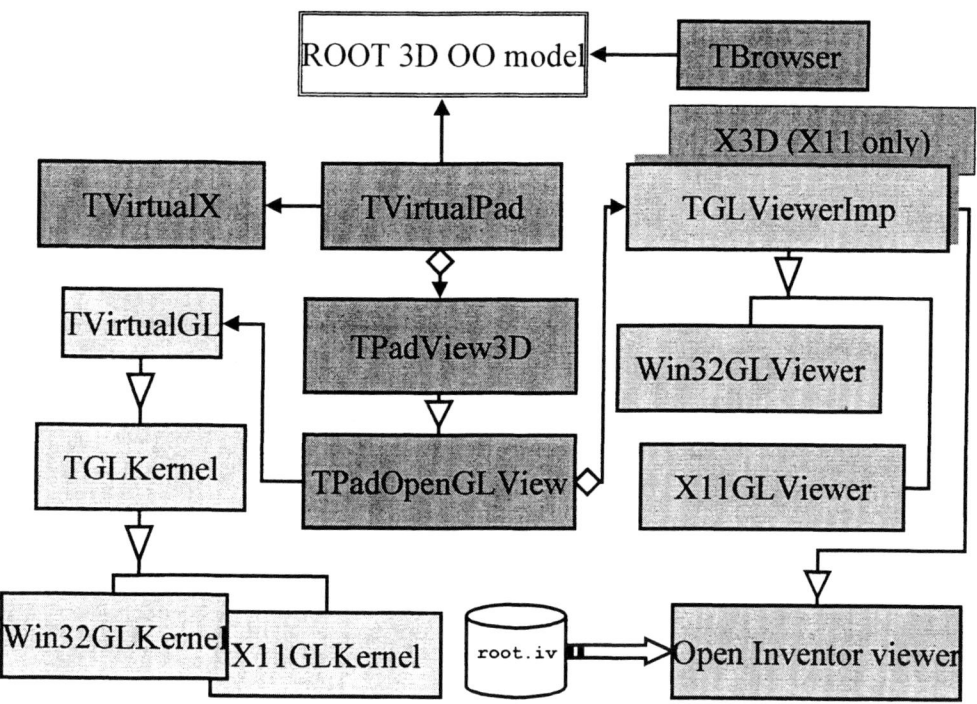

FIGURE 1. UML class diagram of ROOT low-level 3D rendering classes

At present ROOT provides four different viewers to represent its classes. Depending on current needs the user may choose any or all of them to represent his / her 3D object model.

Representing ROOT 3D objects with TBrowser class

The TBrowser class is used to represent complex models with multilevel hierarchical organization, such as HEP detector geometries. It provides an interface to select some subset of the detector before it can be drawn with other types of viewers. Figure 2 shows part of the STAR detector geometry [3] via TBrowser class.

Representing 3D objects with TVirtualPad class

TVirtualPad is another generic class one can employ to represent any object of a class derived from the ROOT TObject base class via TObject::Draw method. It gives just a simple wire frame view of 3D objects. However, it works with any kind of local and remote video terminals and does provide a facility to directly "pick" objects and access any class method of the picked C++ object interactively. One can write the whole picture out with postscript, gif or ROOT file formats. Figure 3 shows a typical "TPad" view of a simulated event within the STAR "forward TPC" used to debug the reconstruction algorithm.

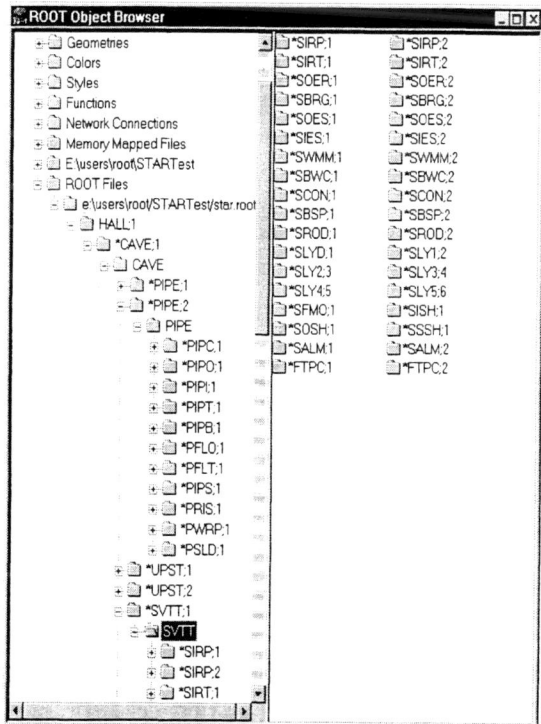

FIGURE 2. TBrowser view of 3D ROOT objects defining STAR detector via TVolume class

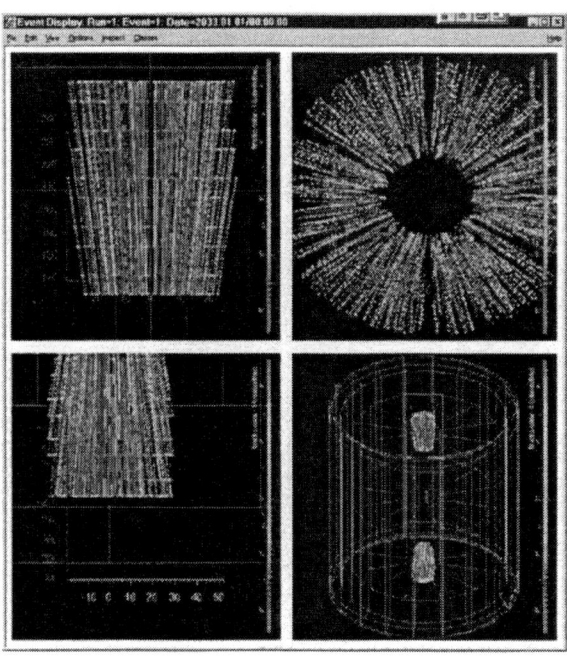

FIGURE 3. "TPad" view of the STAR simulated event

Special ROOT classes to render 3D objects

The ROOT package provides an abstract interface TPadView3D. That can be used to get the high-quality graphics for ROOT 3D objects available via third party software packages. These classes can be used only if the particular hardware /software conditions are satisfied.

UNIX users with X-terminal connections can use X3D view. The X3D class provides very fast rendering via TCP/IP connections but for the UNIX platform only, and it lacks many features one expects from solid 3D rendering packages. It is not available for the Windows platform.

The TGLViewerImp class provides an abstract interface to OpenGL rendering packages. At the moment ROOT provides two kinds of implementation of that interface. One of them is "plain" OpenGL; another one provides the advanced "Open Inventor" view whenever "Open Inventor" is available. The plain OpenGL can render ROOT 3D objects and provide primitive interaction tools. The "Open Inventor" viewer allows using the "Open Inventor "Scene Viewers" classes [4] and merging the views of the ROOT object with those provided from the Open Inventor file. Figure 4 presents the well known NA49

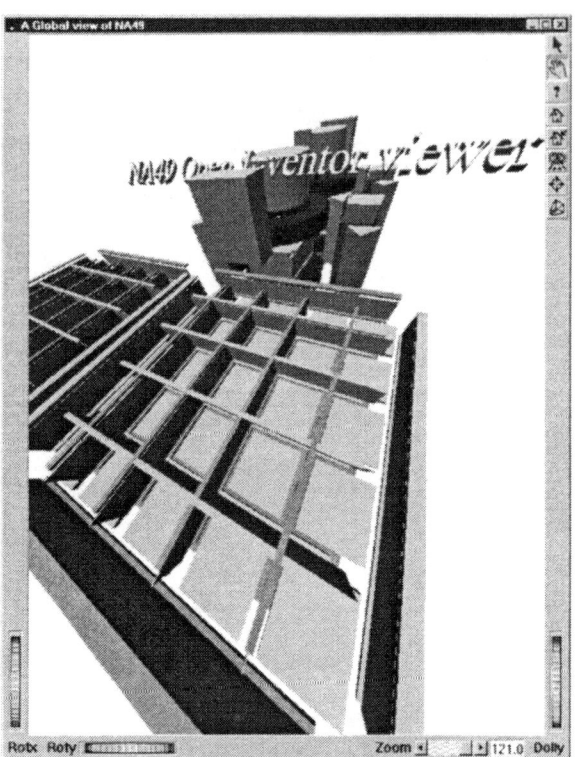

FIGURE 4. ROOT OpenGL view of NA49 detector merged with Text3 "Open Inventor" object.

detector drawn with the Open Inventor viewer and merged with the Text3 class object from the external root.iv file.

REFERENCES

1. Brun, R., Buncic, N., Fine, V., Rademakers, F., "ROOT: An Object-Oriented Framework" in AIHENP'96 Workshop Proceedings, Lausanne: 1996.

2. Fine, V., Nevski P., "OO model of STAR detector for simulation, visualization and reconstruction" in *Computing in High Energy Physics - CHEP'2000*, edited by M.Mazzucato, CHEP'2000 Conference Proceedings 181, Chicago: 2000, pp. 143-146

3. Fine, V., Fisyak Y., Perevoztchikov V., Wenaus T., "The STAR offline framework" in *Computing in High Energy Physics - CHEP'2000*, edited by M.Mazzucato, CHEP'2000 Conference Proceedings 181, Chicago: 2000, pp. 196-200.

4. Wernecke J., *The Inventor Mentor*, Addison-Wesley, 1998.

A C++ Particle Data Table Interface

L.A. Garren

Fermi National Accelerator Laboratory, Batavia, Illinois 60510

Abstract. As a result of discussions within the HEP community, we have written a C++ package which can be used to maintain a table of particle properties, including decay mode information. The classes allow for multiple tables and accept input from a number of standard sources.

INTRODUCTION

For some time, there has been a need for a C++ class embodying the information contained in the Review of Particle Properties(1). We have written HepPDT to fill this need. HepPDT allows access to particle name, particle ID, charge, nominal mass, total width, spin information, color information, constituent particles, and decay mode information. HepPDT is designed to be used by StdHepC++(2), HepMC(3), or any generated particle class. Generated particles will contain a pointer to the particle data information found in the HepPDT particle data table. HepPDT also has simple mechanisms to enable customized decay chains.

HEPPDT DESIGN

HepPDT has been designed to be used by any Monte Carlo particle generator or decay package. It contains only generic particle attributes. In principle, all information which can be found in the Review of Particle Properties(1) can be encapsulated in HepPDT. HepPDT contains particle information such as charge and nominal mass as well as decay mode information. This information is contained in a table which is accessed by a particle ID number. This ID number is defined according to the Particle Data Group's Monte Carlo numbering scheme(4).

HepPDT may be used alone or as part of the StdHepC++(2) package. StdHepC++ provides a standard generated particle class which can be used to communicate among various Monte Carlo generators and decay packages. A StdHep particle contains momentum information, generated mass, information about its generated decay, and a pointer to the appropriate HepPDT particle data. The StdHep particle inherits properties from the HepMC particle class, which also has a pointer to the relevant HepPDT particle data.

Decay information is a crucial part of the particle data in HepPDT. Standard decay information is a list of allowed decay channels with associated branching fractions, decay model names and decay model code. There may also be extra information needed by the decay model (e.g., helicity). A mechanism is provided so that the decay model code can be accessed using the decay data information instead of needing to use a series of if statements based on the decay model name. In addition, users often need the ability to "force" a particle to decay in a certain way. To do this, you must provide custom decay information. Often this information involves the entire decay chain (e.g., $D^{*+} \to D^0 \pi^+, D^0 \to K^- \pi^+$). The design provides for the generated particle to have a pointer to a custom DecayData object. If this pointer is present, it overrides the use of the DecayData associated with the generated particle's ParticleData. To customize the decay chain, the user may create particle aliases which use other special DecayData objects.

Methods are provided to create ParticleDataTable objects from Pythia, Herwig, Isajet, QQ, and EvtGen decay information. Methods are also provided to facilitate creation of custom particle and decay information. A ParticleDataTable object may be created from multiple information sources.

The design requires that ParticleDataTable objects must be fully created before they are used. Multiple data tables are allowed. Although potentially dangerous, we recognize that this is also a powerful option.

Figure 1 shows the interactions of the basic classes.

HEPPDT CLASSES

The ParticleDataTable class contains a map of ParticleData which is keyed on the ParticleID class. Particle

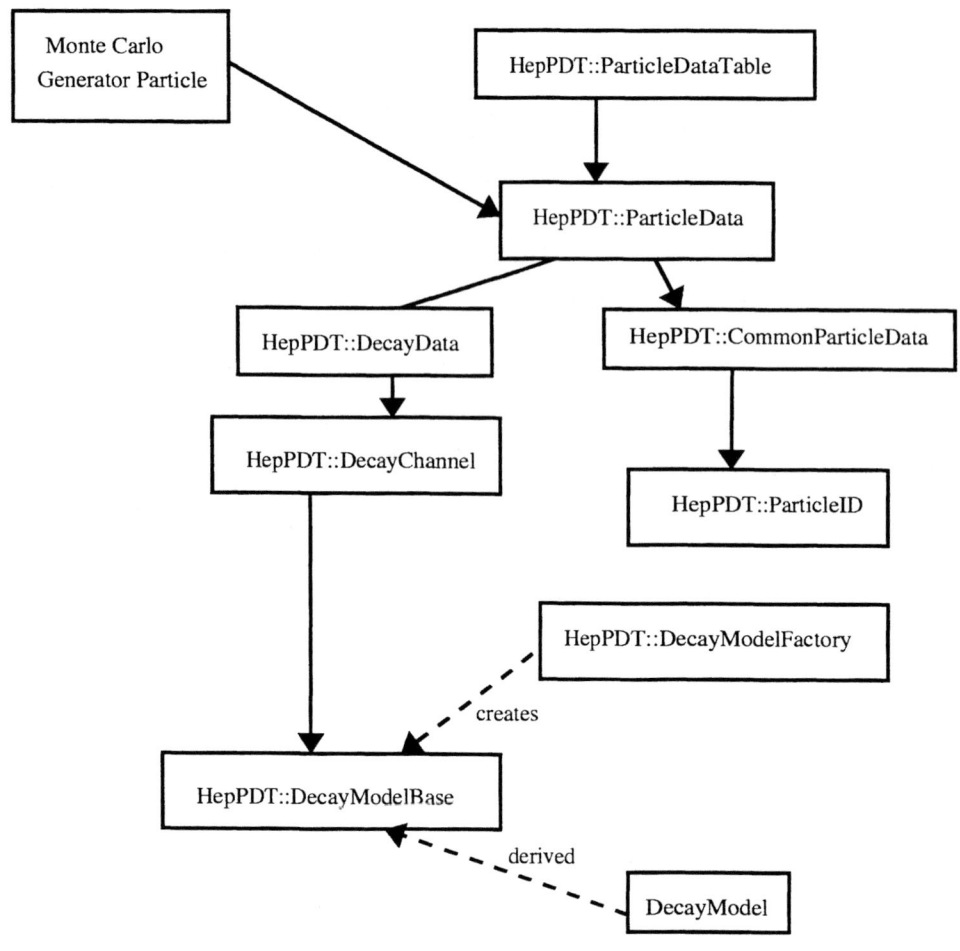

FIGURE 1. HepPDT Classes: Particle information is accessed by a pointer to ParticleData from any Monte Carlo generated particle. CommonParticleData contains particle information such as mass, charge, and total width. Decay information is found in DecayData. The ParticleDataTable contains a map of ParticleData objects, referenced by ParticleID, as well as lists of CommonParticleData and DecayData. ParticleData has indices to CommonParticleData and DecayData, as well as methods to access all relevant information. The DecayModelFactory is used to create DecayModelBase objects which are derived from user DecayModel classes.

ID aliases can be used to add custom DecayData. ParticleDataTable also contains lists of CommonParticleData and DecayData.

The ParticleID class can be used to retrieve all the information that is implied in the particle ID (e.g., charge and quark content). Boolean methods (such as isMeson, isBaryon, hasBottom, and hasTop) are provided for ease of searching for various types of particles.

The ParticleData class has iterators into the lists of CommonParticleData and DecayData. CommonParticleData is extensible and includes particle name, particle ID, charge, mass, total width with cutoffs, spin information, color information, and constituent particles (e.g., quark content). This class is not templated.

The DecayData class is a collection of DecayChannels. A generated particle may use the DecayData information from the ParticleDataTable entry or it may use a customized DecayData that allows, for instance, only a single DecayChannel. Users may add customized DecayData objects to the ParticleDataTable.

Each DecayChannel has a collection of decay channel products (which are pointers to ParticleData), a decay name, a branching fraction, an optional vector of extra decay model parameters, and a pointer to DecayModelBase. We recognize that other information, such as helicity, may be needed by a particular DecayChannel object. Because there are many options, this information is stored as a vector of doubles.

DecayModelBase is the mechanism that allows the user to invoke the actual decay method from this class. Because the decay method must know what kind of generated particle will be created, this class, and by inference the other HepPDT classes, is templated off the generated particle.

The DecayModelFactory provides an interface between the user decay methods and the ParticleDataTable. The user calls the factory before creating the ParticleDataTable object. The factory object is a singleton which registers DecayModels for each decay method. The DecayModelFactory then makes the appropriate DecayModelBase object when it is invoked during DecayData construction.

CONCLUSIONS

HepPDT provides access to all useful particle data properties and is designed to be used with any generated particle. It also contains a factory to allow the user to directly access decay model code instead of needing to use a lookup table or series of if statements based on the decay model name. HepPDT will be part of the StdHepC++ package in CLHEP(5) and is available now at http://www-pat.fnal.gov/stdhep/c++/.

REFERENCES

1. Particle Data Group: Groom, D.E. *et al.*, *The European Physical Journal* **C3**, (2000).
2. StdHepC++: http://www-pat.fnal.gov/stdhep/c++/.
3. HepMC: http://mdobbs.home.cern.ch/mdobbs/HepMC/.
4. Particle Data Group: Groom, D.E. *et al.*, *The European Physical Journal* **C3**, (2000) 205, http://www-pdg.lbl.gov/mc_particle_id_contents.html.
5. CLHEP: http://wwwinfo.cern.ch/asd/lhc++/clhep/.

Parallel Computing on a PC Cluster

X.Q. Luo[1], E.B. Gregory[1], J. C. Yang[2], Y. L. Wang[2], D. Chang[2], and Y. Lin[2]

[1]*Department of Physics, Zhongshan University, Guangzhou 510275, China*
[2]*Guoxun, Ltd, Guangzhou, China*

Abstract. The tremendous advance in computer technology in the past decade has made it possible to achieve the performance of a supercomputer on a very small budget. We have built a multi-CPU cluster of Pentium PC capable of parallel computations using the Message Passing Interface (MPI). We will discuss the configuration, performance, and application of the cluster to our work in physics.

INTRODUCTION

The lattice field theory group at the Zhongshan University has faced the familiar pressures of trying to balance the need for increased computational power against the constraints of an academic research budget. The group's primary research interest is in lattice quantum chromodynamics (QCD), the study of quark interactions. This field lends itself well to numerical simulation, but requires significant computational resources for forefront research. Traditionally this has been the domain of supercomputers. However, in recent years, advances in technology and falling hardware prices have blurred the distinction between the definition of a supercomputer and personal computer. Desktop machines of today are far more powerful than the supercomputers of yesteryear.

A further development is that the fastest computers today are in fact parallel computers, with multiple processors working together on a problem. Parallel computation has its limitations, the biggest being that it is applicable only to problems that can be divided into concurrent tasks. Furthermore, since communication between processors is usually the slowest part of the computation, the separate tasks should ideally be as independent as possible. Fortunately many computational physics problems, including lattice QCD, fall into the category of parallelizable problems. Indeed, many computational problems in the commercial world are suitable for this type of computation as well. Applications of parallel computation include graphics and animation, telecommunications and internet service, and many other fields heavily reliant on computer processing.

It is possible to join multiple cheap, fast PC type computers to build a parallel "supercomputer" with an arbitrarily high aggregate speed. A cluster of this type is called a "Beowulf Cluster" (1) and the idea was pioneered by the United States' National Aeronautics and Space Administration.

CONSTRUCTION

Hardware

One big advantage of a PC cluster over other types of supercomputers is the low cost and easy availability of the hardware components. All the hardware in our cluster is available at retail computer suppliers. This gives great flexibility in both building and the cluster and in any future upgrades or expansions we may choose to make.

Our cluster consists of ten PC type computers, each with two 500 MHz Pentium III processors inside. The logic behind dual CPU machines is that one can double the number of processors without the expense of additional, cases, power supplies, motherboards, network cards, et cetera. Also, the inter-node communication speed is faster for each pair of processors in the same box as compared to communication between separate computers. Each computer has an 8GB EIDE hard drive, 128 MB of memory, a 100Mbit/s ethernet card, a simple graphics card a floppy drive and a CDROM. In practice the CDROM, the floppy drive, and even the graphics card could be considered extraneous, as all interactions with the nodes could be done through the network. However, with these components, all of which are relatively cheap in comparison to the total cost, the operating system installation and occasional maintenance is significantly easier. One computer has a larger hard disk (20 GB), and a SCSI card for interaction with a tape drive. For the entire cluster we have only one console consisting of a keyboard, mouse and monitor.

A fast ethernet switch handles the inter-node communication. The switch has 24 ports so there is ample room for future expansion of the cluster to up to a total of 48 processors. Of course it is possible to link multiple switches or use nodes with more that two processors, so the possibilities for a larger cluster are nearly limitless. The layout of the cluster is illustrated in Figure 1.

Software

The cluster runs on the Linux operating system. Linux is powerful and inexpensive. It easily supports important features like multiple processors. It allows the configuration of a network file server. We have mounted the largest hard disk on to all of the machines in the cluster. Each machine can read and write to it as if it were physically part of that computer. Linux also supports a network information system to share user accounts across the entire cluster. One uses the same account and home directory, no matter which machine he or she logs into. Standard Linux distributions also supply C, C++, and Fortran compilers.

We can use the the cluster for parallel processing by using the message passing interface (MPI)(2), a library of communication functions and programs that allow for communication between processes on different CPUs. The programmer must design the parallel algorithm so that it appropriately divides the task among the individual processors. He or she must then include message passing functions in the code which allow information to be sent and received by the various processors. MPI is one of the most popular standards for message passing parallel programming, and is widely used in the physics community. Therefore we are able to share parallel programs in C, C++, or Fortran with collaborators elsewhere in the world who may even be running MPI on a different platform.

PERFORMANCE

Serial Benchmark

We have run the LINPACK benchmark (3), a standard serial benchmark test on our computers to measure the speed of a single processor. The benchmark showed that a single 500 MHz Pentium III processor is capable of a peak speed between 84 Mflops and 114 Mflops (million floating point operations per second) for single precision arithmetic and between 62 Mflop and 68 Mflop for double precision arithmetic. The peak aggregate speed for the entire cluster of twenty processors, is therefore about 2 Gflops.

Table 1. Comparison of performance of MPI QCD benchmark. Comparison data from Hioki and Nakamura. (6)

Machine	μ-sec/link	MB/sec
SX-4	4.50	45
SR2201	31.4	28
Cenju-3	57.42	8.1
Paragon	149	9.0
ZSU's Pentium cluster	**7.3**	**11.5**

QCD Benchmark

As we primarily developed the cluster for numerical simulations of lattice QCD, we have also performed a benchmark which specifically tests the performance in a parallel lattice QCD code. Lattice QCD simulations are well suited for parallelization (5) as they involve mostly local calculations on a multi-dimensional lattice. The algorithm can conveniently divide the lattice and assign the sections to different processors. The communication between the nodes therefore is not extremely large. Hioki and Nakamura (6) provide comparison performance data on SX-4 (NEC), SR2201 (Hitachi), Cenju-3 (NEC) and Paragon (Intel) machines. Specifically, we compare the computing time per link update in microseconds per link and the inter-node communication speed in MB/sec. The link update is a fundamental computational task within the QCD simulation and is therefore a useful standard. The test was a simulation of improved pure gauge lattice action (1×1 plaquette and 1×2 rectangle terms) on a 16^4 lattice. In each case the simulation was run on 16 processors. The results are summarised in Table 1.

Cost Comparison

We believe that such a parallel cluster of PCs may be the cheapest solution to the problem of developing computing resources for scientific simulations. In 1999, our cluster cost about US$14,000, including all hardware and software. This equates to roughly $7/Mflop. We can compare this to a commercial supercomputer. The Cray T3E-1200E uses 1.2 Gflop processors (4). The basic starting model comes with six processors for a total peak speed of 7.2 Gflops. The cost for the six node model, though, is US$630,000, or $87.50/Mflop. Our home made supercomputer is more than an order of magnitude cheaper.

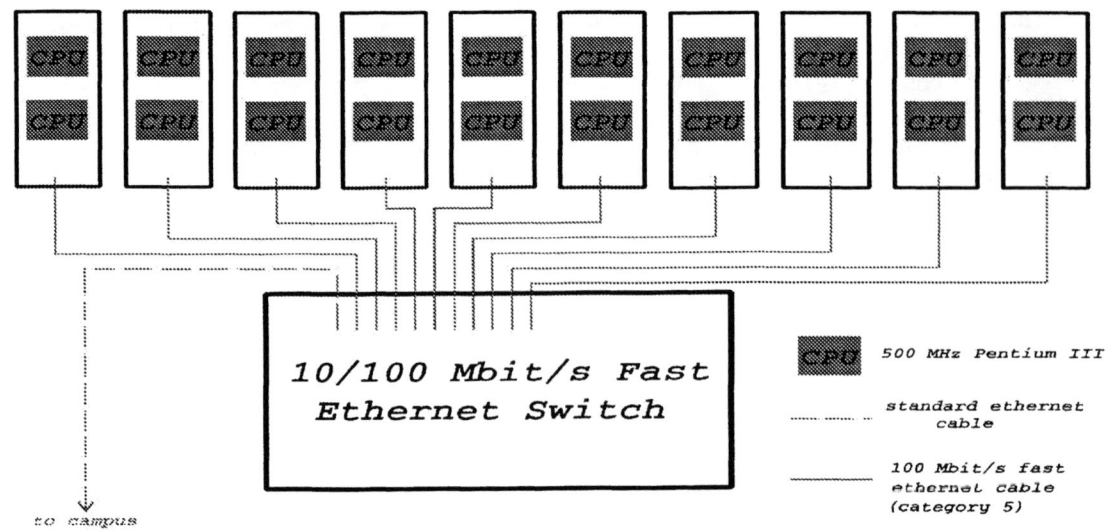

FIGURE 1. The layout of a 10 dual-CPU node cluster.

Of course this is a naive comparison, as the Cray differs in many ways. Notably, faster individual processors means serial jobs will run much faster, and parallel programs will require fewer processors, and hence less inter-processor communication. Furthermore the inter-processor communication is much faster on the Cray.

It is clear, however that for numerical tasks that are easily broken in fairly independent tasks, a farm of PCs is an extremely economical solution by comparison. Additionally, the PC cluster is highly scalable. PCs and their components are so ubiquitous, that expansion of the system is trivial. Nearly anyone with a screwdriver, can upgrade or replace components so it is not nescessary to have a service contract with a commercial vendor.

CONCLUSIONS

We feel that our parallel cluster of PC type computers is an example of an economical way to build a powerful computing resource for academic purposes. On an MPI QCD benchmark simulation it compares favorably with other MPI platforms. It is also drastically cheaper than commercial supercomputers for the same amount of processing speed. PC clusters such as this one have applications in both academia and in commercial enterprises. It is particularly suitable for developing research groups in countries where funding for pure research is more scarce. We believe that our cluster may be the first such facility at an academic physics department in mainland China.

ACKNOWLEDGEMENTS

This work is supported by the National Science Fund for Distinguished Young Scholars (19825117), National Science Foundation, Guangdong Provincial Natural Science Foundation (990212) and Ministry of Education of China. We are grateful for generous additional support from Guoxun (Guangdong National Communication Network) Ltd.. We would also like to thank Shinji Hioki of Tezukayama University for the use of the QCDimMPI code. The C version of the LINPACK benchmark was written by Bonnie Toy.

REFERENCES

1. http://www.beowulf.org/.
2. http://www-unix.mcs.anl.gov/mpi/
3. http://www.netlib.org/benchmark/index.html
4. http://www.cray.com.
5. R. Gupta, *General Physics Motivations for Numerical Simulations of Quantum Field Theory* http://xxx.lanl.gov/abs/hep-lat/9905027.
6. S. Hioki, A. Nakamura. *Nucl. Phys.* **B(Proc. Suppl.)73**, 895,(1999).
7. http://tupc3472.tezukayama-u.ac.jp/QCDMPI/

Some Advance Methods of Statistical Analysis for the Muon g-2 Experiment at BNL

S. I. Redin*

Department of Physics, Yale University, New Haven, CT 06511
and Budker Institute of Nuclear Physics, Novosibirsk, Russia

Abstract. For the muon g-2 experiment at BNL a simple procedure was developed for analytical evaluation of statistical errors and correlations of parameters for the fit of time distribution of decay electrons with 5-parameter function $n(t) = N_o e^{-t/\tau}[1 + A\cos(\omega t + \phi)]$. From 5 parameters $(N_o, \tau, A, \omega, \phi)$ the angular frequency of g-2 oscillations ω is the most important since it is one of two numbers (along with the magnetic field) to be measured precisely (at sub ppm level) in this experiment. It was shown that parameters ω and ϕ are correlated and knowledge of the phase of g-2 oscillations ϕ at the moment of injection of the muon beam into the storage ring can be used to improve the statistical accuracy of frequency ω. An appropriate formula for that case was derived. Methods of statistical analysis were also used to find systematic shifts of parameters of the 5-parameters function in case a small background of arbitrary time dependence is present.

The procedure, developed for statistical analysis for the muon g-2 experiment, was applied for statistical analysis of Breit-Wigner (nonrelativistic and relativistic) and Gaussian distributions.

A simple method to search for presence of small periodic background with known period is also presented. This method was developed for monitoring of the so called AGS flashlets background for the muon g-2 experiment at BNL and presently is routinely being used. It's based on "folding" of the time distribution of decay positrons into one AGS period of 2.694 μsec. For such applications this method supersedes the Fourier analysis method.

INTRODUCTION

One of the most common procedures of data analysis in high energy physics is fitting a histogram or distribution $n_i = n(t_i)$ with a function $f_{\vec{X}}(t)$ with one or several parameters $\vec{X}$, which are usually related to the values being studied in a given experiment. The main objective of the fitting is to find the best estimates for these parameters as well as their statistical errors and correlations. Although statistical errors and correlations come up routinely as an output of MINUIT or ROOT optimization, it is worthwhile to have the ability to estimate them independently of the fit itself. In this presentation we develop a simple procedure to derive a formula for statistical errors and correlations of parameters of a fit as a function of these parameters and statistics (usually $1/\sqrt{N}$ is involved). The procedure includes some integrals and for simple distributions analytical expressions can be obtained.

As a separate issue, in the last section we discuss a simple method, adopted in the muon g-2 experiment, to search for the presence of small periodic background with known period. This method has a clear advantage over Fourier analysis method for such an application.

STATISTICAL ERRORS FOR CORRELATED PARAMETERS

The general procedure for analytical evaluation of statistical errors and correlations can be better understood from the simplest case of 2 correlated parameters, say x and y. Let's assume a chi-square function $W(x,y)$ of a fit with a 2-parameter function has a minimum at x_o, y_o. Then the equation $\Delta W = W(x,y) - W(x_o, y_o) = A(\delta x)^2 + B\delta x \delta y + C(\delta y)^2 = 1$ defines the ellipse of errors, see Fig.1 from (1). Here $\delta x = x - x_o$, $\delta y = y - y_o$. From the definition of statistical errors σ_x, σ_y, also shown in Fig.1, one can derive $\sigma_x^{-2} = A - \frac{B^2}{4C}$, $\sigma_y^{-2} = C - \frac{B^2}{4A}$ (ABC formula).

Equation for multidimensional ellipsoid of errors for a fit with a multiparameter function $f_{\vec{X}}(t)$ would be

$$1 = \Delta W = \sum \frac{[f_{\vec{X}+\delta\vec{X}} - n_i]^2}{\sigma_i^2} = \sum \frac{[f_{\vec{X}+\delta\vec{X}} - f_{\vec{X}}]^2}{n_i} = \quad (1)$$

---

* on behalf of g-2 collaboration

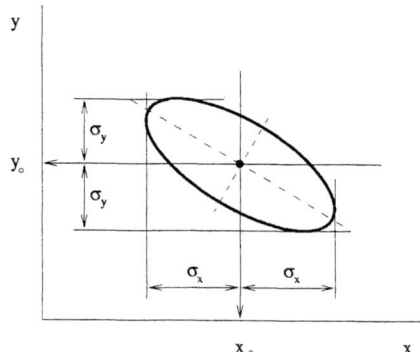

FIGURE 1. Ellipse of errors

$$= \sum \frac{\left(\sum_j \frac{\partial f}{\partial X_j} \delta X_j\right)^2}{f} \to \frac{N}{\int f dt} \int \frac{dt}{f} \left(\sum_j \frac{\partial f}{\partial X_j} \delta X_j\right)^2 = 1$$

In the last step we replace the sum by the integral: $\sum \to \int \frac{dt}{bin}$, where bin is the bin size of the histogram, and then get rid of that bin by using the number of events: $N = \sum n_i \to \frac{1}{bin} \int f_{\vec{X}}(t) dt$.

As an example and check of the procedure, let's find the errors of mean t_o, σ and amplitude A of a Gaussian distribution $f(t) = A e^{-\frac{(t-t_o)^2}{2\sigma^2}}$. From eq.1 find

$$1 = \Delta W = \frac{N}{\int f dt} \int \frac{dt}{f} \left(\frac{\partial f}{\partial t_o} \delta t_o + \frac{\partial f}{\partial \sigma} \delta \sigma + \frac{\partial f}{\partial A} \delta A\right)^2 =$$

$$= \frac{N}{\sigma^2}(\delta t_o)^2 + \frac{3N}{\sigma^2}(\delta \sigma)^2 + \frac{2N}{A\sigma}\delta\sigma\delta A + \frac{N}{A^2}(\delta A)^2 \quad (2)$$

and then use the *ABC* formula to find $\sigma_{t_o} = \frac{\sigma}{\sqrt{N}}$, $\sigma_\sigma = \frac{\sigma}{\sqrt{2N}}$ and $\sigma_A = \sqrt{3}\frac{A}{\sqrt{2N}}$. The formers two equations are well known, of course (1). The procedure gives the correct result.

In this case only one cross term, namely $\delta A \delta \sigma$, does not vanish and hence A and σ are correlated and t_o is statistically independent of A, σ. If, by any reason, parameter A is known and can be excluded from the set of fitted parameters, in eq.2 all terms with δA drop. As a result, σ_{t_o} does not change but σ_σ does: $\sigma_\sigma = \frac{\sigma}{\sqrt{3N}} < \frac{\sigma}{\sqrt{2N}}$. In general, any additional information on one parameter of optimization improves the statistical precision of those parameters which correlate with this one and does not affect the others.

TIME DISTRIBUTION OF DECAY ELECTRONS IN MUON (G-2) EXPERIMENT

In the muon (g-2) experiment at BNL (2) the time distribution of decay electrons is fitted with a 5–parameter function $f(t) = N_o e^{-t/\tau} [1 + A\cos(\omega t + \phi)]$ in order to get precise value for the g-2 angular frequency ω. Other parameters are: amplitude A and phase ϕ of g-2 oscillations, muon lifetime in the storage ring, $\tau \approx 64.4 \mu sec$, and normalization factor N_o. From the 5 parameters N_o correlates with τ, and ω correlates with ϕ. For ω and ϕ exclusively, eq.1 gives

$$\frac{NA^2}{2}\left[(T^2 + 2T\tau + 2\tau^2)(\delta\omega)^2 + 2(T+\tau)\delta\omega\delta\phi + (\delta\phi)^2\right] = 1, \quad (3)$$

where T is a histogram starting time. From eq.3 and the *ABC* formula find $\sigma_\phi = \frac{\sqrt{2}\sqrt{(T+\tau)^2+1}}{\tau A \sqrt{N}}$, $\sigma_\omega = \frac{\sqrt{2}}{\tau A \sqrt{N}}$. The latter can be used for various estimations and, in particular, for choosing an energy threshold E_{thr} of the decay electrons to maximize $A\sqrt{N}$ (both A and N are functions of E_{thr}). The errors of other parameters are: $\sigma_A = \frac{\sqrt{2}}{\sqrt{N}}$, $\sigma_\tau = \frac{\tau}{\sqrt{N}}$ and $\sigma_{N_o} = \frac{N_o}{\sqrt{N}}\sqrt{(T/\tau+1)^2+1}$.

It was known for years that the (g-2) frequency and phase correlate with each other, but only recently it was found "experimentally" that the magnitude of the correlation depends on the choice of time origin. Fig.2 shows result of fit for frequency and phase for the same histogram, first for regular case when $T \approx 70 \mu sec$ (ω and ϕ) and then for shifted time origin such that $T \approx -65 \mu sec$ (ω and ϕ'). The frequency does not change at all, but for the phase obvious correlations with frequency get greatly reduced (ϕ and ϕ' are in the same scale in Fig.2). That can be understood from eq.3: when $T = -\tau = -64.4 \mu sec$, the $\delta\omega\delta\phi$ cross-term vanishes and so the $\omega-\phi$ (and $\tau-N_o$) correlations.

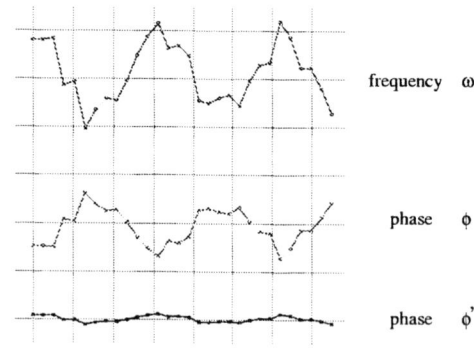

FIGURE 2.

In the muon (g−2) experiment, a beam of polarized muons is injected into the storage ring and at that point the muon polarization vector is directed at a certain small anlge to the direction of motion. Thus, in general, the phase of the (g−2) oscillations is known at the injection time and one can use this knowledge to improve the precision of (g-2) frequency ω. In a special case when the phase is known at some arbitrary time T' with precision σ_F we should add a term $\frac{[\delta(\omega T'+\phi)]^2}{\sigma_F^2} = \sigma_F^{-2}[T'^2(\delta\omega)^2 + 2T'(\delta\omega)(\delta\phi) + (\delta\phi)^2]$ in eq.3. Then after little algebra we find

$$\sigma_\omega^{-2} = \sigma_{\omega\infty}^{-2}\left[1 + \frac{\sigma_F^{-2}}{(\tau\sigma_{\omega\infty})^{-2} + \sigma_F^{-2}}\frac{(T+\tau-T')^2}{\tau^2}\right], \quad (4)$$

where $\sigma_{\omega\infty} = \frac{\sqrt{2}}{\tau A\sqrt{N}}$.

SYSTEMATIC SHIFT DUE TO SMALL BACKGROUND

A procedure, similar to that developed above for the statistical errors and correlations, can be developed for estimation of the systematic shift of fit parameters in case a small background is present in the data. Suppose we have some low level background $h(t)$ admixed to the data, otherwise perfectly described by some multi-parameter function $f_{\vec{X}}(t)$. The background is small enough to observe "by eyes" or even to spoil the chi-squared considerably. Nevertheless fitting the histogram with a function $f_{\vec{X}}(t)$ alone will give us parameters $\vec{X}$, shifted with respect to the "true" parameters $\vec{X}_\circ$: $\vec{X} = \vec{X}_\circ + \delta\vec{X}$.

We can find the systematic shift $\delta\vec{X}$ from the chi-squared minimization requirement $\partial W/\partial X_j = 0$:

$$\frac{\partial W}{\partial X_j} = \frac{\partial}{\partial X_j}\sum_i \frac{\left(f_{\vec{X}_\circ+\delta\vec{X}} - n_i\right)^2}{\sigma_i^2} = 2\sum_i \frac{\left(f_{\vec{X}_\circ+\delta\vec{X}} - n_i\right)}{n_i}\frac{\partial f}{\partial X_j} =$$

$$= 2\sum_i \frac{\left(f_{\vec{X}_\circ+\delta\vec{X}} - f_{\vec{X}} - h\right)}{f_{\vec{X}}}\frac{\partial f}{\partial X_j} = 2\sum_i \frac{\left(\sum_k \frac{\partial f}{\partial X_k}\delta X_k - h\right)}{f}\frac{\partial f}{\partial X_j} =$$

$$= 2\sum_k \delta X_k \sum_i \frac{1}{f}\left(\frac{\partial f}{\partial X_k}\frac{\partial f}{\partial X_j}\right) - 2\sum_i \frac{h}{f}\frac{\partial f}{\partial X_j} = 0 \quad (5)$$

Replacing the sum by an integral we finally get a system of linear equations:

$$\sum_k \delta X_k \int \frac{dt}{f}\frac{\partial f}{\partial X_k}\frac{\partial f}{\partial X_j} = \int dt\frac{h}{f}\frac{\partial f}{\partial X_j} \quad (6)$$

For g-2 distribution $N_\circ e^{-t/\tau}[1 + A\cos(\omega t + \phi)]$ eq.6 gives systematic shifts

$$\delta N_\circ = \frac{1}{e\tau}\int_{-\tau}^{\infty} h(t)\,dt$$

$$\delta\tau = \frac{1}{eN_\circ \tau}\int_{-\tau}^{\infty} t\,h(t)\,dt$$

$$\delta A = \frac{2}{eN_\circ \tau}\int_{-\tau}^{\infty} \frac{h(t)\cos(\omega t + \phi)}{1 + A\cos(\omega t + \phi)}\,dt \quad (7)$$

$$\delta\phi = -\frac{2}{eN_\circ A\tau}\int_{-\tau}^{\infty} \frac{h(t)\sin(\omega t + \phi)}{1 + A\cos(\omega t + \phi)}\,dt$$

$$\delta\omega = -\frac{2}{eN_\circ A\tau^3}\int_{-\tau}^{\infty}\frac{t\,h(t)\sin(\omega t+\phi)}{1+A\cos(\omega t+\phi)}\,dt +$$

$$+\frac{2}{e^2 N_\circ A\tau^4}\int_{-\tau}^{\infty}h(t)\,dt\int_{-\tau}^{\infty}t\,e^{-t/\tau}\sin(\omega t+\phi)\,dt +$$

$$+\frac{2}{e^2 N_\circ A\tau^6}\int_{-\tau}^{\infty}t\,h(t)\,dt\int_{-\tau}^{\infty}t^2 e^{-t/\tau}\sin(\omega t+\phi)\,dt$$

For $\delta\omega$ we have also included two next-to-leading terms, which are important if $h(t)$ is a smooth function. Note that we chose the time origin in such a way that the histogram starting time is $T = -\tau$. That doesn't change result, only makes the calculations easier.

BREIT–WIGNER DISTRIBUTION

In approximation of narrow width the nonrelativistic Breit–Wigner distribution can be written as $f(E) = A\frac{\Gamma^2/4}{(E-M)^2+\Gamma^2/4}$. For statistical errors of parameters M, Γ and A eq.1 gives

$$\frac{2N}{\Gamma^2}(\delta M)^2 + \frac{3N}{2\Gamma^2}(\delta\Gamma)^2 + \frac{2N}{A\Gamma}\delta\Gamma\,\delta A + \frac{N}{A^2}(\delta A)^2 = 1 \quad (8)$$

and from that

$$\sigma_M = \frac{\Gamma}{\sqrt{2N}}, \quad \sigma_\Gamma = \sqrt{2}\frac{\Gamma}{\sqrt{N}}, \quad \sigma_A = \sqrt{3}\frac{A}{\sqrt{N}} \quad (9)$$

As follows from eq.8, parameters Γ and A are correlated and M is independent.

For the Breit–Wigner plus constant background distribution $f(E) = A\frac{\Gamma^2}{(E-M)^2+\Gamma^2/4} + C$ parameter M is still independent but the rest three parameters Γ, A and C correlate with each other. That makes the algebra calculation much more tedious. The result is

$$\sigma_M = \frac{\Gamma}{\sqrt{2N}}\sqrt{\frac{C}{A}}\sqrt{\frac{(1+Z)^3}{3+Z}}$$

$$\sigma_\Gamma = \sqrt{2}\frac{\Gamma}{\sqrt{N}}\sqrt{\frac{C}{A}}(1+Z) \quad (10)$$

$$\sigma_A = \frac{A}{\sqrt{N}}\sqrt{\frac{C}{A}}\sqrt{Z(1+3Z)}$$

$$\sigma_C = \frac{C}{\sqrt{N_c}}$$

where $N \propto A\Gamma\pi/2$ is a number of Breit–Wigner events, $N_c \propto (E_{max} - E_{min})C$ is a number of background events and $Z = \sqrt{1 + A/C}$.

The formula for the relativistic Breit–Wigner distribution $f(s) = A\frac{m^2\Gamma^2}{(s-m^2)^2+m^2\Gamma^2}$ can be written as $f(s) = A\frac{Y^2/4}{(s-X)^2+Y^2/4}$ with $X = m^2$ and $Y = 2m\Gamma$ and we can use eq.9: $\sigma_X = \frac{\Gamma}{\sqrt{2N}}$, $\sigma_Y = \sqrt{2}\frac{Y}{\sqrt{N}}$, $\sigma_A = \sqrt{3}\frac{A}{\sqrt{N}}$. Making inverse transformation, find statistical errors for the relativistic Breit–Wigner distribution: $\sigma_m = \frac{\Gamma}{\sqrt{2N}}$, $\sigma_\Gamma = $

$\sqrt{2}\frac{\Gamma}{\sqrt{N}}\sqrt{1+\frac{\Gamma^2}{4m^2}} \approx \sqrt{2}\frac{\Gamma}{\sqrt{N}}$, $\sigma_A = \sqrt{3}\frac{A}{\sqrt{N}}$, which are exactly same as for the nonrelativistic case, see eq.9.

FOLDING METHOD TO MONITOR THE AGS BACKGROUND

In the first run of the muon (g−2) experiment at BNL in 1997 an accelerator background admixture was found in (g−2) data. It was caused by a proton leakage after extraction of the main pulse out of the AGS into our beam line. It revealed itself as a fence-like structure over the regular (g−2) wiggles, see Fig.3. The time intervals be-

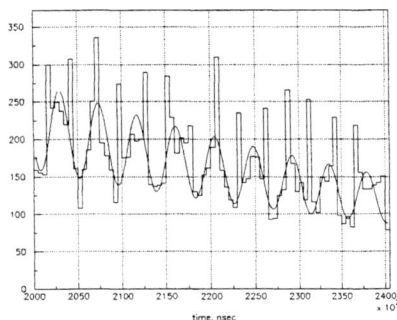

FIGURE 3. Fragment of decay positrons time distribution for the run #883

tween these "flaslets" are 2.7 μsec, i.e. equal to the revolution time for a proton bunch in the AGS, T_{AGS}. That was the main reason to associate these to an AGS proton leakage. The problem was discussed with the AGS experts and various schemes of solution, including fast shutter magnet in the (g−2) beam line, were proposed and implemented. Still, the monitoring for the presence of the AGS flaslet background was and is very important.

A very sensitive method to monitior the AGS background was developed. The idea is to put all the data into one AGS period of 2.7 μsec. To do that, the transformation

$$t' = \frac{t}{T_{AGS}} - \text{integer}\left(\frac{t}{T_{AGS}}\right) \qquad (11)$$

was applied to the data. Fig. 4 shows the result from one of the longest 1997 runs (#1041) without AGS background. The (g−2) oscillations average out completely. The exponential decay with time constant $\tau = \gamma \cdot \tau_\mu = 64.4\,\mu$sec appears as a 4% slope, as shown by the solid line. The distribution has a maximum at $t'=0.15$ which corresponds to the start of data taking.

The distribution of t' for the run #883 is shown in Fig. 5. The highest peak at $t' = 0.90$ corresponds to the 2.7 μsec structure seen in Fig. 3. Also 7 smaller peaks are clearly seen. All together the 8 peaks correspond to the 8 bunches in the AGS. The distribution in t' is very sensitive

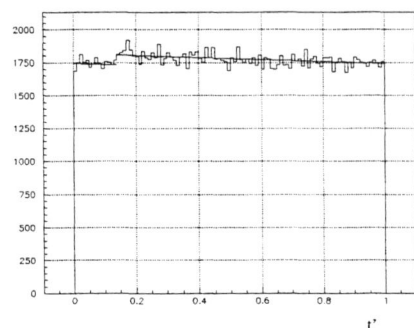

FIGURE 4. Distribution of t' for the run #1041

to the exact value of the AGS period T_{AGS}. Comparing plots for several T_{AGS} with 1 nsec interval, we clearly saw the sharpest structure for $T_{AGS} = 2.694\,\mu$sec. AGS experts confirmed that 2.694 μsec is indeed the best estimate for the AGS period.

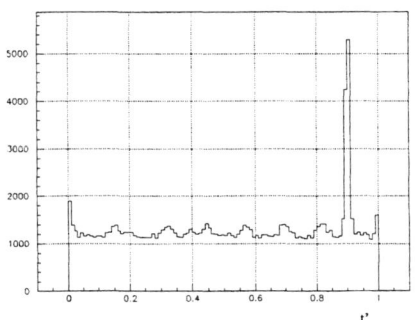

FIGURE 5. Distribution of t' for the run #883

For monitoring the AGS background the folding method (also known in the (g−2) collaboration as a "Redin plot") has some advantages compared to the Fourier analysis method. Fourier analysis works well for sinewave structures. But for the AGS background shown in Fig. 3, which looks like a set of equally spaced delta-functions in time, the Fourier analysis would give a similar set of equally spaced delta-functions in a frequency domain, where the statistical analysis is complicated. Furthermore the 8 beam structure can be better understood from analysis of the t' distribution than from the Fourier analysis.

REFERENCES

1. Review of Particle Physics, Eur. Phys. J. **C15**, 1(1998)
2. R. M. Carey et al., Phys. Rev. Lett. **82**, 1632 (1999); H. N. Brown et al., Phys. Rev. **D62**, 091101 (2000)

Analytical calculation of heavy baryon correlators in NLO of perturbative QCD

S. Groote, J.G. Körner and A.A. Pivovarov

Institut für Physik der Johannes-Gutenberg-Universität, Staudinger Weg 7, 55099 Mainz, Germany

Abstract. We present analytical results for the correlator of baryonic currents at the three-loop level with one finite mass quark. We obtain the massless and the HQET limits as particular cases from the general formula. Calculations have been performed with an extensive use of the symbolic manipulation programs MATHEMATICA and REDUCE.

Baryons form a rich family of particles which has been experimentally studied with high accuracy [1]. A theoretical analysis of these experimental data gives a lot of information about the structure of QCD and the numerical values of its parameters. The hypothetical limit $N_c \to \infty$ for the number N_c of colours which is a very powerful tool for investigating the general properties of gauge interactions was especially successful for baryons [2]. The spectrum of baryons is contained in the correlator of two baryonic currents and the spectral density associated with it. To leading order the correlator is given by a product of N_c fermionic propagators. The diagrams of this topology have recently been studied in detail [3, 4, 5, 6, 7, 8]. They are rather frequently used in phenomenological applications [9]. With the advent of new accelerators and detectors many properties of baryons containing a heavy quark have been experimentally measured in recent years [1]. However, theoretical calculations beyond the leading order have not been done for many interesting cases. In this note we fill up this gap.

We report on the results of calculating the α_s corrections to the correlator of two baryonic currents with one finite mass quark and two massless quarks. We give analytical results and discuss the magnitude of the α_s corrections. The massless and HQET limits are obtained as special cases. Note that the massless case has been known since long ago [10]. The mesonic analogue of our baryonic calculation was completed some time ago [11] and has subsequently provided a rich source of inspiration for many applications in meson physics.

A generic baryonic current has the form
$$j = \varepsilon^{abc}(u_a^T C d_b)\Gamma \Psi_c \qquad (1)$$

where Γ is the Dirac matrix, in the following $\Gamma = 1$.

The correlator of two baryonic currents is expanded as

$$i\int \langle T j(x)\bar{j}(0)\rangle e^{iqx}dx = \gamma_\nu q^\nu \Pi_q(q^2) + m\Pi_m(q^2). \qquad (2)$$

Here we show results for the function $\Pi_q(q^2)$ and compare it with $\Pi_m(q^2)$ [12]. The dispersion relation reads

$$\Pi_\#(q^2) = \frac{1}{128\pi^4}\int_{m^2}^\infty \frac{\rho^\#(s)ds}{s-q^2} \qquad (3)$$

where $\rho^\#(s) = \rho^{q,m}(s)$ are the spectral densities. The spectral density is the real object of interest for phenomenological applications,

$$\rho^\#(s) = s^2\left\{\rho_0^\#\left(1 + \frac{\alpha_s}{\pi}\ln\left(\frac{\mu^2}{m^2}\right)\right) + \frac{\alpha_s}{\pi}\rho_1^\#\right\}. \qquad (4)$$

Here μ is the renormalization scale parameter, m is a pole mass of the heavy quark (see e.g. Ref. [13]) and $\alpha_s = \alpha_s(\mu)$. The leading order two-loop contribution is shown in Fig. 1(a). This topology coincides with water melon diagrams for which a general method of calculation (with arbitrary masses) has recently been developed [5, 6, 7]. The leading order results read

$$\rho_0^q = \tfrac{1}{4} - 2z + 2z^3 - \tfrac{1}{4}z^4 - 3z^2 \ln z, \qquad (5)$$

$$\rho_0^m = 1 + 9z - 9z^2 - z^3 + 6z(1+z)\ln z \qquad (6)$$

with $z = m^2/s$. The next-to-leading order contribution is given by three-loop diagrams with one external momentum. For an arbitrary mass arrangement such diagrams have not yet been calculated analytically. However, if we take the case of one massive line, the result within $\overline{\text{MS}}$-scheme can be obtained analytically and reads

$$\begin{aligned}\rho_1^q &= \tfrac{71}{48} - \tfrac{565}{36}z - \tfrac{7}{8}z^2 + \tfrac{625}{36}z^3 - \tfrac{109}{48}z^4 \\ &\quad - \left(\tfrac{49}{36} - \tfrac{116}{9}z + \tfrac{116}{9}z^3 - \tfrac{49}{36}z^4\right)\ln(1-z) \\ &\quad + \left(\tfrac{1}{4} - \tfrac{17}{3}z - 11z^2 + \tfrac{113}{9}z^3 - \tfrac{49}{36}z^4\right)\ln z \\ &\quad + \left(\tfrac{1}{3} - \tfrac{8}{3}z + \tfrac{8}{3}z^3 - \tfrac{1}{3}z^4\right)\ln(1-z)\ln z \\ &\quad - 2z^2\left(9 + \tfrac{4}{3}z - \tfrac{1}{6}z^2\right)\left(\tfrac{1}{2}\ln^2 z - \zeta(2)\right) \\ &\quad + \left(\tfrac{2}{3} - \tfrac{16}{3}z - 18z^2 + \tfrac{8}{3}z^3 - \tfrac{1}{3}z^4\right)\text{Li}_2(z) \\ &\quad - 12z^2\left(\text{Li}_3(z) - \zeta(3) - \tfrac{1}{3}\text{Li}_2(z)\ln(z)\right) \end{aligned} \qquad (7)$$

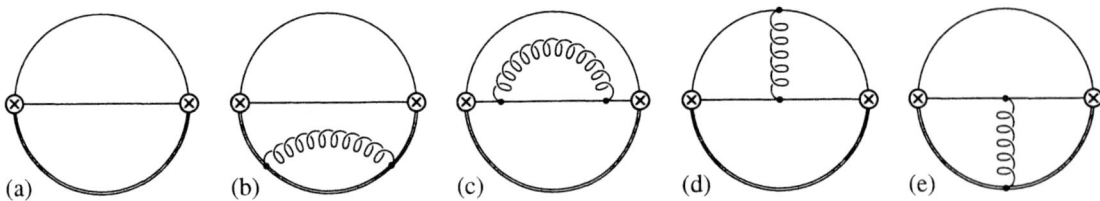

FIGURE 1. The calculated (a) two-loop and (b–e) three-loop topologies

The contributing three-loop diagrams are shown in Figs. 1(b) to (e). They have been evaluated using advanced algebraic methods for multi-loop calculations along the lines decribed in Refs. [6, 11]. This result should be compared to

$$\rho_1^m = 9 + \tfrac{665}{9}z - \tfrac{665}{9}z^2 - 9z^3$$
$$- \left(\tfrac{58}{9} + 42z - 42z^2 - \tfrac{58}{9}z^3\right)\ln(1-z)$$
$$+ \left(2 + \tfrac{154}{3}z - \tfrac{22}{3}z^2 - \tfrac{58}{9}z^3\right)\ln z$$
$$+ 4\left(\tfrac{1}{3} + 3z - 3z^2 - \tfrac{1}{3}z^3\right)\ln(1-z)\ln z$$
$$+ 12z\left(2 + 3z + \tfrac{1}{9}z^2\right)\left(\tfrac{1}{2}\ln^2 z - \zeta(2)\right)$$
$$+ 4\left(\tfrac{2}{3} + 12z + 3z^2 - \tfrac{1}{3}z^3\right)\text{Li}_2(z)$$
$$+ 24z(1+z)\left(\text{Li}_3(z) - \zeta(3) - \tfrac{1}{3}\text{Li}_2(z)\ln z\right). \quad (8)$$

Our method of integration is a completely algebraic one and therefore symbolic manipulation programs can be used for performing the long calculations. Two independent calculations of some steps were done using MATHEMATICA and REDUCE, the latter being rather actively used for high energy calculations (see e.g. Ref. [14]).

The results given in Eqs. (7) and (8) represent the full next-to-leading order solution. Since the anomalous dimension of the current in Eq. (1) is known up to two-loop order [15], the results shown in Eqs. (7) and (8) complete the ingredients necessary for an analysis of the correlator in Eq. (2) within operator product expansion at the next-to-leading order level.

Two limiting cases of general interest are the near-threshold and the high energy asymptotics. With our result given in Eq. (7) both limits can be taken explicitly.

In the massless limit $z \to 0$ the corrections read

$$\rho_1^q = \tfrac{71}{48} + \tfrac{1}{4}\ln z - \tfrac{41}{3}z - 6z\ln z + \ldots, \quad (9)$$
$$\rho_1^m = 9 + 83z - 4\pi^2 z + 2\ln z + 50z\ln z + \ldots. \quad (10)$$

Therefore we obtain

$$\rho^q(s) = \tfrac{s^2}{4}\left\{1 + \tfrac{\alpha_s}{\pi}\left(\ln\left(\tfrac{\mu^2}{s}\right) + \tfrac{71}{12}\right)\right\}$$
$$- 2m_{\overline{\text{MS}}}^2(\mu)s\left\{1 + \tfrac{\alpha_s}{\pi}\left(3\ln\left(\tfrac{\mu^2}{s}\right) + \tfrac{19}{2}\right)\right\}, \quad (11)$$
$$m\rho^m(s) = m_{\overline{\text{MS}}}(\mu)s^2\left\{1 + \tfrac{\alpha_s}{\pi}\left(2\ln\left(\tfrac{\mu^2}{s}\right) + \tfrac{31}{3}\right)\right\}. \quad (12)$$

For the momentum part $\rho^q(s)$ we retain the $O(m^2)$ correction. The relation between the pole mass m and the $\overline{\text{MS}}$ mass $m_{\overline{\text{MS}}}(\mu)$ we have used reads

$$m = m_{\overline{\text{MS}}}(\mu)\left\{1 + \tfrac{\alpha_s}{\pi}\left(\ln\left(\tfrac{\mu^2}{m^2}\right) + \tfrac{4}{3}\right)\right\}. \quad (13)$$

In the near-threshold limit $E \to 0$ with $s = (m+E)^2$ one explicitly obtains

$$\rho_{\text{thr}}^m(m,E) = \tfrac{16E^5}{5m}\left\{1 + \tfrac{\alpha_s}{\pi}\ln\left(\tfrac{\mu^2}{m^2}\right)\right. \quad (14)$$
$$\left. + \tfrac{\alpha_s}{\pi}\left(\tfrac{54}{5} + \tfrac{4\pi^2}{9} + 4\ln\left(\tfrac{m}{2E}\right)\right)\right\} + O\left(\tfrac{E^6}{m^2}\right).$$

We have shown the coincidence with the result of the explicit HQET calculation,

$$m\rho_{\text{thr}}^m(m,E) = C(m/\mu,\alpha_s)^2\rho_{\text{HQET}}(E,\mu) \quad (15)$$

where $\rho_{\text{HQET}}(E,\mu)$ and $C(m/\mu,\alpha_s)$ with

$$\rho_{\text{HQET}}(E,\mu) = \tfrac{16E^5}{5}\left\{1 + \tfrac{\alpha_s}{\pi}\left(\tfrac{182}{15} + \tfrac{4\pi^2}{9} + 4\ln\tfrac{\mu}{2E}\right)\right\}$$
$$C(m/\mu,\alpha_s) = 1 + \tfrac{\alpha_s}{\pi}\left(\tfrac{1}{2}\ln\left(\tfrac{m^2}{\mu^2}\right) - \tfrac{2}{3}\right) \quad (16)$$

are taken from Refs. [16, 17], respectively. Note that the higher order corrections in E/m to Eq. (14) can easily be obtained from the explicit result given in Eq. (7).

Of interest is whether the two limiting expressions (the massless limit expression as given in Eq. (11) and the HQET limit expression in Eqs. (14) and (15)) can be used to characterise the full function for all energies.

For this discussion we compare components of the baryonic spectral function in leading and next-to-leading order. In Fig. 2 and 3 we show the ratio $\rho_1^\#(s)/\rho_0^\#(s)$ for $\# = m$ and $\# = q$, respectively. In the following we shall always use the specific renormalization scale value $\mu = m$ if it is not written explicitly. One can see that a simple interpolation between the two limits can give a rather good approximation for the next-to-leading order correction in the complete region of s. We therefore conclude that in going even one order higher it is very likely

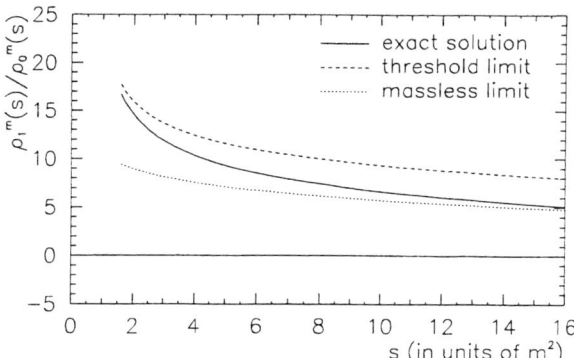

FIGURE 2. The ratio ρ_1^m/ρ_0^m in dependence of s

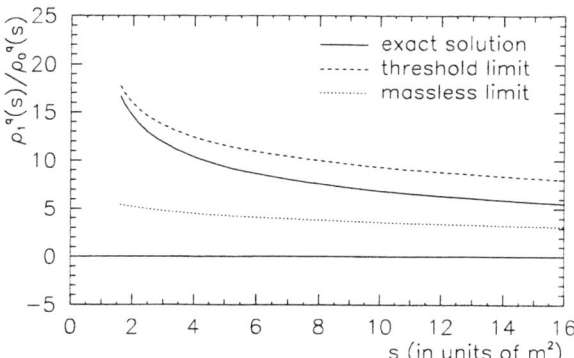

FIGURE 3. The ratio ρ_1^q/ρ_0^q in dependence of s

that the full four-loop spectral density can be well approximated by the corresponding massless four-loop result which can be calculated using existing computational algorithms [18, 19].

ACKNOWLEDGEMENTS

We want to thank Richard Kreckel for presenting our poster at the conference. The present work is supported in part by the Volkswagen Foundation under contract No. I/73611 and by the Russian Fund for Basic Research under contract 99-01-00091. A.A. Pivovarov is an Alexander von Humboldt fellow. S. Groote gratefully acknowledges a grant given by the DFG, FRG.

REFERENCES

1. Particle Data Group, Eur. Phys. J. **C3** (1998) 1
2. E. Witten, Nucl. Phys. **B160** (1979) 57
3. F.A. Berends, A.I. Davydychev, N.I. Ussyukina, Phys. Lett. **426 B** (1998) 95
4. S. Groote, J.G. Körner and A.A. Pivovarov, Phys. Rev. **D60** (1999) 061701
5. S. Groote, J. G. Körner and A. A. Pivovarov, Eur. Phys. J. **C11** (1999) 279
6. S. Groote, J. G. Körner and A. A. Pivovarov, Nucl. Phys. **B542** (1999) 515
7. S. Groote, J. G. Körner and A. A. Pivovarov, Phys. Lett. **443 B** (1998) 269
8. S. Groote and A.A. Pivovarov, Nucl. Phys. **B580** (2000) 459; A.I. Davydychev and V.A. Smirnov, Nucl. Phys. **B554** (1999) 391; N.E. Ligterink, Phys. Rev. **D61** (2000) 105010
9. J.O. Andersen, E. Braaten, M. Strickland, Phys. Rev. **D62** (2000) 045004; S. Narison and A.A. Pivovarov, Phys. Lett. **327 B** (1994) 341; T. Sakai, K. Shimizu and K. Yazaki, Prog. Theor. Phys. Suppl. **137** (2000) 121; S.A. Larin *et al.*, Sov. J. Nucl. Phys. **44** (1986) 690; J.M. Chung and B.K. Chung, Phys. Rev. **D60** (1999) 105001; K. Chetyrkin and S. Narison, Phys. Lett. **485 B** (2000) 145; H.Y. Jin and J.G. Körner, "Radiative correction of the correlator for $(0^{++}, 1^{-+})$ light hybrid currents", Report No. MZ-TH/00-11, hep-ph/0003202
10. A.A. Ovchinnikov, A.A. Pivovarov and L.R. Surguladze, Sov. J. Nucl. Phys. **48** (1988) 358; Int. J. Mod. Phys. **A6** (1991) 2025
11. S.C. Generalis, Report No. OUT-4102-13 (1984), later published as J. Phys. **G16** (1990) 367, see also D.J. Broadhurst, Phys. Lett. **101 B** (1981) 423; D.J. Broadhurst and S.C. Generalis, Report No. OUT-4102-8/R (1982)
12. S. Groote, J.G. Körner and A.A. Pivovarov, Phys. Rev. **D61** (2000) 071501(R)
13. R. Tarrach, Nucl. Phys. **B183** (1981) 384
14. A.A. Pivovarov, Proceedings of the Conference "Pisa AI-HENP 1995", p. 301–306 [hep-ph/9505316]
15. A.A. Pivovarov and L.R. Surguladze, Yad. Fiz. **48** (1988) 1856 [Sov. J. Nucl. Phys. **48** (1989) 1117]; Nucl. Phys. **B360** (1991) 97
16. S. Groote, J.G. Körner and O.I. Yakovlev, Phys. Rev. **D55** (1997) 3016
17. A.G. Grozin and O.I. Yakovlev, Phys. Lett. **285 B** (1992) 254
18. K.G. Chetyrkin and F.V. Tkachov, Nucl. Phys. **B192** (1981) 159; F.V. Tkachov, Phys. Lett. **100 B** (1981) 65
19. K.G. Chetyrkin and V.A. Smirnov, Phys. Lett. **144 B** (1984) 419

Summaries of Recent Computer-assisted Feynman Diagram Calculations

M. Fischler*

Summarizing results submitted by

EE Boos[a], D. Broadhurst[b], S. Eidelman[c], S. Groote[d], R. Harlander[e], S. Heinemeyer[e], T. Ishikawa[f], F. Jergerlehner[g], T. Kaneko[h], A. Kataev[i], J. Körner[d], A. Kotikov[j], C. Maxwell[k], C. Oleari[m], G. Parente[n], A. Pivovarov[d], A. Sidorov[j], O. Veretin[g], and F. Yuasa[f],

Fermi National Accelerator Laboratory, Batavia Il 60510
[a] *SINP, Moscow State Univ.;* [b] *Open University, Milton Keynes;* [c] *Budker Inst. for Nucl. Phys, Novosibirsk;*
[d] *Johannes-Getenberg-Universität;* [e] *Brookhaven Nat'l Lab.;* [f] *KEK;* [g] *DESY;* [h] *Meiji-Gakuin Univ.;* [i] *CERN and INR;*
[j] *JINR, Dubna, Russia;* [k] *Univ. of Durham;* [m] *Univ. Wisconsin;* [n] *Univ. Santiago do Compostela*

Abstract. Recent results from several researchers, in the area of automated high order computation of quantities in QCD, the Standard Model, and other models are summarized.

The AIHENP Workshop series has traditionally included cutting edge work on automated computation of Feynman diagrams. The conveners of the Symbolic Problem Solving topic in this ACAT conference felt it would be useful to solicit presentations of brief summaries of the interesting recent calculations. Since this conference was the first in the series to be held in the Western Hemisphere, it was decided that the summaries would be solicited both from attendees and from researchers who could not attend the conference. This would represent a sampling of many of the key calculations being performed. The results were presented at the Poster session; contributions from ten researchers were displayed and posted on the web.

Although the poster presentation (which can be viewed at *conferences.fnal.gov/acat2000/* placed equal emphasis on results presented at the conference and other contributions, here we primarily discuss the latter, which do not appear in full form in these proceedings.

This brief paper can't do full justice to each contibution; interested readers can find details of the work not presented at this conference in references (1), (2), (3), (4), (5), (6), (7).

Standard Model Higgs Prodution

Robert Harlander has results(1) for $gg \rightarrow H$ to two loops (NNLO) in the heavy top limit. This will be the dominant production mechanism for the Higgs at the LHC, so it is important to improve on the theoretical accuracy. At next-to-leading-order, the theoretical uncertainty

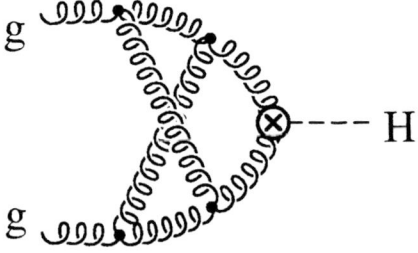

Source: R. Harlander(1)

FIGURE 1. Sample two-loop diagram contributing to $gg \rightarrow H$ at NNLO.

is a factor of 1.5 to 2. The NNLO contributions include diagrams like that shown in figure 1.

When $M_H < M_t$ an expansion in M_H^2/M_t^2 yields an excellent approximation at NLO) (and presumedly at NNLO). The leading term in this expansion may be obtained by using effective Lagrangian for the Higgs-gluon interaction. The coefficient of $H(G_{\mu\nu})^2$ for the effective vertex (marked $\otimes$ in the figure) was previously computed to the needed order in $\alpha_S(9)$.

The NNLO corrections sum several contributions, for example, one-loop amplitudes involving radiation of a single quark or gluon. Some (but not all) of these have been determined. The Harlander calculation computes contribution of the gauge invariant set of corrections (of order α_S^4) involving two loops and no extraneous radiation.

The diagrams which are planar had been reduced, using an integration-by-parts algorithm, to convolutions of one-loop integrals(11). Non-planar diagrams such as the one shown are, as usual, less straightforward. The technique used was one developed by Baikov and Smirnov(10): The recurrence relations for these 2-loop integrals with 3 external legs are related to those for 3-loop integrals with two external legs. Thus such diagrams as Fig. 1 are mapped onto massless three-loop two-point functions.

These calculations were done, using a modification of the program MINCER(13) which is written in the symbolic manipulation system FORM(16). Programmed in this manner, the computer calculation was not very extensive: It completed in a few minutes on a fast processor.

The primary result(1) is a second order correction to the virtual cross section for $gg \rightarrow H$. As an estimate on the magnitude of the corrections, the ratio of time-like to space-like form factor is considered. For 5 light (on the scale of M_H) quarks, the NLO correction to this was 52.8%. The newly computed NNLO correction is found to be 17.2% (and in the same direction). Thus the correction is large, but there is good convergency.

Advanced mathematical techniques and high-order diagrams

A.V. Kotikov(2) uses a Differential Equations Method to do two-loop self-energy diagrams with one non-zero internal mass or external momentum. Combinations of DEM and programs by O.L. Veretin and M. Kalmykov have been used to evaluate that full set of two-loop, two-point onshell master diagrams, as well as three-point two-loop integrals with one and two-mass thresholds in a small-moment expansion.

The Differential Equations Method makes use of the integration-by-parts method(15): when applied to an internal n-point subgraph of a Feynman diagram, IBP generates new diagrams which can be represented as derivatives with respect to masses (or external momenta) of the initial diagram. Thus a differential equation for the initial diagram can be found; this equation has inhomogeneous terms containing diagrams with more trivial topological structure and/or fewer loops or legs. Complicated diagrams may be evaluated by recursively applying this procedure to reduce to known results for simpler diagrams. Some results can be found in reference (2).

In the summary posted at the conference results for two-loop self-energy diagrams in figure 2 were presented. For example,

$$q^2 \cdot I_{125} = -2\log^2 y \log(1-y) - 6\zeta_3 + 6\text{Li}_3(y) - 6\log y \text{Li}_2(y)$$

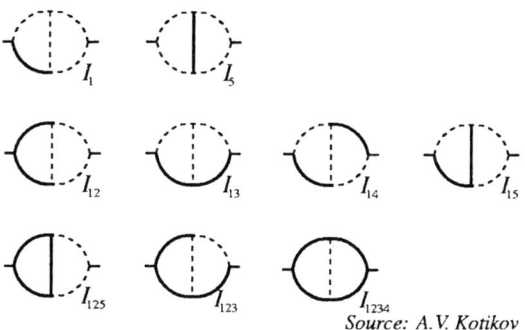

Source: A.V. Kotikov

FIGURE 2. Two-loop self-energy diagrams. Solid lines denote propogators with mass m; dashed lines denote massless propagators.

where

$$y \equiv \frac{1 - \sqrt{q^2/(q^2 - 4m^2)}}{1 + \sqrt{q^2/(q^2 - 4m^2)}}$$

A.L. Kataev, G. Parente and A.V. Sidorov sent results(3) of using the method of Jacobi polynomials to to a next-to-next-to-leading order analysis of Fermilab data for the xF_3 structure function of νN deep-inelastic scattering. Using analytical expressions(14) for the theoretical behaviour of QCD (including $\alpha_S(M_Z)$ and higher-twist terms), these researchers along with A.V.Kotikov(4) use a FORTRAN program realizing the Jacobi Polynomial method, to fit the data. This results in values of $\Lambda_{\overline{MS}}$ normalized to the x-behavior of the nonperturbative contribution, modeled as $h(x)/Q^2$.

They observe that using the Jacobi polynomial method it is possible to reconstruct the structure function to rather high precision, using only ten of its Mellin moments. An interesting physics point is that there is interplay between the effects of the NNLO perturbative QCD corrections and $1/Q^2$-contributions, which result in effective "shadowing" of the power-suppressed terms by the perturbative NNLO effects. If the $1/Q^2$-contributions are fixed through a special model, the NNLO value of $\alpha_S(M_Z)$ is $0.118 \pm 0.002(\text{stat}) \pm 0.005(\text{syst}) \pm 0.003(\text{theory})$.

S. Eidelman, F. Jegerlehner, A.L. Kataev, and O.V. Veretin(5) contributed results of three-loop massive corrections to the Adler D-function of the e^+e^- annihilation process. These were calculated in the Euclidean region, uing a Padé resummation method(12). Massive diagrams with one external momentum were considered, with the Padé resummations realized in a custom FORTRAN program. These were run in several minutes on an Alpha workstation.

At high energies, the perturbative QCD prediction starts to agree with the experimentally motivated behaviour of the Adler D-function only after inclusion of the mass dependence of this 3-loop order α_S^2 term. Thus, these results allow extraction of hadronic shifts to the fine structure constant, from experimental data on the cross section for annihilation of e^+e^- into hadrons, including data obtained at the low-enery e^+e^- in Novosibursk.

D.J. Broadhurst, A.L. Kataev and C.J. Maxwell(6) report on large N_f expansion of scalar correlators and estimates of higher-order QCD corrections to Higgs $\longrightarrow \bar{b}b$ and strange-quark-mass sum rules. These large N_f terms come from a single chain of quark bubble diagrams, and the two-point correlator of the scalar quark current $\bar{\psi}\psi$ was calculated to 20 loops analytically, and up to 100 loops numerically. Such correlators are related to the decay width of the scalar Higgs into quark-antiquark pairs.

The n-loop diagrams were calculated by inserting $n-2$ quark loops in the pair of two-loop skeletons for the scalar correlator. The method entails recurrence relations for $_3F_2$ hypergeometric series, which were implemented in REDUCE. The analysis demonstrates that one must take a twice-cubtracted dispersion relation to avoid an ambiguity of order Λ^2/Q^2 even in the zero-quark-mass limit. Failure to do so leads to explosion of the perturbation series.

Estimates are obtained for order α_S^4 contributions, paying particular attention to terms that result from analytic continuation, which are resummed to all orders in α_S and to leading order in β_0. It is concluded that the perturbative uncertainties in the extraction of strange quark mass are mild, compared with uncertainties related to poor knowlegde of the low-enery hadronic spectral function.

C. Oleari, contributed results(8) obtained in collaboration with C. Anastasiou, E.W.N. Glover and M.E. Tejeda-Yeomans, on two-loop QCD corrections to $q\bar{q} \longrightarrow q'\bar{q}'$. They associate tensor integrals with scalar integrals in higher dimension and with higher powers of propagators, by using the Schwinger-parameter form. Systematic application of the Integration-By-Parts technique, along with recursion relations, is sufficient to reduce these integrals to master intergrals in $D = 4 - 2\varepsilon$. With four external quark lines, the most challenging two-loop topology is the "crossed box" diagram:

The IBP and recursion techniques for reducing these two-loop, four-leg diagrams to master integrals (and in particular integrals for the massless crossed box topology) are discussed in Oleari's parallel session presentation at this conference(17).

They have used these identities to construct MAPLE, MAXIMA, and FORM programs to rewrite the tensor integrals for massless $2 \longrightarrow 2$ scattering directly in terms of the basis set of master integrals.

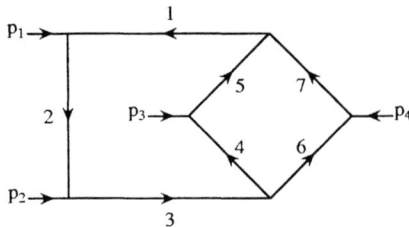

Source: C. Oleari(17)

FIGURE 3. The generic two-loop crossed box.

Then amplitudes are computed by generating the one- and two-loop diagrams using QGRAF. After projecting by tree level and summing over colours, spins and Dirac traces in D dimensions, they identify the scalar and tensor integrals present and replace them with combinations of master integrals. These are then expanded in ε. The expansion can be broken into two parts: one proportional to the Born amplitude $\mathcal{A}^4$, and one that depends on kinematic structures that do not occur at the tree level. The detailed results are rather lengthy for presentation in a summary paper such as this, and can be examined in (8).

Calculations in the MSSM Model

S. Heinemeyer presented work(7) done with G. Weiglein, the developer of the MATHEMATICA program *TwoCalc*. They used that program, along with *FeynArts* (developed by T. Hahn) to compute electroweak two-loop corrections in the Minimal Supersymmetric Standard Model.

The physics motivation is that SUSY particles are too heavy to directly observe in today's colliders, so one must search for indirect effects, by looking at precision observables. The electroweak precision data can be compared with theoretical predictions of the Standard Model and MSSM, to see which model fits better and potentially contradict one or the other. But this tests the theory at the quantum level, and is sensitive to loop corrections. Very high accuracy of measurements and theoretical predictions are needed. In particular, two-loop calculations are necessary to achieve this accuracy on the theoretical side.

The ρ-parameter gives the main contribution to correctins to electroweak observables such as M_W and $\sin^2\theta_W^{\text{eff}}$, and the leading two-loop corrections in MSSM to $\Delta\rho$, which are of order $G_F^2 m_t^4$, are comparable to the accuracy obtained in the Standard Model and to prospective experimental uncertainties. The two-loop results for $\Delta\rho_1^{\text{SUSY}}$ are given in (7).

The contribution of these two-loop diagrams to ΔM_W(MSSM–SM) depends on the Standard Model

$\tan\beta$, and on m_h or M_A (which are related in MSSM). Its dependence on m_h is presented in (7) for several values of $\tan\beta$, and the poster summary displays the dependence of ΔM_W on $\tan\beta$, for several values of M_A.

Beyond the large number of diagrams involved in this computation, the probelm of a proliferation of scales in the MSSM further complicates evaluation of the two-loop corrections. The computation made heavy use of MATHEMATICA-based computer algebra programs: *FeynArts* to generate Feynman diagrams and amplitudes, and *TwoCalc* for reduction of tensor integrals to scalar integrals and evaluation of those integrals. The computing time amounted to about a day on a 500 MHz Pentium.

Results Presented in Parallel or Poster Sessions

The Feynman calculation summary poster was intended to be inclusive: neither all extra-conference material, nor all in-conference presentations. The following summaries were provided based on work presented at this conferences; the detailed papers can of course be found in these proceedings:

S. Groote and A.A. Pivovarov submitted results of a calculation of three-loop QCD diagrams for massive baryonic correlators. These are next-to-leading-order calculations for such processes. Such diagrams are ingredients for QCD sum rules which aim to determine basic baryonic quantities like gorund state energy or residues. The massless contributions were comuputed using REDUCE and MATHEMATICA in a few minutes of computing time; the massive contributions were done by hand and took a few days. These results were persneted as a poster session by R. Kreckel, and appear, co-authored by J.G. Körner, in these proceedings(18).

E.E. Boos presented a scheme(19) for finding gauge-invaiant subclasses of diagrams for a given process. For example, in Bhabha-scattering, the two *s*-channel and two *t*-channel SM diagrams are separately gauge invariant. Aside from the advantage of being able to deal with smaller pieces of a difficult calculation and still produce physically meaningful numbers, this subclassing scheme is important because the precision of computation can be helped by the freedom to use different kinematical variables of integration for different susbsets of diagrams.

F. Yuasa, T. Kaneko and T. Ishikawa presented the Feynman graph selection tool (`grcsel`) in the GRACE system. GRACE(20) is a collection of tools which provides automated generation of Feynman graphs and corresponding helicity amplitudes, phase space integration of the squared amplitudes, and event generation for data analysis. It also contains a facility GRACEFIG for generating figures containing the generated diagrams. The `grcsel` tool(21) can handle tree and 1-loop graphs and supports the Standard and MSSM Models.

State of Computer Techniques

This sample of leading-edge work can provide a perspective on the way advanced computing techniques are being applied to the difficult problem of high-order Feynman graph calculation. Four observations:

- The state of the art has long since passed the point where you could consider doing all these calculations by hand, though there are still some important calculations which have not been done, yet which are not so large as to absolutely require computer assistance.

- There is no clearly established prefered symbolic manipulation system for Feynman integral calculations. Of the ten summaries submitted, five systems (FORM, REDUCE, MATHEMATICA, MACSYMA, MAPLE, and QGRAF) were utilized for calculations, and only one (MATHEMATICA) was used in two cases. Three major programs within systems (*Mincer*, *FeynArts*, *TwoCalc*, three specialized FORTRAN programs, and a major framework GRACE for the non-integration parts of the problem were also used.

- Not everything that could be automated was done via computer. There were a couple of cases where one class of diagrams or one major step was done by hand. This reflects either continuing difficulty in expressing to an automated system the steps done by hand, or a lack of faith that the automated expressions of these steps would be executed correctly. Here, there is room for improvement in the computer tools available.

- Surprisingly, none of the calculations occupied a significant amount of computer time. The running times were generally several minutes and ranged up to one day, and there was no temptation to use any computing platform beyond a simple workstation.

This last observation leads to the conclusion that in principle, the Symbolic Probelm Solving community has the hardware and the software frameworks to attack substantially more complex problems than are currently being pursued. Instead of computing power, one limiting factor (and an area where work being done today will fundamentally advance the field) is continuing development of mathematical techniques and physical insights to do (and to organize) higher-loop calculations with several distinct masses and external momenta. And a related opportunity for improvement is in tools to comfortably program those sophisiticated mathematical techniques in a reliable and readable manner.

Acknowledgments

The author wishes to thank fellow topic conveners J. Vermasseren, V. Ilyin, J. Fleischer, J. Lykken, and D. Perret-Gallix for guidance. And special thanks go to the researchers who made available summaries of their results.

REFERENCES

1. Harlander, R. V., *Physics Letters* **B492**, 74 (2000).
2. J. Fleischer, A. K., and Veretin, O., *Nucl. Phys.* **B547**, 343 (1999).
3. A.L. Kataev, G. P., and Sidorov, A., *Nucl. Phys.* **B573**, 405 (2000).
4. A.L. Kataev, A. V. Kotikov, G. P., and Sidorov, A., *Physics Letters* **B417**, 374 (1998)..
5. S.I. Eidelman, F. Jegerlehner, A. K., and Veretin, O., *Physics Letters* **B454**, 369 (1998).
6. D.J. Broadhurst, A. K., and Maxwell, C., *Nucl. Phys.* **B591**, 1 (2000).
7. Heinemeyer, S., *Two-loop Calculations in the MSSM with FeynArts*, Abstract SPS403, these proceedings.
8. Anastasiou, C. Glover, E., Oleari, C., and Tejeda-Yeomans, M. *Two-loop QCD corrections to* $q\bar{q} \longrightarrow q'\bar{q}'$, Abstract SPS304, these proceedings.
9. Chetrykin, K.G., Kniehl, B.A. and Steinhauser, M. *Nucl. Phys.* **B(510**, 61 (1998).
10. Baikov, P., and Smirnov, V., *Physics Letters* **B477**, 367 (2000).
11. Davydychev, A., and Osland, P., *Phys. Rev.* **D59**, (1999).
12. Fleischer, J., and Tarasov, O., *Z. Phys* **C64**, 413 (1994).
13. S.A. Larin, F. T., and Vermasseren, J., *NIHKEF-H* **91** (18).
14. S.A. Larin, P. Nogueira, T. v. R., and Vermasseren, J., *Nucl. Phys.* **B492**, 338. (1997).
15. Tkachov, F., *Physics Letters* **B100**, 65 (1981).
16. Vermasseren, J., *Symbolic Manipulation with* FORM, CAN, 1991.
17. Oleari, C., *The Tensor Reduction and Master Integrals of the Two-loop Massless Crossed Box*, hep-ph/0012007, presented at ACAT 2000.
18. Groote, S., Körner, J.G. and Pivovarov, A.A. *Analytical calculation of heavy baryon correlators in NLO of perturbative QCD* Abstract P108, these proceedings.
19. Boos, E.E. *Gauge Invariant Classes of Feynman Diagrams and Applications for Calculations*, these proceedings.
20. MINAMI-TATEYA group *GRACE Manual*, KEK Report 92-19.
21. F. Yuasa, T. Kaneko and T. Ishikawa *A Feynman graph selection tool in GRACE system*, these proceedings.

WORKING GROUPS AND PANEL DISCUSSIONS

Experiences Reviewing Scientific C++ Code

Marc Paterno

Computing Division
Fermi National Accelerator Laboratory

Abstract. In this paper I present several issues related to the use of C++ in scientific code, drawing from my experience reviewing large bodies of such code for the Fermilab community, especially for the CDF and DØ experiments at the Fermi National Accelerator Laboratory.

INTRODUCTION

C++ has become the *lingua franca* for scientific programming in the high-energy and nuclear physics (HENP) communities. C++ has many advantages to recommend it: it supports multiple programming "paradigms" (object-oriented, structured, and generic programming), it is (or at least can be) efficient, it is widely available, and there is an international standard defining the language [1]. However, the C++ language is both large and complex, and it is not always obvious how to take advantage of the power it makes available to the user.

Over the past 18 months, my colleague Jim Kowalkowski and I have been part of a number of reviews of medium-to-large C++ projects, mostly for the CDF and DØ experiments at Fermilab. In this paper, I summarize what we have found, through these reviews, to be the most problematical issues with the HENP use of C++.

ABSTRACTION

Useful, Yet Problematic

Support for abstraction is the most valuable feature of C++ in the HENP community. To appreciate this we must remember from where we come: minimally structured Fortran code. While such coding practice was sufficient to the tasks of smaller experiments in the past, the software for current experiments (and those of the future) is of a scale that requires better programming techniques. Good use of those C++ features which support abstraction (*e.g.* object-oriented programming and generic programming) is key to the currently understood better techniques.

Unfortunately, this strength of C++ is a double-edged sword: the most problematic feature of C++, as it is currently used in the HENP community, is this same support for abstraction. When poorly used, abstraction can make code less clear, less maintainable, and less efficient. It is not always easy to determine the right level of abstraction for a given use. For many physicists, previous programming experience does not apply – new techniques have to be learned.

Common Mistakes With Abstraction

The mistakes made with abstraction tend to fall into three categories: too little abstraction, too much abstraction, and poorly organized abstraction. Within each category, one can identify clear subcategories. In this section, I list those errors we have found to be most common, along with a very brief description of how the problem may often be solved.

Too Little Abstraction

Programmers coming to C++ from Fortran 77 often do not have experience in defining clear abstractions to be expressed in code, because Fortran 77 provides so few facilities to help the programmer. One of the first learning tasks faced by such newcomers to C++ is to make use of the features of the language that support abstraction – mostly, the use of classes.

The most extreme form of "missing the abstraction" is the *missing class*. A common symptom of this error is a set of functions that pass around a common set of arguments, or perhaps an array of some basic type. The reader of the code must keep track of which items should be associated to represent a concept. This error is generally not hard to fix: introduce a class to represent the concept. The resulting code is generally easier to understand, and thus easier to maintain.

The next most extreme form of "missing the abstraction" is the *do-nothing* class. This is a class that has nothing but "set" and "get" methods – it is essentially a C-style *struct*. The user of the class must extract values and perform calculations on these values themselves. The resulting code is difficult to read, because the code is full of low-level manipulations. Often a significant improvement can be obtained by inspecting the code for common manipulations, and by putting those manipulations into the class as member functions. The resulting code is more expressive, and thus easier to maintain.

Too Many Abstractions

Some time after becoming aware of the power of abstractions to make code more expressive, many users succumb to the urge to introduce abstractions everywhere. The result can become a problem of too many abstractions.

The most extreme form of "too many abstractions" is the *unused base class*. This occurs when the programmer has introduced a base class to represent each concept, planning for future flexibility, where such flexibility is never needed. The result is a design in which many abstract classes have only one derived class, which all clients use. In this case, the base class adds complexity to the system, but provides no resulting gain in functionality. This sort of over-design makes the code more difficult to understand, and thus more difficult to maintain. This can be improved by eliminating the base class in favor of the sole derived class.

Another problem caused by over-design is that of *very deep class hierarchies*, in which a new layer of inheritance is added for each small increment of functionality – even though those increments (the middle layers of the hierarchy) are not very useful in and of themselves. This sort of over-design makes it much more difficult for users to understand the code, requiring them to look through a long series of headers (or requiring the generation of a great deal more documentation to be read). This extra burden is a disincentive to the reuse of such classes, thus defeating the purpose of the design.

Yet another problem arising from the introduction of needless abstraction is illustrated by the class design shown in Figure 1. In this design there is only one subclass of each base class. Users who get an *AbsComponent** from the *AbsThing* interface have to (dynamic) cast before they can use it to get at function *h( )*. Thus the introduction of the needless abstraction makes the code both more difficult to understand *and* potentially less efficient – the dynamic cast, and the runtime type checking it uses, would not have been necessary if the needless layer of abstraction had not been present.

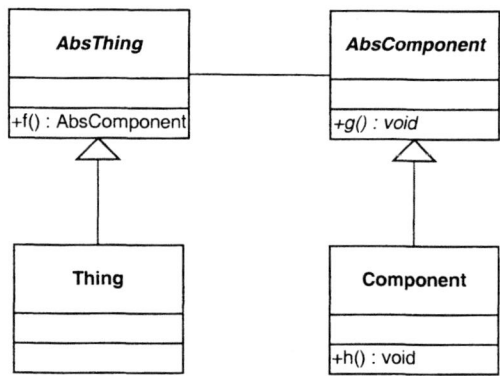

FIGURE 1. UML static diagram showing needless abstraction.

It must be noted that I am not arguing that there is no place for abstract classes and inheritance. I am arguing that it is not advantageous to use an inheritance hierarchy in places where the flexibility it gains is of no benefit, and the complexity it introduces is a burden. Needless complexity makes code harder to understand, and thus harder to maintain.

Poorly Organized Abstraction

The third common problem I have noted in reviews is "poorly organized abstraction." This is a problem not with missing classes, or with unused classes, but with classes that exhibit poor organization.

The first common example of this problem is the *fat interface*, a class with scores of functions that fall into several related groupings. The vast array of functions makes it difficult for the user to determine how to use the class, and forces the user to accept a huge collection of unused functions in order to obtain those few he wants. This introduces what may be a significant physical design burden on the user, who would prefer not to be coupled to those functions for which he has no use. Generally, such a class benefits from being broken up into a set of classes, each of which implements a single, well-defined concept. If some users want the combined interfaces of the entire assemblage, a class that provides the combination can remain as part of the system. This allows the user who wants the entire interface to obtain it, while allowing the user who does not want the entire interface to avoid having his class contain functions which don't make sense for him.

A second form of the problem is the class with multiple purposes, and which contains a member datum whose state determines whether an instance is to perform task *A* or task *B*. This problem is exacerbated when the state of the object must be queried from the outside in order to determine how to use the object. A typical solution for this introduces a distinct class for each distinct purpose.

Abstraction: Summary

It seems quite difficult to find the right level of abstraction, but doing so is clearly important. Wrongly used abstraction results in code that is difficult to understand, difficult to maintain, and difficult to (re)use. Cases of too little abstraction seem easier to repair than cases of too much abstraction. This may be partly sociological, because it is usually those less certain of their command of C++ and object-oriented design that make the first mistake, and those with somewhat more experience who make the second mistake. We must be prepared to review our code and to make improvements if we are to gain the benefits that object-oriented design promises.

An excellent resource for additional methods of improving the design of existing code may be found in reference [2]. An excellent antidote to over-design may be found in the methodology described in reference [3].

LIBRARIES

HENP Libraries

There is a dearth of high-quality scientific C++ libraries. This is mostly a reflection of the fact that many projects were started with early versions of C++, and it has been difficult for these to catch up with modern C++. The lack of use of modern C++ manifests itself in several ways. The result is often not robust, less efficient than possible, and more difficult to maintain than need be.

One of the simplest problems is a lack of *const* correctness. This is generally a result of sloppiness on the part of the library designer, and not an issue with lack of compiler support. Lack of *const* correctness in a library can be especially painful, since user code must either follow suit in ignoring *const* correctness, or must over-use *mutable* data members and *const_cast* operators. The best solution to lack of *const* correctness is to fix the library; this is such a fundamental problem that users should not be forced to work around it.

A second problem is the use of non-Standard language extensions. Since such use tends to make code non-portable, this problem is especially significant. A simple example is the use of the type *long long*, an extension supported by several popular compilers. While this is a legal type in C99 [4], it is not in C++. In this simple example, the solution is clear: create a user-defined type with the appropriate semantics.

A third problem is the use of homegrown classes, rather than use of the classes of the Standard Library. This problem is most often seen in older libraries, especially those created before the Standard Library was widely available, even in draft form. Use of homegrown "string" and container classes causes a problem for code that must inter-operate with different libraries. This is what the Standard Library was designed for, and its use should be preferred to use of homegrown classes.

A fourth problem, the avoidance of templates in systems where templates should be applied, falls into the category of gratuitous language restrictions. This restriction is often a relic left over from early versions of compilers, many of which had poor support for templates. The result is widespread use of macros, with attendant loss of type safety, or the introduction of needless base classes, to provide a type that can be held in a homogenous container. The resulting code is often littered with casts, and is both less efficient and more difficult to maintain than appropriately templated code. Modern compilers provide much improved support, and the effort required to update old codes to conformance with the modern Standard is well worthwhile.

A final category of problem is one that appears at first glance to be a voluntary restriction, but which really amounts to a language extension. This is the use of compiler-specific directives to "turn off" C++ exception handling. Since exceptions are a part of the Standard, and are used by the Standard Library, this immediately results in non-Standard behavior – code that is mandated by the Standard to throw an exception cannot do so. One of the worst results of the decision to use such a language extension is that constructors no longer have a way to signal failure. Instead, one encounters classes with methods to query the status of created objects, to determine if they are in a "good" state. Careful coding would then require that this status be checked in many different places – leading, once again, to less-maintainable code, either because the code is cluttered with tests, or because many necessary tests are not performed.

Use of the Standard Library

One way to take advantage of the power of modern C++ is to take advantage of the Standard Library. I have found that use of the Standard Library in code I have reviewed is broad (many programmers use parts of it) but shallow (few programmers take advantage of more than a few features).

The containers of the Standard Library are used frequently, except where the use of some other library's containers have taken over. The most common problem is that users have some trouble choosing the right container for their purpose.

The iterator classes of the Standard Library are used almost as widely, but are generally used in a simplistic manner. They are commonly used in loops, but rarely are used to denote a range of items. Instead, programmers usually pass around a collection (or a reference to one), which is less useful for generic programming purposes.

The algorithms of the Standard Library are chronically under-used. Many programmers have told me that the idiom of using function objects combined with the Standard algorithms seems unnatural, and that they are happier with explicit loops – even though their code is more likely to contain errors (such as off-by-one mistakes) or inefficiencies. Rarer still are user-invented generic algorithms, even though appropriate use of generic algorithms would help reduce the degree of coupling between libraries, and help the overall physical design of many projects.

CONCLUSION

C++ is a powerful tool for expressing concepts in code. As such, it is also large and complex. The HENP community is still learning to take advantage of the features of the language. Nonetheless, we have made vast improvement over our previous *lingua franca*, barely-structured Fortran.

The major problems we still face are several, and stem from both the innate complexity of our task, and the fact that many programmers are still catching up with the Standard.

REFERENCES

1. ISO, *ISO/IEC 14882:1998 – Programming Languages – C++*. This is the international standard for the C++ programming language.

2. Fowler, Martin, *Refactoring: Improving the Design of Existing Code*, Reading, Massachusetts: Addison Wesley Longman, Inc., 1999.

3. Beck, Kent, *Extreme Programming Explained*, Reading, Massachusetts: Addison Wesley Longman, Inc, 2000.

4. ISO, *ISO/IEC 9899:1999 – Programming Languages – C*. This is the international standard for the C programming language.

Reflections on a Decade of Object-Oriented Programming in Accelerator Physics

Leo Michelotti

Fermi National Accelerator Laboratory[†]
P. O. Box 500, Batavia, IL 60510, USA

Let me begin by stating the obvious: there is neither enough space in a three page paper nor was there enough time in a fifteen minute talk to do justice to the title printed above. Before too many more years pass, someone should undertake a serious psycho-sociological study of scientific C++ programming in the 20th century. There is substantial paydirt to be sifted, enough to generate a score of Ph.D. theses on human behavior and a half dozen volumes on the philosophy of science. Invaluable anecdotal information can be mined on the processes of software acceptance by physicists. If such books are ever written, I would readily contribute a few chapters on sundry proven prescriptions for failure. In anticipation, a few personal reminiscences are scrawled here.

MOTIVATED TO OOP

Three unrelated circumstances led me, in 1988, to search feverishly for new ways of programming. The first was my use of a remarkable graphics engine, the Evans & Sutherland PS390, to display projections of four-dimensional orbits. The PS390 was programmed using a Phigs-like graphics protocol. Even though everything was done with FORTRAN subroutine calls, the underlying graphics model was object oriented. One constructed a data processing net *by combining simple objects*: translators, rotators, scalars, and so forth. Some of these objects could be linked to external peripherals, such as knobs, which enabled real-time manipulation of both data and display.

At the same time, a minor revolution was taking place in the way that (a few) accelerator physicists were doing calculations. Rapidly increasing computing power, coupled with some clever ideas, encouraged the introduction of old mathematical objects – Lie algebras, exponential maps, differential algebras – into new problem domains. A desire to program these algebraic manipulations within a natural syntax moved me to search for a language that permitted operator overloading.

The third circumstance was a strong personal aversion to the "monolithic code paradigm," in which accelerator physics calculations are carried out by submitting an input deck to a single program expected to do everything. As an alternative, I thought it would be useful to have a collection of *simple objects that could be combined to do complicated things*, as was done in Phigs. At the same time, I didn't believe that physicists should be creating their own programming languages, which some were doing. If a *real* language could be used both for describing beamlines and doing calculations, then there would be no need of specially formatted input decks or their parsers; the language itself would be the parser. Physicists could then write their own applications using this toolkit, rather than relying on the monolithic codes.

Such was the goal, anyway. A few requirements for its achievement were: (a) clean separation of physics from geometry and/or hardware considerations; (b) local, not global, coordinates to be used throughout; (c) default "propagators" – objects containing the physics – supplied for each type of beamline element, but specialized ones could be plugged in if desired; (d) automated algebraic algorithms to be implemented expressively; (e) an ordered hierarchy of libraries to make calculator objects work seamlessly with beamline objects; (f) identical coding to be used for tracking and analysis.

It was in this frame of mind that I first ran across *The C++ Programming Language*, by Bjarne Stroustrup: a small paperback, only 15 mm in width, recommended to me by my colleague Michael Allen. What followed is summarized briefly in the next section.

[†] Operated by the Universities Research Association,Inc, under contract with the U.S. Department of Energy

A TELEGRAPHIC PERSONAL TIMELINE

Phase 1, 1988-1992: Beginnings. Terrible compilers and no debuggers. Learning a new approach to programming in surprisingly uncooperative sociological and technical environments. Rapidly evolving "vision." Continually rewriting software: version after version. Spectacularly unsuccessful attempts at proselytizing. Redeeming feature: C++ was a simple, powerful, inexpensive language that allowed me to program the way I wanted. Wrote first versions of BEAMLINE (modeling accelerator components) and MXYZPTLK (algebraic classes) libraries by 1989. (Answer to the one and only MXYZPTLK FAQ: it's pronounced ZIE-tul.)

Phase 2.1, 1992-1993: Communication. First contacts with like-minded people. Learn quickly that writing robust OOP/C++ classes useful to others is *very* challenging. Begin to profess: "Using good classes is easy; writing them is hard."

Phase 2.2, February 16-19, 1994: The Workshop. I convene a "Workshop on C++ Classes for Design, Analysis, Modelling, and Control of Accelerators and Detectors." Approximately 30 people show up to share ideas. Many exchanges; few points of agreement. Presentation of my BEAMLINE/MXYZPTLK libraries leads to the one almost unanimous consensus that I don't know what I'm doing. (After the workshop, Christoph Iselin returns to CERN and delivers a seminar suggesting that writing MAD 9 be replaced with developing a set of class libraries. This idea is not received enthusiastically.)

Phase 2.3, 1995-1997: Collaboration. At the 1994 Workshop, Paul Kunz suggested that accelerator physicists devise a common interface, a set of header files that "everyone" could agree on. Shortly afterward, a collaboration, called CLASSIC, is formed, loosely under John Irwin's leadership, to do that. Several collaboration meetings are held; I attend one and kibitz the others. Christoph Iselin, arguably CLASSIC's most prolific member, writes MAD 9 in C++. Others go in various directions and write their own class libraries. The collaboration no longer exists in any meaningful sense.

Phase 3, 1996-1999: Withdrawal. Concluding that an international software collaboration of physicists is like an orchestra of manic musicians, all playing their own compositions, I return to doing calculations in support of local projects. Joel Butler hides me in the Computing Division (CD) for awhile, and I improve the BEAMLINE/MXYZPTLK libraries by adding new classes, new functionality, and making everything more user-friendly. Afterward, an ambitious online modelling (OLM/OAM) effort is begun: connecting accelerator models with control consoles. At a critical juncture, 2/4 of the OLM/OAM team leave Fermilab. Elliott McCrory and Jean-Francois Ostiguy program heroically, salvaging what can be, but the project's long term viability is doomed.

Phase 4, 1996-1998: FPCLTF (pronounced ZOOM). CDF and D0 decide to rewrite software in C++. Informally attending brain-storming sessions at D0, I eventually join the FPCLTF, a laboratory supported, organized, C++ support group, and learn more good programming techniques. Unfortunately, CD chooses to support only detector-related HEP activities.

Phase 5, 1999-present: To oblivion and beyond. OLM/OAM disintegrates quietly from lack of support. I join FNAL's NLC team, upgrading the BEAMLINE/MXYZPTLK libraries to study linacs. We now have excellent compilers and debuggers, but C++ is no longer a simple language. Controls Department, never having accepted C++, is moving toward JAVA.

PROGRAMMING ISSUES

Using C++ so early in its development meant running headlong into problems and anticipating solutions that did not yet (or still do not) exist. Some of them are presented here.

Run time type identification (RTTI). From the beginning, long before RTTI was part of the language, it was essential to be able to clone an object given only its address, without knowing its type. To accomplish this, classes inherited from an abstract base class which contained a pure virtual function, *.Clone()*: defined as returning a pointer to an instance of the base class, implemented in the derived classes by returning a pointer to a new copy of the object. This left responsibility to the invoking module for (a) recasting properly and (b) deleting the object. (Such responsibilities are considered dangerous and offend purists.) Querying an object's type was done similarly: the base class contained a pure virtual function *.Type()* which, as implemented in the derived classes, returned a string identifying the class.

Ownership. An instance of `class beamline` contains a set of elements – quadrupoles, dipoles, cavities, and so forth – any one of which may be other beamlines. I made an early, but probably correct, decision that a `beamline` should not own its elements. (This is opposed to the STL philosophy, which came later.) Thus, a `beamline` does not have responsibility for its objects, and their handles are available for outside influence. (The same object can even appear in several positions.)

Garbage Collection. Large containers of volatile objects are created and destroyed on the fly, and copy constructors reproduce data unnecessarily. My earliest attempt to deal with these problems involved inheriting from a base class, `Junk`, which possessed a virtual destructor. All dynamic, volatile objects were derived from `Junk`, and their pointers were stored in an `Attic` object. When the number of `Junk` objects reached some critical value, a percentage of the oldest were deleted.

This was, of course, a foolishly risky scheme, which I abandoned after learning the "envelope-letter idiom" from Coplien's *Advanced C++ Programming Styles and Idioms*. This fundamental procedure was an eye-opening introduction to the values of reference counting. Nothing would be deleted until it was no longer needed; garbage collection occurred on the fly; data existed only in one place; no large objects need ever be copied.

A serious problem remains: the inefficiency of dynamically creating and destroying large numbers of objects. To get around that, we pre-create pools of objects to draw from (and replace). However, this works well only for objects of well-defined size.

Persistence. Persistence is important! The C++ language does not provide it; perhaps it never will. Instances of `class beamline` and their elements must persist, available for use by multiple applications. We tried various pickling and swizzling techniques: early and simple solutions utilized streaming and virtual "writeTo" functions. However, essential tasks of providing object pointers and function pointers were not handled correctly. Our most recent solution uses Joseph Hesse's `EZSave` and seems to be adequate for now. In this context, a "Factory" class is almost indispensable. (Presumably, the best solution is an object oriented data base.)

Ferrets and Visitors. At first, adding functionality to `class beamline` meant continually modifying basic header files and source code. However, many calculations proceeed in a sequential manner, processing one beamline element at a time. An early CLASSIC idea was the "Ferret" or "Algorithm," a base class for objects that would propagate through beamlines and do specific calculations by changing state as they traversed the elements. This was superseded by the visitor pattern, which we adopted enthusiastically.

Barnacles. Different `calculator` objects (e.g., visitors) may be able to use information generated by earlier calculations. Without invoking centralized organization, how can a later `calculator` know whether the earlier calculation has been done and, if so, where to find the information? `Barnacle` is a base class for a "Post-It note," if you will, that could be attached to individual beamline elements. A calculator would attach its own information, in an object derived from `Barnacle`. The information can be retrieved later via a keyword search. (This was something akin to an STL multimap, except that the data could be anything whatsoever and need not be of uniform type across all the `Barnacles`.)

Functors. The goal of separating physics from other considerations, while allowing for dynamic specialization, was achieved with the exceedingly useful concept of functors. Default "propagator" functors are provided for every beamline element class. Specialized ones can be written by users – e.g., to employ higher order symplectic integrators. Propagator states can be adjusted by control objects; they can be bound to elements dynamically. Functors satisfied all the required criteria.

Dynamic Polymorphism. At times, it is convenient for a variable to "change its type" in the midst of a program – e.g., from a `double` to a `polynomial` – without changing its name or address. This can be mimicked using the envelope-letter idiom, wherein an object is only a container for a pointer to data. Double dispatch provides a mechanism for handling operator overloading in algebraic classes. There are several ways to do this; I chose to maintain "doubly indexed function tables."

Sages. People misuse software, sometimes by thinking incorrectly about physics (e.g., trying to calculate a closed orbit with RF cavities energized). Without resorting to the monolithic code paradigm, I tried to make calculations more foolproof by introducing `Sage`, a base class for objects that know how to use correctly lower level calculator and beamline classes to do a particular category of calculations . In fact, this comes dangerously close to writing monolithic code and fundamentally violates the original intent of the BEAMLINE/MXYZPTLK. The libraries were not supposed to be foolproof; they were meant to be used by those who understood at least the physics and mathematics of what they were doing. There should have been no need for a `Sage`, but human nature dictated it.

Broadcasting; event handling. It is frequently necessary to inform certain categories of objects about events that have occurred, such as modifications to a beamline. This information must be broadcasted to all existing objects possessing this "need to know." Again, here is a problem not solved by the language itself. It is the sort of problem that arises continually in graphics applications, and the solution can be found therein. Basically, a unary central message server must exist which maintains a registry of all objects responding to particular kinds of events. Objects register (deregister) with this server via their constructors (destructors). (Jean-Francois Ostiguy has become a local expert on this; I am indebted to him for these observations.)

NON SEQUITUR CONCLUSION

There are a minimum of two distinct levels at which C++ can be used: object-based programming (OBP, using objects, which is easy) and object-oriented programming (OOP, creating object classes, which is hard). Physicists who have broken free of the monolithic code paradigm have a natural inclination to peer into nooks and crannies, leading to an inverted population of OOPers over OBPers. At one time this was unavoidable, because classes did not yet exist. That is no longer the case, but the natural inclination is hard to overcome.

C++ In Scientific Application: A Case Study

Walter E. Brown

Computing Division
Fermi National Accelerator Laboratory
Batavia, Illinois 60510-0500

Abstract. This paper describes a project, SIunits, designed and implemented in C++. Addressing a problem that the author has been working on for nearly twenty years, SIunits could not have been implemented without the generic programming and meta-programming capabilities afforded by C++.

HISTORY

Galileo provided us the basis of all scientific computation when he wrote, "We must measure what is measurable, and make measurable what is not so." However, mensuration in other contexts has a long history. For example, Leviticus 19:35 enjoins us, "Ye shall do no unrighteousness in judgment, in measures of length, of weight, or of quantity."

Clearly, there has been significant progress in accuracy. Consider, as a very simple example, the value of pi. In I Kings 7:23 we read, "And he made a molten sea, ten cubits from brim to brim ..., and a line of thirty cubits did compass it round about." The ratio of these data yields $\pi = 3$, an error of less than 5%.

It is noteworthy that widespread use of standardized units came only much later. Among the significant provisions of the Magna Carta, we find the declarations, "Throughout the kingdom there shall be standard measures of wine, ale, and corn.... Weights [also] are to be standardized similarly."

ACCEPTED FIRST PRINCIPLES

A beginning student in any scientific discipline is taught, almost at once, the critical importance of calculations involving physical quantities. Often termed "dimensional analysis" or "quantity algebra," the appropriate techniques are invariably stressed to students early on. For example, Halliday & Resnick (1970) emphasize the importance of such discipline in such terms as:

- "In carrying out any calculation, always ... attach the proper units to the final result, for the result is meaningless without this..."
- "You should check the dimensions of all the equations you use."
- "One way to spot an erroneous equation is to check the dimensions of all its terms...."

If these are indicative of the accepted and expected methodology to be used in hand calculation of physical quantities, consider the formulation of analogous statements bearing on computer-based calculations. For example, we might have, as a parallel to the last of the above (differences italicized):

- "One way to spot an erroneous *program* is to check the *data types* of all its *objects*."

This clearly suggests that it is likely to be useful for computer-based calculations to carry out both units checking and type checking.

TYPE CHECKING AUGMENTS TRADITIONAL UNITS CHECKING

Type checking is far from a new concept. Aristotle, for example, wrote, "There is no transfer into another kind, like the transfer from length to area and from area to solid." Indeed, there is considerable evidence that type checking is still needed.

As a representative case in point, a correspondent (who has requested his identity be withheld) writes, "[A] famous bug in this shop is `2*pi+r*w` as the area of a strip of a cylinder, which escaped notice *for years....*" [emphasis added]. This incident provides compelling evidence of the subtle nature of computer-based calculation. It also strongly hints at the value of computer-based type checking, for such checking would have discovered that the above expression attempts to sum incommensurate quantities and is thus inconsistent with a purported area calculation.

NUMERIC PROGRAMMING TODAY

In the very early stages of this project, we had undertaken a study to get a feel for the degree to which modern coding practices apply these principles with respect to numeric programming involving physical quantities. Extensive inspection of code samples in 1998 showed:

- Heavy use of native numeric types (*e.g.*, `double`), and
- Occasional use of synonymous types (*e.g.*, CLHEP's `HepDouble`).

However, it is easy to demonstrate that such practices are wholly inadequate. We have merely to consider the meaning of such expressions as:

- Avogadro's number + speed of light?
- Distance + Energy?
- Mass + Momentum?

Alas, arbitrary expressions involving such generally meaningless combinations will typically compile without complaint!

As further evidence of the need for computer-based checking, the following excerpt from the <u>Washington Post</u> (Oct. 1, 1999) discusses the failed Mars Climate Orbiter mission, graphically demonstrating that unit errors are both costly and very hard to find.

> NASA's Mars Climate Orbiter was lost ... because engineers failed to [convert] from English units to metric, an embarrassing lapse that sent the $125 million craft fatally close to the Martian surface.... [T]he error had affected the orbiter mission from its launching almost 10 months and 416 million miles before its ... failure. And yet the problem was never caught and corrected....

We believe the present unfortunate state of affairs is a relic of past limitations. In particular, early programming languages (such as Fortran and Cobol) provided inadequate expressiveness. Data types, for example, were limited to those implemented in the computer hardware. This seemed adequate, however, because these types were of sufficiently general utility. Frankly, we just didn't know what we were missing!

Programmer-defined data types came later. Such languages as Pascal and Modula had enough features to validate the feasibility and utility of the then-novel "data abstraction" methodology. However, the same languages lacked generic programming and meta-programming capabilities.

Today, we can do much better. Using contemporary technology afforded by C++, we have expressive power resulting from such newer features as classes, namespaces, and templates.

The early, experimental "data abstraction" methodology has evolved, as well. We now have object-oriented techniques to apply to software design and programming. Generic programming and meta-programming technology has become available. Finally, static type checking has become the norm.

Static type checking is a well-known and well-understood concept in computer science. When consistently applied, static type checking provides:

- Succinct, intelligible, consistent (!) program documentation;
- Information that a compiler could exploit to produce more efficient code; and
- Important programmer feedback via early (compile-time) error detection.

SIUNITS

It is against this background that the current project, known as **SIunits**, has evolved. The following five goals were established from the outset:

- Application of contemporary technology to computation involving physical concepts,
- Convenient (near-trivial) expression,
- General utility based on existing standards,

- Nomenclature from our problem domain, and
- No run-time performance penalties!

These design goals have been realized via application of modern C++ coding practices.

SIunits is based on *le Système internationale d'unités* (SI), the international standard (having the force of treaty) that codifies accepted practice in dealing with physical quantities. SI specifies base and derived quantities as well as corresponding units of measure; SIunits has adopted these specifications and enforces them via compile-time static type checking. This effectively forbids all incommensurate expressions, exactly as desired! Further, any implied quantities (such as might arise mid-computation) are generated as needed without programmer intervention.

In addition, the SIunits project has met its remaining goals, too. Among its other features, SIunits attaches no run-time overhead for typical usage:

- No extra memory space per object;
- No extra execution time for user functions;
- Small initialization costs for I/O, etc.

In addition, SIunits includes a spectrum of five models, a feature first suggested by a colleague, Dr. Mark Fischler. In the relativistic model (where the speed of light is one), for example, it is possible to add meters to seconds; it is otherwise (i.e., in the standard model) highly unlikely to be a meaningful expression, and thus highly likely to reflect a conceptual and attendant computational error. Models are upwardly compatible.

SIunits PROGRAMMING

Within SIunits, physical quantities (often known as dimensions) are implemented as types. Of particular interest, and consistent with SI, SIunits provides the seven base quantities (Length, Mass, Time, Current, Temperature, AmountOfSubstance, and LuminousIntensity). It is from these seven that all remaining quantities are derived. Indeed, 22 derived quantities are given special names by SI; these include, for example, Energy, Force, Entropy, etc. These and many, many more derived quantities are provided by SIunits.

Units behave as right-hand constants. Of special note are the 7 base units (meter, kilogram, second, Ampere, Kelvin, mole, candela) and the 22 composite units (joule, watt, hertz, katal, …) specified by SI, as well as the various multiples and submultiples (mega-, nano-, etc.). All these are found in SIunits, too.

The following code provides some of the flavor of programming with SIunits.

```
// --- Preliminaries:
#include "siStdModel"
using namespace si;
typedef  Length<float>  Len;

// --- Good Instantiations:
Len d1;    // implicitly 0.0F
Len d2( 2.5*meter );
Len d3( 1.2*centi_*meter );
Len d4( 5*d3 );

// --- Bad Instantiations:
Len d6( d2*d3 );      // no!
Len d7( 3.5 );        // no!
Len d8 = 3.5;         // no!
Len d9;               // ok, but …
    d9 = 3.5;         // no!
// However:
Len d10( d2+d3 );     // ok
```

CONCLUDING VIEWPOINTS

It seems common, in the professional scientific community, to hear assertions such as, "I did enough dimensional analysis in school. I know the units my work involves, so I don't need to repeat my analyses for every calculation." This attitude is certainly one of convenience; it is equally certainly unprofessional.

As our correspondent has written, "[T]hese are really insidious bugs, very hard to find by proofreading ... because your brain knows what the expression is supposed to be [so] you tend to see the intent and not the error." Given the convenience of SIunits, there appears no longer any excuse for such a lackadaisical attitude. When first presented with SIunits, a potential user quickly remarked, "[SIunits] is the first really good reason I've seen to switch from Fortran to C++."

In brief, we note that SIunits could not have been carried out absent modern C++ metaprogramming and generic programming support. This support is rapidly improving as vendors provide increasingly Standard-compliant C++ compilers. Alas, the concepts underlying C++ seem well understood and well applied by only a small percentage of the scientific community.

The Development of the ROOT Data Analysis System

René Brun

CERN – CH-1211 Geneva 23, Switzerland

Abstract. The previous generation of data analysis systems such as PAW had been designed as stand-alone executable modules to be used in the final stage of data analysis. This conventional model has been used very successfully to display histograms or generate histograms on the fly from some ntuple files. It had the advantage to be totally independent of any framework used in the simulation, reconstruction or physics analysis steps. With the advent of Object-Oriented programming, this simple model appears to be insufficient. The ROOT system has been designed to provide all the facilities in a system like PAW, but in addition has the ambition to provide a complete and coherent framework that can be used at all stages of data processing and well matched to the OO paradigm.

INTRODUCTION

The PAW model

When PAW was created in 1985, histograms were the lowest common denominator for data types. The HBOOK system had been used for many years and was a de-facto standard. The introduction of the so-called row-wise ntuples was immediately successful and called for one more step. Column-Wise Ntuples(CWN) were introduced in 1989 to optimize the access time to larger and larger files. CWNs were very similar to tables in relational data bases. An ad-hoc query mechanism was introduced and was improved multiple times to support more and more complex queries. The data types, however, were pretty simple, integers, floats and characters. The CWNs and histograms were created from data stored in sequential files with systems such as ZEBRA.

ZEBRA was designed in 1983 to support dynamic data structures of simple entities called banks. A bank was like a C structure but limited to the basic types, integers, floats and characters. ZEBRA was particularly useful to describe large collections of banks to build hierarchical or graph-like structures. The concept of structural and reference links was powerful enough to describe the type of physics event data in medium or large size detectors. CWNs were files created by each individual physicist. When the PAW system became more stable, CWNs were used as micro data summary files (microDSTs). Because PAW was a system available on most platforms and well supported, CWNs became quickly the standard data format for the final analysis stage.

However, the PAW CWNs were not designed to describe the complex event structures that ZEBRA was able to describe. In addition, the CWN internal description tables could not scale to the Gigabyte range. Despite all these limitations, PAW has been THE standard data analysis system used by most experiments in the past decade. This success has been such that finding a suitable alternative has taken more time than initially expected in the new world of software based on OO languages.

MOVE TO OO PROGRAMMING

The first attempts to introduce the OO concepts in data analysis were made in 1989. These attempts were unsuccessful for many reasons:

- OO languages were not mature and not widely available.

- The performance of compilers could not compete with the Fortran compilers.

- The introduction of a new system requires that this new system provides all the facilities of the previous system in a stable and proven environment. This is a chicken and the egg problem. Convincing the silent majority to adopt a new tool or framework is a challenging work.

- The problem of Object Persistency had been totally underestimated or a wrong solution assumed.

- Last, but not least, the introduction of a new system could only be done with the strong support of a major laboratory with a long term commitment.

A major problem: Object Persistency

In the middle nineties, it was naively assumed that the solution to Object Persistency could be provided by commercial Object Oriented Data Base management systems. Systems like BOS or ZEBRA had been used successfully to manage large volumes of data, but these systems were perceived as difficult to use and to maintain. For a few years, Object Persistency was left in the hands of computer scientists or people with some experience in data management but a limited experience in data analysis. As a result, progress in understanding this difficult problem has been slow and highly controversial. Solutions to Object Persistency should have been thought with data analysis in mind. On the contrary, too much emphasis was put on the use of OODBMS systems to store event raw data or results of the first pass reconstruction. A few experts were designing systems to create large data bases. Processing data in the data base was assumed to be an easy task. The official policy was to delegate this function to a set of commercial tools. An unfortunate confusion was made between a data analysis system and a 3-D interactive visualization system. As a result, too much emphasis was put on the pure visualization side of the problem and tools capable of making queries in large OODBMS were not developed.

Towards the ROOT framework

Following our many years of experience with the development of the PAW system, we decided in 1995 to start the design and the implementation of a system capable of doing at least the same thing in an OO context, but also to serve as a complete framework from data taking to data analysis. During a few months, we learnt the basic ingredients of an OO system by implementing several variants of a histogramming package. We quickly implemented a rudimentary I/O sub system and also some very basic collection classes. It became rapidly clear to us that a more ambitious persistency mechanism had to be developed. There was no point in developing a system supporting only the PAW CWNs in a world dealing with classes and complex object hierarchies. OODBMS could have been the solution to our problem, but we were convinced that the corresponding proposed commercial tools were not appropriate for a flexible data analysis environment.

The User Interface Problem

One of the main tools in PAW was the command interpreter KUIP. Many man years had been invested in this package. Designed initially as a simple command line interface, KUIP was further developed to include many features of a programming language with control statements, loops, local and global variables. We had substantially underestimated the effort to develop a coherent system. With a growing number of users, it was clear that all the features of a programming language were requested at the interpreter level. We had seen so many KUIP macros with hundred or even thousands of lines that we were absolutely convinced that a modern data analysis system requires all the power of a high level programming language. It was out of question to develop our own language. It was also out of question to use a scripting language different from the main programming language. Because most users were going to write large scripts to analyze large data sets, we were convinced that the scripts had to be written in the same programming language than the main language used in the other stages of data acquisition, simulation and reconstruction.

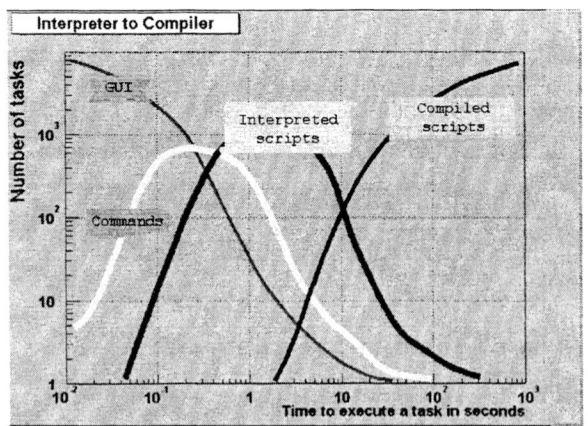

FIGURE 1. Estimated frequency of tasks as a function of the time taken to execute the task.

With today's desktop machines, it takes between a few seconds and one minute to compile and dynamically link a realistic analysis script. With computers becoming faster and faster, one may hope that in a few years from now, dynamic compilation and linking will become affordable for an increasing number of tasks. Having the same language for the interpreted and the compiled codes will be a tremendous advantage. On the other hand, nobody will trust results produced by a pure interpreted language. An interpreted language is fundamental for tasks that must be executed rapidly, such as short scripts edited very frequently or all the tasks called via the graphical user interface. Our goal was to combine the advantages of an interpreted and/or compiled language in one single framework. To achieve this goal, we had to develop a powerful object persistency system with as few limitations as possible to support the main stream proposed OO language C++. We were lucky to find an existing C++ interpreter CINT capable of parsing the complex C++ header files and to support a very large subset of the language interactively. We developed an extended Run Time Type Information (RTTI) used in the I/O system but also in many other places including the Graphical user Interface. This RTTI goes far beyond the C++ RTTI and looks more like the Introspection mechanism in Java.

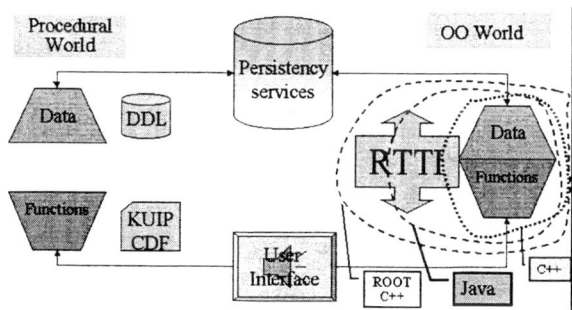

FIGURE 2. The ROOT RTTI is used by the persistency services and by all user interfaces.

A dictionary describing the classes (data and functions) is a fundamental requirement. It is the heart of a framework and is used to implement automatic or hand-coded converters, the automatic class schema evolution system, browsers, inspectors, the GUI context menus and the automatic documentation tools.

In the past, we developed ad-hoc dictionaries for data and functions, for example the Adamo DDL, the Zebra DZDOC for data dictionary or the KUIP Command Definition File for the functional part. When using an OO language, it is an advantage to use one unique dictionary to describe the data and function part of a class because the function arguments, and their return types require the same description than the data members.

BUILDING A MODULAR SYSTEM

Modularity is a buzzword with different meanings.. A modular system is sometimes presented as a system with many small and independent components. In general such systems do not have an object bus and the communication between the components is left to the application using these components. Systems with a deep hierarchy of components may be difficult to maintain because of too many interdependencies between the top level and low level modules. Is a system with well defined interfaces a modular system? Probably not, because too much emphasis is put on the interfaces at the expense of the object bus. In such systems, the interfaces may have long argument lists instead of well designed collections and object folders. An end user will see a system as modular if the structure is easy to understand, while a system developer will put more emphasis on the maintenance aspects, probably the two aspects being strongly related. A modular system can also be seen as a system easy to integrate into another system. The truth is that modularity is difficult to achieve, in particular in a rapidly growing system.

Modularity and Dependencies in ROOT

One of the main problems we had to face in building a complex and large system like ROOT (version 2.25 has about 650,000 lines of code) was to minimize the dependencies between shared libraries. ROOT consists of about 25 shared libraries or DLLs. A library may be dependent on another library if linking one library forces the linking of another library even if the application does not have a direct reference to this library.

The early versions of the ROOT system had many inter-library dependencies. This problem was rightly pointed out by many users as something to be fixed. We did this. In the current system, only a small set of base libraries (libCore) is required when creating e.g., Histograms in batch mode. Besides the decoupling of the graphics system many more abstract layers were introduced to decouple other parts of the system: histogram from its painter, the tree storage system from its query mechanism (treeplayer), fitting from Minuit, etc. Following this reorganization none of the lower level libraries depend anymore on higher level libraries. These changes improved besides modularity also overall system performance.

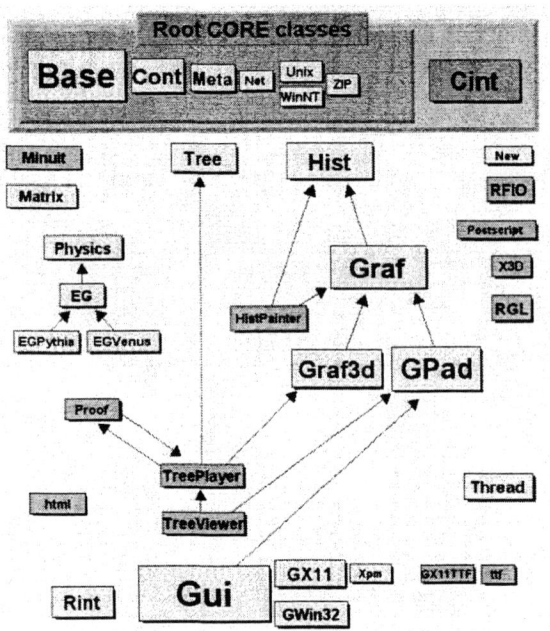

FIGURE 3. The ROOT shared libraries. Arrows indicate dependencies. All libraries depend on the Core libraries.

Abstract Interfaces

One of the ways to improve modularity is to use abstract interfaces. Abstract base classes can be referenced by the low level libraries. Only applications using the real implementation of the derived classes will be forced to link with the corresponding library. For example, the libCore references the TVirtualPad abstract class (pad/canvas graphics interface), but only applications doing real graphics will need to link with the graphics libraries libGpad, libGraf.

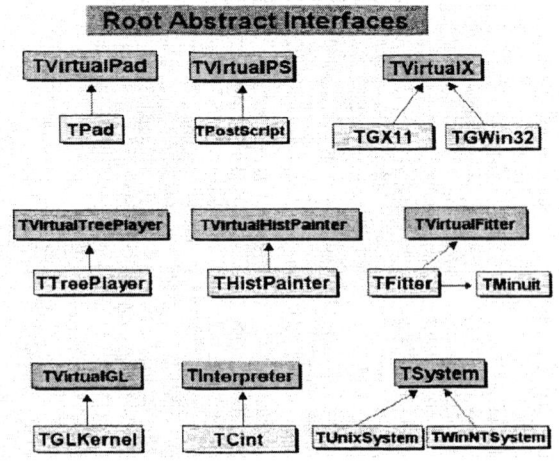

FIGURE 4. Some of the ROOT abstract classes and derived classes.

Quality Assurance

In a rapidly growing system, it is inevitable that the implementation of new functionality introduces new bugs or side-effects. By making early and frequent releases, users are introduced in the development loop and give a lot of feedback. Data analysis systems cannot be developed by a committee or a closed circle of developers. If the original idea and the system design is correct, users will soon use the system and will be happy to give feedback or even to contribute to some parts of the code.

There is a delicate compromise to be made between adding too much functionality, getting too many users, porting to new systems, writing the documentation, answering questions, etc. and minimizing the number of bugs.

In the figure below, we show our monitoring of the number of bugs that we fixed as a function of time. Our experience with ROOT or any previous system indicates that it takes at least five years to reach some stability. Thanks to several thousand users, ROOT is now close to reaching a stability regime.

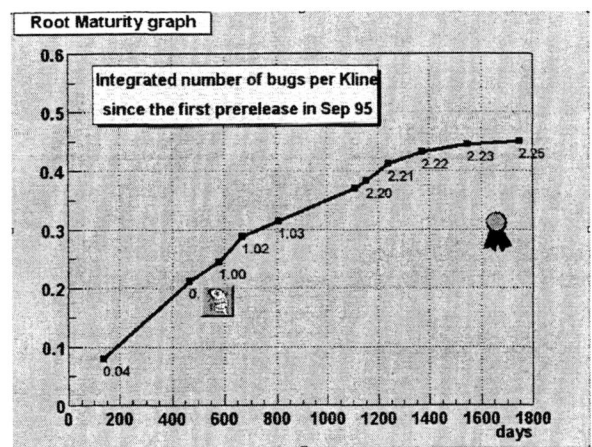

FIGURE 5. Monitored number of fixed bugs per thousand lines of code in each ROOT release during the five years of development.

REFERENCES

1. ROOT: http://root.cern.ch/
2. PAW: CERN Program Library Q121
3. KUIP: CERN Program Library I202
4. ADAMO: CERN Program Library Q190
5. ZEBRA: CERN Program Library Q100

Analysis Environment Challenges

Lassi A. Tuura*

Northeastern University, Boston, Massachusetts 02115

Abstract. This paper discusses some of the challenges modern high-energy physics analysis tools are or will be facing shortly. We take a brief look at the environment in which the tools will need to operate and the requirements this implies. We then explore a handful of concepts that may prove helpful in responding to the challenges. Some general ideas on implementation are also given.

INTRODUCTION

Experimental particle physics has high data processing demands due to the need to transform large volumes of data into highly distilled statistical analyses. To present results to others, large samples of data must be scanned, drilled into, reduced, viewed in different ways, and fitted to models. This process involves much of the experiment software. Sometimes reconstruction needs to be redone with changed parameters and improved algorithms. It is not uncommon to find that a part of the "final analysis" needs to be migrated back the chain, for example to the offline reconstruction or the triggers.

With the growing volume and complexity of the data gathered as well as the rising user-interface expectations, pressure on the analysis tools is increasing. Existing tools no longer perform satisfactorily. This paper takes a look at some of the key challenges from the software technology point of view, concentrating on the more important issues and requirements. These are followed by a discussion of concepts that may prove useful to improving the tools, including a few ideas on implementation.

The purpose of this paper is to motivate discussion on where future analysis tools should be going. Hopefully this discussion will lead us to better understanding of the problems we are trying to solve, and hence better tools for physics analysis.

ANALYSIS ENVIRONMENT CHARACTERISTICS

It could be said that physics analysis is largely an iterative process of reducing data sets to more interesting subsets and deriving new quantities from the data at hand, as illustrated by Figure 1. The outcomes are highly distilled statistical analyses intended for human interpretation and discussion, usually histogram plots in two or three variables of a large number of measurements fitted to a distribution function. New quantities are usually derived either to increase the abstraction level ("this trajectory is a muon particle track"), to summarise data ("the highest momentum track in this event has p_T of 65.5 GeV"), or to derive statistical quantities from repeated measurements ("classify the momentum of these million tracks into a range from 50 to 100 GeV in steps of 0.5 GeV"). Much of this work can be done in an interactive analysis and presentation tool on very high-level entities such as simple scalars and 4-vectors. This tendency is clearly manifested by the past focus on tools operating on simple summary information such as DSTs, N-tuples and tag databases.

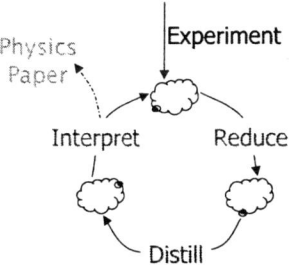

FIGURE 1. Generic physics analysis process

However, it could also be said that the above description is, if not incorrect, at least an immensely limited view of physics analysis. Analysis involves *a lot* more than just the interactive tool. We can learn from the "PAW revolution": PAW (1) provided a new user interface and new, more powerful ways of working with data through N-tuples. Better access to data did not stop with PAW, only today we can do much more and better than just N-

---

* This work is supported by US National Science Foundation

tuples. For example, ROOT (2) has added trees, and CMS uses a complete object model (3).

Consequently experiments are making their data accessible in new ways by exploiting object models that have proved themselves very powerful. Besides this advances in networking, distributed computing and mass-storage systems have opened new possibilities of automation and integration to the collaborations. It is now quite reasonable to assume that one can in many cases navigate transparently from one part of the data to the other even when the destination objects will need to be first fetched from tape or transferred from a remote site. For example, we are now growing to expect that one can navigate from a N-tuple back to the data it describes. Similarly, we are no longer taking for granted that an analysis job should run on a single computer, or even any particular computer at all.

Where we are particularly lacking is new user interfaces to match these advances. The new object and computing infrastructure models require corresponding advances from analysis tools. To accomplish this, the next-generation analysis tools will need considerably deep links with the rest of the infrastructure in the experiments. This is bad news as it will be much harder to develop experiment-independent tools.

ANALYSIS TOOL REQUIREMENTS

Let us now turn to specific challenges new tools will need to address. Here we explicitly assume we are reaching *beyond* the present functionality: all the existing capabilities in data analysis and presentation are expected to be retained in one form or another, including familiar operations on N-tuples, histograms, fitting and plotting.

We observe that the range of analysis activities is great, with no clear boundaries. They include for example:

- Batch analysis;
- Interactive analysis;
- Setting up software configuration management tools, application frameworks and reconstruction packages;
- Data store operations such as replicating entire data stores; copying runs, events, and event parts across stores; not only copying but also doing something more complicated (filtering, reconstruction, ...);
- Browsing data stores, selections and other collections down to object detail level;
- 2D and 3D visualisation;
- Moving code across final analysis, reconstruction and triggers.

Each of these has considerable range in itself. People have different preferences for interactive analysis style, and those tastes vary according to the task at hand. Sometimes a "pointy-clicky" graphical interfaces is called for, at other times an interface more powerful if also more cryptic in the spirit of Emacs (4), yet at other times a command line interface is the best. As another example, analysis often develops from a quick interactive cut to a more elaborate expression to some quite complicated code that is best ran in the reconstruction framework. By this time one needs to know how to compile and incorporate code into the reconstruction framework—for example, how to check out a workspace package from the source code management system, copy the sources into it, build it, and tell the framework to load and run it at an appropriate time.

Today all this is largely possible but involves far too many tools and the physicist has to painfully learn to master each one. This usually results in a collection of "magic incantations"—commands and code fragments not understood by the author but somehow causing the system to carry out the desired task—not a coherent environment. The transition from one environment to another usually implies pain (the all too familiar "Oh dear, I must now convert this KUMAC into a FORTRAN reconstruction package and start using data in a completely different format" scenario). The other extreme, a single environment that solves all the problems, is not very appealing either as experience has too often showed that such environments fall short on quite a few important accounts and escaping outside the environment is often painful. The concept of a coherent, seamless and well-integrated environment *is however extremely appealing*. In our opinion one of the greatest challenges is how one can develop such analysis tools without dictating a single environment to the experiments.

A Concrete Example

Let us consider a relatively simple concrete example to illustrate these points. Let us assume I wish to replicate and share some of my experiment's data for analysis tasks and GRID-style[1] use. While I have tools to do so, I immediately stumble on the following:

- Do I understand my experiment's world-wide configurations well enough to use the tools confidently? (That is, can I be sure that I am not doing something utterly silly?)

---

[1] Please refer to other talks in the conference and references in them.

- How do I find the data store nearest to me in the first place?
- If I want a private working store that shares the experiment data at the same time, exactly what should I do?
- What if I do not want just a plain file copy, but want only a copy of the reconstructed data for the calorimeter from a certain sample that includes events in tens of files?
- What if I want to share my analysis settings and results with my colleague for verification?

There are answers to all these questions in each experiment. Some of the answers are even well documented. The challenge to us is: is there a *coherent* answer? To what degree can the answer be applied from one experiment to another?

CONCEPTS TO GET US THERE

We will now suggest a couple of concepts that might prove helpful in responding to the above challenges. These are simply ideas; the author certainly has not yet tested all of them in analysis tools. We do however include some examples below how existing (mostly unrelated) applications are already employing these ideas. Some analysis tools, such as JAS (5), are also pursuing some of these concepts already. In other words, these concepts are not novel. Most of them can be considered industry-standard software design practice. As we apply the concepts to our domain, hopefully something useful will emerge.

First we observe that a *uniform integrated user interface* to the whole task range would be most useful. It should encourage a coherent look and feel within reasonable limits, both for command line and graphical use. It could take a form of a *tool suite* or a work bench, together with a sufficiently large set of common building blocks. It should be possible to host tasks that are most naturally done via a graphical user interface[2] inside a coherent, standard framework—a document-centric model such as those found in most of today's office applications would probably feel most natural to many users.

Secondly, a cohesive solution to a whole slew of the tasks, including the more complex ones, is required. *Wizards* seem to be emerging as the popular solution: most find them easy to use as they guide one through each task by explaining the choices, giving sensible defaults and by making the steps clear. In our environment most of the wizards ought to leave behind a repeatable "trail" such as a script so that someone preferring to see actual code can replicate the wizard's actions once the steps are more familiar. It would also allow the wizard's actions to be postponed and carried out in a batch environment.

Another quite evident idea is to represent the data store or parts of it[3] as a *directory structure*—and allow one to browse it as such, much like one views the file system with modern file browsers. It is not a long step from there to provide a conceptual *home directory* where one ends up upon "logging into" the system. It should be easy, if not the default, to put objects relating to one's analyses under this home directory, including setups for the framework and reconstruction packages, parameters and so forth. Data store browsers should be able to show, inspect and interact with any object in the directory structure, just like modern file system browsers can not only show the file system structure and basic file information, but can also spawn applications for any file type within themselves (6).

Finally, it should be easy to access and interchange analysis setups and objects with different individuals and physics groups. It should be possible for one to keep track of configurations, input and output data selections, and so forth. Following again prevailing user interface standards, perhaps a *desktop* would be most appropriate, allowing one to put on it shortcuts or symbolic links to the objects and "directories" of interest. Standard shortcuts for common things would naturally be part of this.

For all this we must bear in mind that one size never fits all. It is the analysis tools that must adapt, not the physicists and the experiments. Hence in all the above there is an implied requirement for the implementations to be modular.

Examples In Today's Applications

Figures 2 and 3 show some of these ideas in action. It is not difficult to extrapolate them to the physics analysis.

The left side of Figure 2 shows a recent version of Microsoft Outlook (7). On the left there is a quick access bar to common items. Next to the bar there is a hierarchical browser. On the right side are various levels of detail on the item selected in the hierarchical browser. In the figure that happens to be a mail folder and a message from the folder, but it could also be a several different views of the calendar—and many other things.

The right side of Figure 2 shows the Windows 2000 file save dialog above and an IGUANA (8, 9) database browser prototype below. The file save dialog presents

---

[2] Here the user preference matters just as much as the task itself.

[3] Not just event data, but also other data and possibly even software components in the running program.

again the standard "root" locations on the left for an easy access to items of interest to me personally. One can imagine such a dialog to be offered when saving any analysis object such as a histogram, event collection or a N-tuple. The IGUANA prototype demonstrates how easy it is to work towards these concepts: this simple prototype can only browse histograms and doesn't offer any of the conveniences a full-fledged browser ought to offer, but already has much to offer in that direction. Similar examples can be found in other recent analysis tools.

Figure 3 shows a combination of co-operating views. DDD (10) is a graphical debugger front-end—it only provides a graphical user-interface on top of command-line debuggers; exactly what it does is mostly irrelevant here. In DDD one can click around in the data to expand and collapse the view (top window), view data and issue commands in the source window (in the middle; note the tooltip when mouse hovers above a variable name), execute common commands from the floating command tool (on the right), or type in commands on the command line (bottom window). The command line reflects actions carried in the other windows and vice versa; the visualisation window reflects changes in the program being debugged. As far as the user is concerned, none of the views is more important than the others; one just uses whatever window is most convenient. What is not visible is that each of the views can be disabled and enabled separately, or split off into separate windows. DDD can also run in a command line mode without any of the windows—the requested windows will then pop up when needed.

Road Map

How might we implement such concepts? Few experiments can afford to develop a new interactive analysis tool, let alone coherent tool set for the entire range of analysis tasks. If as a community we want these tools, we need to partition the problem and co-operate. We need to define categories such as GUI, event and detector visualisation, and data analysis and presentation. In each category we need to use existing modules where possible, and write our own only when nothing suitable exists. All the components should be integrated to form a user-friendly and productive environment.

In this manner we can build a pool of components out of which we can draw to construct customised applications with relatively little effort. For this work out the pool needs to be *truly modular* which means we must consider *all* component dependencies, not just the obvious ones. We can get a good idea of the immediate dependencies of each component by analysing how all the features provided by that component can be *tested* (11). The result will be the union of the source and binary dependencies (the bare minimum required to build and run the test) plus logical dependencies from the ways the components must be used together.

Architectural Issues

The single key to a successful analysis environment we consider to be an architecture that is modular where it matters the most. Not everything needs an abstract interface—it may be better to make a strategic choice to use a particular product if it can be contained and completely replaced in something like 6–9 months.[4] In other words, the point is to *assess and bound risks*, not achieving total security of any change.

We suggest the following as guiding design principles:

- Use Model-View-Controller and similar patterns to partition the domain;
- Use layering to separate front-ends and back-ends;
- Ensure a standard for visual components to facilitate integration.

Abstract interfaces are required for all data access, and very narrow interfaces are required to link the analysis and visualisation subsystem to the experiment's core frameworks. Let us illustrate this principle with event selection using summary data: a concept that is implemented in many ways, such as tags, N-tuples, DSTs, and B-tree indices. N-tuple is both an access paradigm and a storage method, and historically the emphasis has been on the latter. It would be necessary to shift the focus to the *access and query interface* that continues to provide the look and feel of the proven access method with natural modern extensions. The implementation should be allowed to vary, and could be changed to take advantage of better ways of storing and organising events, indexing schemes provided by the data stores, or even computation on demand by the core framework and tape reads. Other interfaces should be provided to control the aspects outside querying, while the primary query interface should remain unaware of and unaffected by such concerns. As far as the client code cares, a physical N-tuple should not even need to exist.[5] For example, a database query result could be represented as a N-tuple: it could be iterated over one row at a time, and the previous row might vanish once it has been discarded.

---

[4] For example, it would be better to use OpenInventor (12) than to invent one's own 3D library.
[5] Obviously the client *user* might care and might want to direct the system to make the query results persistent.

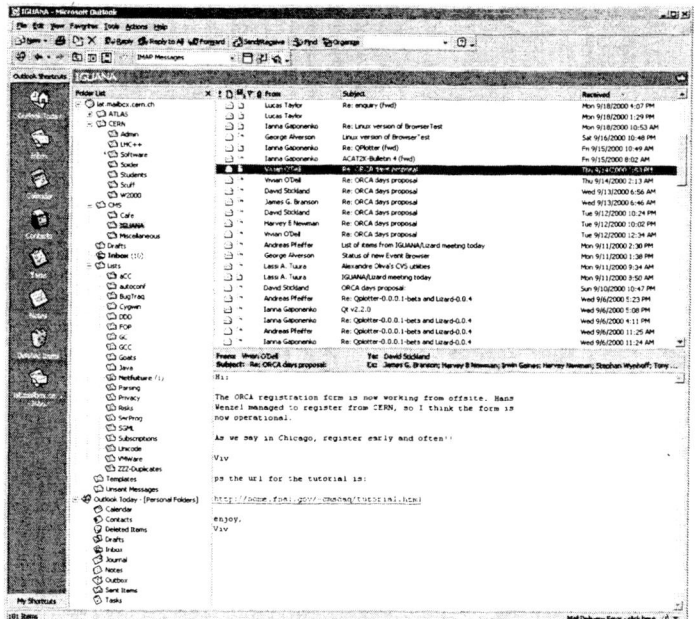
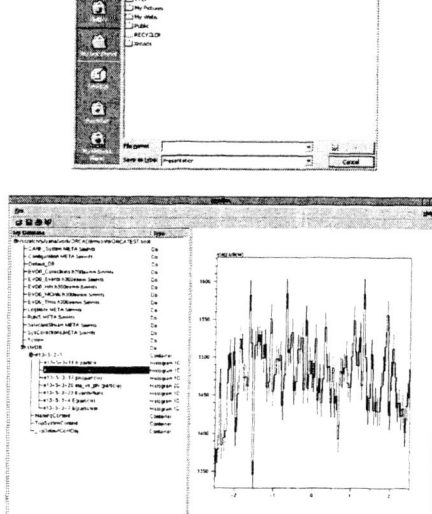

FIGURE 2. Modern user interface concepts: flexibility in Microsoft Outlook (left), and quick access to important items in the Windows 2000 file save dialog (top right). Also an IGUANA prototype on the way to such concepts (bottom right).

CONCLUSIONS

Analysis environments include much more than just the interactive analysis and presentation tools. So far the focus has been largely on the latter, but as the experiment complexity grows we need to be able to drill down and interact with the data in new ways. Consequently a good, coherent, solid user interface for the whole range of tasks all the way from batch mode operation to the quick point-clicky jobs is required. Building such a tool suite from scratch is neither affordable nor wise. Instead we need to use existing components from our own community, open source and commercial packages.

For this to be practical, each component must be given a very clearly defined responsibility: a simple clear mission statement. Abstract interfaces are useful means in this to carry out some of the domain partitioning and layering. This way risks can be bounded should a particular technology or a component fail. Abstract interfaces will also provide natural points to hook in many of the exciting new ideas that we will see. Finally, they have the benefit of allowing people to co-operate without disturbing each other too much.

Some of these ideas are already being explored by many projects, but we still lack a comprehensive solution with a vision of catering to the complete range of analysis environment tasks.

FIGURE 3. DDD illustrating how multiple different views can be combined for a flexible and productive interface to interact with the underlying system, ranging from very graphical (top) to a basic but powerful simple window (middle) to command line (bottom). Command line reflects actions in other windows.

ACKNOWLEDGEMENTS

I would like to thank my colleagues earlier in ATLAS and now in CMS, and IGUANA project in particular, for the many insightful discussions that have lead to the development of these concepts.

REFERENCES

1. Brun, R. et al, *PAW Physics Analysis Workstation*, 1989. CERN/CN Long Write-Up Q121.
2. ROOT Home Page, http://root.cern.ch/
3. Innocente, V., *CMS Software Architecture: Software framework, services and persistency in high level trigger, reconstruction and analysis*, 1999. CMS IN/1999-034.
4. GNU Emacs Home Page, http://www.gnu.org/software/emacs/emacs.html
5. Java Analysis Studio Home Page, http://www-sldnt.slac.stanford.edu/jas/
6. Microsoft Internet Explorer Home Page, http://www.microsoft.com/windows/ie/
7. Microsoft Outlook Home Page, http://www.microsoft.com/office/outlook/
8. Alverson, G., Gaponenko, I., Taylor, L., Tuura, L., *The CMS Interactive Graphical User Analysis (IGUANA) "Functional Prototype" Software*, 2000. CMS IN/2000-052.
9. Alverson, G., Gaponenko, I., Taylor, L., *IGUANA – Interactive Graphical User Analysis*, 1999. CMS IN/1999-042.
10. GNU DDD Home Page, http://www.gnu.org/software/ddd/ddd.html
11. Lakos, J., *Large Scale C++ Software Design*, Addison-Wesley, Reading, MA, 1996.
12. Wernecke, J., *The Inventor Mentor*, Addison-Wesley, Reading, MA, 1994.

CDF Monte Carlo'2000

P. Murat, CDF Collaboration

Fermilab Batavia IL 60510

Abstract. We present status of CDF Run II Monte Carlo project which is being developed for about 3 years. We discuss general architecture of the project, its major components, performance and future prospects.

INTRODUCTION

Detector simulation code written for any detector reflects specifics of this detector. The heart of the CDF Run II detector (1) is its excellent tracking system. The CDF tracker consists of the Si vertex detector (SVX) with 8 layers of double-sided silicon, inner of which is positioned at $R \simeq 1.5$ cm from the beam and the outer - at $R \simeq 30$ cmq. SVX provides resolution of about $7 \oplus 21/P_t$ microns in impact parameter. It is positioned inside the central drift chamber (COT) which extends up to R=135cm in radius and provides up to 96 space measurements with the single-hit resolution of about 150 microns. The COT is surrounded by the layer of TOF counters with the resolution of 120ps which significantly improve particle ID capabilities of the detector. CDF has modest but adequate sandwich calorimeter with EM resolution of $\frac{13.5\%}{\sqrt{(E)}}$ and HAD resolution of $\frac{80\%}{\sqrt{(E)}}$ and a robust muon system which covers the calorimeter with 4-8 layers of drift chambers. Luminosity is monitored by 2 hodoscopes of the Cherenkov counters with the timing resolution of about 50ps. To simulate the detector performance adequately, the CDF Monte Carlo does detailed hit-level simulation of the tracker and muon systems, and, to improve the performance uses fast parametrizations to simulate the calorimeter response.

OVERALL DESIGN

CDF Analysis Framework, AC++ (2), is written in C++. One of the key features of AC++ is its modularity, meaning that each data processing algorithm is implemented as a software module and one can easily plug the algorithms in to and out of the event loop. Following this scheme the detector simulation code is implemented as an AC++ module and every MC event generator also has a corresponding module as well. All the event generator modules and the detector simulation module are combined into a common CDF MC executable. Data structures produced by the event generators have unified format and the detector simulation takes these structures as input. Such organisation significantly simplifies debugging and also allows to avoid writing out intermediate files with the event generator output. However the functionality of writing out the files with the events generated but not simulated in the detector is fully preserved. When the reconstruction code becomes more stable, we plan to run event generation, detector simulation and reconstruction in a single job.

GEOMETRY MODELLER / TRACING ENGINE

CDF MC is using GEANT3 (3) as an underlying geometry modelling and tracing engine. To interface GEANT3 to CDF offline framework, we put a thin layer of C++ code on the top of GEANT3. Currently this interface is given by TGeant3 class (4), which is substantially shared with the ALICE collaboration. Digitization code has been implemented in C++ from the very beginning. The choice of GEANT3 which has its algorithms debugged over the decades, allowed CDF to build its detector simulation framework on a very solid foundation. Presently we are making an effort to unify geometry description in the simulation and the reconstruction code by using GEANT4 (5) geometry classes on top of GEANT3. The goal of this effort is to use GEANT3 tracing utilities for extrapolating the muons in very complex geometry of the CDF muon system and also to provide a migration path to GEANT4.

PARAMETRIZED CALORIMETRY SIMULATION

CDF has a conventional iron(lead)/scintillator sandwich calorimeter. To simulate its performance we are using GFLASH (6) - a parametrized calorimetry simulation code developed within H1 collaboration. GFLASH is based on a parametrized shower model, both longitudinal and transversal profiles are modelled as functions of the incident particle energy and properties of the tracking medium. Important feature of GFLASH implementation is that it is fully integrated with GEANT3: it uses GEANT3 geometry and other data structures, such that given a tracking volume it is up to user to decide whether to simulate shower development in this volume using fast GFLASH or detailed GEANT3 algorithms. Another feature of GFLASH is that it has many handles to tune the performance. We found that with the proper choice of constants it is even possible to use GFLASH to simulate the response of the CDF showermax detectors - the scintillation strips in the plug calorimeter and the gas chambers in the central calorimeter.

It is also worth to comment on a common limitation of the fast calorimetry simulations: because of the approximations used they usually do not work well in complex inhomogeneous geometries with the calorimeter-type compartments separated in space.

EVENT GENERATORS

CDF is using various event generator (EG) packages:

- widely used in HEP event generators, such as ISAJET, PYTHIA, HERWIG, which generate events with equal weights

- VEGAS-type generators, such as VECBOS, DYRAD, WGRAD, which on output produce events with different weights

- different generators of minimum-bias events

- single particle gun, used for debugging purposes

Specifics of the event generator packages is that all of them are written in FORTRAN and require special interfaces to C++. In addition, for each event generator package there exists an AC++ module, interfacing the generator to the CDF offline framework. All the EG modules have unified interface to the particle decay packages TAUOLA (7) and QQ(8) and independent of which primary EG package is used to generate the events, the same switches allow to use QQ and TAUOLA to decay B's and τ's.

RANDOM NUMBER GENERATORS

As large detector simulation programs are using random numbers for many different purposes and to ensure that the code is producing the same results on different platforms it is very important to manage the random numbers on a consistent basis.

CDF Monte Carlo is using a single random number (RN) manager, which maintains multiple RN streams and is responsible for saving their state to a data stream when requested. Thus it is possible to resimulate once simulated events starting from the last one for which the random numbers have been saved and get the identical results.

Many external to CDF packages, such as GEANT3 and MC event generators, are using their own implementations of RN generators. To guarantee the reproducibility of the results, all the external RN generators have been identified and substituted with the implementations based on the CDF RN manager, such that all the random numbers in CDF Monte Carlo code are coming from the streams in a single place. No exception from this rule is allowed.

OUTPUT FORMATS

The typical output of the detector simulation program in HEP is the simulated event data (hits, digits) and also so-called "MC truth" information which links the hits, generated in the detector, to their parent particles. The CDF detector simulation writes the event data out in the format identical to that of CDF DAQ system. The driving requirement here is that CDF Run II DAQ is 100% digital and it should be possible to reconstruct trigger decisions for the simulated events and for the real data using the same procedure.

PERFORMANCE AND MOCK DATA CHALLENGES

We expect the bulk of the Monte Carlo calculations in Run II to be performed on the Linux-based PC farms, so we tested the performance of the code using 500-750MHz PIII Linux boxes with 256-512 MBytes of memory and 100MHz bus. With all the code compiled in debug mode, the size of the simulation executable on disk is about 100 MBytes. When the debug symbols are stripped, the size of the executable goes down to about 45 MBytes. The executable size in memory is about 50 MBytes at startup, 10-20 MBytes are allocated dynamically during the execution. Time necessary to simulate one event ranges from about 1sec for the single 45 GeV electron events to about 10 sec for the $t\bar{t}$ events (all the code compiled in debug mode). We expect further improvement in timing by a factor of about 2 when the code is compiled in optimized mode. This level of performance is adequate for generating any volume of MC statistics necessary for developing and debugging the reconstruction code and for studying the physics signals in various channels.

We'd like to emphasize that simulating a $t\bar{t}$ event in the CDF tracker takes about 2-3 sec, thus to improve the overall performance further one has to use significantly different approach to tracing or much simpler geometry model.

In fall of 1999 and in summer of 2000 the CDF Run II Monte Carlo system has been used to generate input data for the 2 Mock Data Challenges. It turned out to be very stable - millions of events and Terabytes of data were simulated without a single crash.

CONCLUSIONS

In conclusion, CDF experiment has developed and stable Monte Carlo system, which provides detailed hit-level simulation of the tracker and the muon detectors and relies on fast parametrized algorithms to simulate the calorimeter response. Combining detailed and parametrized algorithms allowed to achieve quite impressive level of performance - a $t\bar{t}$ event is fully simulated in about 10 sec on 750MHz Linux box. The system is very stable such that millions of events corresponding to the Terabytes of data are simulated without a single crash and the CDF Monte Carlo code is ready to be tuned when the first Run II data will become available in spring'2001.

REFERENCES

1. The CDF II Collaboration. The CDF-II Detector Technical Design Report, FERMILAB-Pub-96/390-E

2. http://www-cdf.fnal.gov/upgrades/computing/projects/framework/framework.html

3. CERN Program Library Long Writeup W5013, PM0062. An html-based version is available from http://wwwinfo.cern.ch/asdoc/geant_html3/geantall.html

4. http://purdue-cdf.fnal.gov/CdfCode/source/geant_i/TGeant3.h

5. home page of GEANT4 project is http://wwwinfo.cern.ch/asd/geant4/description.html

6. G.Grindhammer, S. Peters, The Parameterized Simulation of Electromagnetic Showers in Homogeneous and Sampling Calorimeters, NIM A290(1990)469-488, also hep-ex/0001020

7. S. Jadach, Z. Was, R. Decker, J.H. Kuehn, The tau decay library TAUOLA: version 2.4 , Comp. Phys. Commun. 76 (1993) 361

8. QQ home page: http://www.lns.cornell.edu/public/CLEO/soft/QQ/index.html

Computational Challenges For Large-Scale Astrophysics Calculations

B. Fryxell

ASCI Center for Astrophysical Thermonuclear Flashes, University of Chicago, Chicago, IL

Abstract. Numerical astrophysics is filled with a wide range of computational challenges. Many of these relate to developing a code for solving a given problem, or preferably a code that can address a large class of problems. Once the code is built, it is necessary to test it to make sure that the results are believable. Direct validation of astrophysics codes is virtually impossible, so other techniques must be employed. Finally, there are many challenges associated with performing the simulations, in particular managing the huge quantities of data produced and performing analysis and visualization of the results.

INTRODUCTION

Working in numerical astrophysics usually involves trying to perform the largest simulations possible with current technology. This means pushing every aspect of computation to the limits. It is only natural that many obstacles will have to be overcome along the way. Many of the challenges faced by computational astrophysicists are common to all fields of large-scale computing. However, the extreme conditions found in astronomical objects offers many challenges unique to this field.

The following section will discuss issues related to developing a general purpose high-performance astrophysics code. In many cases, the physics and numerics required for dealing with astrophysical objects is in direct conflict with achieving good performance on parallel computers, so compromises must be made. The next section will be devoted to techniques that can be used to verify and validate astrophysics codes, including comparison with standard test problems and with laboratory experiments. This will be followed by an analysis of a hypothetical astrophysics calculation, which illustrates the challenges presented by the immense amounts of data generated by a typical astrophysics simulation.

CODE DEVELOPMENT ISSUES

Frequently, astrophysics calculations are characterized by enormous ranges in length and time scales. It is not unusual for important length scales to vary by ten or more orders of magnitude in a single object. For example, in a Type Ia supernova, the initial radius of the star is 10^9 cm, while the thickness of the nuclear burning front near the center of the star can be less than 1 mm. Resolving both length scales using a uniform grid would require 10^{10} grid points in each spatial dimension. It is unlikely that we will ever have computers capable of dealing with this situation. Another common example is caused by the extremely small viscosity in most astrophysical fluids, which permits the formation of fully developed turbulence. These flows are characterized by a cascade of energy into small length scales that cannot be resolved on a typical grid.

There are a number of techniques available for dealing with multiple length and time scales. One is Adaptive Mesh Refinement (AMR), which concentrates grid points in regions of the flow where they are most needed. This method can significantly reduce time to solution and decrease memory usage, allowing higher-resolution simulations to be performed. Unfortunately, the increase in dynamic range of length scales provided by AMR is highly problem dependent and does not approach the needs discussed above. To make matters worse, the benefit comes at the cost of code complexity and poorer performance on parallel computers. Data must continually be redistributed between processors as the refinement pattern changes, increasing the ratio of communication to computation. In addition, these communication patterns are highly irregular and difficult to predict, making efficient implementation on parallel computers difficult.

Another effective way to improve the dynamic range of length scales in a simulation is front tracking. This technique can prevent unphysical mixing due to numerical diffusion at sharp interfaces. In the supernova example described above, the burning front could be replaced by an interface, so that the code would not need to resolve the width of the front. In this case, a physical model of how the flow changes across the front would also be required. Front tracking also adds significantly to the complexity of the code and can be difficult to implement efficiently on parallel computers. The grid points that contain the front are much more expensive to update than those in smooth portions of the flow, and this can lead to load balancing problems. As the front moves through the grid, one must continually redistribute data so that each processor has the same amount of work. This is similar to the problems encountered with AMR.

A third way to simulate flows with multiple length scales is the use of sub-grid models. This has been used primarily for turbulent flows. The approach is to replace all the physics that takes places on a length scale smaller than the grid spacing by a physical model. The main problem with this method is that the optimal model is problem dependent, and reliable sub-grid models are available only for relatively simple flows. For example, no good sub-grid models exist for magnetized flows.

In addition to challenges associated with multiple length and time scales, many astrophysical calculations also have to deal with a wide range of flow velocities. Efficient solution of low Mach number flows requires very different techniques from supersonic flows. It is not uncommon for both regimes to appear in a single simulation. Low Mach number, incompressible flows are usually calculated with some variety of spectral technique, while the supersonic regime is most often calculated using finite difference or finite volume shock capturing methods. One possibility would be to follow the low Mach number portion of the calculation with an incompressible spectral code and then switch to a compressible code after the Mach number becomes too large. However, it is not clear at this time that there is an intermediate region in Mach number space where the two codes give identical answers. In addition, the data structures used by the two types of code are very different. Because of these issues, mapping smoothly from one code to the other without creating a sudden jump in the solution may not be possible.

An alternate approach would be to use a purely implicit or an implicit/explicit hybrid algorithm for following the fluid flows. These techniques would allow treatment of flows of any Mach number within a single code. However, they are extremely difficult to program and maintain. They also require global communications and are therefore less efficient on parallel computers. For a hybrid method, the need for load balancing would provide an additional complication.

A third set of issues centers around the wide variety of physical processes that take place in astrophysical objects. The physics can be highly localized and non-uniform. In the supernova example described earlier, nuclear burning occurs only in the small number of grid points located near the flame front. It would be a waste of computer resources to call the nuclear reaction module at other grid points. However, without the use of load balancing techniques, processors containing only non-burning regions would have to wait while other processors were performing the burning calculations,, destroying parallel efficiency. Unfortunately, load balancing will require frequent redistribution of data, thereby increasing communication costs. In addition, many of the physical processes are non-local and require global communication. Examples are radiative transfer and self-gravity, which requires the solution of Poisson's equation.

An additional challenge in constructing a code involves the rapidly changing computing environment that we must deal with. New computer architectures are constantly being developed. The time scale for developing a complex code can be comparable to or longer than the life time of a given computer architecture. Development of portable codes rather than ones designed for a particular type of computer is essential. The emergence of Beowulf clusters requires incorporation of latency tolerant algorithms. Clusters of Shared Memory Processors (SMP's) may require the use of hybrid programming algorithms, such as using threaded code within a single box and MPI communication between boxes. This is a severe challenge to portability. Even without computer architecture changes, hardware and software is frequently unstable and subject to frequent changes. It is not unusual for "improvements" in operating systems or compilers to reduce the performance of a code or cause it to stop working entirely.

CODE VERIFICATION AND VALIDATION

In order to do believable science with complex simulation codes, it is first necessary to test them as thoroughly as possible with an intense verification and validation effort. There are many ways to test a code, but none of them can definitively determine if a code is giving the correct answer for a real astrophysics problem. Nevertheless, by careful use of a combination of these methods, it is possible to obtain some degree of confidence in the results of the code.

The most obvious test of an astrophysics code would be to validate it by comparing to observations of the event being simulated. Unfortunately, direct validation of astrophysics codes is extremely difficult. The stars we wish to simulate are so far away that they appear as tiny points in the sky, even using the most powerful telescopes. Even if the surfaces of these objects could be resolved, we could not see into their interiors, where much of the interesting physics is taking place, since the stellar gases are completely opaque. There are a number of events, such as supernovae, where one could try to reproduce observations of light curves and spectra. However, this gives only indirect clues to the physical processes that cause the event. Even if a model is found that reproduces the observations, it is unclear if the result is unique. Perhaps a completely different model would give similar agreement with observations.

The first step in checking a code is to run a standard set of verification tests. These are important in finding obvious flaws in a code, but there usefulness is fairly limited. Many problems with analytic solutions exist, but generally these test only the simplest aspects of the code. Few good test problems exist for multi-dimensional flows, and multi-physics tests are virtually non-existent. Convergence tests, in which the resolution of the calculation is increased to see if the answers change, can also be useful, but it is possible for a code to converge to the wrong solution or to converge to multiple solutions as the resolution is increased. To make matters worse, due to the complexity of astrophysical objects, it is almost never possible to reach convergence for a complete simulation. Another way to find errors in a code is to compare the results with those obtained from other codes. The problem with this approach is that, if the codes disagree, it can be difficult to determine which, if any, of them is producing the correct solution.

One additional approach to testing codes is validation through the use of laboratory experiments. Although it is impossible to obtain the physical conditions in a laboratory that one encounters in stellar astrophysics, except perhaps in high-energy laser experiments, much of the important physics, such as fluid flow, obeys simple scaling laws. In other words, the same physical phenomena can occur in both astrophysical objects and in laboratory experiments, but with different length and time scales. The chief difficulty with this approach results from the poor resolution of diagnostics in many laboratory experiments. It is usually possible to get agreement with the experiments for relatively large length scales, but little information can be obtained on the small scale phenomena, which are most likely to be in error.

DATA MANAGEMENT

Another major challenge that must be faced in performing large-scale astrophysics calculations is how to deal with the enormous quantity of data that is produced. In many types of large-scale computing, the result of a very long calculation can be just a few important numbers. This is not the case in most astrophysics simulations. A complete analysis of a simulation will require writing to disk the entire data set, not only at the final time, but also at a large number of intermediate stages. We are interested not only in the final result, but also how the system evolved into that state.

For a typical high-resolution three-dimensional astrophysics calculation, we would like to use a grid size of perhaps 1000^3 for a total of 10^9 grid points. If we store 25 variables at each grid point and limit the precision of the output data to 4 Byte words, each output file will require 0.1 TBytes of disk space. A 30 second movie of the results at 30 frames per second requires about 1000 output files – a total disk space of 100 TBytes for a single calculation. This will obviously put a severe stress on mass storage systems.

Even using parallel I/O, on currently available computers, the best throughput that could be hoped for is 100 MBytes per second. At this rate, it would take more than 10 days to write all of the data to disk. To retrieve all of the data from a remote computer at current network speeds would require more than a year. Data could also be retrieved by mailing tapes, but writing the tapes would also be time consuming and labor intensive. The tapes would also require significant storage space at the local site. Clearly, an alternate approach is needed, which will probably require analysis and visualization of the data at the remote site.

These large data sets also put a strain on visualization and data analysis capabilities. Full three-dimensional volume rendering of the data will require a large parallel computer. Even trying to visualize two-dimensional slices through a three-dimensional data set will be beyond the capability of a typical desktop workstation.

ACKNOWLEDGMENTS

This work was supported by the Department of Energy under Grant No. B341495 to the Center for Astrophysical Thermonuclear Flashes at the University of Chicago and by NASA/Goddard Space Flight Center.

DZero Monte Carlo Simulation

G. E. Graham*

Department of Physics, University of Maryland, College Park, Maryland 20742-4111 [†]

Abstract. The DZero Monte Carlo is discussed. In particular, the generation and GEANT simulation programs are described. The DZero Parametrized Monte Carlo System (PMCS) is also discussed. Finally, the DZero Monte Carlo mass production system is described.

INTRODUCTION

Monte Carlo simulation at DZero is quickly gearing up in preparation for Run II. In this note, we will discuss the various components of the DZero Monte Carlo; including the available generators, the GEANT-based simulation program D0gstar, the digitization and minimum bias/pile-up simulation D0sim, and the DZero Parametrized Monte Carlo System (PMCS). In addition to the core software effort, a Monte Carlo mass production effort for the GEANT-based Monte Carlo using a new set of tools and the SAM system will be discussed.

Core Software

Software at DZero is written in C++ and organized into modules using cvs in order to support a multi-developer environment. All production executables at DZero make use of a framework(1) which further organizes DZero code into packages. Each package roughly corresponds to a logical unit of work, e.g.- event reconstruction in the central preshower, geometry management, or pythia Monte Carlo event generation. At runtime, packages are registered within the framework and are invoked through known hook methods in a well defined order on a set of events. The ordering of the hook methods are determined at initialization time by user specification or by probing of package dependencies by the framework itself.

Each event comprises a collection of chunks(2). Each chunk contains a logical unit of data along with parentage information, so that it is possible to identify later the exact package which created the chunk and its runtime configuration. Runtime configuration parameters are tracked by a custom built database(3).

Monte Carlo data is stored at DZero using the SAM system(4). SAM is a distributed file management and storage system produced by the Fermilab Computing Division. SAM retains parentage information at the file level. Events are recorded using D0OM(5) (DZero Object Model) for persistancy. Though events are stored in files, random access is still possible through the use of event keys in each file header. D0OM is loosely based on ODMG standards, and supports the CERN DSPACK and EVPACK I/O packages.

Monte Carlo Generators

There are several physics generators available in the DZero software environment.

- Pythia, version 6.129.
- Isajet, version 7.44.
- Herwig, version 6.1
- Custom generators for single particles and cosmics

The generators can all use the TAUOLA package for correct tau decay dynamics and the QQ libraries for the correct B decay dynamics. They also make use of the STD-HEP interface and Monte Carlo particle numbers, but the format(MCPP) is internal to DZero.

GEANT-based simulation : D0gstar

D0gstar is a simulation program for the DZero detector in Run II based upon GEANT 3.21. The GEANT 3 FORTRAN library code is made available to the DZero framework by wrapping the FORTRAN in C++ where

---

* for the DZero collaboration
[†] Present Address : Fermi National Accelerator Laboratory, Batavia, IL 60510

needed. Most of the DZero sub-detector systems are included, and the remaining will be integrated soon. These include the Forward Proton Detector which requires the use of a double precision version of GEANT 3, and the luminosity monitor.

The performance of D0gstar was estimated on a 200 MHz Origin 2000 using Pythia generated $t\bar{t}$ events including all decays with no minimum bias. On average, D0gstar required 110 seconds to process each event and wrote 0.9 MB of output. These numbers are preliminary, as issues of output content and optimization are still being resolved.

Digitization, minimum bias and pile-up : D0sim

D0sim is a multi-purpose program which does the digitization of the hits data provided by GEANT. Other packages are included with D0Sim which do the pile-up simulation and the minimum bias simulation. Minimum bias events need to be added in the current crossing and also for a number of previous and future crossings to simulate luminosity effects. This simulation takes into account hardware baseline subtraction done by the DZero calorimeter electronics(6).

The performance of D0sim was also estimated on a 200 MHz Origin 2000 using Pythia generated $t\bar{t}$ events including all decays and processed by D0gstar with no minimum bias. On average, D0sim required 80 seconds to process each event and wrote 1.0 MB of output. In addition, tested with a Monte Carlo sample of Isajet generated QCD events with a lower limit on P_T of 5 GeV/c, it was found that D0Sim required about 1 additional minute of running time plus one additional megabyte of output per event for each additional minimum bias event added in the crossing of interest. These numbers are also preliminary, as issues of output content and optimization are still being resolved.

Parametrized Monte Carlo System : PMCS

DZero also has a parametrized Monte Carlo System (PMCS) for generating large numbers of events quickly. PMCS is logically divided into three tracks.

- FAST track : Simulation of high level physics objects only.

- MEDIUM track : Simulation of high level physics objects plus some important detectors.

- SLOW track : Simulation of detectors.

As other DZero code, PMCS is organized in packages. This plus the above organization has facilitated the development of PMCS. While the management focus is on the FAST track packages to simulate high level physics objects, PMCS can still easily accept contributions of parametrized detector simulations at any time from developers outside of the normal PMCS development team.

The output of PMCS packages is in the form of standard DZero chunks compatible with other DZero tools such as analysis programs and visualization tools. The lower level detector packages in PMCS should also be suitable for providing input to some of the standard reconstruction packages.

The performance of PMCS was also estimated on a 200 MHz Origin 2000 using Pythia generated $t\bar{t}$ events including all decays, processed by D0gstar with no minimum bias. On average, PMCS and Pythia together created and processed about 22 events per second with Pythia accounting for about 50 percent of the processing time.

EVERYBODY WORKING TOGETHER : MC_RUNJOB PACKAGE

The mc_runjob package consists of about 10K lines of Python code which sets up and produces shell scripts to run most D0 offline executables. All of the interesting input parameters are settable through the mc_runjob interface, and repetitive tasks such as random number seed generation are handled automatically. Furthermore, it is possible to configure and chain together multiple executables in a single job. Thus generation, simulation, digitization, reconstruction, and analysis can be configured together through a united interface. Intermediate files are retained and handled automatically. Furthermore, and perhaps more importantly, mc_runjob is script driven so that an mc_runjob script amounts to specification of a processing chain. This aids immensely the tasks associated with large scale production and makes possible completely automated large scale Monte Carlo production.

mc_runjob is modular so that the addition of Python code to handle new executables or tasks is greatly simplified. Furthermore, the inheritance hierarchy is organized to permit the use of a graphical user interface that is automatically generated from stored information about each executable only; so that minimal maintenance is needed to extend the GUI to new executables once the basic Python code is written.

MONTE CARLO PRODUCTION

DZero is able using the above tools to generate massive amounts of Monte Carlo data. This Monte Carlo data is used to flex the generation and storage tools and to estimate the physics performance of the DZero detector. In the production phase ending April 2000, well over one million Monte Carlo events were generated at five centers around the world(7). Currently, more centers are coming online and 600K events have been generated already for the Fall 2000 production. Capacity is already approaching 100K events per day equivalent no minimum bias.

An automatic Monte Carlo request processing system taking advantage of the mc_runjob system and SAM is envisioned for the start of RunII. Tables in the SAM Oracle database have been completed which describe

- production requests for Monte Carlo
- production input parameters
- generator and detector card files
- physics processes
- processing chain (for input to mc_runjob)

The system as currently planned will accept Monte Carlo processing requests via a WWW interface and generate events with the appropriate DZero generator at FNAL using mc_runjob. The resulting generated events are then stored into SAM for tracking purposes. Later, an automatic process retrieves the generated events from SAM and deposits them in a remote-site specific area with instructions for further processing in the form of an mc_runjob script. Remote sites are responsible for retrieving these generator files and instructions, performing necessary processing remotely, and returning the results back into SAM.

REFERENCES

1. J. Kowalkowski, et al. "D0 Offline Reconstruction and Analysis Control Framework," Proceedings of the International Conference on Computing in High Energy and Nuclear Physics (CHEP) 2000, INFN, Padua

2. May, B. and Paterno, M. "The D0 Event Data Model," Proceedings of the International Conference on Computing in High Energy and Nuclear Physics (CHEP) 1998, ANL, Chicago, IL

3. Paterno, M. "Run Control Parameters at D0," http://cdspecialproj.fnal.gov/d0/rcp/D0LocalGuide.htm

4. L. Lueking, et al. "The Data Access Layer for D0 Run II: Design and Features of SAM," Proceedings of the International Conference on Computing in High Energy and Nuclear Physics (CHEP) 2000, INFN, Padua

5. Greenlee, H. and Snyder, S. "The D0 Object Model for Persistent Data," Proceedings of the International Conference on Computing in High Energy and Nuclear Physics (CHEP) 1998, ANL, Chicago, IL

6. The DZero Collaboration. "D0 Upgrade Technical Design Summary," D0 Internal Memo #2962 and references therein.

7. G.E. Graham. "The D0 Monte Carlo Challenge" Proceedings of the International Conference on Computing in High Energy and Nuclear Physics (CHEP) 2000, INFN, Padua

Large-Scale Simulations of Clusters of Galaxies

P. M. Ricker[a], A. C. Calder[a], L. J. Dursi[a], B. Fryxell[a], D. Q. Lamb[a], P. MacNeice[b],
K. Olson[a,b], R. Rosner[a], F. X. Timmes[a], J. W. Truran[a], H. M. Tufo[a], and M. Zingale[a]

[a]*ASCI Flash Center, University of Chicago, Chicago, IL 60637*
[b]*NASA/Goddard Space Flight Center, Greenbelt, MD 20771*

Abstract. We discuss some of the computational challenges encountered in simulating the evolution of clusters of galaxies. Eulerian adaptive mesh refinement (AMR) techniques can successfully address these challenges but are currently being used by only a few groups. We describe our publicly available AMR code, FLASH, which uses an object-oriented framework to manage its AMR library, physics modules, and automated verification. We outline the development of the FLASH framework to include collisionless particles, permitting it to be used for cluster simulation.

SIMULATING CLUSTERS

Clusters of galaxies are the largest gravitationally bound objects in the universe. They consist mainly of dark matter and diffuse, hot plasma, with galaxies themselves contributing only a few percent of the total mass. Clusters have attracted attention in recent years because they are large enough to serve as a representative sample of the universe; they provide strong contraints on cosmological models. Clusters are also interesting from an astrophysical point of view. The intracluster medium (ICM), at densities $\sim 10^{-4} - 10^{-2}$ cm^{-3} and temperatures $\sim 10^{7-8}$ K, is collisionally ionized and emits X-rays, primarily via bremsstrahlung [21]. The radiative cooling time can be short enough to produce cooling flows [11]. Observations of Faraday rotation show that the ICM is magnetized, with $B \gtrsim 1$ μG [7]. With the diffuse, nonthermal radio [15] and X-ray [14] emission seen in some clusters, this suggests that clusters are sites for cosmic-ray acceleration [2]. Magnetic fields may also help to suppress diffusion in the ICM [6]. Galaxies orbiting in cluster potentials experience tidal and ram-pressure stripping and help to stir the ICM [22]. Star formation and supernovae may affect the abundances of heavy elements in the ICM as well as its global energetics [10, 17]. Many elements of a complete model for the ICM are known, but we still cannot answer such questions as: what is the source of entropy and metals in the ICM? what happens to the gas that cools below X-ray temperatures in cooling flows? how robust are cooling flows? how is energy partitioned among thermal and nonthermal particle populations, magnetic fields, and turbulent motions?

Cluster mergers play a key role in these phenomena. Mergers strongly affect the ICM, producing long-lived

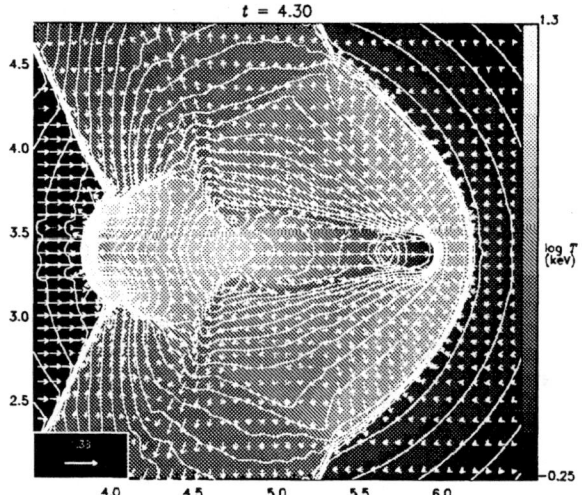

FIGURE 1. Gas density (log contours), temperature (shading), and velocity (arrows) in a merger of clusters with mass ratio 1:3 (Ricker & Sarazin 2000, in prep.). Times and lengths are in Gyr and h^{-1} Mpc, respectively, with Hubble constant $h = 0.6$.

distortions in X-ray images and temperature maps. The energies they release ($\sim 10^{64}$ erg) are easily enough to heat the ICM to 10^8 K. These events are complex, nonlinear, multiphysical, and three-dimensional; thus numerical simulations are appropriate tools for studying them.

Cluster simulations with multiphysics (at least hydrodynamics in addition to dark matter) have reached $\gtrsim 10^6$ particles or zones, simulating a cluster and its environment down to kpc scales [12, 16]. Cosmological simulations with 10^9 particles, yielding 10^5 clusters, have been performed [8], but thus far these have only included dark matter. To understand recent observations, we must resolve scales needed for cosmological context ($\gtrsim 10$ Mpc) and galaxies ($\lesssim 1$ kpc) – a dynamic range of $> 10^4$ – with hydrodynamics, cooling, magnetic fields, and nonequilibrium plasma effects. Star formation and supernova feedback will remain unresolved for the forseeable future and must be included phenomenologically.

Because shocks are important in the ICM, Eulerian shock-capturing methods such as the Piecewise-Parabolic Method (PPM) [9] are very desirable. Fig. 1 shows an example merger calculation performed using the COSMOS *N*-body/hydro code [20]. COSMOS uses PPM on a single nonuniform grid. This 3D calculation covered a dynamic range of ~ 300 on 128 processors of the San Diego Cray T3E, requiring 10,000 node-hours. To add new physics and increase dynamic range, the cost of such calculations must be reduced significantly while retaining the shock-capturing properties of single-grid Eulerian schemes.

ADAPTIVE MESH REFINEMENT

The computational issues involved in studying gravitational clustering with multiphysics involve the coupling of small and large scales through gravity and hydrodynamics, requiring large dynamic range, and the presence of short-range source terms that upset load balancing.

Block-structured adaptive mesh refinement (AMR) methods address these issues by placing fine grids only where they are needed to resolve fine features [3]. An example of a freely available AMR package is PARAMESH [18]. PARAMESH manages an octree (in 3D) data structure whose nodes are uniformly gridded meshes ('blocks'); Fig. 2 shows an example. Each block is a factor of two more refined than its parent. Refined blocks are placed according to user-defined criteria; interpolation is used to obtain their initial and boundary data from coarser blocks. PARAMESH distributes blocks among processors using a work-weighted space-filling curve, keeping spatially adjacent blocks on the same processor when possible and balancing the computational load.

Long-range coupling can be handled in AMR using multilevel relaxation techniques [4]. The coarse-grid solution is obtained on one processor, while finer levels use neighbor-to-neighbor communication.

Short-range forces produce highly clustered distributions of work. The mixed material representations in PPM-based cluster simulations (Lagrangian particles, Eulerian gas) require different domain decompositions. AMR can solve both of these problems by weighting blocks appropriately, e.g., by source terms or particle content. Blocks also can be evolved on different timesteps and weighted inversely by their timestep.

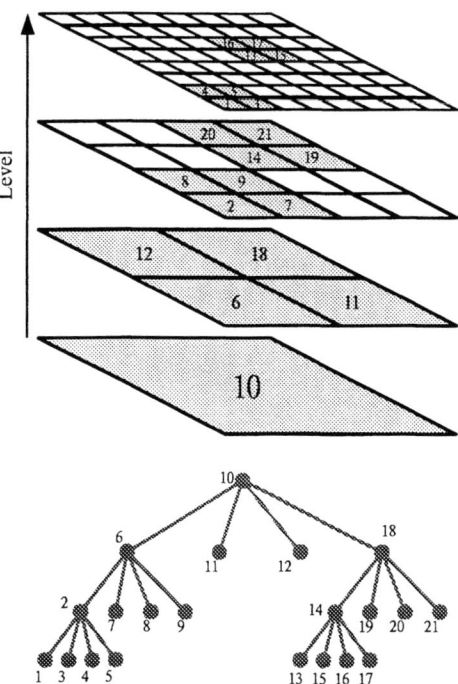

FIGURE 2. Example 2D mesh managed by PARAMESH [18].

AMR techniques are widely used in cosmological *N*-body and smoothed-particle hydrodynamics codes, but few groups have used them with Eulerian schemes [19]. AMR codes are difficult to construct, and most cosmological codes are proprietary. However, during the coming year we expect to see several AMR codes useful for cosmology emerge, some of which are freely available.

THE FLASH CODE

We are developing FLASH, an adaptive-mesh astrophysical simulation code based on PARAMESH [23, 13]. FLASH is coded mainly in Fortran 90 and uses the Message-Passing Interface (MPI). It is highly portable and scales to thousands of processors. We have recently been awarded the Gordon Bell Prize for achieving 0.24 TFlops with FLASH using 6,420 processors of ASCI Red on a cellular detonation problem relevant to Type Ia supernovae [5]. We intend for FLASH to evolve into a community simulation framework; the code is publicly available at http://flash.uchicago.edu/.

Many astrophysical problems require multiple physical processes and a wide range of scales. Each physical process requires a different numerical method and different tests. Exploiting AMR also requires complicated mesh management libraries. Such complex software is best managed using a framework.

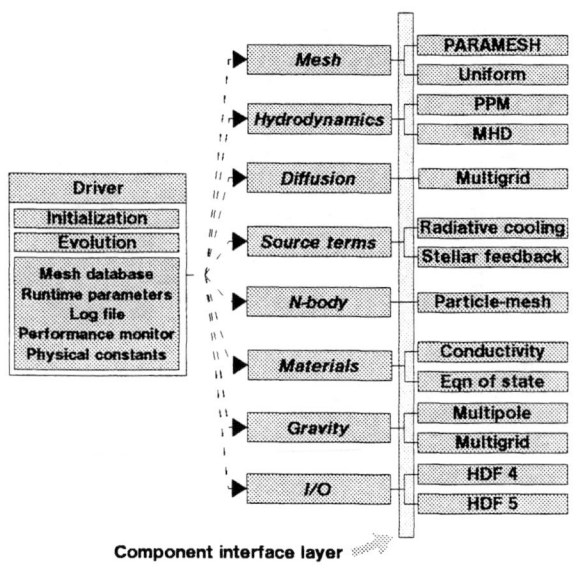

FIGURE 3. The framework of the FLASH code [13], showing components useful for cluster simulation.

ules appropriate for supernova problems, including a partially degenerate equation of state and nuclear reaction networks. Modules for front tracking, implicit diffusion, magnetohydrodynamics, and collisionless particles are under active development by our group. With these new components FLASH will be capable of simulating individual clusters with multiphysics and a dynamic range of $\gtrsim 2,000$ per dimension during the coming year.

This work was supported by DOE under Grant No. B341495 to the ASCI Flash Center at the University of Chicago. Calculations were performed using the resources of the San Diego Supercomputer Center.

REFERENCES

1. Armstrong, R., et al., Argonne Natl. Lab. MCS preprint P759-0699, 1999
2. Berezinsky, V. S., Blasi, P., & Ptuskin, V. S., *Astroph. J.* **487**, 529-535 (1997)
3. Berger, M. J., & Oliger, J., *J. Comput. Phys.* **53**, 484-512 (1984)
4. Briggs, W. L., Henson, V. E., & McCormick, S. F., *A Multigrid Tutorial*, 2d ed., SIAM, Philadelphia, 2000
5. Calder, A. C., et al., in *Proc. Supercomputing 2000*
6. Chandran, B. D. G., et al., *Astroph. J.* **525**, 638-650 (1999)
7. Clarke, T. E., Kronberg, P. P., & Böhringer, H. B., in *Cluster Mergers and their Connection to Radio Sources*, 24th IAU, Joint Discussion 10, 2000
8. Colberg, J. M., et al., *Mon. Not. R. Astron. Soc.* **319**, 209 (2000)
9. Colella, P., & Woodward, P. R., *J. Comput. Phys.* **54**, 174-201 (1984)
10. Dupke, R. A., & White, R. E., *Astroph. J.* **537**, 123-133 (2000)
11. Fabian, A. C., *Ann. Rev. Astr. Ap.* **32**, 277-318 (1994)
12. Frenk, C. S., et al., *Astroph. J.* **525**, 554-582 (1999)
13. Fryxell, B., et al., *Astroph. J. Suppl.* in press (2000)
14. Fusco-Femiano, R., et al., *Astroph. J. Lett.* **513**, 21-24 (1999)
15. Kempner, J. C., & Sarazin, C. L., *Astroph. J.* accepted (2000)
16. Lewis, G. F., et al., *Astroph. J.* **536**, 623-644 (2000)
17. Lloyd-Davies, E. J., Ponman, T. J., & Cannon, D. B., *Mon. Not. R. Astron. Soc.* **315**, 689-702 (2000)
18. MacNeice, P., et al., *Comput. Phys. Comm.* **126**, 330-354 (2000)
19. Norman, M. L., & Bryan, G. L., in *Numerical Astrophysics*, eds. S. M. Miyama et al., Kluwer, Boston, 1999, p. 19
20. Ricker, P. M., Dodelson, S., & Lamb, D. Q., *Astroph. J.* **536**, 122-143 (2000)
21. Sarazin, C. L., *X-Ray Emission from Clusters of Galaxies*, Cambridge U. P., Cambridge, 1988
22. Stevens, I. R., Acreman, D. M., & Ponman, T. J., *Mon. Not. R. Astron. Soc.* **310**, 663-676 (1999)
23. Rosner, R., et al., *Comput. Sci. Eng.* **2**, 33-41 (2000)

Object-oriented languages provide several features useful for building simulation frameworks. Encapsulation allows us to interchange solvers that need conflicting internal data structures; inheritance allows us to abstract common features of different types of solvers; and polymorphism allows us to switch between solvers.

Component frameworks scale better with increasing complexity by providing standard ways for components to describe themselves to each other. Such frameworks are commonly used in business, but they have not seen wide use in science, because they impose unacceptable overhead and lack features needed for scientific applications. An appropriate scientific component standard, such as that being developed by the Common Component Architecture (CCA) Forum [1], is still several years away.

The FLASH framework is object-oriented and makes use of some component ideas. Its class structure appears in Fig. 3. The driver maintains mesh data in a static container class and instantiates objects from various classes of physics solvers. The solvers are divided into different classes by their level of coupling and by differences in solution method (e.g., hyperbolic solvers for hydrodynamics, elliptic solvers for radiation and gravity). The AMR library is also treated as a class. Solver and mesh objects access mesh data through methods supplied by the mesh container class. The component interface layer, for which we are developing a standard, will consist of F90 module wrappers implementing an interface that is abstractly specified in an interface definition language (IDL).

FLASH includes hydrodynamics using PPM, self-gravity using multigrid and multipole methods, and mod-

CMS Monte Carlo Status, Performance and Future plans

Hans Wenzel *

Fermi National Accelerator Laboratory, P.O.Box 500, Batavia, IL 60510-0500

Abstract. In this article we will discuss the current status of the CMS software. A first major Monte Carlo production was conducted in spring 2000. We will discuss the experiences we collected during this production. A Distributed Monte Carlo production is currently in progress (Fall 2000).

INTRODUCTION

CMS is an experiment under construction for the LHC pp collider at CERN. The LHC is designed to operate at $\sqrt{s}$= 14 TeV. The nominal luminosity of $10^{34} cm^{-2} s^{-1}$ and bunches crossing every 25 nsec result in 10^9 pp interactions per second. The first collisions are expected in 2005. The LHC will not commence with its nominal luminosity but a low-luminosity start-up phase of several years with 10% of the nominal value in the first year, 33% in the second year and 67% in the third year (1). LHC experiments will have to deal with larger data sets, more (and wider distributed) computing power with collaborators more spread out than ever[1] before experienced in high energy physics[2].

The first major Monte Carlo production was conducted in spring 2000. Distributed Monte Carlo production is happening right now (Fall 2000). The highest priority for the production effort was the Higher Level Trigger validation to be performed using full simulation and reconstruction of high luminosity LHC events. Other goals of the Monte Carlo production were to provide 'realistic' simulation data to study trigger and physics performance of CMS, to provide data in order to develop reconstruction and calibration algorithms and to provide feed back for realistic simulation of network/CPU usage and access patterns (MONARC)(2).

CMS relies on a distributed computing model with 1/3 of the computing and data storage at tier 0 (CERN), 1/3 at tier 1 (CERN, Fermilab,INFN, IN2P3, ??) and 1/3 at tier 2 centers. For this to work entirely new tools must be developed and exercised. These includes tools for data base management, load balancing, data replication, monitoring, production farm management and production system replication. The current production efforts are the initial step to develop and use these tools and to estimate if the computing model will scale up to the size and complexity expected for LHC experiments. The major challenges for LHC software and computing today include:

- Events are big (raw event is 2MB).

- at the design luminosity of $10^{34} cm^{-2} s^{-1}$ approximately 17 minimum bias events per crossing are produced. With the bunch crossing time so short (25 nsec) a realistic detector digitization has to take into account the fact that events from different bunch crossings contribute to the digitization. At least 9 crossings (-5 to +3) contribute to calorimetry digitization [3]. Typically information from more than 150 (9 × 17) minimum bias events is needed for each signal event. Thus for 1 million signal events you would need to generate > 150 million minimum bias events. This is impossible with current available CPU, storage etc.. Instead we include the minimum bias events in the digitization step and recycle simulated minimum bias events. For that we created a minimum bias data set consisting of a few hundred thousand events. For each signal event we randomly select the necessary number of minimum bias events in a straightforward manner. Problems can arise when one single minimum bias event by itself would trigger the detector. One would get this trigger many many times. Therefore it is necessary to filter the minimum bias events, but remember to take into account the removed events. The size of the 150 required minimum bias events which have to be read in for each signal event is greater than 50MB. This results in a massive data movement problem. The

---

* representing the CMS collaboration
[1] 1700 physicists of 150 institutes in 32 countries
[2] there are two possible attitudes: optimistic- great opportunity or pessimistic- the greatest computing nightmare ever experienced!

[3] Muon digitization is affected by even more crossings

advantage of this approach is that it is easy to study the same signal events at different luminosities and also with and without pile-up.

- The CMS Tracker is immersed in a high magnetic filed of 4 Tesla. The tracks of low momentum charged particles loop in the magnetic field and can persist for many crossings. This requires track finding in a very complex environment.

- The total tracker material adds up to 1 radiation length resulting in lots of bremsstrahlung for the electrons. This makes matching tracks to calorimeter clusters a non-trivial task.

One CMS event requires of order 100-400 times more computing to digitize than a Tevatron Run II event.

CMS SOFTWARE

CMS decided to use object oriented technology and C++ has been chosen as the programming language. This allows the use of many modern C++ standard tools (e.g. standard library) (3). The Software and runtime environment is managed and distributed using SCRAM [4]. The source code is organized into modules using CVS in order to support a multi-developer environment. Currently the Monte-Carlo events generators and the detector simulation are still written in Fortran.

The Framework

The basic framework is called CARF[5]. To achieve maximum flexibility CARF implements an "implicit invocation architecture" (Action on Demand) (4) where Modules register themselves at creation time and are invoked when required. In this way only those modules really needed are loaded and executed. CARF pins Database objects (Objectivity) in memory for the duration of an event. The user accesses C++ objects and for the user it is invisible if the object is transient or in the Database. No "Objectivity handles" appear in the user code. All modules see the same persistent data and each modules can change its local copy of the data, but one can not unintentionally change persistent data nor the data which is viewed by another module in the same reconstruction job.

Monte Carlo generators and Detector Simulation

There are several physics generators available in the CMS software environment. The current production (fall 2000) uses Pythia (version 6.152), Isajet (version 7.48) and in some cases the TAUOLA package for correct tau decay dynamics. The results of the generators are stored as HEPEVT ntuples.

For the simulation of the CMS detector response we use CMSIM (version 120) (5) which is written in FORTRAN and based on GEANT3. The plan is to retire this package, as soon as OSCAR[6] is available and is sufficiently tested. OSCAR is written in C++ and based on GEANT4 at this time all sub-detector geometries are in and OSCAR is now being interfaced to CARF to use persistent storage. FAMOS is a fast parametrized simulation written in C++ and fully integrated into CARF. A first release is underway. FAMOS is required for the Physics TDR II planned for 2003. A fast simulation ,for example, will allow us to explore SUSY parameter space.

Reconstruction

The reconstruction framework called ORCA [7] is written in C++ and is fully embedded in the CARF framework. ORCA provides a sub-detector reconstruction skeleton into which new developers can implement realistic objects and algorithms which can be used for final detector optimizations, trigger studies or global detector performance evaluation. The ORCA sub-systems relevant for production include:

- **OOHits:** is the first step of OO reconstruction. OOHits reads CMSIM files which are in Zebra format and writes the simulated hits into an Objectivity database. This process is completely IO limited and does not require a lot of CPU.

- **OODigis:** adds minimum bias events (> 150) of several crossings. The Pile-up is selected randomly from a sufficiently large database of minimum bias events. OODigis then performs the full digitization and stores the resulting objects in Objectivity.

- **RecReader:** this is where the final steps of reconstruction is done including: calorimetric clustering, jet finding, muon segment finding, track finding and primary vertices finding using information from the silicon pixel detectors.

---

[4] Software Configuration, Release And Management.
[5] CMS Analysis and Reconstruction Framework

[6] Object oriented Simulation for CMS Analysis and Reconstruction
[7] Object Reconstruction for CMS Analysis

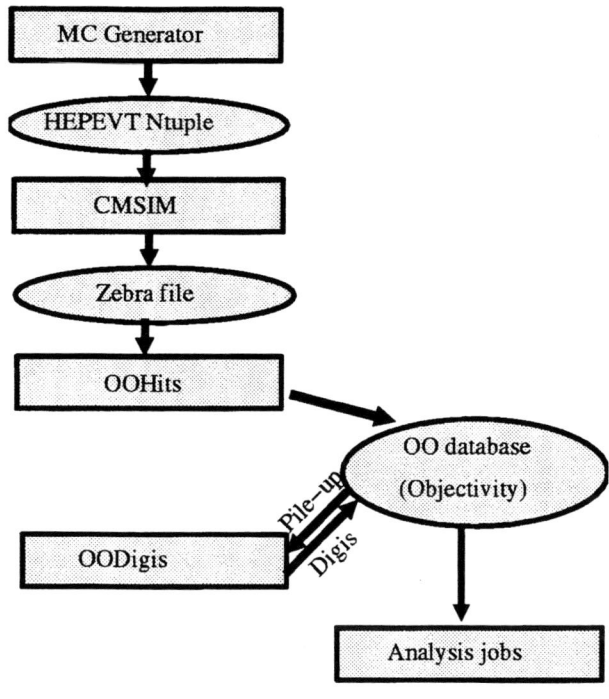

FIGURE 1. Steps and datasets involved in a CMS production job.

SPRING 2000 PRODUCTION

The first major Monte Carlo production was conducted in spring 2000. The steps involved are shown in Figure 1. The first steps of production (MC generation and the simulation) were partly done in regional centers outside of CERN. The Zebra files of the simulation were then transfered to CERN. The OO reconstruction (OOHits and OODigis) was done at CERN. The OO processing was done on a Linux farm. 70 CPU's were dedicated to projects of the muon physics group. Typically these jobs required 90 seconds/event and resulted in 20MB/event output. An additional 70 CPU's were dedicated to projects of the jets and missing E_t physics group with typically 60 seconds/event and 35MB/event. 30 Linux computers were used as dedicated database servers (running the Objectivity AMS server) to deliver the minimum bias events to the digitization processes. The results were written to a Objectivity database on a high speed SUN server. The database access patterns were similar to those expected for an analysis facility with 140 jobs reading asynchronously and chaotically from 30 servers writing into one server. The best reading rate achieved out of Objectivity was ≈70MB/sec and a continuous 50MB/sec reading rate could be sustained.

There was not enough disk space to keep the entire data base, nor even the minimum bias sample on disk therefore non disk-resident data needed to be staged from tape (MSS). OOHits was completely limited by the IO performance into and out of MSS, OODigis was IO limited in reading Pile-up events from dedicated database servers. Analysis jobs by physicists are again limited by IO performance staging data in and out of MSS. This is due to limited amount disk space (the ratio Tape/Disk=6).

FALL 2000 PRODUCTION

Production started in October 2000 and is still ongoing. Eight regional centers in Europe and the US are participating. Most centers do the only generation and simulation steps and transfer the resulting Zebra files to CERN using Perl scripts using scp. Some of the regional centers (e.g. Fermilab) are in the process will of setting up their production farms to do OO reconstruction (OOHits and OODigis). To move the Objectivity databases to and from CERN, the GDMP tools are used (6)

NEAR FUTURE

After we finish the current production (Fall 2000), a series of milestones with ever increasing functionality has to be met. To answer the fundamental questions of the collaboration on trigger and physics performance it is necessary to develop and test many new features in the computing and software. Some of the goals include:

- The use of the production data to develop algorithms and trigger strategies to achieve a final factor of 10 reduction in level 2 trigger rates. The results will be included in the final Trigger/DAQ TDR due in spring 2001.

- In 2002 a computing and software TDR and a 5% [8] data challenge are planned.

- In 2003 we plan to have the final Physics TDR.

- In 2004 we plan to perform a 20% data challenge.

---

[8] Here the percentage numbers refer to the level of complexity not necessarily the total amount of data.

REFERENCES

1. *The Tracker Project, Technical Design Report*, CERN/LHCC 98-6, CMS TDR 5.
2. http://monarc.web.cern.ch/MONARC.
3. http://cmsdoc.cern.ch/cmsoo/cmsoo.html.
4. M. Shaw *Some Patterns for Software Architecture*, Proceedings of the Second Workshop on Pattern anguages For Programming, Addison- Wesley, Reading Masachusetts, 1994
5. http://cmsdoc.cern.ch/cmsim/cmsim.html
6. http://www.globus.org.

Large Scale Cluster Surveys and Distributed Computing

James Annis

Experimental Astrophysics Group, Fermilab, Batavia, IL 60510

Abstract. The Sloan Digital Sky Survey is among the first of the new generation of large scale sky surveys that enabling the construction of revolutionary new cluster catalogs. I describe here some of the experiences of creating a SDSS cluster catalog, some of the computational challenges we encountered, and what we can expect in the near future.

ENTERING THE AGE OF TERABYTE SCALE SKY SURVEYS

There is only one sky: on it are 40,000 sq-degrees. Take an image of the sky with 1" pixels and you have 520 billion pixels. Call it 10 bytes/pixel and this image weighs in at 5 Terabytes. Astronomy is beginning to take these images and is moving to do so at a variety of wavelengths. Soon we will be doing it as a function of time. We will have access to huge amounts of high quality data, and we will be learning new techniques to analyze it all.

The Sloan Digital Sky Survey [1] is one of the first of the large scale surveys. It is an experiment to map 10,000 square degrees of the the Northern Celestial Hemisphere and to obtain spectra of a million of the brightest galaxies. It covers only one quarter of the sky, but it does so with 0.4" pixels and through 5 bandpasses. By the completion date of 2006, The SDSS will have have generated 15 Terabytes of images, 250 million objects strong object catalogs, and billion pixel scale spectral catalogs.

The first year data collection is complete and we now have experience with the analysis of terabyte scale astronomical data sets. What have we learned? The data volumes are large: a 200 sq-degree catalog is 12 gigabytes. The data rates are high: the data comes in at a rate of 200 sq-degrees a month. The data are complex: there are 120 parameters per object, a radial profile, 5 bandpass atlas images, and often detailed spectroscopy. And the data are reprocessed and re-calibrated on 6 month time scales, so is a moving target. All of these lessons will apply to the forth-coming surveys: the collection of astronomical data is riding the same Moore's Law wave that the analysis of astronomical data rides; and recalibration of data is notoriously fascinating to those who take and analyze the data because it is so useful.

CLUSTER SURVEYS IN THE TERABYTE AGE

Clusters of galaxies are useful tools for studying the growth of structure in the universe as well as being laboratories for the effects of environment on galaxy evolution. Using the large scale, high quality, and uniform data of the SDSS for finding clusters is a pleasure scientifically, but it does present certain technical problems that are generic to any cluster finding algorithm.

We can use the Chak-Bey[1] (Annis et al 2001) algorithm as an example. This algorithm takes a definite view on what constitutes a cluster: a luminous elliptical galaxy with quite uniform properties surrounded with a sequence of lower luminosity galaxies whose colors are remarkably similar to each other. It takes the view that in physical units clusters all look the same, modulo the number of galaxies on the sequence. Given that, one can predict with great precision their appearance at any given distance. In fact for any observed galaxy on the sky, one can ask if it could be that luminous central elliptical at any possible redshift. One maximizes the likelihood that it is by calculating the likelihood at all redshifts and choosing the maximum. This algorithm as a by-product produces a precise (1%) photometric redshift measurement.

Cluster finding is intrinsically an N^2 problem. The number of galaxies in the catalog is N. Checking the distance to all the galaxies surrounding it takes N^2 time. Doing so for every redshift z is then $N^2 R_z$ time, where R_z is the resolution in z, typically 10^2 to 10^3. Fighting the N^2 slowness involves minimizing N. Chak-Bey does so via a variety of heuristic ways. Primarily these take advantage of the natural tree structure of the data (imaging run, camera column, field number) to minimize N: clusters are only searched for in a given field and the surrounding 8

---

[1] Chak-Bey: red road in classic Mayan; the E/S0 ridgeline of clusters presents a red road to efficiently locating them.

fields, not the entire imaging run. Secondarily, one can use work at a given z for the entire field at a time, applying color cuts to bring down N before calculating the distances.

It is possible to transpose from the natural tree structure of the data into kd-trees built on position, magnitude and color. Kd-tree algorithms are $N \log N$ and are directly applicable to Chak-Bey in particular and cluster finding in general. There are other computer science developed algorithms of interest: k^{th} nearest neighbor searches and pyramid data models using multi-scale statistical summaries of underlying data points from immediately to mind. In the future of multiple Terabyte scale datasets, any algorithm that is not $N \log N$ may well be infeasible.

Bottlenecks

The complicated nature of our data in conjunction with our desire for standard data formats lead to our first bottleneck: I/O. Reading the FITS binary tables that contain the data on 200 sq-degrees of sky takes 14 hours. This corresponds to 1/3 Mbyte/sec, clearly CPU dominated, traceable to a remarkably limited FITS reader. The solution is to make compact ASCII catalogs containing only the parameters of interest. Reading the compact catalog for 200 sq-degrees takes 5 minutes, 30 Mbytes/sec, and takes Chak-Bey back into the CPU dominated regime.

The second bottleneck is CPU time. Chak-Bey takes 40 CPU hours per 200 sq-degrees of sky on a 500 Mhz PIII CPU; taking the inherent photo-z aspect of the finder to its limits took this to 200 hours. Both are too long for an efficient code/run/debug cycle. Our solution to this is a cluster of commodity computers, now themselves commodity items thanks to the rise of the Internet economy. Observational astronomers are behind the curve in learning how to use clusters; fortunately we are fast studies. A modest cluster of 6 dual CPU compute nodes brings the compute time down to an acceptable 3.5 hrs.

Both these issues are made more problematic given the necessary calculation of the selection function[2]. The calculation is itself straightforward: add a cluster to the data and run the cluster finder to locate it, a large number of times. This is, of course, more stressful to the computational system than cluster finding is, by the ratio of $N * A_{test}/A_{total}$, where N is the total number of selection function tests run, A_{test} is the area of the sky used in the test, and A_{total} is the total area of the sky used in cluster finding. Since $A_{test}/A_{total} \approx 50$ and $N \approx 1000$, one sees the problem.

---

[2] The probability of selecting a cluster of given mass and at a given redshift

TURNING THE BOTTLENECK INTO A RIVER: DISTRIBUTED ARCHITECTURES

Distributed Cluster Finders

Locating quasars is not a spatial problem: one can decide on an object by object basis which ones are quasars. Cluster finding, on the other hand, is inherently spatial. At a minimum it is performed in the 2 dimensional space of celestial positions; often it has a third, redshift dimension. But cluster finding is only mildly spatial, as clusters have a limited extent. This is what makes the speedup of tree structure codes possible; it also makes possible the construction of distributed cluster finding codes.

The insight is the extent of the largest cluster one looks for. The angular extent depends both on the redshift and the bandwidth of the cluster finding filter. One looks for clusters in a given area; required to do so is a buffer area about the main area, of size equal to the largest extent. Chak-Bey typically uses 500 kpc; at $z = 0.1$ this comfortably fits in a single field. The 8 fields directly about the main field are used to ensure that any clusters at the edge of the main field are not missed. This is helped by the adaptive nature of Chak-Bey: the cluster likelihood measurement is made at the positions of galaxies, and the question of whether there is a cluster in this field devolves to the question of whether the central galaxy of the cluster is in this field.

The price paid for this ability to distribute is twofold. First, one must duplicate the data: the data in the buffer area will be duplicated at least once. This can be minimized by making the central area as large as convenient, and is in any case not a big problem. Second, one must have code to join the answers together at the end. In fact, this is an example of a hash machine, the computational analog of the hash-sort algorithm.

Datawolves: Specialized Clusters

The next step is to specialize the computing: build special purpose machines. Based on our experience with the analysis of terabyte scale astronomical data sets, we are rapidly converging on the Datawolf design: if Beowulfs are computational clusters optimized for efficient message passing parallel processing, Datawolves are clusters optimized for extremely large data sets.

The first design is for a database cluster: the SDSS science archive database SX (Szalay et al. 1999) has a master/slave architecture that allows a full, high speed scan of the entire data set to be distributed over the machines of a cluster. We expect aggregate throughputs approaching 1

Gbyte/sec. The results of the query are then sent out over TCP/IP to the querying source.

The second design is for an analysis cluster: the Terabyte Analysis Machine (Annis et al. 2000) is designed either to bring large amounts of CPU power to bear on the data stream from the database cluster, or, if the data are locally available, to do its own high speed data scan. In fact, we are exploring the possibilities of making database repartitioning and re-indexing available as a facility class instrument for high speed problem solving through advanced algorithms. We expect the general use of TAM to be for bringing raw compute power to bear on the SDSS dataset.

Distributed Computing

The vision of specialized clusters is closely associated with the vision of a grid of high-horsepower computing resources available to specialized users with specialize problems. For the most CPU bound problems, however, there is no real need for tightly bound clusters.

It may well turn out that there are many terabyte scale problems that are distributable enough that extremely loosely coupled clusters will be of great interest. We are thinking of the example of Seti@Home, or of the less celebrated but more useful approach of Condor[3]. Once one has the joining/hash machine code in place, there is no real reason to depend on the workload of a single cluster. We expect this to be an evolutionary pathway.

At this point, we have a virtual data engine. The galaxy catalogs form the starting point; the database cluster extracts the relevant galaxies and sends them on to the analysis cluster, where the cluster finding codes run, and the resultant cluster catalog is reported. To turn this into a truly on demand process will take significant work, a task being pursued by the GriPhyN collaboration[4].

Versioning

Inevitably, the virtual data engine approach brings up the pedestrian topic of versioning. Of course the science code must be versioned: quality science codes have a evolutionary life of their own that can extend a decade or more, and one *must* be able to go back to the code that produced a given catalog. Less obvious is the fact that the catalogs themselves must be versioned as both the finder code changes, and *as the input data changes*. Using a galaxy catalog, for instance, means that one has implicitly used choices of star/galaxy separation and limiting magnitude; as these change so will the cluster catalog. Even less obvious is the fact that cluster catalog creators will have to provide cross-catalog identification tables. The users of the catalogs will have spent considerable energy examining single objects; they have a right to know its true name in the next catalog, or at very least its fate.

THE FUTURE: MULTIPLE-SIGNATURE TECHNIQUES

A cluster is $\sim 10^{15} M_\odot$ mass in ~ 1 Mpc^3 volume. This is not directly observable, of course, but there are many observational signatures: the star light from galaxies is visible in optical/infrared imaging, the thermal bremsstrahlung from cluster plasma is visible via x-ray imaging, the inverse Compton scattering of cosmic microwave background photons by the cluster plasma electrons (the Sunyaev-Zeldovich effect) is seen via sub-mm/radio imaging, and weak gravitational lensing induced distortion of the background galaxies by the cluster surface mass is visible by deep optical/infrared imaging. Next generation techniques will use as many of these signatures as are available. We are currently exploring techniques that place ROSAT All Sky Survey (Voges et al 1999) photons on an equal footing as SDSS galaxies for example, and are looking to incorporate weak lensing signals.

Given that in the future, the likelihood that any given place on the sky contains a cluster will be determined using galaxies, x-ray photons, SZ decrement pixels, and lensing signals in unison, we can expect to be developing cluster finder architectures that must connect to many database clusters to extract the data, and find many analysis machines upon which to run.

REFERENCES

1. York, D. et al., *Astronomical Journal*, 120, 1579 (2000)
2. Annis, J. et al., 2001. in preparation.
3. Annis, J., Garzoglio, G., Ruthsmandorfer, K., and Stoughton, C., *ACAT 2000*, these proceedings.
4. Szalay, A., Kunszt, P., Thakar, A., Gray, J., Slutz, D., and Brunner, R.J. 1999. *astro-ph/9912382*. (http://arXiv.org/abs/astro-ph/9912382)
5. Voges, W. et al. *Astronomy & Astrophysics*, 349, 389 (1999)

---

[3] http://www.cs.wisc.edu/condor

[4] http://www.griphyn.org

Simulation Packages in Accelerator Studies for the ν Factory

V. Daniel Elvira

Fermi National Accelerator Laboratory, P.O. Box 500, Batavia, IL 60510

Abstract. The muon collider and the neutrino source are very challenging projects. Recently, a feasibility study for a 20 GeV muon storage ring yielding on the order of 10^{19} muon decays per year has been successfully completed by the muon collider/neutrino source collaboration. Simulation tools are crucial to support these conceptual design efforts. In this paper, we will present different software packages used in simulations related to the neutrino factory study.

INTRODUCTION

Theoretical arguments indicate that new physics, beyond the standard model, associated with the electroweak gauge symmetry breaking and fermion mass generation will emerge in parton collisions at or approaching the TeV energy scale.

The LHC (hadron collider under construction at CERN) will provide hard parton-parton collisions with typical center-of-mass (COM) energies of a few TeV. The route toward TeV-scale lepton-antilepton colliders is less clear. In circular e^+e^- colliders the energy loss increases as E^4/ρ, where E is the particle energy and ρ the radius of the orbit. Linear colliders avoid this problem but would be very long (30-40 km) to attain the TeV energy scale. In addition, radiation during beam-beam interaction would limit the precision of the COM energy (1). For a lepton with mass m the radiative energy losses in a circular orbit are inversely proportional to m^4. This means that the energy loss problem could be solved by using muons ($m_\mu \approx 207 m_e$), therefore enabling higher energies to be reached and smaller rings to be used (2, 3).

A muon collider would be a unique facility for neutral Higgs boson studies through the s-channel resonance production, since the cross section is proportional to m_μ^2 (more than 40,000 times larger than in a e^+e^- collider). Measurements can also be made of W boson and Top quark pair production as well as, eventually, supersymmetric particles. The muon storage ring could also be used as an intense source of neutrinos of well-defined flavor to study the ν-mass hierarchy, the mixing matrix driving flavor oscillations, and CP-violation.

Both the muon collider and the neutrino source are very challenging projects. Recently, a feasibility study for a 20 GeV muon storage ring yielding on the order of 10^{19} muon decays per year has been successfully completed by the muon collider/neutrino source (ν/μ) collaboration (4). Simulation tools are crucial to support these conceptual design efforts. In this paper, we will present different software packages used in simulations related to the neutrino factory study (5). In particular, we will discuss the packages used to simulate the accelerator section which would perform ionization cooling, the main challenge in the design of both the neutrino source and the muon collider.

DESIGN AND TECHNICAL CHALLENGES

The neutrino source is based on an intense proton beam driven onto a target. The outgoing pions are captured by a solenoid and then decay into muons which are cooled (beam size in position-momentum space is reduced), accelerated, and transfered to a storage ring. A fraction of the muons will decay in a straight section, which will produce an intense, well collimated neutrino beam.

The design of the target is an important technical issue to address. The cooling section or channel is, however, the main challenge to face. Accelerator physicists define "emittance" to measure the size of a beam in position-momentum space. The simplest expression for a muon beam normalized emittance, assuming there are no correlations between the six coordinates defining the state of a muon, is:

$$E_N = (\sigma_x \sigma_{p_x} \sigma_y \sigma_{p_y} \sigma_{ct} \sigma_E)/m_\mu^3.$$

The cooling factor is defined as the ratio of emittances at the end and the beginning of the channel. The large emittances at the end of the capture solenoid, are a manifestation of the pions coming out of the target at large angles. As a consequence, the beam equations are not linear. In other words, theoretical predictions based on linearized envelope equations or linear transport matrices are not a good approximation to the problem. The use of solenoids as focusing elements is a unique feature of the neutrino source design. Conventional quadrupole based focusing cannot be used due to the large emittances involved.

Ionization cooling is the principle used to reduce the beam size. The muons are directed to low Z absorbers (typically liquid hydrogen or lithium hydride). The muons loose momentum both longitudinally (z-direction) and transversely. An electric field, applied as the beam goes through r.f. cavities, replenish the muon momentum only in the longitudinal direction. A cooling effect should be observed for muons in the plane transverse to the beam direction. Muons are heavy enough so that the cooling effect produced by energy loss is greater than the heating effect due to multiple scattering.

The unprecedented technical challenges in the design of the new generation of accelerators are very demanding on the software packages used for simulations. They should not only support different types of r.f. cavities, magnetic fields, and provide very precise tracking capabilities over long distances. They must also allow the introduction of absorbers with complex geometry and good modeling of dE/dx, straggling, and multiple scattering.

ACCELERATOR SOFTWARE

Many beam physics software packages were used to simulate different elements of the neutrino factory. We will concentrate only on those extensively used for the cooling simulations. For completeness, we cite in this section several other packages utilized in the neutrino source design:

- MARS(6) is a Monte Carlo code for simulation of hadronic and electromagnetic cascades, muon and low-energy neutron transport in shielding and in accelerator and detector components. It was used for pion production and target simulations in the neutrino source design.

- MAD(7) (Methodical Accelerator Design) is a computer program to design and calculate the properties of particle accelerators. COSY INFINITY(8) is an arbitrary order beam dynamics simulation and analysis code. It allows the study of accelerator lattices, spectrographs, beamlines, electron microscopes, and many other devices. MAD and COSY INFINITY were used for lattice design of the neutrino source re-circulator linear accelerator and muon storage ring.

- MAFIA(9) is a multi-purpose ECAD system designed to solve all kinds of electromagnetic problems. It was used to generate electromagnetic field maps of r.f. cavities which were then interfaced to other simulation packages.

SOFTWARE FOR COOLING SIMULATIONS

For conceptual design, fast and simple programs (mostly privately developed by a single author) were used. Once we agreed on the basic design of a cooling channel, it was necessary to perform a detailed simulation, which included a realistic modeling of the geometry, the electromagnetic fields, and the physics processes. Two packages were supported: ICOOL(10) and DPGEANT (based on GEANT3(11)). Here are the basics about both packages:

- **ICOOL** is being developed by Rick Fernow at Brookhaven National Laboratory specifically for muon cooling studies associated with the ν/μ project. It is written in Fortran, and includes a well targeted library maintained by the ν/μ collaboration. It comes as a pre-built executable and is user-driven via a simple ASCII file, where generic accelerator elements like solenoids, dipoles, r.f. cavities, etc can be defined. The ability for modeling these elements is therefore limited. Their geometry is also restricted to 2D (longitudinal and transverse directions). No graphic capabilities are available.

- **DPGEANT** is based on the popular HEP detector simulation kit tool developed at CERN (GEANT3). It has been upgraded by Paul Lebrun at Fermilab to perform double precision tracking and support electric fields. It is written in Fortran, and includes an extensive physics library maintained by CERN. The user has to define the beam line elements, that is both the volume shapes and electromagnetic field maps. This implies more flexibility but also more work. The geometry of the accelerator elements can be defined in 3D but is limited to a set of pre-defined shapes. Primitive graphics capabilities are available.

Both packages simulate the physics processes associated with dE/dx, delta rays and Moliere multiple scattering (ICOOL is based on the GEANT3 implementation). ICOOL was developed on a PC running Windows, but it also runs on SGIs, SUNs and CRAYs. GEANT3 has been ported to virtually all computers used in HEP. Typically, we use DPGEANT on a PC running LINUX. The speed of the two packages is approximately the same.

A COOLING EXAMPLE USING GEANT4

GEANT4(12) is the object oriented C++ version of the CERN GEANT package. It is the result of a collaborative effort of more than 100 physicists around the world. It offers several obvious advantages like an interface with CAD systems, more basic shapes (new shapes and physics processes can be added with no changes in the tracking code), a more complete set of EM processes, better hadronic physics, and modern 3D visualization. In addition, double precision is built-in. Other reasons to migrate to GEANT4 are that C++ is a language better suited than Fortran for large software packages with a large number of contributors, and that the new generation of students and computing professionals are not proficient in Fortran anymore (or willing to use it). In addition, CERN will probably discontinue support for GEANT3 at some point in the future (new accelerator projects are long term). Good visualization and geometry flexibility is not just a luxury but a need when studying complex cooling channels like the helical wiggler which seeks to achieve emittance exchange using wedge absorbers inside a solenoid plus a dipole. The principle behind this channel is the use of a dipole which rotates in the transverse plane along the longitudinal direction to produce an increasing p_z dispersion as a function of the radial distance (r) to the center of the beam (large p_z corresponds to large radius). In this way, the longitudinal emittance is reduced by the effect of the absorber being increasingly thick as a function of r. Figure 1 (side view) shows a muon (trace in red) moving along a helical channel. The yellow and green wedges are the absorbers and the blue disks are an ideal representation of r.f. cavities. Open Inventor (13) allows direct manipulation of the objects on the screen, plus perspective rendering via the use of light.

CONCLUSIONS

The complex problems we are facing in the design of the next generation of accelerator machines call for large and multidisciplinary software packages which take elements from accelerator, particle, and even plasma physics. At Fermilab, we are migrating toward GEANT4 for detailed simulations, and developing classes of beam physics related objects (solenoids, r.f. cavities, etc) to facilitate the application of this tool kit to accelerator physics.

I would like to thank my colleagues from the ν source/μ collider collaboration for their contributions to this work.

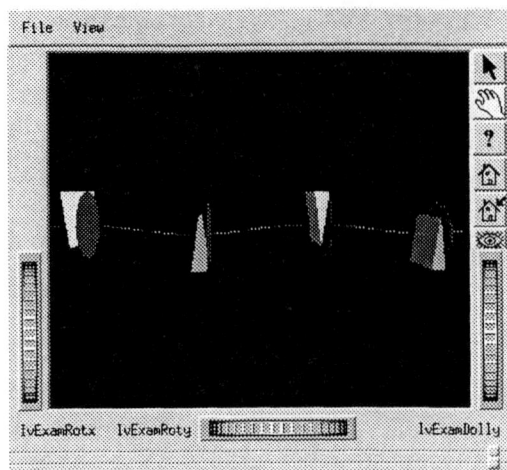

FIGURE 1. Side view of a helical cooling channel (Open Inventor).

REFERENCES

1. Tigner, M., in *Advanced Accelerator Concepts*, edited by J. S. Wurtele, AIP Conf. Proc. # 279 (AIP, New York, 1993), p. 1.

2. Ankenbrandt, C., *Status of muon collider research and development and future plans*, Phys. Rev. Special Topics - Accelerators and Beams, **2**, 081001 (1999).

3. HTTP://WWW.FNAL.GOV/PROJECTS/MUON_COLLIDER/HISTORY.HTML

4. The Muon Collider/Neutrino Source Collaboration, *A Feasibility Study of a Neutrino Source based on a Muon Storage Ring.* Submitted to Phys. Rev. Special Topics - Accelerators and Beams. FERMILAB-PUB-00-108-E, Jun 2000. 158pp.

5. Lebrun, P., *Computational Needs for Muon Colliders and Muon Storage rings*, ICAP2000 conference, Darmstadt, Germany (2000).

6. HTTP://WWW-AP.FNAL.GOV/MARS

7. HTTP://WWWSLAP.CERN.CH/FCI/MAD/MAD_HOME.HTML

8. HTTP://WWW.BEAMTHEORY.NSCL.MSU.EDU/COSY

9. HTTP://WWW.CST.DE/MAFIA/MAINPAGE.HTM

10. R. C. Fernow, ICOOL: *A Simulation Code for Ionization Cooling of Muon Beams*, Particle Accelerator Conference, 1999, New-York, paper # THP31. HTTP://PUBWEB.BNL.GOV/PEOPLE/FERNOW/ICOOL/README.HTM

11. HTTP://WWWINFO.CERN.CH/ASDOC/GEANT_HTML3/GEANTALL.HTML

12. HTTP://WWWINFO.CERN.CH/ASD/GEANT4/GEANT4.HTML

13. Open Inventor. Registered trademark of Silicon Graphics Inc.

ATLAS Simulation: Status, Performance, and Future Plans

Frederick C. Luehring (for the ATLAS Collaboration)

Physics Department, Indiana University, Bloomington, IN 47405, USA

Abstract. From the earliest stages, simulation has played a key role in designing the ATLAS detector for LHC. For more than ten years the detector simulation was done using the FORTRAN-based GEANT3 program. Work is now underway to move the simulation to the object-oriented GEANT4 program.

INTRODUCTION

Simulation has played a key role in designing the general-purpose ATLAS detector for CERN's Large Hadronic Collider (LHC). Millions of fully simulated Monte Carlo events were used to design the ATLAS detector. The original FORTRAN-based simulation is now being superceded by object-oriented code written with the GEANT4 [1] detector simulation tool. We are validating the GEANT4 physics models with testbeam data and investigating using XML to describe our detector geometry.

THE ATLAS DETECTOR

ATLAS is one of two large, general-purpose detectors at LHC and consists of:

1. A three-part charged particle tracking system in a 2 T magnetic field: an inner system of silicon pixels, a system of silicon strips, and an outer system of straw tubes using transition radiation for particle identification.

2. Two calorimeters: an electromagnetic liquid argon calorimeter using "accordion" shaped electrodes/lead plates, and a hadronic calorimeter consisting of iron and scintillator tiles centrally and a liquid argon system for $|\eta| > 1.5$ where the radiation dose is higher.

3. An outer muon detection system using drift-tubes and an air core toroidal magnet system

The ATLAS collaboration consists of over 1800 participants from approximately 170 institutions.

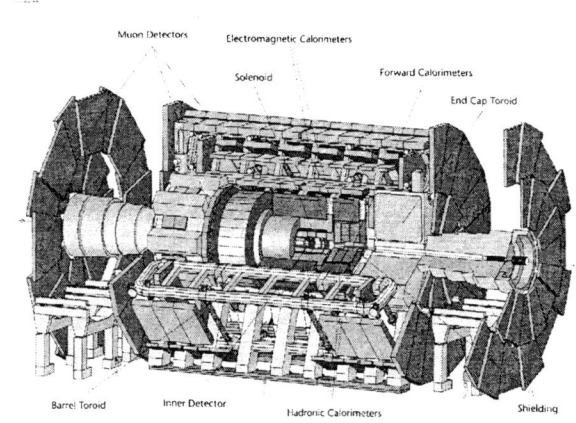

FIGURE 1. The ATLAS detector.

PREVIOUS SIMULATION EFFORT

In the early 1990s the first full detector simulation of ATLAS was written using the FORTRAN-based GEANT3 [2] program. DICE (as this simulation was called) used the standard particle physics software of that time: ZEBRA [3] for memory management and the CERN-written tool CMZ for source code management. DICE was a batch system and used ASCII files containing 80-column "datacards" for control. It ran mainly on IBM mainframes and DEC VAX minicomputers. With the advent of RISC based technology, DICE was ported to all major types of RISC machines, with the HP RISC machine being the main development platform. As the ATLAS detector design took shape, more detail was needed in the simulation and the total number of lines of source code rapidly expanded. The LHCC review panel asked for a series of technical design reports (TDRs) to be written to validate the design of the ATLAS detector that also required many changes to the simulation. The batch-based DICE system made rapid development of new code difficult.

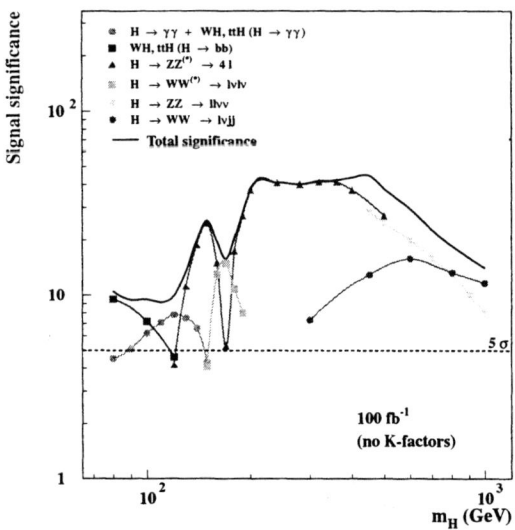

FIGURE 2. Calculated ATLAS sensitivity for the discovery of a Standard model Higgs boson. An integrated luminosity of 100 fb^{-1} is assumed.

In 1995, a new version of the simulation, DICE95 [4], was introduced. It had a preprocessor with its own language (AGE) to ease defining the detector geometry, materials, hits, and digitizations within GEANT3. There was a new interactive version of the simulation called ATLSIM that allowed dynamic linking of small parts of the simulation without having to rebuild the entire program. ATLSIM used the CERN-written KUIP [5] user interface that allowed the user to interact with ATLSIM on a command line or to use a macro scripting language called KUMAC. The new system greatly reduced development time and provided a nice environment to implement and debug the changes in the simulation. A new version of the batch mode program was also introduced at this time.

The ATLSIM/DICE95/AGE system has some limitations. First, the code is not written using Object-Oriented (OO) techniques. The code is so complex that only a few experts can understand it. Finally, key parameters describing the detector are coded into the source files instead of being maintained in a database.

The FORTRAN and GEANT3-based software was used to generate millions of fully simulated events. These simulations formed the basis of the ATLAS Detector and Physics Technical Design Report (Physics TDR) [6]. Figure 2: shows a key result of the Physics TDR studies: the ATLAS detector can discover the standard model Higgs particle with a significance of $\geq 10\sigma$ for masses up to 1 TeV with an integrated luminosity of 100 fb^{-1}. The events for the Physics TDR were fully reconstructed using the standard ATLAS reconstruction program ATRECON without using any of the Monte Carlo truth information. The simulation had about 16 million volumes representing detector elements and the surrounding service material. Table 1 shows the CPU consumption of the GEANT3, simulation including the tracking phase and the calculation of digitizations (simulated detector readouts).

TABLE 1. CPU time need for simulation of ATLAS detector with GEANT 3.21 (400 MHz Pentium II).

Event Type	Timing in Inner Detector	Column Header Goes Here
Single 10 GeV electron	6	100
Single 10 GeV pion	4	60
Minimum-bias event ($|\eta| < 3.0$)	190	2500

Many of the events were generated with the full background of pile-up events. When running at design luminosity, the LHC particle bunches cross every 25 ns, producing an average of 23 interactions per crossing. With the expected ~80% LHC duty factor and pp cross-section at 14 TeV this leads to ~720 million particle interactions per second. Simulating the effect of these background hits was the most challenging part of producing accurate results. To introduce the effect of the pile-up hits, we generated the background events individually. We then combined the hits from a large number of background events with the hits from an event-of-interest during the calculation of the detector response (digitization). The amount of computer time spent generating the pile-up events was greatly reduced by storing a few thousand background events in a file, then randomly reading 100-150 background events from the file. As long as sufficient care is taken in randomizing the selected events, it is then possible to reuse the background events many times without generating new events. In the ATLAS implementation of GEANT3, it was still necessary to store all the hits in memory to combine them, which limited the luminosity that could be simulated.

CONVERSION TO OO

ATLAS has decided to convert the simulation to GEANT4, using object oriented (OO) software engineering methods. This decision is conditional on validating that, using GEANT4 code; we can model the physics in such a way as to reproduce the results of the ATLAS testbeam program. To expedite the conversion process to GEANT4, the wrapping of FORTRAN routines with C++ is allowed. In addition, it has been decided to store parameters describing the detector geometry and materials in a software-neutral database. This will allow all parts of the ATLAS offline software to use the same source of information about the detector instead of having to manually enter information into each program separately.

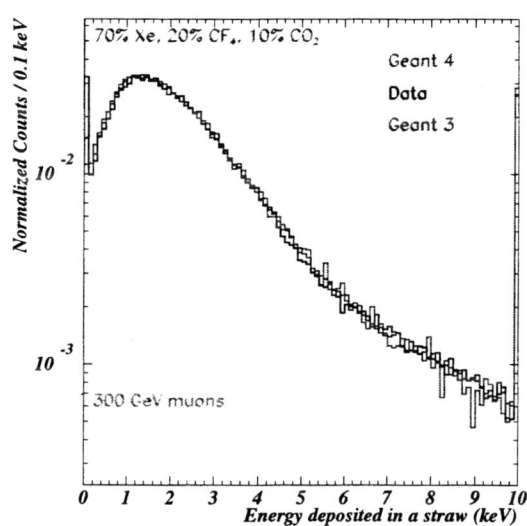

FIGURE 3. Comparison of ionization energy deposit in a straw tube for GEANT3 and testbeam data to GEANT4 results.

For the past year, the ATLAS offline software group has conducted a comprehensive program to compare testbeam results for all ATLAS subdetector systems with GEANT4 simulations of each testbeam setup. This is one of the first real checks that GEANT4 produces accurate results, and has been strongly supported by the GEANT4 team. Figure 3 shows a preliminary comparison of the ionization deposition in an ATLAS straw tube for GEANT3, GEANT4, and actual testbeam data

In designing the new simulation it became obvious that we needed a systematic way of storing information about the detector dimensions and materials. We decided to develop a generic detector description language called ATLAS Generic Detector Description (AGDD) using XML. XML was selected because it is a commercial standard and many compatible tools (such as parsers and editors) are commonly available.

CONCLUDING REMARKS

ATLAS has many simulation tasks to perform over the next 18 months. We must finish comparing the GEANT4 physics models with our testbeam data to ensure valid results before running large Monte Carlo event productions. We must implement the model of the full ATLAS detector in GEANT4, integrating the full detector simulation within our analysis framework. We need to design hits and digitizations for the GEANT4 simulation. Our first mock-data challenge requires the full chain of event-generation, simulation, and reconstruction to be operating with the new software. Meeting the November 1, 2001 deadline for this will be a challenge.

REFERENCES

1. S. Gianni and G. Folger, *Introduction to Geant4*, http://wwwinfo.cern.ch/asd/geant4/G4UsersDocuments/Welcome/IntroductionToGeant4/html/index.html.

2. Application Software Group, *Geant Detector Description and Simulation Tool*, CERN Program Library Long Writeup W5013, Geneva: CERN, 1993.

3. Application Software Group, *The Zebra System*, CERN Program Library Long Writeup Q100/Q101, Geneva: CERN, 1995.

4. A. Artamonov et al., *Dice-95*, Atlas-Soft/95-14c, Genva: CERN, 1995.

5. Application Software Group, *KUIP: Kit for a User Interface Package*, CERN Program Library Long Writeup I102, Geneva: CERN, 1993

6. The ATLAS Collaboration, *ATLAS Detector and Physics Performance Technical Design Report*, ATLAs TDR 14, CERN/LHCC 99-14, Geneva: CERN, 1999

RAPPORTEUR TALKS

Artificial Intelligence

Harrison B. Prosper

Department of Physics, Florida State University, Tallahassee, Florida, USA

Abstract. I summarize the ideas that emerged from the parallel sessions on Artificial Intelligence and I offer a critical assessment of what we learnt.

INTRODUCTION

In 1956, at a conference in Vermont organized by John McCarthy, the term "Artificial Intelligence," (AI to the cognoscenti) was coined. The grand vision of the pioneers was to create, in due course, machines that combined phenomenal computing power with human-like characteristics. In his classic science fiction novel "2001, A Space Odyssey" Arthur C. Clarke envisioned such an awesome machine, the HAL 9000. But, unfortunately for the crew of the Jupiter Mission, that machine, in spite of its enormous intellectual power, displayed the severest of human failings: it became murderously psychotic! Today, forty-five years after the brainstorming in Vermont, machines have indeed grown enormously more powerful. But the real success of AI, to date, has been not so much in making HAL-like machines but rather in solving difficult real-world problems.

Some of these problems are indeed extremely hard. An interesting example is the optimal scheduling of computer jobs in a distributed computing system, as described in the contribution by Newman and Legrand (1). Here, the term artificial intelligence has some resonance. The problem addressed by the authors is daunting: to build an autonomous system that schedules computer jobs in a a dynamic distributed computing environment that could contain millions of interacting components subject to thousands of time-dependent constraints. On the face of it, to have any chance of a near-optimal solution, this problem appears to demand a very high level of rapid and intelligent decision-making. The complexity is surely on a scale that is likely to overwhelm intelligence of the natural sort.

For the most part, the contributors to the parallel sessions on Artificial Intelligence had a more prosaic goal: the optimal analysis of multi-dimensional data in a more-or-less automatic manner. I shall focus on these contributions and refer the reader to the paper on job scheduling (1) and to the contribution by Berg (2) dealing with a generalization of the Clopper-Pearson method for constructing binomial confidence limits.

The remaining contributions can be usefully grouped into the following categories.

- **Classification**
    - Selecting W-pair events (DELPHI) (3)
    - Searching for single top events (DØ) (4)
    - Searching for a low-mass Higgs boson events (5)
    - Searching for instanton-induced processes (HERA) (6)
    - Electron/jet discrimination (ATLAS) (7)
    - Electron and τ-lepton identification (DØ) (8)
    - Rule induction (9)
- **Parameter Estimation**
    - Measuring the top quark mass (10)
    - Calorimeter energy estimation (ATLAS) (11)
- **Feature Recognition**
    - Vertex finding (ZEUS) (12)

Below I briefly consider each of these problem categories separately and end with a few conclusions.

CLASSIFICATION

The problem addressed by the contributors in this category is the optimal classification of objects, specifically, high energy events or particles. Each object is described by a feature vector $\mathbf{x} = (x_1, x_2, \ldots, x_N)$ of N measured quantities x_i. The goal is to classify the object based on the value of its feature vector. The solution to this problem is easy to state: use the *Bayes decision rule*, namely, given **x** assign the object to the class k with the highest posterior probability

$$P(k|\mathbf{x}) = \frac{f(\mathbf{x}|k)p(k)}{\sum_k f(\mathbf{x}|k)p(k)}, \quad (1)$$

where $f(\mathbf{x}|k)$ and $p(k)$ are the density function and the prior probability of class k, respectively. This rule minimizes the probability of making incorrect classifications.

From the viewpoint of classification, the principal difference between the parallel session contributions is the algorithm used by the authors to approximate the posterior probability $P(k|\mathbf{x})$. Most authors (for well-founded reasons (13, 14, 15, 16)) approximated this probability using feed-forward neural networks (17). Koblitz (6), however, found it computationally more efficient to approximate this probability with a kernel density method that, in the neigborhood of a given point, counts the number of feature vectors of each class. The key feature of this method is an efficient (range searching) algorithm to determine which points lie within a given neighborhood.

Stepanov (9) noted the persistent perception, by some, of neural networks as "black boxes." In view of the considerable body of mathematical work that underpins neural networks and the great conceptual clarity that has ensued, I am inclined to conclude that this perception is a problem of sociology. Stepanov proposes to circumvent this (sociological) problem by using an interesting semi-automatic rule induction system that, in effect, approximates the posterior probability with a decision tree. The claimed advantage of this system is that a decision tree, constructed directly from the components of the feature vectors, provides an *explanation* for every classification it makes. This is true and could be important. For instance, a medical practitioner presumably has an interest in understanding how an expert system has arrived at a medical diagnosis. However, for the sorts of applications for which the rule induction system is targeted it is not clear what is gained by having a decision tree, rather than a single cut on a single discriminant.

Of course, it is often instructive to understand why an object was classified the way it was; why an electron and not a jet. But my claim is that one can gain an understanding of the "reasons" for a decision by examining how the single cut has distorted the distribution of each component or, for more insight, the distributions of different sub-sets of the components taken two or even three at a time. Where a rule induction system could be extremely valuable is, for example, in the alarm reporting system of a large and complex experiment such as DØ, where there is great interest in understanding why a particular class of fault was reported in response to a feature vector representing the active state of the experiment. A rule induction system is precisely what is needed in such an expert system.

It comes as no great surprise that every author reported, in some cases significant, improvement in classification performance relative to more traditional methods. For example, Chakraborty (8), who talked about electron and τ-lepton identification, reported a factor of 10 improvement, with respect to previous work by the DØ collaboration, in the correct classification of objects as background rather than as electrons, while achieving a classification probability of 90% for electrons.

One issue that was addressed by Seixas (7), in the context of electron/jet discrimination, is that of dimensionality reduction. All things being equal, it is advantageous to reduce the dimensionality of the feature vectors, if for no other reason than to decrease the computational burden. (For a counter example, see Dudko (4) who showed that even with moderately large dimensions it is still possible to make progress provided one proceeds with due caution.) The authors faced a severe computational problem: a feature vector of dimension 496, one dimension per calorimeter cell! However, by a judicious use of the method of Principal Component Analysis (PCA) (17) the 496 dimensions of the original feature vectors were reduced to as few as 9, while maintaining efficient electron/jet discrimination.

PARAMETER ESTIMATION

A remarkable property of feed-forward neural networks, which in the final analysis are nothing more than highly non-linear mathematical functions, is that they are universal approximators (16). In principle, they can be used to learn any mapping of the form $f : \Re^n \to \Re^m$. Another useful property, not specifically tied to neural networks, is their Bayesian interpretation (13, 14, 15). In the paper presented by Beri (10), explicit use was made of the fact that the outputs of feed-forward neural networks, when appropriately trained, approximate class posterior probabilities. The authors exploited the Bayesian interpretation to construct an algorithm to estimate the top quark mass using events in which a full kinematic reconstruction of the top quark 4-vectors is not possible.

In many situations, the chief concerns during the training phase of a neural network are simply to guard against over-training, that is, "over-fitting" and to ensure convergence. But, as Seixas (11) showed, the training strategy itself can be put to good use. The goal was to improve the estimation of the energy deposited in the ATLAS calorimeter by reducing the non-linearity in its response, while maintaining acceptable energy resolution. This was achieved by dividing the training into two phases. In the first, the network target output is set to the beam energy while during the second, the target output is set to a linear sum of the calorimeter cell energies. The use of a neural network, trained in this novel way, reduced the nonlinearity by almost a factor of 4, and caused no significant degradation in the energy resolution.

FEATURE RECOGNITION

The application described by Etzion (12) provided a beautiful example of feature recognition that incorporates parameter estimation. Etzion described a hierarchy of neural networks that recognizes tracks in the ZEUS vertex detector. The inspiration for this architecture is the hierarchically processing of information in the primary visual system. The system studied consists of a "retina" of inputs followed by three layers of neural networks. The outputs of the neural networks in the first and second layers of the hierarchy provide the inputs to the subsequent network. The first layer finds line segments, the second combines segments into arcs and the last combines arcs into tracks. This application, in which data are compressed into more and more useful forms, is an excellent example of the possible advantage of trying to solve a problem by mapping it to simpler and simpler subproblems, each solved by its own "expert."

CONCLUSIONS

All classification methods are approximations to the same problem: the calculation of class posterior probabilities, $P(k|\mathbf{x})$. Given a fixed and finite set of resources, time, computational, intellectual, it remains a matter of experimentation to decide which method of approximation is the best for a given application since experience suggests that no single method is uniformly the best over the set of all possible problems. Therefore, provided that they are mathematically well-founded, one should embrace them all.

That being said, however, it is nonetheless useful to have a *default* method that is known to work well in circumstances of general interest. It seems that neural networks are emerging as the default method for classification, parameter estimation, functional approximation and feature recognition. And for truly difficult problems, such as the problem of optimal job scheduling in a large and complex distributed system, the use of neural networks of one sort or another may prove to be the only feasible approach that stands any chance of succeeding.

It has been amply demonstrated at this workshop, in the many excellent contributions at the parallel sessions, that neural networks work well in practice. But, as is true of anything that one wishes to do well, it is necessary to exercise a fair measure of non-artificial intelligence to render that statement true!

REFERENCES

1. Newman, H.B., Legrand, I.C., these proceedings.
2. Berg, B.A., these proceedings.
3. Becks, K.-H., Buschmann, P., Drees, J., Müller, U., Wahlen, H., these proceedings.
4. Dudko, L., these proceedings.
5. Tentindo-Repond, S., Bhat, P.C, Prosper, H.B., these proceedings.
6. Carli, T., Koblitz, B., these proceedings.
7. Vassali, M.R., Seixas, J.M., these proceedings.
8. Chakraborty, D., these proceedings.
9. Stepanov, N., these proceedings.
10. Beri, S., Bhat, P.C., Kaur, R., Prosper, H.B., these proceedings.
11. da Silva, P.V.M, Seixas, J.M., Seixas, J., these proceedings.
12. Dror, G., Etzion, E., these proceedings.
13. Ruck, D.W. *et al.*, *IEEE Trans. Neural Networks* **1 (4)**, 296 (1990).
14. Wan, E.A., *IEEE Trans. Neural Networks* **1 (4)**, 303 (1990).
15. Blum, E.K., and Li, L.K., *Neural Networks* **4**, 511 (1991).
16. Hornik, K., Stinchcombe, M., and White, H., *Neural Networks* **2**, 359 (1989); Hornik, K., *Neural Networks* **6**, 1069 (1993);
17. Bishop, C.M., *Neural Networks for Pattern Recognition*, Clarendon Press, Oxford, 1998; Haykin, S., *Neural Networks: A Comprehensive Foundation*, Prentice Hall, Inc, Upper Saddle River, New Jersey, 1999.

Status of Neural Network Hardware in High Energy Physics

Bruce Denby

Université de Versailles and
Laboratoire des Instruments et Systèmes, Paris, France

Abstract. This paper examines the current status of hardware implementations of neural networks in high energy physics experiments, as reflected in the applications presented at ACAT 2000, Fermilab, October, 2000.

INTRODUCTION

It has now been 8 years since the first simple test of neural network hardware in high energy physics [1]. Today, the technique is well established in running experiments and has been proposed for LHC. Currently, efforts are hampered by a lack of available commercial NN products, obliging physicists to examine other solutions for implementing NN in their experiments. We present here some of the more recent NN applications as well as some possible solutions for future developments.

USE OF NN IN THE TRIGGERS OF RUNNING EXPERIMENTS

Two currently running experiments, Dirac at CERN and H1 at DESY use NN hardware directly in the trigger.

The Dirac Experiment at CERN

Neural network hardware is used in the first level trigger of the Dirac experiment at the CERN PS [2]. The experimental setup is illustrated in Fig. 1. The experiment is designed to detect pionium production on a copper target with a 34 GeV proton beam and to measure the lifetime of pionium.

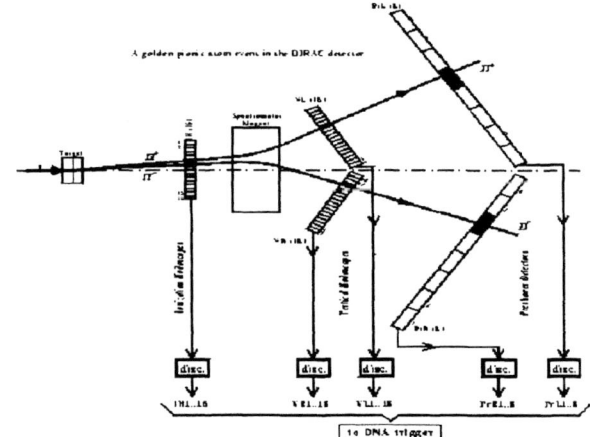

FIGURE 1. Experimental setup of the Dirac experiment at CERN. A proton beam impinges on a copper target. Two–track event are analyzed in a dipole magnet followed by hodoscopes.

Tracks produced in the collisions are detected in hodoscopes following a dipole magnet. Hodoscope positions are read out as binary words. As shown in Fig. 2, the hodoscope information passes first into a data formatting and event selection card, which essentially reduces the dimensionality of the hodoscope data and reformats it, and then is passed into 4 neural network cards. All of the electronics was produced at the University of Basel.

Fig. 3 shows the architecture of the Basel NN cards. The unique feature of these cards is that the synaptic multiplications and the transfer function evaluations are all subsumed into memory lookups. This enables the neural networks to produce a result in only 60 ns. Full level-1 trigger processing occurs in only 210 ns.

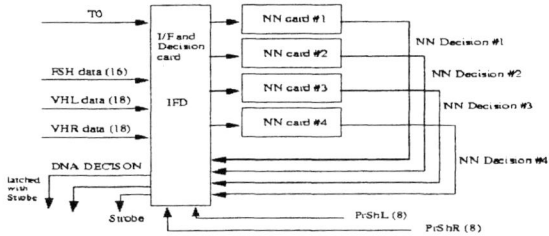

FIGURE 2. Trigger electronics schematic of the Dirac experiment at CERN showing preprocessing and reformating card and the 4 neural net cards.

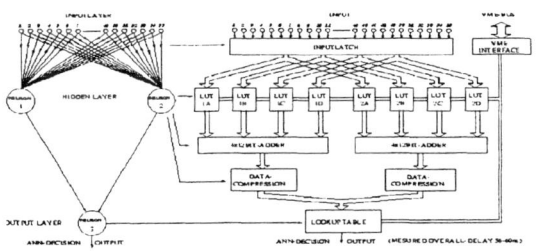

FIGURE 3. Schematic of the basel NN cards. Synaptic multiply and activation function are performed in memories. The MLP structure is 55-2-1.

The Dirac trigger achieves an overall background reduction factor of 2.5 and is 95% efficient for the pionium events under study. The trigger is functioning normally as an integral part of the Dirac experiment.

The H1 Experiment at DESY

The H1 experiment of the HERA accelerator at DESY in Hamburg has adopted a completely neural architecture for its level 2 trigger [3]. An initial 10 MHz data rate off the detectors is reduced to 1 kHz by a hardwired level-1 trigger with a 2.3 microsecond latency. Then the level-2 trigger, which performs its function in under 20 microseconds, takes effect, reducing the background by an additional factor of 20.

The background is predominantly from beam-gas events.

The hardware chosen is the CNAPS chip from Cromemco in Germany, which is a 64 processor SIMD architecture. Twelve separate CNAPS cards are used, each one tuned to a particular physics channel. The networks have typically several tens of variables at the input, several tens of hidden units, and a single output unit which simply signals whether the corresponding physics process is believed to be present or not.

It is important to realize that in online NN applications, any preprocessing must also be done online. In the case of H1, this is done in data distribution bus DDB boards constructed at MPI in Munich using FPGA technology. The entire procedure must be completed in 4 microseconds which is a challenging electronics task. The overall system is illustrated in Fig. 4.

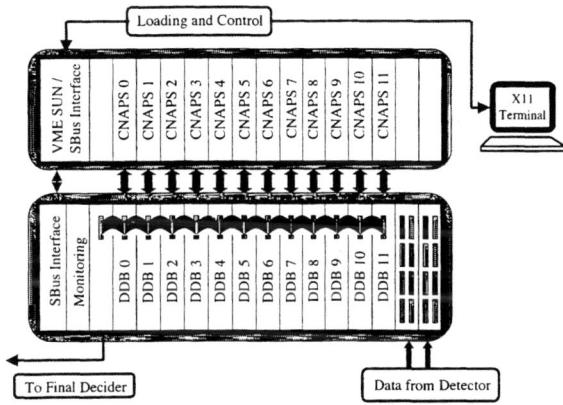

FIGURE 4. Schematic Diagram of the H1 NN trigger including the DDB preprocessing cards and the 12 corresponding CNAPS NN cards.

The H1 NN trigger has been in operation for 5 years and has enabled the H1 collaboration to produce physics results which would otherwise have been unattainable, particularly in the realm of vector meson photoproduction (phi, psi, etc.). The additional acceptance for these channels is due to the intelligent triggering of the NN hardware.

FUTURE DEVELOPEMENTS FORESEEN

Apart from the currently running NN trigger applications mentioned above, there are also a certain number of new projects that are being planned using these techniques. We shall speak here about the upgrade to H1 and a possible application to a muon trigger in ATLAS at CERN, and conclude with some new approaches to NN hardware.

The H1 Upgrade

For its next run, the H1 experiment will be upgrading its NN trigger [3,4]. The number of CNAPS cards will be extended to 16 and a new DDBII preprocessing system will be built. The DDBII is based on intelligent preprocessing achieved in only 4 microseconds, and are implemented in FPGA using the Virtex family of Xylinx chips. The overall plan is to execute the preprocessing in 4 steps: clustering in the individual detectors; matching of corresponding clusters between devices into physics objects; ordering of phyics objects by energy or other criteria; and finally construction of the network in put variables. The procedure is shown schematically in Fig. 5.

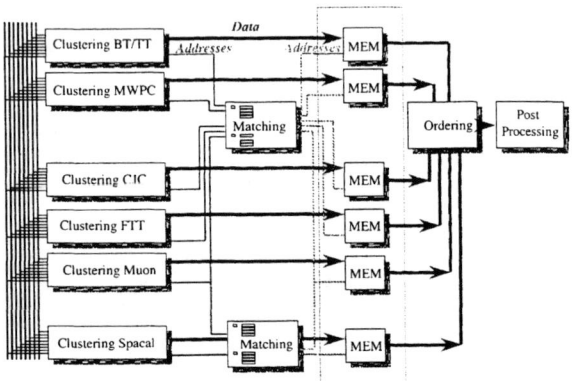

FIGURE 5. Schematic of the DDBII architecture for the H1 upgrade. The circuit will be implemented with FPGA's.

The DDBII's have shown, in simulation, to be able to improve rather dramatically the ability of H1 to study certain processes, in particular photoproduction of phi's.

A Possible Muon Trigger for ATLAS

As the magnetic field in the ATLAS experiment is rather non-uniform, the mapping between raw track parameters (slopes and intercepts in two planes) and physics quantities such as the 3-vector and charge of a muon is a complicated function which varies with position around the ATLAS detector. In an experimental simulation [5], it was shown that this mapping can be learned by a 4-7-7-4 multilayer perceptron architecture with very good accuracy. Given the parallel nature of neural networks, it is interesting to ask the question as to whether or not this architecture could also be realized at the trigger level to providean 'intelligent' muon trigger. This will be addressed in a preliminary way in the next section.

Future Hardware Approaches

It is worth noting that the CNAPS chips is no longer manufactured. Many of the other sorts of NN chips such as the ETANN which used to exist have also over the years been discontinued. One may ask the question today, if one wanted to build an NN trigger, what hardware should be used?

One approach, as already cited [2], is to implement the NN, both synaptic multiply and transfer function using memory lookups. Another method, proposed by a group in Paris [6], is to use some the new of FPGA's coming out today to build an NN architecture directly. Such a circuit, named MAHARADJAH, is illustrated in Fig. 6.

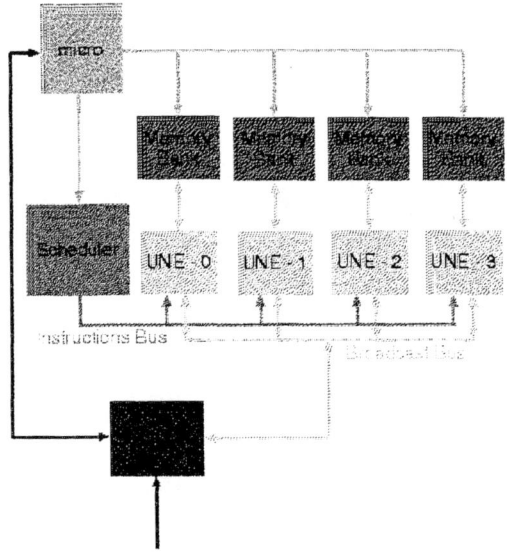

FIGURE 6. Architecture of the MAHARADJAH system for evaluating NN in realtime. Implementation is in FPGA's.

The architecture consists of four UNE units containing multiply-accumulators, in communication with memory banks holding synaptic weights and lookup tables of activation function. Control is via a simple microcontroller. Such a circuit can equal the performance of a Pentium III or a Sparc Ultra workstation while running at only 40 MHz. As a test, the architecture was benchmarked against the ATLAS 4-7-7-4 network presented in the previous section.

The predicted execution time was of the order of only 2 microseconds, which would be quite acceptable at trigger level and is thus very interesting.

CONCLUSIONS

Neural network triggers have been incorporated into a certain number of running experiments and are being used to produce physics. In the case of H1, the NN trigger has been instrumental in obtaining certain classes of physics. For the future, H1 will be upgraded and along with it the NN trigger, which will include a new, more intelligent preprocessor based on FPGA's. Additionally, there is a possibility of a neural muon trigger for ATLAS.

In a general sense, neural network hardware is difficult to find; a number of longstanding products are no longer available. In future, it should be possible to turn for example to full-FPGA solutions such as the MAHARADJAH architecture cited above, or other digital approaches such as memory lookups.

ACKNOWLEDGMENTS

The author is indebted to the authors of the presentations which served as the basis for this rapporteur's summary.

REFERENCES

1. C.S. Lindsey et. al,, *Nucl. Instrum. Meth.* **A317** (1992) 346.

2. Sotirios Vlachos, these proceedings.

3. Christophe Kiesling, these proceedings.

4. Jean-Christophe Prévotet, these proceedings.

5. Erez Etzion, these proceedings.

6. Bertrand Granado, these proceedings.

Innovative Software Algorithms and Tools Parallel Sessions Summary

Irwin Gaines

Fermi National Accelerator Laboratory, P.O. Box 500, Batavia, IL 60510

Abstract. A variety of results were presented in the poster and 5 parallel sessions of the Innovative Software, Algorithms and Tools (ISAT) sessions. I will briefly summarize these presentations and attempt to identify some unifying trends.

INTRODUCTION

An exciting thread running through many of these presentations is the quiet revolution of enabling technologies. New programming languages (C++ and Java), programming methodologies (object orientation), GUIs, open source techniques, and Grid technologies are being more and more widely used and are enabling better science. Most of the presentations emphasized the innovations that resulted from creative use of these new tools and technologies, and so rather than seeing religious wars about the technology choice, discussions focused on effective use of a variety of techniques, resulting in very productive sessions.

PRESENTATIONS

I have divided the presentations up into several broad categories for discussion. In all cases the reader is referred to the full writeups of each talk for more details.

Accelerators and Control Systems

Four papers discussed applications involving accelerator (and experiment) control systems. Two papers from Fermilab presented, first, the online modeling of the Fermilab accelerator using accurate beam physics computation classes, integrated into the control system with database lookup for device information. Class libraries are provided for Lie Algebra and Automatic Differentiation, and a beamline library for lattice construction and hierarchy and analysis. The use of Object Oriented C++ was stressed. Examples were given of an online comparison between a model of the Main Injector 8 GeV line and the readings form the beam position monitors.

A second Fermilab talk discussed an offline beamline matching application based on open source software, including many of the same class libraries described in the previous talk, but with lots of additional functionality not available online due to the limited online user interface. The system is designed to provide graphical feedback, with full user specification of all aspects of the beamline element matching problem, adaptable for special needs, and running on both Unix and Windows platforms. The work benefited both from the use of C++ and from an open source philosophy making it easier to share code.

Another talk in this area described a runtime tailorable architecture used in the KEK control system, wit an authoring interface for developers and component based tailorability by integration for end users. Java and BML (beans markup language) were used). Finally, the KLOE integrated dataflow (KID) in an experiment data acquisition system was presented, again with differing developer (C based library) and user (Uniform Resource Identifier [URI] interpreter) views.

Shared Code

Two talks discussed sharing code. Tony Johnson described the FreeHEP initiative, a library of Hep-wide Java software, some HEP specific and some general purpose. The libraries are fully open source, all available in cvs, web browsable, freely reusable, with a preview area for half baked ideas. General purpose code includes 2D vector graphics, Studio (a common base GUI for application development), Java access to C++ objects, and java package dependency tools. HEP specific packages include the hep.physics package (including 3- and 4-vectors, jet finding and event shape routines), and XML particle property database, hep.io packages including StdHEP and ROOT input output, an interface to the AIDA package of abstract interfaces for data analysis, and Hep 3D event display utilities. This project is a fruitful exploration of techniques for sharing code using open source methods, and for extensive use of Java.

Jon Thaler presented the Lattice package that he has written for CLEO. This project makes object relationship into an object; allows adding and removing connections between objects, associating data with each connection, and a complete set of data access methods. Specifications are to be flexible with minimal impact on existing code. Mappings are persistent over read/write cycle and optimized for data access at expense of slower data insertion. Jon provides both a generic implementation and a CLEO specific interface derived by inheritance. He provides many powerful data access methods. Other experiments at the talk were quite interested in making use of the package.

Reconstruction and pattern recognition techniques

Five papers described innovative techniques for track finding, vertex finding, and data analysis. Topics included vertex finding using hits before tracking, faster track finding using z finders and hit filters, simultaneous vertex and track fitting, reuse of tracking code in different experiments, and a bias free technique to search for new physics.

Different talks focused on mathematical methods, innovative algorithms, or computing techniques (like object orientation) that provide clear advantages. One new approach was to reduce the enormous combinatorics present in hadron collider experiments by taking advantage of the differences between physics and pileup events to clean up the event before starting tracking. First, a z finder algorithm uses rapid one dimensional histograms generated by pairs of hits to identify the primary z vertex. Then, a hit filter algorithm is able to quickly identify hits coming from high Pt tracks originating at the primary vertex. Efficiencies and performance were shown for complex Atlas events, with excellent efficiency above 2 GeV Pt.

Andrew Haas showed a new elastic fitting approach which simultaneously fits hits, tracks, primary and secondary vertices. He takes advantage of continuing optimization as the fitting procedure can reassign hits to tracks and tracks to vertices. He also uses an improved "ghostbuster" track seeding algorithm to better choose triplets of hits to use for candidate tracks. He applied these techniques to the problem of tracking and verticizing in the new D0 central fiber tracker.

Sijin Qian presented a Kalman Filter track finding and fitting program that has been successfully reused in two experiments and several different environments. OO/C++ is now mature enough for development and debugging by non experts. A stable framework in which to implement code is important. The use of C++ and object orientation makes long distance collaboration easier, and makes reuse possible

Bruce Knuteson described SLEUTH, a quasi model independent new physics search strategy. Its motivation is too many competing candidate theories, so there is a need to perform generic searches and a need to quantify "interestingness" of a few strange events a posterior. His strategy is to use final state variables and use exclusive final states (since presence of extra objects in an inclusive search changes interpretation); He assumes the existence of standard object definitions: e, mu, tau, jet, gamma, b, c, missing Et, W, Z. He creates a general rule to select variables for any final state (there are lots of final states); for example, to look for SUSY, look for final state particles with large Pt, and use as variables sums of Pt (of leptons, jets, gamma/W/Z) and missing Et.

For an algorithm he puts variables into the unit box, searches for interesting regions, finds fraction of hypothetical experiments that would see something this interesting; starting with background, he transforms these events to get uniform distribution in unit box. Voronoi diagrams divide box up using density of events; search for region with greatest excess of events; generate lots of pseudo experiments to get probability of the excess in the data interesting region. Results: he finds Monte Carlo leptoquarks with > 3.5 sigma in 80% of experiments (but sees no new physics in 40/80 D0 final states yet)

Using distributed resources

One ISAT parallel session took place on a day emphasizing grids and worldwide computing, which included plenary talks, a VLSC (very large scale computing) parallel session, and a working group on worldwide computing. The ISAT session focused on use of grids to perform real life scientific calculations and analysis.

Harvey Newman discussed Worldwide Distributed Analysis and Data Grids for Next-Generation Physics Experiments. He considered some of the upcoming challenges of new generation experiments, and surveyed some of the R&D projects working on Grid solutions, including PPDG and GriPhyN. Data Grids offer the possibility of better global resource use and faster turnaround, building information and security infrastructures, and coordinated use of computing, data handling and network resources.

Consistent with the overall aims of the ACAT conference, Harvey sees both an opportunity and an obligation for HEP/CS collaboration, with high energy and nuclear physics as both an early adopter and leading developer of DataGrid technology

Bill Allock focused on Grod Components, reporting on Protocols and Services for Distributed Data-Intensive Science coming from the Globus project. There are two major categories of tools: data transport and access tools, typified by GridFTP; and replica management tools, which maintain mapping between logical collections of files at different physical locations. Replica manager components include replica catalog definition, a low level replica catalog API (application programmer interface), and a high level reliable replication API. These tools, which are available now are the start of a full set of grid tools to provide transparent and efficient access to widely distributed collections of data.

Myron Livny reported on the Condor-G project, a computation management agent for multi institution grids. He emphasized the current real life use of grid techniques for science. He described the solution of the NUG30 computing problem which was able to deliver 11 CPU years of compute time in 7 days using an international array or processors. The project started with existing Condor tools to manage collections of workstations, and added the G to Condor by adding GSI based authentication services, adding specific support for the Grid universe, and adding glide-in tools. Heavy use was made of existing Globus Grid tools.

Iosif Legrand described a set of modeling tools to simulate the performance of large scale distributed systems. The tools were originally developed in the context of the MONARC project to model architectures for LHC (Large Hadron Collider) experiment computing. He described the design and development of the tool, the elegant GUI to specify simulation parameters and to publish results, validation of the tool against measurements on real systems, and the use of the tool to predict behavior in a large scale CMS high level trigger production exercise.

As was true in previous sessions, these talks demonstrated that the grid is already a real entity, and making use of grid tools and techniques is enabling new or better science.

Miscellany

Two final papers defied classification. One described a Component based architecture supporting scientific workflow (for support of the CRISTAL project constructing the CMS electromagnetic crystal; calorimeter). Software components were described using the standard UML modeling language. The second discussed mathematical techniques for image matching, used to search the internet for specific content.

CONCLUSIONS

A major conclusion is that high energy and nuclear physicists have absorbed many of the new computing technologies and are making effective use of them to further their science. Further, it is clear that the computing challenges represented by experiments in these fields are an attractive proving ground for computer scientists to show off their wares. Conferences such as this one provide useful meeting grounds where the problems and the technologies can confront one another, and point the way for specific collaborations between experimental physicists and computer scientists.

Panel Discussion: C++ in Scientific Computing[*]

Walter E. Brown, James Kowalkowski, and Mark Fischler

*Fermi National Accelerator Laboratory
Batavia, IL 60510-0500*

Abstract. We summarize the issues and concerns voiced by the ACAT conference attendees regarding their experiences using the C++ programming language and attendant methodologies.

INTRODUCTION

Because of the widespread acceptance of the C++ programming language in the physics community, this panel session sought to explore the ways in which the programming paradigms supported by C++ have impacted the community.

Two sessions were held, totaling nearly four hours, over a two-day period. In the first session, three speakers provided prepared presentations to serve as catalysts for subsequent discussions. Leo Michelotti provided *Reflections on a Decade of OOP in Accelerator Physics*, Marc Paterno shared his *Experiences from Reviewing Scientific C++ Code*, and Walter Brown presented *C++ in Scientific Application: A Case Study*. Each of these talks is separately documented elsewhere in these Proceedings.

Following the prepared talks in the first session, and continuing through the second session, panelists responded to comments and questions from attendees. Moderated by Walter Brown, the panel was composed of the following members of the physics computing community at Fermilab:

Amber Boehnlein, DØ
Elizabeth Sexton Kennedy, CDF
Jim Kowalkowski, Computing Division
Leo Michelotti, Beams Division
Marc Paterno, Computing Division

In addition, the original designer and developer of C++, Bjarne Stroustrup, served as a distinguished member of the panel and also gave a plenary address on modern C++ (*Speaking C++ as a Native*) between the two panel sessions.

The remainder of this document will summarize the main themes and viewpoints that were expressed. We extend thanks to all of the panelists and attendees for their participation.

PREPARED QUESTIONS

To provide a starting focus, the following questions were posed by the moderator, and attendees were invited to comment:

- What features of the C++ core language and of its library have you found most useful and/or most problematical?

- How has C++ aided or hindered your transition to object-oriented and generic programming methodologies?

- How have your software designs been affected by the wide variations in the quality of C++ compilers, libraries, and debugging tools?

- In the context of your experience with C++, to what extent are the startup costs and ongoing overhead of:

---

[*] This written summary is largely based on the extensive notes taken by Jim Kowalkowski. as well as on his concluding presentation.

- o   O-O design/implementation, and
- o   Generic programming techniques

now paying off with respect to the benefits such methodologies provide?

Attendees and panelists addressed these and other concerns in a variety of contexts during the subsequent discussion. The following summary outline is organized by topic, not chronologically.

DISCUSSION SUMMARY

Overview

Essentially all attendees' comments fit into the following recurring topics, each of which is separately summarized below.

- Frustrations with the language
- Performance
- Desired C++ features
- Use of language and library subsets
- C++ is difficult

Frustrations with the Language

Issues:

a) Decomposing a problem is difficult to do properly

b) Requires design time

c) Results are not available quickly

d) Relatively few experts are available to help

e) Appropriateness of algorithm/data separation

Responses:

a) Good design yields flexible, maintainable, and correct code

b) Prototyping is very important; distinguish it from production code

c) Striking a balance in separating concepts is difficult; extremes are usually bad

Performance

Issue:

Fortran code apparently executes faster than C++

Responses:

a) Performance testing is difficult; one must compare very similar things

b) Cannot ignore performance when designing and implementing code

c) Well designed code is generally easier to optimize

d) Correctness is more important than speed: extremely fast programs that produce erroneous results, even part of the time, are still incorrect!

Desired C++ Features

Desiderata:

a) Persistence

b) Reflection (enhanced object-querying)

c) Garbage collection

d) Compiler optimization hints

Responses:

a) The technology for some of these issues seems not mature enough yet; certainly there has been no Committee consensus to date (standards require this)

b) Not pushed hard by groups in the Committee

c) Everyone is welcome to join the Standards Committee and/or to submit and defend proposals; our community seems to be underrepresented

Use of Language and Library Subsets

Issues:

a) Exceptions (cost)

b) Templates (compiler support)

c) Algorithms are "weird"; they're expressed in a different style than many are accustomed to

d) Feature fear in general

Responses:

a) Take a fresh look at exceptions and templates; much improved compiler support nowadays

b) Turning off exceptions is not using Standard C++

c) Algorithms allow direct expression of concepts in code

d) Algorithms are efficient

   i) Much effort has gone into codifying the best available algorithms

   ii) Their use allows many opportunities for optimization

e) Avoiding features misses out on some of the real strengths of C++; the weirdness will go away with increased familiarity; language features are there for a reason

f) Code to the Standard, not to a subset, because you'll end up reinventing the features you're avoiding

C++ is Difficult

Issues:

a) Abundant Fortran expertise not yet available with C++

b) C++ is difficult to read

Responses:

a) Fortran skill and code was, in hindsight, really not very good

b) Poor C++ is hard to read, however good C++ code matches the concepts and is straightforward to read and maintain

c) Rely on others' code, concentrate on your application; there's no need to understand every detail

CONCLUSIONS

A few major items emerged from the nearly four hours of discussion. Much discussion centered on various proposed experiment-wide subsets of C++, typically to ease the transition for inexperienced experimenters. However, the panel generally agreed that this was poor policy. "Code to the Standard, not to any subset" seemed the prevailing wisdom.

There was also significant discussion regarding language and library features perceived to be missing from C++. The panel concluded that the scientific community is under-represented on the C++ Standards Committee, and encouraged attendees' institutions to join and participate in order to make their needs known.

Panelists generally encouraged a fresh look at modern, standard C++. Support for object-oriented programming, generic programming, as well as traditional procedural programming is a strength of the language. Programmers should use those features that are most appropriate for solving their problem.

Advanced Analysis Environments - Summary

Suzanne Panacek

Fermi National Laboratory, PO Box 500, Batavia Illinois 60510, USA

Abstract. This is a summary of the panel discussion on Advanced Analysis Environments. René Brun, Tony Johnson, and Lassi Tuura shared their insights about the trends and challenges in analysis environments. This paper contains the initial questions, a summary of the speakers' presentation, and the questions asked by the audience.

INTRODUCTION

The panel was started with a set of questions to start the discussion.

1. What is the role of Java in physics analysis?

2. Will programming languages be relevant? We all thought that the answer to this question was "yes, the programming languages will be relevant.

3. Can commercial products help meet our needs in this area?

4. What is the role of modularity and abstract interfaces?
   Modularity is very important however not well defined.

5. Do we want an all encompassing framework or a collection of configurable tools?

THE SPEAKERS

The three speakers were: René Brun from CERN, Tony Johnson from SLAC, and Lassi Tuura from Northeastern University in Boston.

René is a leading architect of ROOT the large HEP Analysis Framework written in C++. René is also an architect of PAW and many other large analysis software packages.

Tony Johnson is a leading architect of JAS which is a HEP Analysis Framework written in Java and geared towards the Java user.

Lassi Tuura is part of the Iguana team that is currently evaluating analysis tools for the Atlas collaboration.

René Brun: Future of Analysis Environments: Personal Views

René observed two trends in data storage for physics analysis. The first is to store everything in an object data base such as Objectivity. He noted that many experiments following this path have abandoned it, and he would not recommend it for a PAW like analysis.

The second trend is to put the write-once data into an object store as is done by the ROOT streamers. In addition, the run catalogs, and calibration data is stored in a relational data base. For example ROOT and Oracle or ROOT and Objectivity.

René summarized the basic requirements for a framework. One being that the interpreted code can call the compiled code and vice versa. He also addressed Automatic Code generation, noting that in ROOT 40% of the code is automatically generated. He listed several options on how to integrate ROOT and Java.

Languages for data analysis need to be powerful in interpreted and compiled mode. A scripting language is not the solution. He showed a graph to illustrate the time to execute a task vs. the number of tasks. This shows the different situation for using interpreted vs. using compiled scripts.

He sees Oracle as a likely commercial product to be of use. He does not see commercial GUI components, or special fitting algorithms as playing a strong role. He believes in open source and strong discussion groups.

René, described the trend in ROOT to go towards a GRID like architecture with it's parallel processing capability PROOF.

Modularity is not well defined. René asked many interesting questions concerning modularity. He explained how ROOT addressed modularity, and showed a historical graph on how the physics analysis tools evolved from libraries to a framework.

Tony Johnson: The Role of JAVA

Tony's presentation focused on two startup questions. 'What is the role of Java in physics analysis?' and 'What is the role of modularity and abstract interfaces?'

He explained that Java is a great language for analysis because it is clean, modern, Object Oriented, relatively simple, and platform independent. It also is a mainstream language with a set of standards. In addition, large research teams are working on development and performance. Java is simple and lets the physicist spend time solving physics problems rather than working around language problems.

Java solves three problems we see in data storage, it has the ability to represent complex data structures, provides persistence, and access to named data at run time (RTTI). C++ is lacking built-in persistence and RTTI.

Tony showed a graph illustrating the large advances made in the performance of Java. It is now at 60% of C++ performance. No performance improvement for C++ is expected, but Java performance continues to improve.

Tony showed a list of Java applications. They are: WIRED, JAS, LCD – reconstruction/analysis, and FreeHEP – a library of components.

Tony showed how Java has built-in support for the GRID. He also thinks that computer languages will be relevant.

Tony made several points about modularity. He sees it as fundamental to Object Oriented design. As an example he showed us how JAS is divided into modules. He also showed a slide on AIDA, the Abstract interface for Data Analysis.

Tony saw no paradox in having both a framework and a collection of tools.

Lassi Tuura: Analysis Environment Challenges

Lassi started by noting we can do better than just ntuples. For example ROOT trees, and CMS's full object model. He noted that many experiments are making big jumps by using objects rather than just data. The user interfaces still need to catch up. He sees a need for several ways a user will interface with an analysis tool: Batch, interactively, data store operations, browsing, 2D and 3D visualization, and moving code across final analysis.

Lassi showed a slide on the challenge of distributing your data store. He noted the need for uniform integrated interface to the whole task range. He mentioned MS Outlook as a user interface that could be mimicked for patching together system analysis tools.

Lassi showed a slide on his recommendation for implementation. Divide and conquer the problem into categories, share existing modules, integrate the tools into a user friendly environment, and make applications by choosing from a module pool.

He showed us a couple of slides on modularity. His opinion was to use modularity where it matters, but not everything needs to be modular.

In his summary he reminded us that as complexity grows we need to be able to interact with the data in many new ways. Building all from scratch is not feasible and we need to help people co-operate and not disturb each other too much.

Lassi talked about the three-tier architecture and the use of wizards.

He showed us a list of pros and cons on modularity and interfaces. He stated that modularity is good, but has a cost associated with it. Bad interfaces can make it very awkward. A good interface clearly defines a mission.

Lassi stated that in spite of what language is used great concepts will survive in almost any language. However, the realities are that a PAW analysis will not run on a C++ object, and finding someone to port the FORTRAN code is difficult.

Lassi also showed us illustrations of the modularity in CMS and IGUANA.

Questions From The Audience

Question: Why can we now bring code to the data when it has failed in the past?

The panelists thought that new technology (Java) is key to making it happen now. However, it will not happen until there is a user demand for it. Also, the solution may include a combination of moving data and code rather than just moving code to data.

Questions: Memory leaks, Modularity

Is Java really memory leak free? The answer is yes, due to multi-level garbage collection.

Can software be modular if user classes must inherit from a base class? Yes, one can provide a standard behavior in other ways.

Questions: What is the vision for "your" tool in 5 years from now?

For ROOT the emphasis is on PROOF and Schema evolution. In JAS the emphasis is to integrate JAS components with FreeHEP, leveraging reuse and open source. In CMS/IGUANA we beg, borrow, and steal tools that meet CMS's requirements.

Questions: What about the Open Source model?

Many people contribute to ROOT (three contributions last week). RTTI and a working product are essential to enable and inspire contributors. JAS is moving to integrate with FreeHEP and both projects are now open source.

Summary: Panel Discussion on Large-scale Astrophysical Calculations

Robert Rosner

Depts. of Astronomy & Astrophysics and Physics, The University of Chicago, Chicago, Illinois 60637

Abstract. I summarize the key points of the discussion following the astrophysics presentations in the panel discussion sessions covering large-scale simulations in high energy, accelerator and astrophysics.

INTRODUCTION

Large-scale computations come in two flavors in astronomy and astrophysics: First, modern astronomical data sets are now largely digital in nature, and have reached levels of size and complexity that call for large-scale storage and archiving facilities, and require significant computational resources in order to analyze the data. There is a serious shift in how observational data is gathered and organized: as the sky is being digitized over a broad energy range (from the radio to gamma rays), we are seeing a shift to the "Virtual Observatory", in which the notion of "observing" is coming to mean the remote mining of a multi-wavelength digital data repository. Second, large-scale parallel computational facilities have now reached the point that numerical simulations can now be used both to model astrophysical systems with considerable fidelity, and to explore the physics of astrophysical systems by using numerical simulations as a "laboratory" for conducting astrophysically-relevant "experiments"; such calculations produce "fully instrumented" data sets, allowing heretofore unrivaled detailed study of physically interesting regimes. (What has changed to make this possible is that the dynamic range of 3-dimensional numerical simulations can now be sufficiently large that the Reynolds number of flows can be pushed into the turbulent regime, so that it is now possible to start the exploration of the fully nonlinear behavior of such systems). The Panel provided a hint of the broad range of astronomical problems that can now be tackled via very large-scale computations.

Numerical simulations

Two of the computational talks focussed on the extremes of modern astrophysical simulations: Bruce Fryxell [1] discussed calculations related to thermonuclear flashes associated with degenerate stars, and Paul Ricker [2] related experiences with simulations which coupled fluid dynamics to gravitational N-body calculations (in the context of structure and cluster formation). Characteristic of these types of calculations is that parallel communications are both local and global; and that the latter type of communications are unavoidable in most astrophysics calculations involving gravity, magnetic fields, or radiation transport.

Data Bases, Data Mining, and Exploratory Data Analysis

The data-oriented talks focussed on two of the essential aspects of modern astronomical data bases: first, Alex Szalay [3] showed how one goes about to create a data base whose size can be in the terabyte range, yet allows multiple users effective and practical access, based on his experiences with Object-Oriented data base programming of the SDSS[1] science data base. As an illustration of what can be done with such a data base, Jim Annis [4] discussed the problem of finding structure in the SDSS data. The computations encountered in this domain involve primarily local parallel communications, and therefore scale extremely well on massively-parallel machines.

THE ISSUES

Our discussions centered on the various issues which we must deal with in developing application tools for large-scale computations. These issues largely reflect the fact that software for our computational hardware (e.g., large parallel computers) remains immature, especially in the commercial market; that the half-life of much of this

---

[1] SDSS = Sloan Digital Sky Survey

hardware is very short, measured in 1-2 years; and that standards (in computer architecture, in code architecture, and – in general – in the computing and coding infrastructure) are far from settled.

Language paradigms

The first issue discussed revolved around the role of object-oriented (OO)-based programming. It quickly became clear that OO-based programming is becoming more pervasive, primarily because the programming community in astrophysics is realizing the advantages to be gained: code re-use via modular design, ease of code maintenance, greater emphasis on clean code design (and whence increased code reliability). As for the specifics of what language to use, there remains considerable diversity: many hydrodynamicists continue to rely on Fortran (though in its more modern flavors, e.g., F90/F95), while the data-intensive astrophysicists are focusing on the use of C/C++. Some sentiments were expressed that in the ideal programming environment, code agnosticism should prevail; and that proper "glue" components (e.g., wrappers, written in languages such as Python) would be able to take care of interfacing between modules written in different languages.

Code architecture paradigms

The second issue raised focused on the programming paradigm for structuring large codes. Two contending philosophies vie for the favor of the coder: "frameworks" and "toolkits". Frameworks tend to define all data structures and interfaces, and building a new code basically entails constructing the modules for computing the desired physics, and "hooking" them to the framework backbone. For example, frameworks such as Lawrence Livermore's *Samrai* which are designed for computational physics/fluid dynamics provide services such as mesh generation, adaptive mesh refinement, various types of time stepping algorithms, solvers for various types of partial differential equations, input/output and visualization tools, all packaged to work seamlessly on a parallel machine. In contrast, toolkits focus their energies on defining a rich set of interfaces, and thus rely on a very lightweight code framework. Modules such as mesh generators, adaptive mesh codes, visualization tools and so forth are then all regarded as peers; this is the code architecture paradigm represented by the Common Component Architecture (CCA) working group. Obviously, the choice between these two paradigms depends entirely on the particulars of the application: In my view, application codes with well-defined, restricted application domains are likely to benefit from the rigor and consistency of frameworks; while application codes that are likely to have varied (and difficult to define in advance) application domains are most likely to benefit from the flexibility of toolkits. Toolkits also have the (current) advantage of broader support – frameworks tend to be supported by small individual development groups, with little commonality between competing frameworks.

Unique issues to simulations

Certain issues that were discussed were clearly unique to the world of numerical simulation. For example, until very recently, the dominant parallel programming paradigm has been the use of message-passing (primarily based on MPI), which works well for single processor/node architectures such as that of Beowulf clusters. However, several computer vendors (IBM perhaps the most prominent) have pushed towards a clustered SMP parallel architecture, based on multiple processors per node (with the ASCI White IBM machine currently at 16 processors/node); in this case, the use of threads (and the OpenMP paradigm) has become much more important in order to achieve decent parallel efficiency. Since MPI does not currently support threads, we are facing the (hopefully short-term) possibility of once again dealing with non-portable codes.

The verification and validation of codes has become a much more visible stumbling block for code development. Modern code development efforts deal with verification (meaning determining whether a code contains bugs, and properly solves the equations it is supposed to solve) by enforcing strict code development procedures (including version control and automated regression tests). Validation (meaning determining whether a code correctly solves physical problems purported to be approximated by the equations defining the code) in contrast remains a research issue: typically, it is not trivial to arrange laboratory (or other) experiments so that the characteristic control parameters overlap with those governing the numerical simulations; and for those cases in which such an overlap can be constructed, it remains unclear exactly what is meant by "agreement" between code and experiment: local measures are very likely inappropriate (because of the high sensitivity of nonlinear systems to initial conditions), and global measures (e.g., energy spectra) may be insufficient.

Unique issues to Data Mining and Exploratory Data Analysis

Similarly, the world of large-scale data set analysis must also face a number of unique problems. Perhaps the most pressing (and most often mentioned) is the quest for a stable, bug-free, feature-rich OO-based data base. The core problem seems to be that the commercial market for such data bases remains very limited, so that the commercial vendors tend to be insufficiently capitalized to be able to afford the level of problem-resolution (viz., bug-fixing) and feature extension that modern data base efforts such as the SDSS require. A related problem is the fact that the expertise for constructing such databases does not reside in the astronomical community; the expertise instead resides in computer science departments and in the commercial world. The resulting sociological challenge of getting computer scientists, physicists, and astronomers to work together effectively remain a problem common to all large-scale code development efforts in astronomy.

Boundaries and frontiers

Some discussion focused on the current limitations to computing. The traditional measure of code performance – most sensibly measured by "wall clock time" – continues to play a key role (for example, few of the simulators expressed willingness to give up more that 10-20% in performance as a trade-off for clean modular design!). However, it is also clear that other metrics are becoming important: As codes become more complex, the metrics of "time-to-code-completion" and "time-to-solution" (meaning, the time from starting code construction to the time a solution is finally obtained) become critical. These additional code metrics have played an important role in promoting code modularity and code (module) re-use, since both serve to speed the code development process. Modularity also promotes code portability (since the machine-dependent pieces can be readily isolated), as well as promoting code longevity (since new techniques, or new machine architectures, are far more readily accounted for within the context of a modular code architecture). An interesting observation was that for many forefront computational hydrodynamics problems, run time was no longer the dominant limiting factor: available memory size and archival storage capacity, and I/O and (internet) communication speeds seem to have become critical limiting factors for state-of-the-art hydrodynamic simulations.

Where are the current unsolved bottlenecks? Some examples identified by the various speakers and the audience include:

- Parallel I/O: currently, parallel I/O works relatively poorly, and a number of machines do not yet implement a standard parallel I/O (such as the HDF5 standard).

- Visualization: Visualization tools need to scale, as data sets reach the terabyte scale. Commercial tools simply do not perform well in such applications. Research is progressing in the direction of multi-resolution visualization, but this has not yet reached the desktop.

- Debuggers and performance analyzers: These programming tools also need to scale. Tools such as Argonne National Laboratory's *Jumpshot* go a long way to addressing this problem (for performance analysis), but much more needs to be done.

DISCUSSIONS AND CONCLUSIONS

The panel discussions made clear that the recent huge increase in computing capabilities (especially on the memory size end of things) have led to entirely new ways of using computations in astrophysics: The "virtual observatory" is no longer a dream, but is coming into being; and the use of numerical simulations as an "experimental" tool for astrophysics theorists is becoming commonplace.

Certain "rules of thumb" came up during our discussions that seem to be well-worth highlighting:

- Nothing is for the ages/design for obsolescence: Any given application code is not likely to survive for long. For this reason, allowing for code evolution (via modular design and code component re-use) is a very worthwhile goal.

- Don't "over-optimize": The breakeven point for optimization is usually well in the past, precisely because code life is usually short.

- Getting correct results tops all: No amount of coding elegance or algorithmic cleverness will replace the most important point of all – the application code must produce correct results. For this reason, the common error of short-changing code verification and code validation is a terrible blunder.

ACKNOWLEDGMENTS

I would like to thank the organizers for providing the venue for interesting discussions, Rajendran Raja for collaborating on co-organizing this panel session; and the Department of Energy ASCI/Alliances grant at the University of Chicago for partial support.

REFERENCES

1. Bruce Fryxell, these Proceedings.
2. Paul Ricker, these Proceedings.
3. Alex Szalay, these Proceedings.
4. James Annis, these Proceedings.

Summary of HEP and Accelerator physics part of panel discussion on Large Scale Simulations

Rajendran Raja

Fermilab

Abstract. I present a summary of the discussion that took place after the presentations in the panel discussion session covering large scale simulations in high energy, accelerator and astrophysics.

INTRODUCTION

In the panel discussion on large scale simulations, Pavel Murat described the CDF Monte Carlo (1), which now has an extensive GEANT component. This was followed by Gregory Graham(2) of DØ who described the status of simulations software of that experiment. Hans Wenzel(3) described the CMS Monte Carlo and Fred Luehring(4) the Atlas Monte Carlo. The accelerator talk by Daniel Elvira(5) covered muon Cooling and transportation code for the Muon Collider/Neutrino factory project.

The discussion period in the sessions was lively and included insights into the problems common to astrophysics and particle physics as well as those common to particle physics experiments.

SUMMARY OF DISCUSSION SESSION

One of the items that stood out during the joint astrophysics HEP panel discussion is the commonality of clustering algorithms between HEP and Astrophysics. In HEP, one needs to cluster calorimeter cells with energy deposited in them to form jets. The clustering should ideally be done in three dimensions- in pseudo-rapidity, azimuthal angle and depth. In astrophysics, an identical problem occurs when trying to cluster stars found by the Sloan Digital Sky Survey for instance, where one has to associate stars into clusters also in three dimensions. Some joint algorithm development and/or sharing would be helpful. Perhaps some joint working groups can be set up to investigate this matter further.

The other problem that became evident in HEP simulation and reconstruction is the absence of the use of a common database to describe geometry of the detectors in simulation and reconstruction. Almost all the HEP experiments mentioned above use one geometry database to simulate the detector and another one to reconstruct the data. Much time and effort is spent in trying to ensure that the information contained in either is the same as the other. What is lacking is a well defined suite of routines that will serve the purposes of both simulation and reconstruction that use the same set of numbers to describe the geometry. Such a database should also be able to handle alignment information and calibrations.

The question then arose as to why the full power of C++ is not utilized to provide generic reconstruction programs (generic track finders and, calorimeter reconstructors). Already we use GEANT3 and GEANT4 as generic simulation tools. So why should one not strive for a suite of programs that will contain the optimal algorithms for tracking in non-uniform magnetic fields that handle multiple scattering errors correctly and produce the best estimate for the momenta of tracks. Why is the wheel being re-invented so many times?

The answer to this question seems to be that once the geometry data base tool is developed, then perhaps we would get to a situation where generic reconstruction programs will be available. But such programs are much easier to write and develop than GEANT3/4 and so re-inventing the wheel would never be fully abolished, especially since it is so much fun and is also educational.

ACKNOWLEDGMENTS

The author would like to acknowledge the organizers for putting together a stimulating conference and would like to thank Bob Rosner for his efforts in co-organizing the panel discussion session.

REFERENCES

1. Pavel Murat, these proceedings.

2. Gregory Graham, these proceedings.
3. Hans Wenzel, these proceedings.
4. Fred Luehring, these proceedings.
5. Daniel Elvira, these proceedings.

Worldwide Computing Working Group Summary

Irwin Gaines

Fermi National Accelerator Laboratory, P.O. Box 500, Batavia, IL 60510

Abstract. The worldwide computing working group brought together a number of experts in this rapidly emerging field. A combination of presented talks and discussion identified common interests and issues to be pursued in the future.

INTRODUCTION

The Worldwide Computing Working Group session came at the end of a day that had already featured plenary talks and parallel sessions in both the Very Large Scale Computing and Innovative Software Algorithms and Tools tracks. This session gathered experts in a somewhat more informal setting, allowing more time for interchange of views and discussions of the points raised by the speakers. A panel helped to stimulate the discussion as well as responding to audience questions.

PRESENTATIONS

Four presentations by physicists from both sides of the Atlantic with experience in the challenges of large scale distributed computing led off the program.

Luciano Barone – Management of Large Scale Data Production for the CMS Experiment

Luciano described the experience of the CMS experiment in managing a large distributed Monte Carlo production exercise. The current production involves the generation, simulation, digitization and reconstruction of several million events at sites located a CERN, in the US, Italy, Russia, Finland and elsewhere. A combination of older (Fortran-Zebra based, Geant-3, Pythia) and more modern (C++, Objectivity object database) software is being used. Data relocation (from production sites to central locations and distributing data to remote sites for analysis) is an important issue.

The current production exercise is viewed as a transition between older ad hoc management and full use of Grid tools. There is some attempt at standardization of platforms and operating systems, with a common set of scripts for job submission and file transfer. The GDMP tool (which was reported on by A. Samar in the parallel sessions) is used for file replication. Issues needing improvement include:

- scripts are not sufficiently generic, but remain somewhat site specific and manpower intensive. Increased robustness would also help. More practice at this large scale is needed;

- information service needs a clear definition and implementation; and

- file replication management tools are just starting to appear and need careful evaluation.

All of these areas could benefit from general purpose Grid tools.

Finally, a case study was given of using Objectivity database files as targets for file replication management between centers. The study concluded that even the early GDMP tool contained the basic needed functionality for production file replication, but that an information service would allow easier synchronization of files and optimized data access during analysis.

Harvey Newman – Issues for Grids and Worldwide Computing

Expanding on his earlier parallel session talk on Data Grids for Next Generation Experiments, Harvey presented a vision of Grids as an enabling technology for future science. Among the broader issues motivating work on grids are:

- a required new level of intersite cooperation and resource sharing, including security and authentication across worldwide boundaries;

- developing methods for effective collaboration between physicists and computer scientists; and

- being ready to adapt to coming revolutions in network, collaborative and internet information technologies.

After highlighting some of the expected coming revolutions in information technology and citing some of the current gird projects, he went on to identify issues in worldwide computing needing immediate further work:

- integration of grid prototypes for end-to-end data transport;

- starting to integrate grid systems with experiment specific software frameworks;

- deriving strategies for data caching, query estimation, co-scheduling, load balancing and workload management among centers, including the use of modeling tools; and

- transparent interfaces for replica management.

Finally, after describing some current work in these areas, he went on to discuss moving beyond traditional architectures through the use of mobile agents: autonomous, goal driven adaptive software objects.

Fabrizio Gagliardi – EU Datagrid

Fab described the motivation and plans for the European Union (EU) Datagrid project. Motivation for the project includes:

- the perfect match between the Grid concept and the proposed LHC computing model;

- the critical need to address the coming shortage of locally available computing resources; and

- the need to coordinate ongoing national Grid research efforts in several EU countries (including the UK, Italy, France, the Netherlands, and others).

The main objectives include:

- middleware for fabric and grid management;

- large scale testbeds;

- production quality HEP demonstrations; and

- demonstrations in other sciences.

The project is funded and underway, with high energy physicists as active participants in many of the 12 work packages, including grid workload management, grid data management, grid monitoring services, fabric management, mass storage management, network services, integration testbed, and high energy physics demonstrations.

Paul Avery – The GriPhyN Project

Paul discussed the GriPhyN, an NSF funded R&D project to build the foundation for Petabyte scale virtual data grids. GriPhyN brings together physics from several scientific disciplines in a strong partnership with computer scientists to design and implement production scale grids with common infrastructure, tools and services based on existing foundations.

Motivations for datagrids include both physical and political reasons:

- optimizing data distribution and proximity;

- making efficient use of networks;

- providing for scalable growth;

- unifying all computing resources as part6 of the grid;

- leveraging use of local resources; and

- overcoming the problems of managing a single massive central facility.

Th GriPhyN research agenda includes:

- virtual data technologies;
- planning and scheduling;
- execution management;
- performance analysis; and
- a virtual data toolkit.

The initial R&D phase of the GriPhyN project is funded and work is underway.

DISCUSSION

Discussion focused around managing and coordinating the diverse grid R&D activities. Are all important areas of research being covered? Is there any mismatch between Europe and the US?

The management of the various projects is in good communication with many people participating in several of the projects. There are general grid forum activities in both the US and Europe, so it is felt that there is sufficient coordination of work. It is very useful to keep the experiments involved, and to provide real services that they can test out against real world problems.

Monique Werlen and Denis Perret-Gallix (IAC Chair)

Perspectives on the workshop series

Monique Werlen[a] and Denis Perret-Gallix[b]

[a] *LAPTH, BP 110, F-74941 Annecy-le-Vieux, France*
[b] *CNRS-JAPON, 3-9-25 Ebisu, Shibuya-ku, Tokyo 150, Japan*

Abstract. After a short history of the AIHENP workshop series started in 1990, we will sketch how, through the 7 meetings, we have been monitoring the computer revolution which is drastically transforming the way we perform basic research and, furthermore, how this series has been anticipating some of the major happenings. We will end up by proposing a critical review of the main goals of the workshop series and a short list of some current budding techniques and their possible applications to basic research.

SHORT HISTORY OF AIHENP

The ACAT'2000 workshop, 7th issue of the AIHENP series, marks the 10th anniversary of the series. The AIHENP acronym was meant to stand for Artificial Intelligence for High Energy and Nuclear Physics. But what does ACAT mean?: A CATapult to the future of Computing? A CATacombs for deadlocks? A CAThedral for the Holy GRID? ...No, simply, "Advanced Computing and Analysis Techniques in Physics Research". This new acronym puts more emphasis on the profound and essential evolution of the data analysis techniques.

Six workshops have been organized until this one hosted by Fermilab. The first one was held in Lyon (1) (March 1990), followed by La Londe-les-Maures (2) (January 1992), Oberammergau (3) (October 1993), Pisa (4) (April 1995), Lausanne (5) (September 1996) and Heraklion (6) (April 1999). Each had its own personality and uniqueness thanks to the deepest involvement of the local organizers.

The goal of the first workshop in the city hosting the French High Energy and Nuclear Physics computing center (CCPN) was to focus on technologies born from the conceptual methods developed in the field of Artificial Intelligence and to discuss their uses in high energy and nuclear physics research. "AI" was selected in a slightly provocative way to attract the attention on our willingness to confront real applications to the sixties smoking dreams. But the needs for more embedded autonomy and adaptability in the software were already a reality. With the beginning of the big LEP experiments, the implementation on on-line real-time and on off-line data analysis software of more intelligence was becoming quite inescapable.

Essentially three sensitive fields had been identified:

- Software engineering: In French: "Génie Logiciel" but "génie" also means genius which should go well along with artificial intelligence! More seriously, at this time, only simple text editors were available. They were providing no support for the large software developments needed for the on-line and the data analysis of those long term and complex experiments. For the first time, in our field, it was made clear that some sort of "intelligent" support was necessary for the software development of these multi platform, multi language, multi user, multi developer, multi country undertakings.

- AI algorithms: A few people were experimenting how basic research could benefit from the implementation of the most promising AI developments. Here the field was quite mined, due to the poor delivery of AI, but we will see how the focus of interest moved from the nineties to the new century and how, nowadays, some of those algorithms are central to fore front physics research.

- Symbolic algebraic Manipulation and automatic calculations: The computation of signal and background processes for LEP and the extensive need for multi-purpose event generators had started the development of automatic calculation based on algebraic languages or on diagrammatic numerical computation. This was the first meeting, world wide, of the different groups involved in this now burgeoning technique.

It was also quite clear, 10 years ago, that the more the computing technology was progressing, the more the plain physicists were at pain in this intrusive jungle with two serious problems that we haven't fully solved yet:

- A loss of manpower in the software development: Many physicists could not efficiently contribute to the software development as they did not have the basic knowledge of the new programming paradigm.
- A loss of physics vista on the computer programs: The computing experts were losing ground on the physics matters for being too much involved on the programming technicalities.

So this series was seen as get-together of experimentalists, theorists and computer scientists for hand to hand presentations and discussions expecting to soften the pain of some of us and bring perspectives to others.

Monitoring the computing revolution

Back in the seventies, high energy experiments were being performed without computers. The data taking was essentially performed by reading a set of scalers, counting particles crossing cleverly arranged detectors gated by triggers based on the timing coincidence of fast devices. The data analysis was done through the interpretation of histograms drawn by hand representing the scaler contents. Final results like cross-section were calculated by applying geometrical detector acceptance and dead-time correction factors.

No computers ... life was easier ... more time to think ... physicists were doing physics. Why did we have to introduce computers in this wonderland story? Because they were there ! or at least the first mini-computers were being assembled. The industry was looking for guinea-pigs to launch the largest ever revolution since the wheel (see the workshop series Logo). Is this actually true ? No, we do not think so, we think that Mother Nature was even more provoking in this happening.

Nature was requiring a push to higher energy interactions; although looked for, the intermediate boson, later on called Z, was not found at 29 GeV on a fixed target experiment! Higher energy collision studies could only be achieved by colliders and 4π detectors which could not anymore be registered by scalers. Furthermore, the higher the energy the more complex the final state, the triggering and detection system. This did drive a complete change in the event read-out and recording technique. Thanks to the newly available micro-electronics and computers technologies, the modern HEP could actually set in.

But, in passing, that the way, in a more general stand, basic research progresses: It attempts to approach the Truth by glancing in the unknown on top of the shoulders of the latest technological achievements.

Since 1990, when the first workshop of the series was organized, we have seen an evolution in the main topics of the workshop, a sliding of the main focus of the sessions.

In the software engineering, at first, we had descriptions of CASE systems and the technicalities of the software design, presentations of "new" languages (Ada, Prolog, Eiffel) and an introduction to interactive analysis.

Then, the focus moved on to the object-oriented paradigm and the starting of the web, then C++ became the center of most of the developments, finally parallel computing got a lot of attention due to the exploding CPU power needs and this year the GRID, the world-wide net-computation entered in the dance.

In AI, the emphasis was first put on the expert systems for the diagnosis of the hardware equipment, but rapidly the main buzz words became neural-networks and genetic algorithms. This session due to the overwhelming interest in neural-nets turned out to be the largest session of the workshop in terms of contributions and attendance. This year we were shown that the latest algorithms are now so powerful that even Higgs can be "detected" by neural-nets (7)!:)

In symbolic manipulation, we started with descriptions of symbolic manipulation languages and their ability to mimic mathematician activities. In addition, some pioneering work on automatic calculations were presented. Then, advanced calculations (1-loop) and physics results produced by automatic procedures were shown. Finally, the question of dealing with those lengthy computations raised the need for more advanced computer technologies like automatic formula optimization and parallelism. Other uses of symbolic languages were exemplified for computing super-algebra string oriented theories.

Anticipating the wave

It is always comforting for the organizers of a workshop to "discover", even if it went confidentially, that some major event did actually occur. For example in January 1992, in La Londe-les Maures, Tim Berners-Lee, associated with J.F. Groff, R. Cailliau and B. Pollermann, presented the " World-Wide Web: An information Infrastructure for High Energy Physics" (8) probably one of the very first presentations of the WEB in front of a wide audience (250 people).

Sometimes new ideas or technologies have a hard time imposing themselves. This has been the case for the automatic calculation of HEP processes. Although most of the symbolic manipulation language designers were high-energy physicists, it was not recognized by the rest of the community that those techniques although not "replacing" the theorists, could alleviate some of their calculation burden. At best this was considered as a learning tool for beginners. It has been therefore a big pleasure to acknowledge the participation in this series of those famous designers. G. Gonnet, A.C. Hearn, J. Vermasseren, S. Wolfram and even M. Veltman (9) (1999 Nobel prize winner) have in some way been backing up this activity. These workshops have contributed to the takeoff of the automatic computation systems which are now fully embedded in the event generators used at colliders.

As already mentioned, during these last 10 years, one has seen the rising importance of neural-net selection in data analysis and in on-line triggering. This workshop has served as a permanent forum for the development of these new techniques.

BACK TO THE FUTURE

What will be the future of this series? Well, what you, participants, will make of it. The Advisory committee will be keen at enforcing the basic goals which are the foundation of this series, but the content, the "substantifique moelle", you, the actors on the field, you will make it happen. We would like to hear from you any proposals or ideas to help us improve this series.

Panel sessions for in depth discussions

This is an efficient way to provoke discussions between people. We all did appreciate the panel sessions organized at ACAT'2000, regretting maybe that they were in parallel sessions. We intend to have more of those in the next workshop.

A catalysis of ideas and concepts

A single idea is rarely good by it-self, it finds its real application through its synergy with other concepts. Let's take the WEB story as an example. Everything started with the idea of creating hyper-link between documents on different computers, but the full power of the WEB came with the development of the graphic interface. We remember using the very first text version of the WEB at CERN, we were not convinced that this tricky way of typing numbers to refer to such and such keywords would ever initiate the so-called Internet Revolution.

What are the new ideas now popping up and about which we would like to have discussions in the future issues of this series? Well this is not easy to say and others in this workshop will address this question, but if we can add our pinch of salt, here is what we personally would like to track in the current developments.

- Virtual Reality: So much have been said about it in the simulation and game playground. But what can this new technology bring to us, actually? First of all this is really the next step in event scanning. The ability to manipulate a complex HEP event as we would with an object of "art", "going" to the interaction point and by Lilliputian steps inspecting all secondary vertices or looking in the particle trajectories using our personal, human, fuzzy logic to estimate, "by eyes", the momentum of the particle or to identify and correct wrong pattern matching, is probably a big plus. This technology can also be useful for the maintenance of the huge electronics and computer environment of the LHC experiments. Expert Systems have been of a great help already to send alarms and to indicate the series of actions to be performed. But imagine wearing those goggles and gauntlet and being guided directly by the computer, like a jet pilot to its target, to the right red button on the module 16 of the bay 15 in the rack 2587 of the room 12 to do the correct manual reset needed by the on-line to resume ... a plus no ? More applications include the help in the design of the detector by visual inspection, the simulation and the guidance for on site detector repairs .... But additional developments are probably needed to realize efficient and fully operational systems.

- Nanotechnology, molecular machines, nanobots are fashionable and strong expectations are put on this budding technology. Matter will become software, and vice versa (we would say). "So some day soon, we could download hardware from the Net just like we download software today." (10) The ability to build molecule by molecule (or atom by atom) electronic devices just as one writes on an magnetic disk by arranging magnetic grains would make possible the building of specific circuit dedicated to a very precise application. It would be possible to create "in hardware" the software algorithm most appropriate for a triggering signal and take the full benefit of the orders of magnitude faster execution time. Moreover this could be built on-line, on demand, when needed. The gap between software and hardware would vanish, as all software could turn into hardware in the running of the program. This give just a taste of what could be the next revolution, but needless to say that many workshops will go before those ideas will be transformed into down to earth applications.

Table 1. Comparison of performances for the salesman problem with 75 cities

algorithm	result	nb of iterations
Particle Swarm algo.	535	3480
Genetic algorithm	545	80000
Evolutionary programming	542	325000
Simulated Annealing	580	173250
Optimal solution	535	

Source: Dorigo and Gambardella, BioSystems 43,73 (12)

- Swarm intelligence, bio-genetic research, ...: Those optimization algorithms are based on the observation of the social behavior of insects (11). An ant colony selects the shortest path to a target thanks to the deposited pheromone. It is actually an example of collective problem solving as opposed to centralized control. Salesman problem algorithms based on this principle give quite good results (12) (see table 1). Traffic routing for Internet packet or transportation vehicles issues can also be addressed by this algorithm. But most interestingly, the particle swarm algorithm (13) can be used in place of the learning algorithms needed by the neural-nets and may evolve both the weights and the structure it-self of the net. Last, but not least those systems are well suited for parallel implementation.

- New Internet Protocol: New Internet protocols like IP v6 are being discussed, increasing dramatically the number of addresses. This would make possible for each important part of an experimental detector to have a different IP address and therefore to embed its own local intelligence wherever it would be located. This is an important aspect when building widely distributed intelligence for an electromechanical equipment.

Promoting new initiatives

The workshop is also a place for informing the participants of new important programs. The GRID project has been largely discussed this year, but other projects will be discussed in the future including Internet II, a high speed Internet dedicated to R&D labs and universities.

Establishing Standards

We would like to advocate here, again, the need to establish standards. Although standards are not always possible or even desirable for some developments, we have to do our best to reach minimum agreement on common standards every time it is possible, just not to re-invent the wheel (especially if you happen to make it squared). Look at the mess of maintaining the same C/C++ package on different computers with different compilers. The big successes in the software history are the cases where many people have contributed to develop, to enhance and to extend the package. This cooperative development has been made possible thanks to a clean definition of the basic standards. Standards may come 'de facto' after long test and trial or may be the output of committees and again they take time to be set. But one has to go through these painful steps to benefit later from the well designed basics. We have to learn to think globally, collectively and cooperatively from the start of the package design.

Open to all fields of research

Although this workshop series started with the application field built in in its acronym (HENP: High Energy and Nuclear Physics), we soon opened it to other fields of research, first in physics, astrophysics and then even more widely to space science, telecommunication, biology. This has never been really successful in the sense that the majority of the speakers and of the attendance always did come from HENP. It is therefore a pleasure to acknowledge this year the participation of astrophysicists and particle accelerator physicists. But in future issue, we would like to have it even more attractive to other research fields as long as they are facing similar problems or using similar approaches or because they bring to particle physics new view points. For example:

- the need for large computing power requiring the use of supercomputers or large scale farm-like systems is found on many problems: in molecular modeling, bio-computing, 3D animation or Internet queries.

- the handling of huge amounts of data like for the LHC experiments is also critical for the human genome or the Internet storage.

- the security and reliability of the computer systems are key points in all systems.

Different research fields like meteorology, fluid dynamics or bio-computing may use similar algorithms for the simulation of processes, the parameter optimization or the multi-dimensional integration. Moreover, we in HEP can learn from discoveries made in other domains like in the brain research or, as we have already seen, from the study of the social behavior of ant colonies.

Permanent link

Finally, we want this workshop to be a permanent link between developers, users and the industry through:

- email, newsgroups and a decent WEB page (this will be done in the coming month $http://www.lapp.in3p3.fr/aihep/$)
- Intermediate video or net conferences
- Tutorials
- Products show-room and demos

ACKNOWLEDGMENTS

The ACAT'2000 has been without contest a really enjoyable and exciting meeting where many new ideas have been discussed and, we hope, new collaborations have been formed. On behalf of the International Advisory Committee, we want to express our deepest thank to Pushpalatha Bhat, Matthias Kasemann and to all the members of the local committee for this smooth organization.

Several proposals have been put forward for the organization of the next workshop. We are happy to announce that the international advisory committee has enthusiastically accepted the Moscow application. We wish our Russian colleagues the same success in 2002.

REFERENCES

1. *New Computing Techniques in Physics Research. Proc. 1st International Workshop on Software Engineering, Artificial Intelligence and Expert Systems in High Energy and Nuclear Physics Lyon (March 19-24 1990)*, ed. by D. Perret-Gallix and W. Wojcik, CNRS, Paris 1990.
2. *New Computing Techniques in Physics Research II. Proc. 2nd Int. Workshop on Software Engineering, Artificial Intelligence and Expert Systems in High Energy and Nuclear Physics, La Londe-les-Maures (Jan. 13-18 1992)*, ed. by D. Perret-Gallix, World Scientific, Singapore 1992.
3. *New Computing Techniques in Physics Research III. Proc. 3rd Int. Workshop on Software Engineering, Artificial Intelligence and Expert Systems in High Energy and Nuclear Physics, Oberammergau, Bayern, Germany (Oct. 4-8 1993)*, ed. by K.H. Becks and D. Perret-Gallix, World Scientific, Singapore 1993.
4. *New Computing Techniques in Physics Research IV. Proc. 4th Int. Workshop on Software Engineering, Artificial Intelligence and Expert Systems in High Energy and Nuclear Physics, Pisa, Italy (Apr. 3-8 1995)*, ed. by B. Denby and D. Perret-Gallix, World Scientific, Singapore 1995.
5. *New Computing Techniques in Physics Research V. Proc. 5th International Workshop - AIHENP'96 Lausanne (Sep. 2-6 1996), Nucl. Inst. Meth.* **A389 (1-2)** (1997), ed. by M. Werlen and D. Perret-Gallix.
6. *New Computing Techniques in Physics Research VI. 6th Int. Workshop on Software Engineering, Artificial Intelligence and Expert Systems*, ed. by G. Athanasiu and D. Perret-Gallix, Parisianou S.A., Greece (2000).
7. ALEPH Collaboration, "Observation of an excess in the search for the standard model Higgs boson at ALEPH", CERN-EP/2000-138, *Phys. Lett.* **B** in press. Bhat P.C., Gilmartin R. and Prosper H.B., "Strategy for discovering Higgs at the Tevatron", *Phys. Rev.* **D62**, 074022 (2000).
8. Groff J.-F., Berners-Lee T.J., Cailliau R., Pollermann B., "World-Wide-Web: An Information Infrastructure for High-Energy Physics", in (2), pp. 157-164.
9. Veltman M., "Symbol Manipulation: Early History and Present Prospects", in (3), pp. 451-458.
10. Ellenbogen, J.C., "Matter as Software", presented at the *Software Engineering and Economics Conference, McLean, 2-3 April 1997*; Gerd Binnig interviewed by Otis Port in Business week on-line 99_35.
11. Bonabeau E., Dorigo M. and Theraulaz G., *Nature* **406**, 39 (2000).
12. Dorigo M. and Gambardella L.M., *BioSystems* **43**, 73 (1997).
13. Kennedy J. and Eberhart R., "Particle Swarm Optimization", in *Proc. of the 1995 IEEE Int. Conf. on Neural Networks, Perth* IV pp. 1942-1948.

Technology Show

Participating Companies

Cisco Systems
Silicon Graphics
Kuck & Associates
Objectivity
Platform Computing
Waterloo Maple
Wolfram Research

More of ACAT 2000 in Pictures

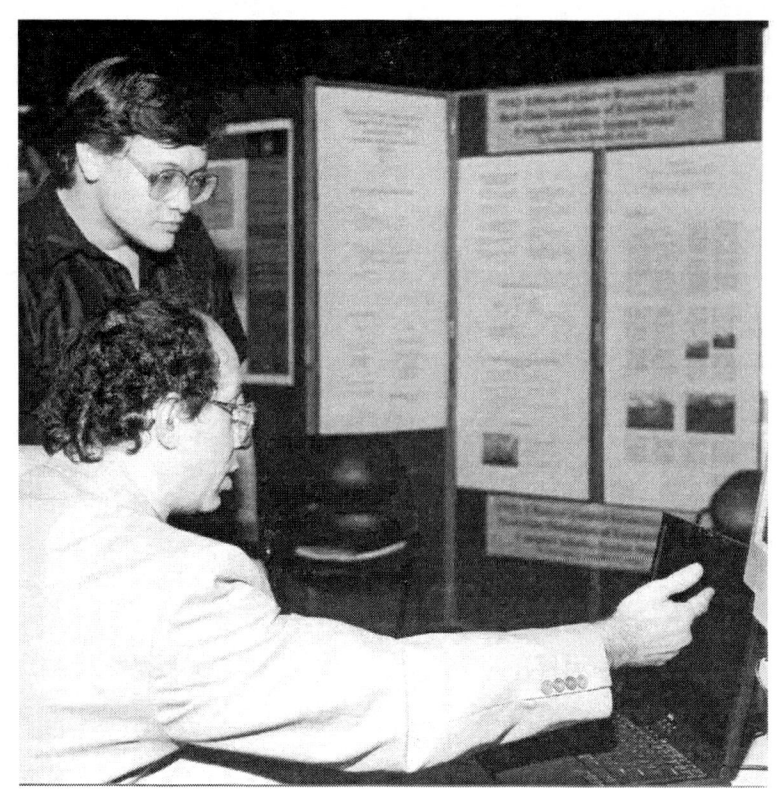

Workshop Program

Monday October 16, 2000
08:00 Registration

Plenary Session, Ramsey Auditorium - **Chair: Matthias Kasemann**
09:00 *Welcome*
    Michael S. Witherell (Director, Fermilab)
09:10 *Workshop Organization*
    Pushpa Bhat (Fermilab)
09:20 *Key Note Address - Information Technology: Transforming our Society and our Lives*
    Ruzena Bajcsy (NSF)
10:00 *Knowledge Discovery through Machine Learning*
    John Moody (Oregon Graduate Institute)

10:45 Coffee Break

Plenary Session, Ramsey Auditorium - **Chair: Joe Lykken**
11:15 *Automatic Program Generation from a computer Algebra System: Efficiency and Correctness*
    Gaston Gonnet (ETHZ)
12:05 *Modern Calculation of higher order QFT perturbation Theory*
    Dmitri Shirkov, Slava Ilyin (Moscow State Univ.)

12:35 Lunch Break

Parallel Sessions 14:00 – 15:30
Innovative Software Algorithms & Tools - Session I, Chair: Leo Michelotti
Online Monitoring and Controls

  ISAT101 *Tailorable Software Architectures in Accelerator Physics Research*
        I. Mejuev, A. Kumagai, E. Kadokura
  ISAT102 *Online Modeling of Accelerators and Beamlines at Fermilab*
        E. McCrory, L. Michelotti, J. Ostiguy
  ISAT103 *A Software Tool for the Online Analysis and Monitoring of the HADES Spectrometer*
        P. Finocchiaro, D. Vasiliev
  ISAT104 *A Beamline Matching Application based on Open Source Software*
        J. Ostiguy

Very Large Scale Computing – Session I, Chair: Edward Boos
Monte Carlo and Simulation Tools

  VLSC101 *CompHEP-PYTHIA interface: integrated package for the collision events generation based on exact matrix elements*
        V.A. Ilyin, A.E. Pukhov, A.N. Skachkova
  VLSC103 *Integration of GRACE and PYTHIA*
        K. Sato, S. Tsuno, J. Fujimoto, et. al.
  VLSC104 *Parton shower method for QED radiative correction*
        Y. Kurihara, J. Kurihara, T. Ishikawa, et. al
  VLSC105 *Algorithm for Computing Excited States in Quantum Theory*
        X. Luo, H. Jirari, H. Kroger, et. al.
  VLSC106 *Simple Scaling for Faster Tracking Simulation in Accelerator Multiparticle Dynamics*
        J. MacLachlan

VLSC107 *Adaptive Mesh Simulations of Astrophysical Detonations Using the ASCI Flash Code*
B. Fryxell, M. Zingale, F. Timmes, et. al.

Symbolic Problem Solving - Session I, Chair: Jerome Fleischer
Feynman Diagram Algorithms and Tools

SPS101 *HELAC: a package to compute electroweak helicity amplitudes*
A. Kanaki, C.G. Papadopoulos
SPS102 *O'Mega: Compact Factorized Representations of Large Sets of Feynman Tree Diagrams*
T. Ohl
SPS103 *A Feynman-graph selection tool in GRACE system*
F. Yuasa, T. Kaneko, T. Ishikawa
SPS104 *Automatic Construct Lagrangian and Deduce Feynman Rules for Supersymmetry Model*
J. Wang

15:30 Coffee Break

Panel Discussions 16:00 – 18:00
Use of C++ in Scientific Computing – Session I, Conveners: W. Brown, M. Fischler

Opening Remarks and Questions

Experiences Reviewing Scientific C++ Code
Marc Paterno
Reflections on a Decade of Object-Oriented Programming in Accelerator Physics
Leo Michelotti
C++ in Scientific Applications: A Case Study in Genericity
Walter Brown
Speaking C++ As a Native
Bjarne Stroustrup

Advanced Analysis Environments - Conveners: R. Brun, T. Johnson

Future of Analysis Environments: Personal Views
Rene Brun
The Role of JAVA
Tony Johnson
Analysis Environment Challenges
Lassi Tuura

Large Scale Simulations - Session I Conveners: R. Raja, R. Rosner

CDF Monte Carlo- Status, performance and Future Plans
Pasha Murat
Astrophysics simulations--Large Scale fluid simulations
Bruce Fryxell
D0 Monte Carlo- Status, performance and future plans
Gregory Graham
Astrophysics Simulations--N body /fluid simulations of cluster formation
Paul Ricker
CMS Monte Carlo- Status, performance and future plans
Hans Wenzel
Discussion

18:00 **Welcome Reception** Wilson Hall Atrium

Poster Session 18:00

P101	*EMS: a Framework for Data Acquisition and Analysis*	
	J. Nogiec, J. Sim, K. Trombly-Freytag, et. al.	
P102	*Effects of Limited Resources in 3D Real-Time Simulation of Extended Echo Complex Adaptive System Model*	
	D. Dominiak, F. Rinaldo, M. Evans	
P103	*High Performance visual display for HENP detectors*	
	M. McGuigan, G. Smithe, J. Spiletic	
P104	*ROOT OO model to render multi-level 3-D geometrical objects via an OpenGL layer*	
	V. Fine, R. Brun, F. Rademakers	
P105	*StdHepC++ and the Particle Data Table*	
	L. Garren	
P106	*Parallel Computer on a PC Cluster*	
	X. Luo, E. Gregory, J.C. Yang, et. al.	
P107	*Some advance methods of statistical analysis for the muon g-2 experiment at BNL*	
	S.I. Redin	
P108	*Analytical results for finite mass baryon correlators in next-to-leading order*	
	S. Groote, J.G. Korner, A.A. Pivovarov, et. al. presented by R. Kreckel	
P109	*Summaries of Recent Computer-assisted Feynman Diagram Calculations*	
	Compiled by M. Fischler	

Tuesday October 17, 2000

Plenary Session, Ramsey Auditorium - **Chair: Harrison Prosper**

09:00 Speaking C++ as a Native
       Bjarne Stroustrup (AT&T Research Labs)
10:10 Computation in Lattice QCD
       Paul Mackenzie (Fermilab)
10:45 Coffee Break

Parallel Sessions 11:15 - 12:45
Artificial Intelligence - Session I, Chair: Jose Seixas
Online Applications of Neural Networks

AI101	*Momentum reconstruction of particles in the forward muon trigger system of the ATLAS detector*	
	G. Dror, H. Abramowicz, E. Etzion, et. al.	
AI102	*The neural network first level trigger for the DIRAC experiment*	
	S. Vlachos	
AI103	*Intelligent preprocessing for Neural Network in the H1 experiment*	
	JC.Prévotet, B.Denby, P.Garda, et. al.	
AI104	*An electronic system for simulation of neural networks with a micro-second real time constraint*	
	E. Chorti, B. Granado, B. Denby, et. al.	
AI105	*A neural network trigger for heavy quark studies*	
	V. Papavassiliou	

Innovative Software Algorithms & Tools - Session II, Chair: Heidi Schellman
Physics Analysis and Reconstruction Algorithms

ISAT201	*The FreeHEP and HEP Libraries for Java*
	T. Johnson, M. Donszelmann, C. Loomis, et. al.
ISAT202	*Stable Algorithm for Extraction of Asymmetries - from the Data on Polarized*

          Lepton-Nucleon Scattering
          G. Nikolai
ISAT203 *Sleuth: A Quasi-Model-Independent Search Strategy for New Physics*
          B. Knuteson, D. Toback
ISAT204 *Faster tracking in hadron collider experiments*
          N. Konstantinidis, H. Drevermann
ISAT205 *More performance and implementation of an OO track reconstruction model in different OO frameworks*
          I. Gaines, S. Qian
ISAT305 *Hermite Polynomials in problem of searching for global Maximum*
          Y. Yatsunenko

Symbolic Problem Solving - Session II, Chair: V. Ilyin
Symbolic Manipulation via Function Objects

SPS201 *Algorithms in Computer Algebra for Feynman diagrams calculation*
          A. Kryukov
SPS202 *A Class Library of Function Objects*
          J. Boudreau, M. Fischler, P. Maksimovic
SPS203 *FunctionalObjects.h: Using Symbolic Syntax in C++ Programs*
          R. Nolty
SPS204 *Large scale symbolic computing with GiNaC*
          C. Bauer, A. Frink, R. Kreckel

12:45 Lunch Break

Plenary Session, Ramsey Auditorium - **Chair: Ruth Pordes**
14:00 *High Performance Networks*
       Bob Aiken (Cisco Systems)
14:35 *General Trends in Computing*
       Rex Tanakit (Silicon Graphics)
15:10 *NMR Quantum Computation*
       Yael Maguire (MIT)

15:30 **Tech Show & Reception**

Panel Discussions   16:00 – 18:00
Use of C++ in Scientific Computing - Session II, Conveners: W. Brown, M. Fischler

      Follow-up discussions with the panel

Wednesday October 18, 2000

Plenary Session,   Ramsey Auditorium - **Chair: Al Goshaw**
09:00 *Advanced Analysis Techniques*
       Pushpa Bhat (Fermilab)
09:30 *Statistical Techniques in HEP*
       Louis Lyons (Oxford University)
10:00 *The H1 Neural Network Trigger Project*
       Chris Kiesling (Max Planck Institute)

10:30   Coffee Break

Plenary Session,   Ramsey Auditorium - **Chair: John Womersley**
11:00 *Very Large Scale Computing in Accelerator Physics*

        Robert Ryne (Los Alamos Lab)
11:30   *Feynman Diagram computation in electroweak theory*
        Georg Weiglein (CERN)

12:30   Lunch Break

Parallel Sessions   13:30 - 15:15

Artificial Intelligence - Session II, Chair: Sotrios Vlachos
Applications in Data Analysis I

AI201	*Selection of W-Pair-Production in DELPHI with Feed-Forward Neural Networks* K. Becks, P. Buschmann, J. Drees, et. al.
AI202	*Search for Electroweak Single Top Quark Production at DZero Using Neural Networks* L. Dudko
AI203	*An Hybrid Training Method for Neural Energy Estimation in Calorimetry* P.V.M. Da Silva, J.M. Seixas, J. Seixas
AI204	*Principal Component Analysis for Neural Electron/Jet Discrimination on Highly Segmented Calorimeters* M.R. Vassali, J.M. Seixas

Innovative Software Algorithms & Tools - Session III, Chair: Wolfgang Rolke
Pattern Recognition Techniques

ISAT301	*Simultaneous Tracking and Vertexing with Elastic Templates* A. Haas
ISAT302	*Vertex reconstruction before tracking in magnetic field* Y. Yatsunenko
ISAT303	*Singular Value Decomposition (SVD) to simplify features recognition analysis in very large collection of images* F. Guillon, D.J.C. Murray, P. DesAutels
ISAT304	*Evolutionary and Genetic Algorithms in Computer Vision* D. Li
ISAT305	*Hermite Polynomials in problem of searching for global Maximum* Y. Yatsunenko

Very Large Scale Computing - Session II, Chair: Steve Wolbers
Analysis Farms and DAQ systems

VLSC201	*Introduction to Matrix Distributed Processing 1.0* M. Di Pierro
VLSC202	*The SDSS-TAMS Experience with GFS* J. Annis, K. Ruthsmandorfer, C. Stoughton
VLSC203	*A Large Linux-PC Farm for Online Event Reconstruction at HERA-B* A. Gellrich, H. Leich, U. Schwanke, et. al.
VLSC204	*Client and Event Driven Data Hub System at CDF* B. Kilminster, T. Vaiciulis, K. McFarland, et. al.
VLSC205	*Results From the Terascale Supernovae Collaboration* D. Swesty
VLSC206	*The COMPASS Computing Farm* M. Lamanna

Symbolic Problem Solving - Session III, Chair: Monique Werlen
Symbolic Techniques for Feynman Diagrams

SPS301	*Optimization of symbolic evaluation of helicity amplitudes*

	V.A. Ilyin, A.E. Pukhov, P. Cherzor
SPS302	*A Feynman Diagram Analyser DIANA*
	M. Tentyukov, J. Fleischer
SPS303	*Calculation of two-loop selfenergies in the Standard Model*
	J. Fleischer, O.V. Tarasov
SPS304	*The tensor reduction and master integrals of the two-loop massless on-shell crossed box*
	C. Oleari

15:15 Coffee Break

15:45 *Special Fermilab Colloquium*
  Science and Mathematica: Some Personal Perspectives
  Stephen Wolfram (Wolfram Research)

18:30 Banquet @ The Mid-America Club, Chicago -
  After-dinner Speaker: Edward Kolb (Fermilab & U. Chicago)

Thursday October 19, 2000
Plenary Session, Ramsey Auditorium - **Chair: Irwin Gaines**
09:00 *Grid Computing*
  Ian Foster (Argonne National Lab)
09:45 *Very Large Scale Computing in Astrophysics*
  Alex Szalay (Johns Hopkins Univ.)
10:15 *Large Scale Molecular Dynamics Simulations of Materials on Parallel Computers*
  Aiichiro Nakano (Louisiana State Univ.)

10:45 Coffee Break

Parallel Sessions   11:15 – 12:45

Artificial Intelligence - Session III, Chair: Bruce Denby
Applications in Data Analysis II

AI301	*Particle Identification Using Neural Networks*	
	D. Chakraborty	
AI302	*Vertex reconstruction of ep interactions with the ZEUS Central Tracking Detector*	
	G. Dror, H. Abramowicz, D. Horn	
AI303	*Measuring the Higgs Boson Mass using Neural Networks*	
	S. Tentindo-Repond, P. Bhat, H. Prosper	
AI304	*Top Quark Mass Measurements Using Neural Networks*	
	S. Beri, P. Bhat, R. Kaur,, et. al.	

Very Large Scale Computing - Session III, Chair: Harvey Newman
Grid Architectures

VLSC301	*A DataGrid Prototype for Distributed Data Production in CMS*
	M. Hafeez, A. Samar, H. Stockinger
VLSC302	*Object level physics data replication in the Grid*
	K. Holtman
VLSC303	*SAM for D0 - a fully distributed data access system*
	V. White, I. Terekhov
VLSC304	*A Dynamic State Machine for the Atlas Software Framework*
	P. Calafiura
VLSC305	*The StoreGate, a Data Model for the Atlas Software Architecture*
	S. Rajagopalan, Atlas Data Model Group

VLSC306 *The Particle Physics Data Grid Project*
M. Livny

Innovative Software Algorithms & Tools - Session IV, Chair: Rene Brun
Common Libraries

ISAT401 *Event Bookkeeping for CLEO-3*
J. Thaler
ISAT402 *KID - KLOE Integrated Dataflow*
I. Sfiligoi
ISAT403 *A Component Based Architecture to Support Scientific Workflow Management.*
N. Baker, P. Brooks, Z. Kovacs
ISAT404 *Brain Computer Interface*
E.J.X. Costa, E.F. Cabral

12:45 Lunch Break

Parallel Sessions   14:00 – 15:30

Artificial Intelligence - Session IV, Chair: Chris Kiesling
Theoretical Aspects and Other Topics

AI401 *Clopper-Pearson Bounds for Neural Networks*
B. Berg
AI402 *Ring recognition method based on the elastic neural net*
I. Kisel, S. Gorbunov, E. Konotopskaya, et. al.
AI403 *Experimenting with rule induction algorithms in HEP data analysis*
N. Stepanov
AI404 *A multivariate discrimination technique based on range searching*
T. Carli, B. Koblitz
AI405 *Using Ensembles of Neural Networks in HEP Analyses*
C. Bhat, P. Bhat
AI406 *A Dynamic Self-Organising Job Scheduling System*
H. Newman, I.C. Legrand

Innovative Software Algorithms & Tools - Session V, Chair: Luciano Barone
Grid and Distributed Computing Techniques

ISAT501 *Worldwide Distributed Analysis and Data Grids for Next-Generation Physics Experiments*
H. Newman
ISAT502 *Using CORBA for remote participation in large physics experiments*
B.U. Niderost, A.A. Gerritsen, W. Lourens, et. al.
ISAT503 *Protocols and Services for Distributed Data-Intensive Science*
B. Allcock, A. Chervenak, I. Foster, et. al.
ISAT504 *Condor-G: A Computation Management Agent for Multi-Institutional Grids*
I. Foster, M. Livny, T. Tanenbaum, et. al.
ISAT505 *Simulating Distributed Systems*
I. Legrand

Symbolic Problem Solving - Session IV – Chair: Mark Fischler
Multiloop Calculations and Results

SPS401 *Gauge invariant classes of Feynman diagrams and applications for calculations*
E. Boos
SPS402 *A Parallel Version of the Symbolic Manipulation Program FORM*
D. Fliegner, A. Retery, J.A.M. Vermaseren

SPS403 *Calculations in the MSSM with FeynArts*
         S. Heinemeyer
SPS404 *Virtual corrections to Higgs production in gluon-gluon fusion up to order α_s^4*
         R. Harlander

15:30 Coffee Break

Panel Discussions   16:00 - 18:00
World-wide Computing - Conveners: P. Avery, I. Foster

> *Management of Large Scale Data Production for the CMS Experiment*
> Luciano Barone
> *World Wide Computing*
> Harvey Newman
> *World Wide Computing*
> Miron Livny

Large Scale Simulations II – Conveners: R. Raja, R. Rosner

(Chair: Robert Rosner, University of Chicago)

> *Astrophysics simulations- large-scale cluster surveys*
> Jim Annis
> *Very large databases*
> Alex Szalay
> *Accelerator Physics Simulations- Muon cooling, transport etc*
> Daniel Elvira
> *Atlas Monte Carlo- Status, performance and future plans*
> Fred Luehring
> *Discussion*

Computer-Assisted High-Order Calculations – Convener: T. Ohl

Friday October 20, 2000
Plenary Session, Ramsey Auditorium -- **Chair: Pushpa Bhat**
Rapporteur Talks
08:45 *Artificial Intelligence - Offline Applications*
        Harrison Prosper (Florida State Univ.)
09:00 *Status of Online Neural Networks*
        Bruce Denby (Univ. of Versailles)
09:15 *Symbolic Problem Solving*
        Jos Vermaseren (NIKHEF)
09:45 *Innovative Software Algorithms & Tools*
        Irwin Gaines (Fermilab)
10:15 *Very Large Scale Computing   & Panel on World-Wide Computing*
        Paul Avery (Univ. of Florida)

10:45 Coffee Break

Plenary Session, Ramsey Auditorium -- **Chair: Steve Holmes**
Rapporteur Talks (Cont'd)
11:15 *Perspectives on the Workshop Series*
        Monique Werlen (LAP-TH)
11:45 *Use of C++ in Scientific Computing*
        Jim Kowalkowski (Fermilab)

12:05 *Advanced Analysis Environments*
   Suzanne Panacek (CERN)
12:25 *Large Scale Simulations I*
   Bob Rosner
12:40 *Large Scale Simulations II*
   Rajendran Raja
12:55 *Vote of Thanks*

WORKSHOP PARTICIPANTS

NAME	AFFILIATION	EMAIL
Agodi, Attilio	University of Catania	attilio.agodi@ct.infn.it
Allcock, Bill	Argonne National Laboratory	allcock@mcs.anl.gov
Amendolia, Salvator Roberto	U. of Sassari and INFN Pisa	amendolia@pi.infn.it
Amundson, James	Fermilab	amundson@fnal.gov
Andronico, Giuseppe	University of Catania & INFN Sez. CT.	giuseppe.andronico@ct.infn.it
Annis, James	Fermilab	annis@fnal.gov
Avery, Paul	Univeristy of Florida	avery@phys.ufl.edu
Azemoon, Tofigh	CERN	Tofigh.Azemoo@cern.ch
Baczy, Ruzena	National Science Foundation	rbajcsy@nsf.gov
Baker, Nigel	UWE Bristol	bakern@ecid.cig.mot.com
Barone, Luciano	INFN and University of Rome	luciano.barone@roma1.infn.it
Berg, Bernd	Florida State University	berg@hep.fsu.edu
Beri, Suman B.	Panjab University	sumanberi@yahoo.com
Bhat, Chandrashekhara	Fermilab	cbhat@fnal.gov
Bhat, Pushpalatha	Fermilab	pushpa@fnal.gov
Boehnlein, Amber	Fermilab	cope@fnal.gov
Boos, Edward	Moscow State University	boos@theory.npi.msu.su
Boudreau, Joseph	Fermilab	boudreau@fnal.gov
Brown, Walter	Fermilab	wb@fnal.gov
Brun, Rene	CERN	Rene.Brun@cern.ch
Buckley-Geer, Elizabeth	Fermilab	buckley@fnal.gov
Calafiura, Paolo	Berkeley National Laboratory	pcalafiura@lbl.gov
Canal, Philippe	Fermilab	pcanal@fnal.gov
Chakraborty, Dhiman	Fermilab	dhiman@fnal.gov
Collins, John	University Hamburg/ DESY	collins@phys.psu.edu
Connolly, Brian	Fermilab	connolly@fnal.gov
Costa, Ernane Jose X.	Universidade de Sao Paulo	ernane@lcs.poli.usp.br
Cranshaw, Jack	Fermilab	cranshaw@fnal.gov
Denby, Bruce	Universite De Versailles	Denby@Ieee.org
Denisov, Dmitri	Fermilab	denisovd@fnal.gov
Di Pierro, Massimo	Fermilab	mdp@fnal.gov
Djurabekov, Ulugbek	Applied Laser Physics Institute	dhltas@online.ru,usd@mp.silk.org
Dominiak, Dana	Illinois Institute of Technology	dominiak@webfootgames.com
Dubois-Felsmann, Gregory	California Institute of Technology	gpdf@caltech.edu
Dudko, Lev	Moscow State University	dudko@fnal.gov
Eisert, David	Univeristy of Wisconsin - Madison	deisert@src.wisc.edu
Etzion, Erez	Tel-Aviv University	erez@lep.tau.ac.il
Filho, Sebastiao	Universidade de Sao Paulo	biscuit@brfree.com.br
Fine, Valeri	Brookhaven National Laboratory	fine@bnl.gov
Firestone, Alexander	National Science Foundation	afiresto@nsf.gov
Fischler, Mark	Fermilab	mf@fnal.gov
Fisyak, Yuri	Brookhaven National Laboratory	fisyak@bnl.gov
Fleischer, Jochem	University of Bielefeld	fleischer@physik.uni-bielefeld.de
Foster, Ian	Argonne National Laboratory	itf@mcs.anl.gov
Fryxell, Bruce	University of Chicago	fryxell@flash.uchicago.edu
Gaines, Irwin	Fermilab	gaines@fnal.gov
Gallagher, Hugh	University of Minnesota	gallag@hep.umn.edu

Garda, Patrick	University of Paris	patrick.garda@lis.jussieu.fr
Garren, Lynn	Fermilab	garren@fnal.gov
Gellrich, Andreas	Humboldt-University, Berlin	Andreas.Gellrich@desy.de
Gerhards, Ralf	University of Hamburg/DESY	Ralf.Gerhards@desy.de
Gonnet, Gaston H.	Institute for Scientific Computing	gonnet@inf.ethz.ch
Goradia, Shantilal	University of Notre Dame	shantilalg@juno.com
Goshaw, Alfred	Fermilab & Duke University	goshaw@fnal.gov
Granado, Bertrand	University of Paris	Bertrand.Granado@lis.jussieu.fr
Guillon, Francis	Source Works Consulting Inc.	guillon@sourceworks.com
Gupta, Ambreesh	University of Chicago	agupta@hep.uchicago.edu
Haas, Andrew	University of Washington	haas@yahoo.com
Hagopian, Sharon	Florida State University	hagopian@hep.fsu.edu
Harrison, Prosper	Florida State Univeristy	harry@fnal.gov
Heikkinen, Aatos	Helsinki Institute of Physics	miheikki@pcu.helsinki.fi
Heinemeyer, Sven	Brookhaven National Laboratory	Sven.Heinemeyer@desy.de
Holtman, Koen	California Institute of Technology	koen@hep.caltech.edu
Huang, Yimei	University of Michigan	yimeih@umich.edu
Ilyin, Viacheslav	Moscow State University	ilyin@theory.npi.msu.su
Jain, Supriya	Tata Institute (TIFR)	sjain@iris.hecr.tifr.res.in
Jerzy, Nogiec	Fermilab	nogiec@fnal.gov
Johnson, Tony	Stanford Linear Accelerator Laboratory	tony_johnson@slac.stanford.edu
Jones, Christopher	Cornell University	cdj@mail.lns.cornell.edu
Kasemann, Matthias	Fermilab	kasemann@fnal.gov
Kiesling, Christian	Max-Planck-Institute for Physics	cmk@mppmu.mpg.de
Kilminster, Ben	Fermilab	bjk@fnal.gov
Knuteson, Bruce	University of California at Berkeley	knuteson@fnal.gov
Koblitz, Birger	MPI, University of Hamburg	koblitz@mail.desy.de
Kolb, Edward	Fermilab and University of Chicago	rocky@fnal.gov
Konstantinidis, Nikolaos	CERN	n.konstantinidis@cern.ch
Kostin, Mikhail	Cornell University	kostin@mail.lns.cornell.edu
Kostritski, Alexander	Institute for High Energy Physics	kostritsky@fnal.gov
Kovalev, Andrew	Fermilab	kovalev@fnal.gov
Kowalkowski, Jim	Fermilab	jbk@fnal.gov
Kreckel, Richard	Johannes Gutenberg University	richard.kreckel@uni-mainz.de
Kryukov, Alexander	Moscow State University	kryukov@theory.npi.msu.su
Kurihara, Yoshimasa	Joint Institute for Nuclear Research	kurihara@suchi.kek.jp
Lamanna, Massimo	CERN	Massimo.Lamanna@cern.ch
Lediaev, Laura	Fresno State University	lediaev@fnal.gov
Legrand, Iosif	California Institute of Technology	Iosif.Legrand@cern.ch
Leyderman, Elena	University of Puerto Rico	elena@hpcf.upr.edu
Linde, Timur	University of Chicago	t-linde@uchicago.edu
Litvine, Vladimir	California Institute of Technology	litvine@fnal.gov
Livny, Miron	University of Wisconsin	miron@cs.wisc.edu
Lopez, Angel	University of Puerto Rico at Mayaguez	angel@fnal.gov
Lueking, Lee	Fermilab	lueking@fnal.gov
Luo, Xiang-Qian	Zhongshan University	stslxq@zsu.edu.cn
Lykken, Joseph	Fermilab	lykken@fnal.gov
Lyons, Louis	Oxford University, Nuclear Physics Lab	llyons1@physics.ox.ac.uk
Mackenzie, Paul	Fermilab	mackenzie@fnal.gov
MacLachlan, James	Fermilab	maclachlan@fnal.gov
Maguire, Yael	Massachussetts Institute of Technology	yael@media.mit.edu
Mahmood, Akhtar	University of Texas - Pan American	mahmooda@panam.edu

Malzacher, Peter	GSI	P.Malzacher@gsi.de
Mans, Jeremiah	Princeton University	jmmans@princeton.edu
May, Edward	Argonne National Laboratory	may@anl.gov
Mazzanti, Paolo	INFN, Bologna	mazzanti@fnal.gov
McBride, Patricia	Fermilab	mcbride@fnal.gov
McCrory, Elliott	Fermilab	mccrory@fnal.gov
Mejuev, Igor	PFU Limited	mejuev@almond.kek.jp
Merritt, Wyatt	Fermilab	wyatt@fnal.gov
Michelotti, Leo	Fermilab	michelotti@fnal.gov
Moody, John	Oregon Graduate Institute	moody@cse.ogi.edu
Mueller, Uwe	Bergische Universitaet / GH Wuppertal	mueller@whep.uni-wuppertal.de
Murat, Pasha	Fermilab	murat@fnal.gov
Nakano, Aiichiro	Louisiana State University	nakano@bit.csc.lsu.edu
Nevski, Pavel	Brookhaven National Laboratory	nevski@bnl.gov
Newman, Harvey	California Institute of Technology	Harvey.Newman@cern.ch
Niderost, Beat	Utrecht University	B.U.Niderost@phys.uu.nl
Nolty, Robert	California Institute of Technology	nolty_r@caltech.edu
Odaka, Shigeru	KEK	shigeru.odaka@kek.jp
Ohl, Thorsten	Darmstadt Tech University	ohl@hep.tu.darmstadt.de
Oleari, Carlo	University of Wisconsin	oleari@pheno.physics.wisc.edu
Olson, Douglas	Lawrence Berkeley National Laboratory	dlolson@lbl.gov
Onofri, Enrico	University of Parma and INFN	enrico.onofri@unipr.it
Ostiguy, Jean-Francois	Fermilab	ostiguy@fnal.gov
Panacek, Suzanne	Fermilab	spanacek@fnal.gov
Papadopoulos, Costas	Institute of Nuclear Physics	costas.papadopoulos@cern.ch
Paterno, Marc	Fermilab	paterno@fnal.gov
Perevoztchikov, Victor	Brookhaven National Laboratory	perev@bnl.gov
Pordes, Ruth	Fermilab	ruth@fnal.gov
Prevotet, Jean-Christophe	Universite Pierre et Marie Curie	prevotet@lis.jussieu.fr
Qian, Sijin	Brookhaven National Laboratory	Sijin.Qian@cern.ch
Rajagopalan, Srini	Brookhaven National Laboratory	srinir@bnl.gov
Rajaram, Durga	Illinois Institute of Technology	durga@fnal.gov
Rajendran, Raja	Fermilab	raja@fnal.gov
Redin, Sergei	Brookhaven National Laboratory	redin@bnl.gov
Ricker, Paul	University of Chicago	ricker@flash.uchicago.edu
Rosner, Robert	University of Chicago	rrosner@flash.uchicago.edu
Rolke, Wolfgang	University of Puerto Rico at Mayaguez	w_rolke@rumac.upr.clu.edu
Rubin, Howard	Illinois Institute of Technology	rubin@iit.edu
Ryne, Robert	Los Alamos National Laboratory	ryne@lanl.gov
Sallach, David	University of Chicago	sallach@uchicago.edu
Samar, Asad	CERN	asad.samar@cern.ch
Sang, Max	Max Planck Institute for Physics	max.sang@cern.ch
Schermerhorn, Betsy	Fermilab	ecs@fnal.gov
Seixas, Jose	UFRJ - Fed. Un. of Rio de Janeiro	seixas@lps.ufrj.br
Sexton-Kennedy, Liz	Fermilab	sexton@fnal.gov
Sfiligoi, Igor	INFN LNF - Italy	Igor.Sfiligoi@lnf.infn.it
Sim, James	Fermilab	sim@fnal.gov
Skow, Dane	Fermilab	dane@fnal.gov
Stepanov, Nikita	CERN	nikita.stepanov@cern.ch
Stroustrup, Bjarne	AT&T Research Labs	bs@research.att.com
Szalay, Alex	Johns Hopkins University	szalay@pha.jhu.edu
Tentindo Repond, Silvia	Fermilab	silvia@fnal.gov

Terekhov, Igor	Fermilab	terekhov@fnal.gov
Thaler, Jon	University of Illinois-Urbana-Champaign	jjt@uiuc.edu
Toback, David	Fermilab	toback@fnal.gov
Trombly-Freytag, Kelley	Fermilab	kfreytag@fnal.gov
Turner, Kathleen	Department of Energy	kathy.turner@science.doe.gov
Tuura, Lassi	CERN	lassi.tuura@cern.ch
V. M. DA Silva, Paulo	Insttitute of Superior Technico/UTL	vitor@lps.ufrj.br
Vaniachine, Alexandre	Lawrence Berkeley National Laboratory	avvaniachine@lbl.gov
Vermaseren, Jos	NIKHEF-Amsterdam	t68@nikhef.nl
Vititoe, David	USAF Academy	david.vititoe@usafa.af.mil
Vlachos, Sotrios	CERN	S.Vlachos@cern.ch
Votava, Margaret	Fermilab	votava@fnal.gov
Walbridge, Dana	Fermilab	dgcw@fnal.gov
Wang, Jian-Xiong	Institute of High Energy Physics	jxwang@hptc5.ihep.ac.cn
Wang, Ming-Jer	Fermilab	ming@fnal.gov
Watson, William	Jefferson National Laboratory	watson@jlab.org
Weiglein, Georg	CERN	Georg.Weiglein@cern.ch
Wellner, Rich	Fermilab	wellner@fnal.gov
Werlen, Monique	CNRS-JAPON	Monique.werlen@cern.ch
White, Greg	Stanford Linear Accelerator Laboratory	greg@slac.stanford.edu
White, Vicky	Fermilab	white@fnal.gov
Witherell, Michael	Fermilab	witherell@fnal.gov
Wolbers, Stephen	Fermilab	wolbers@fnal.gov
Womersley, John	Fermilab	womersley@fnal.gov
Yarba, Julia	Fermilab	yarba_j@fnal.gov
Yatsunenko, Yuri	Joint Inst. for Nuclear Research	yuyatsu@fnal.gov
Yeh, G. P.	Fermilab	gpyeh@fnal.gov
Yuasa, Fukuko	KEK	fukuko.yuasa@kek.jp
Zdrazil, Marian	Fermilab	zdrazil@fnal.gov
Zhou, John	Fermilab	johnzhou@fnal.gov

ACAT Participants

AUTHOR INDEX

A

Allcock, W., 161
Annis, J., 229, 323

B

Bajcsy, R., 7
Baker, N., 155
Bauer, C., 185
Becks, K.-H., 80
Belyaev, A. S., 211
Berg, B. A., 104
Beri, S. B., 101
Bhat, P. C., 22, 98, 101
Boos, E. E., 199, 211
Boudreau, J., 179
Brooks, P., 155
Brown, W. E., 294, 345
Brun, R., 264, 297
Buschmann, P., 80

C

Calafiura, P., 250
Calder, A. C., 223, 316
Campbell, T. J., 57
Carli, T., 110
Carpenter, L., 247
Chakraborty, D., 92
Chang, D., 270
Chervenak, A., 161
Cherzor, P. S., 190
Chorti, A., 76

D

da Silva, P. V. M., 86
Denby, B., 36, 73, 76, 338
DesAutels, P., 143
Di Pierro, M., 226
Dominiak, D. M., 258
Drees, J., 80
Drevermann, H., 130
Dror, G., 67, 95
Dubinin, M. N., 211
Dudko, L., 83
Dursi, L. J., 223, 316

E

Elvira, V. D., 326
Etzion, E., 67, 95
Evans, M. W., 258

F

Fent, J., 36
Fine, V., 261, 264
Fischler, M., 179, 280, 345
Fleischer, J., 193
Fliegner, D., 202
Foster, I., 51, 161
Frink, A., 185
Fröchtenicht, W., 36, 73
Fryxell, B., 223, 310, 316
Fujimoto, J., 214

G

Gaines, I., 133, 342, 357
Garda, P., 36, 73, 76
Garren, L. A., 267
Garzoglio, G., 229
Gellrich, A., 232
Graham, G. E., 313
Granado, B., 36, 73, 76
Gregory, E. B., 270
Grindhammer, G., 36, 73
Groote, S., 277
Guillon, F., 143

H

Haas, A., 137
Haberer, W., 36
Hafeez, M., 241
Heinemeyer, S., 205
Holtman, K., 244

I

Ilyin, V. A., 190, 211
Ishikawa, T., 176, 214

J

Janauschek, L., 36, 73
Jirari, H., 217

K

Kadokura, E., 119
Kalia, R. K., 57
Kanaki, A., 169
Kaneko, T., 176
Kaur, R., 101
Kesselman, C., 161
Kiesling, C., 36, 73
Kilminster, B., 235
Knuteson, B., 128
Kobler, T., 36, 73
Koblitz, B., 36, 73, 110
Kodiyalam, S., 57
Kokkas, P., 70
Konstantinidis, N., 130
Körner, J. G., 277
Kovacs, Z., 155
Kowalkowski, J., 345
Kreckel, R., 185
Kröger, H., 217
Kryukov, A. P., 211
Kumagai, A., 119
Kurihara, Y., 214

L

Lamanna, M., 238
Lamb, D. Q., 223, 316
LeGoff, J-M., 155
Legrand, I. C., 113, 164
Lin, Y., 270
Luehring, F. C., 329
Lueking, L., 247
Luo, X. Q., 217, 270
Lyons, L., 31

M

MacLachlan, J. A., 220
MacNeice, P., 223, 316
Maksimović, P., 179
Matsunaga, H., 235
McClatchey, R., 155
McCrory, E. S., 122
McFarland, K., 235
McGuigan, M., 261
Mejuev, I., 119

Michelotti, L., 122, 291
Moriarty, K., 217
Müller, U., 80
Murat, P., 307
Murray, D. J. C., 143

N

Nakano, A., 57
Nellen, G., 36
Nevski, P., 261
Newman, H. B., 113, 164
Nogiec, J. M., 255
Nolty, R., 182

O

Odaka, S., 214
Ogata, S., 57
Ohl, T., 173
Oleari, C., 196
Olson, K., 223, 316
Ostiguy, J.-F., 122

P

Panacek, S., 348
Papadopoulos, C. G., 169
Paterno, M., 287
Perret-Gallix, D., 360
Pivovarov, A. A., 277
Prévotet, J.-C., 36, 73
Prosper, H. B., 98, 101, 335
Pukhov, A. E., 190, 211

Q

Qian, S., 133

R

Rademakers, F., 264
Raja, R., 355
Redin, S. I., 273
Rétey, A., 202
Ricker, P. M., 223, 316
Rinaldo, F., 258
Rosner, R., 223, 316, 351
Ruthsmandorfer, K., 229

S

Samar, A., 241
Sato, K., 214
Savrin, V. I., 211
Schellman, H., 247
Schmidt, S., 36, 73
Seixas, J., 86
Seixas, J. M., 86, 89
Sfiligoi, I., 152
Sherstnev, A. V., 211
Shichanin, S. A., 211
Shimojima, M., 235
Shimojo, F., 57
Sim, J., 255
Skachkova, A. N., 211
Smith, G., 261
Spiletic, J., 261
Steinacher, M., 70
Stepanov, N., 107
Stockinger, H., 241
Stoughton, C., 229
Stroustrup, B., 11

T

Tauscher, L., 70
Tentindo-Repond, S., 98
Tentyukov, M., 193
Terekhov, I., 247
Thaler, J. J., 149
Timmes, F. X., 223, 316
Trombly-Freytag, K., 255
Trumbo, J., 247
Truran, J. W., 223, 316
Tsuno, S., 214
Tuecke, S., 161
Tufo, H. M., 223, 316
Tuura, L. A., 301
Tzamariudaki, B., 73
Tzamariudaki, E., 36

U

Udluft, S., 36, 73

V

Vaiciulis, T., 235
Vashishta, P., 57
Vassali, M. R., 89
Vermaseren, J. A. M., 202
Veseli, S., 247
Vlachos, S., 70
Vologdin, A. N., 211
Vranicar, M., 247

W

Wahlen, H., 80
Walbridge, D., 255
Walsh, P., 57
Wang, Y. L., 270
Weiglein, G., 45
Wenzel, H., 319
Werlen, M., 360
White, V., 247

Y

Yang, J. C., 270
Yatsunenko, Y. A., 140, 146
Yuasa, F., 176

Z

Zingale, M., 223, 316

Related Titles from AIP Conference Proceedings

574   Modeling Complex Systems: Sixth Granada Lectures on Computational Physics
Edited by Pedro L. Garrido and Joaquín Marro, July 2001, 0-7354-0013-X

573   Computing Anticipatory Systems: CASYS 2000 – Fourth International Conference
Edited by Daniel M. Dubois, June 2001, 0-7354-0012-1

568   Bayesian Inference and Maximum Entropy Methods in Science and Engineering: 20th International Workshop
Edited by Ali Mohammad-Djafari, June 2001, 0-7354-0004-0

567   Bayesian Inference and Maximum Entropy Methods in Science and Engineering: 19th International Workshop
Edited by Joshua T. Rychert, Gary J. Erickson, and C. Ray Smith, May 2001, 0-7354-0003-2

553   Disordered and Complex Systems
Edited by Peter Sollich, A. C. C. Coolen, L. P. Hughston, and R. F. Streater, February 2001, 1-56396-983-1

548   Fundamental Issues of Nonlinear Laser Dynamics
Edited by Bernd Krauskopf and Daan Lenstra, December 2000, 1-56396-977-7

523   Gravitational Waves: Third Edoardo Amaldi Conference
Edited by Sydney Meshkov, June 2000, 1-56396-944-0

517   Computing Anticipatory Systems: CASYS'99—Third International Conference
Edited by Daniel M. Dubois, June 2000, 1-56396-933-5

To learn more about these titles, or the AIP Conference Proceedings Series, please visit the webpage http://www.aip.org/catalog/aboutconf.html